COMPREHENSIVE ANALYTICAL CHEMISTRY

ELSEVIER SCIENCE B.V.
Sara Burgerhartstraat 25
P.O. Box 211, 1000 AE Amsterdam, The Netherlands

First edition 2000

Library of Congress Cataloging in Publication Data
A catalog record from the Library of Congress has been applied for.

ISBN: 0-444-89951-0

∞ The paper used in this publication meets the requirements of ANSI/NISO Z39.48-1992 (Permanence of Paper).

Transferred to digital printing 2005

Wilson & Wilson's

COMPREHENSIVE ANALYTICAL CHEMISTRY

Edited by

D. BARCELÓ

*Department of Environmental Chemistry, CID-CSIC,
Jordi Girona 18-26, E-080034 Barcelona, Spain*

COMPREHENSIVE ANALYTICAL CHEMISTRY

ADVISORY BOARD

Wilson & Wilson's

COMPREHENSIVE ANALYTICAL CHEMISTRY

VOLUME XXXIV

DISCRETE SAMPLE INTRODUCTION TECHNIQUES FOR INDUCTIVELY COUPLED PLASMA MASS SPECTROMETRY

DIANE BEAUCHEMIN
D. CONRAD GRÉGOIRE
DETLEF GÜNTHER
VASSILI KARANASSIOS
JEAN-MICHEL MERMET
TODD J. WOOD

2000
ELSEVIER
AMSTERDAM – LAUSANNE – OXFORD – NEW YORK – SINGAPORE – SHANNON
TOKYO

CONTRIBUTORS TO VOLUME XXXIV

Diane Beauchemin
Department of Chemistry, Queens University, Kingston, Ontario, Canada K7L 3N6

D. Conrad Grégoire
Geological Survey of Canada, 601 Booth Street, Ottawa, Ontario, Canada K1A 0E8

Detlef Günther
Laboratorium für Analytische Chemie und Spurenanalytik, Eidgenössische Technische Hochschule Zürich, CH-8092 Zürich, Switzerland

Vassili Karanassios
Guelph-Waterloo Center for Graduate Work in Chemistry, University of Waterloo, Waterloo, Ontario, Canada N2L 3G1

Jean-Michel Mermet
Laboratoire des Sciences Analytiques, Université Claude Bernard-Lyon 1, F-696222 Villeurbanne Cedex, France

Todd J. Wood
Guelph-Waterloo Center for Graduate Work in Chemistry, University of Waterloo, Waterloo, Ontario, Canada N2L 3G1

WILSON AND WILSON'S

COMPREHENSIVE ANALYTICAL CHEMISTRY

VOLUMES IN THE SERIES

Vol. IA	Analytical Processes Gas Analysis Inorganic Qualitative Analysis Organic Qualitative Analysis Inorganic Gravimetric Analysis
Vol. IB	Inorganic Titrimetric Analysis Organic Quantitative Analysis
Vol. IC	Analytical Chemistry of the Elements
Vol. IIA	Electrochemical Analysis Electrodeposition Potentiometric Titrations Conductometric Titrations High-Frequency Titrations
Vol. IIB	Liquid Chromatography in Columns Gas Chromatography Ion Exchangers Distillation
Vol. IIC	Paper and Thin Layer Chromatography Radiochemical Methods Nuclear Magnetic Resonance and Electron Spin Resonance Methods X-Ray Spectrometry
Vol. IID	Coulometric Analysis
Vol. III	Elemental Analysis with Minute Sample Standards and Standardization Separation by Liquid Amalgams Vacuum Fusion Analysis of Gases in Metals Electroanalysis in Molten Salts
Vol. IV	Instrumentation for Spectroscopy Atomic Absorption and Fluorescence Spectroscopy Diffuse Reflectance Spectroscopy
Vol. V	Emission Spectroscopy Analytical Microwave Spectroscopy Analytical Applications of Electron Microscopy
Vol. VI	Analytical Infrared Spectroscopy
Vol. VII	Thermal Methods in Analytical Chemistry Substoichiometric Analytical Methods
Vol. VIII	Enzyme Electrodes in Analytical Chemistry Molecular Fluorescence Spectroscopy Photometric Titrations Analytical Applications of Interferometry

Vol. IX	Ultraviolet Photoelectron and Photoion Spectroscopy
	Auger Electron Spectroscopy
	Plasma Excitation in Spectrochemical Analysis
Vol. X	Organic Spot Tests Analysis
	The History of Analytical Chemistry
Vol. XI	The Application of Mathematical Statistics in Analytical Chemistry
	Mass Spectrometry
	Ion Selective Electrodes
Vol. XII	Thermal Analysis
	Part A. Simultaneous Thermoanalytical Examination by Means of the Derivatograph
	Part B. Biochemical and Clinical Applications of Thermometric and Thermal Analysis
	Part C. Emanation Thermal Analysis and other Radiometric Emanation Methods
	Part D. Thermophysical Properties of Solids
	Part E. Pulse Method of Measuring Thermophysical Parameters
Vol. XIII	Analysis of Complex Hydrocarbons
	Part A. Separation Methods
	Part B. Group Analysis and Detailed Analysis
Vol. XIV	Ion-Exchangers in Analytical Chemistry
Vol. XV	Methods of Organic Analysis
Vol. XVI	Chemical Microscopy
	Thermomicroscopy of Organic Compounds
Vol. XVII	Gas and Liquid Analysers
Vol. XVIII	Kinetic Methods in Chemical Analysis
	Application of Computers in Analytical Chemistry
Vol. XIX	Analytical Visible and Ultraviolet Spectrometry
Vol. XX	Photometric Methods in Inorganic Trace Analysis
Vol. XXI	New Developments in Conductometric and Oscillometric Analysis
Vol. XXII	Titrimetric Analysis in Organic Solvents
Vol. XXIII	Analytical and Biomedical Applications of Ion-Selective Field-Effect Transistors
Vol. XXIV	Energy Dispersive A-Ray Fluorescence Analysis
Vol. XXV	Preconcentration of Trace Elements
Vol. XXVI	Radionucleide X-Ray Fluorescence Analysis
Vol. XXVII	Voltammetry
Vol. XXVIII	Analysis of Substances in the Gaseous Phase
Vol. XXIX	Chemiluminescence Immunoassay
Vol. XXX	Spectrochemical Trace Analysis for Metals and Metalloids
Vol. XXXI	Surfactants in Analytical Chemistry
Vol. XXXII	Environmental Analytical Chemistry
Vol. XXXIII	Elemental Speciation – New Approaches for Trace Element Analysis
Vol. XXXIV	Discrete Sample Introduction Techniques for Inductively Coupled Plasma Mass Spectrometry

Contents

Series Editor's Preface xv
Preface xvii

Chapter 1. Current status of ICP-MS 1
1.1 Introduction 1
1.2 Description of the technique 3
1.2.1 Fundamentals of the ICP source 3
1.2.2 Instrumentation 15
1.2.2.1 Quadrupole-based ICP-MS 16
1.2.2.2 High resolution 33
1.2.2.3 TOF-ICP-MS 42
1.2.3 Sample introduction 43
1.2.3.1 Common systems with pneumatic nebulizers . . 44
1.2.3.2 Limitations 52
1.2.3.3 Alternative high-efficiency nebulization system 53
1.2.3.4 Other sample introduction systems 59
1.3 Advantages 68
1.4 Limitations 74
1.4.1 Precision 74
1.4.2 Spectroscopic interferences 78
1.4.3 Non-spectroscopic interferences 86
1.5 Present capabilities 97
1.6 Extended capabilities with mixed-gas plasmas 137
1.6.1 Modifications required for the addition of a foreign gas to the plasma 138
1.6.1.1 Gas proportioner or T-piece on a gas line . . . 141
1.6.1.2 Sheathing device 144
1.6.2 Physical properties of the resulting plasmas 144
1.6.2.1 Addition to the outer gas 144
1.6.2.2 Addition to the aerosol carrier gas 148
1.6.2.3 Addition to other or all gas flows 151
1.6.3 Features and limitations of each approach 152

1.6.3.1 Addition to the outer gas 153
1.6.3.2 Addition to the aerosol carrier gas 155
1.6.3.3 Addition to other or all gas flows 163
1.6.4 Applications 165
Acknowledgement 170
References. 170

Chapter 2. Flow injection techniques 213
2.1 Introduction 213
2.2 Principles. 217
2.2.1 Definition and measurement of dispersion 217
2.2.2 Key experimental FI parameters 220
2.2.3 Key components 225
2.2.4 Measurement conditions. 233
2.3 Limited dispersion systems (D = 1 – 3) 238
2.3.1 Instrumental set-up 238
2.3.2 ICP-MS Instrument optimization 240
2.3.3 Features 242
2.3.3.1 Aqueous carrier 242
2.3.3.2 Gaseous or gas-segmented carrier 248
2.3.4 Calibration strategies 252
2.3.4.1 On-line internal standardization 253
2.3.4.2 On-line standard additions 253
2.3.4.3 On-line isotope dilution analysis 259
2.3.5 Applications 259
2.4 Medium dispersion systems (D = 3–10). 267
2.4.1 Instrumental set-up 267
2.4.2 Features 270
2.4.3 Calibration strategies 273
2.4.3.1 Gradient calibration. 273
2.4.3.2 On-line standard additions 274
2.4.3.3 On-line isotope dilution analysis 275
2.4.4 Applications 278
2.5 Large dispersion systems (D > 10) 281
2.5.1 Instrumental set-up 281
2.5.2 Features 284
2.5.3 Applications 286
2.6 Reduced dispersion systems (D < 1). 286
2.6.1 Instrumental set-up 287
2.6.1.1 Preconcentration on a microcolumn 287
2.6.1.2 Vapour generation. 295
2.6.1.3 Liquid–liquid extraction 297
2.6.1.4 Precipitation/dissolution 299
2.6.1.5 Anodic stripping voltammetry (ASV). 300

2.6.2 Optimization 300
2.6.2.1 Preconcentration on a microcolumn 301
2.6.2.2 Vapour generation. 302
2.6.2.3 Liquid–liquid extraction 305
2.6.2.4 Precipitation/dissolution 306
2.6.2.5 Anodic stripping voltammetry (ASV). 306
2.6.3 Features 307
2.6.3.1 Preconcentration on a microcolumn 308
2.6.3.2 Vapour generation. 310
2.6.3.3 Liquid–liquid extraction 312
2.6.3.4 Precipitation/dissolution 312
2.6.3.5 Anodic stripping voltammetry 313
2.6.4 Applications 314
2.7 Other possibilities 321
Acknowledgement 322
References. 322

Chapter 3. Electrothermal vaporization sample introduction for inductively coupled plasma–mass spectrometry 347
3.1 Introduction 347
3.2 Instrumentation 351
3.3 Analytical domain of ETV-ICP-MS 357
3.4 Principle of operation of ETV-ICP-MS 361
3.5 Background spectral features 367
3.6 Mechanism of vaporization of analyte 375
3.6.1 Vaporization of boron in ETV-ICP-MS 381
3.6.2 Vaporization of tungsten in ETV-ICP-MS 383
3.7 Chemical modification, physical carriers and mass transport efficiency 386
3.7.1 Chemical modifiers and physical carriers 386
3.7.1.1 Sodium chloride 395
3.7.1.2 Magnesium nitrate 396
3.7.1.3 Nickel nitrate 398
3.7.1.4 Palladium nitrate 400
3.7.2 Analyte transport efficiency 402
3.8 Non-spectroscopic interferences. 408
3.9 Applications 413
3.9.1 Ultratrace analysis of Arctic snow 417
3.9.2 Trace analysis of single zircons for REE, U and Th . . . 419
3.9.3 Direct analysis of solids introduced as slurries. 423
3.9.4 Determination of Cd isotope ratios in sheep's blood and organ tissue 425
3.10 Concluding remarks 428
References. 428

Chapter 4. Laser ablation for inductively coupled plasma-mass spectrometry . . . 445
4.1 Laser ablation . . . 445
4.2 Different types of laser ablation-based analysis . . . 446
4.3 Specific use of LA-ICP-MS . . . 447
4.4 Laser principles . . . 449
4.4.1 Laser parameters . . . 451
4.4.2 Laser spot size . . . 452
4.4.3 Laser beam processing . . . 453
4.4.3.1 Focusing devices . . . 454
4.4.3.2 Focus positioning devices . . . 456
4.4.3.3 Spatial filtering and homogenizing devices . . . 456
4.4.4 Interaction laser–material: UV laser vs IR laser . . . 456
4.5 Laser ablation systems . . . 457
4.5.1 Specification . . . 457
4.5.2 Ablation cell and transportation system . . . 458
4.5.3 Ablation strategies . . . 462
4.6 Signal acquisition and data processing . . . 462
4.7 Sample preparation . . . 463
4.8 Calibration . . . 464
4.9 Elemental fractionation . . . 466
4.10 Figures of merit . . . 467
4.11 Applications . . . 469
4.12 Spark and arc ablation . . . 469
4.13 The future . . . 470
References . . . 480

Chapter 5. Direct sample insertion . . . 503
5.1 Introduction . . . 503
5.2 Instrumentation for DSIs . . . 506
5.2.1 DSI-drive mechanism and DSI-designs . . . 507
5.2.2 ICP torch . . . 512
5.2.3 Sample-carrying probes . . . 512
5.2.4 Spectrometers . . . 514
5.2.4.1 Spectrometers for DSI-ICP-AES . . . 514
5.2.4.2 Spectrometers for DSI-ICP-MS . . . 514
5.3 Effect of operating conditions on analyte signal behaviour . . . 515
5.3.1 Insertion sequence . . . 517
5.3.2 Sample probe geometry, composition and physical condition . . . 518
5.3.3 Sampling depth and gas flow rates . . . 521
5.3.4 Drying and charring (or ashing) . . . 522
5.3.5 Optimization . . . 523
5.3.6 Plasma power and cup temperature . . . 523

5.4 Mixed-gas DSI-ICP-AES . . . 527
5.4.1 Experimental observations on mixed-gas DSI-ICP-AES . 528
5.4.2 Survey of analyte emission with mixed gases and µl volumes . . . 529
5.5 Chemical modification . . . 532
5.6 Further extension of the analytical capability of the ICP using DSIs . . . 539
5.6.1 Reductions of spectroscopic interference in ICP-MS . . . 539
5.6.2 Reductions of non-spectroscopic interference in ICP-MS . . . 542
5.6.3 Pre-concentration and sample clean-up . . . 542
5.7 Analytical performance characteristics . . . 544
5.7.1 Liquid samples . . . 544
5.7.2 Solid samples . . . 546
5.7.3 Slurry DSI-ICP-AES . . . 550
5.7.4 Benefits of tailoring the plasma environment to the application . . . 551
5.7.5 The plasma as a programmable reactor . . . 553
5.8 Applications . . . 553
5.9 Conclusions . . . 553
Acknowledgements . . . 555
References . . . 555

Index . . . 561

Series Editor's Preface

Inductively coupled plasma–mass spectrometry (ICP-MS) has been one of the most exciting emerging techniques in recent years and it has certainly improved the way of analysing multielements and isotopic determinations. The number of manufacturers has grown accordingly and nowadays it is a standard and routine method in many laboratories devoted to trace metal and speciation analysis. This book has a special interest in the area of ICP-MS since it covers a wide range of sample introduction techniques, such as flow injection electrothermal vaporisation or direct sample insertion. The book contains five chapters and it is written by recognised specialists in this area of expertise.

In a way this volume is complementary to the previous volume in this series on 'Elemental Speciation', edited by J.A. Caruso, K.L. Sutton and K.L. Ackley. With the publication of both volumes, the *Comprehensive Analytical Chemistry* series has covered extensively the area of elemental analysis and speciation and these two volumes are destined to become reference works for all researchers in these fields.

Finally, I would like to congratulate the authors for the excellent job carried out with this new volume in the series, and thank them for the time and effort put into the preparation of their chapters. Without their co-operation and engagement this volume would certainly not have been possible.

D. Barceló

Preface

Papers on inductively coupled plasma mass spectrometry (ICP-MS) began to appear in the literature in 1974. The technique, however, did not evolve into a commercial product until 1983 with the introduction of the Sciex ELAN 250 and the VG PlasmaQuad. Since that time, the number of instruments sold has increased rapidly because of the advantages of ICP-MS for the determination of both the trace elemental and isotopic composition of samples. In 1992, there were over 450 instruments installed world-wide. At the time of writing, several thousand ICP-MS instruments were in use supplied by at least ten different companies in North America, Europe, Asia and Australia. Analysts can now choose from a wide variety and types of ICP mass spectrometers including instruments based on quadrupole time-of-flight and magnetic sector mass spectrometers as well as those equipped with collision and reaction cells. There are now several instruments available incorporating multi-collector detection systems, which can produce isotope ratio data equivalent to that produced by thermal ionisation mass spectrometers. It is therefore not surprising that the growth in the number of ICP-MS publications appearing in the scientific literature has increased exponentially since its introduction!

This book mainly concentrates on developments that took place from 1990 to the present, although earlier work is cited when appropriate. This volume is a response to the ever-growing interest in coupling various sample introduction techniques to ICP-MS in order to extend the use and applicability of the technique.

The book aims at motivating current and potential ICP-MS users to think about the great power of this multielement technique. It critically evaluates what has been accomplished so far and what could be done to extend the capabilities of the technique for the discrete analysis of any type of sample, be it aqueous, gaseous or solid. The authors' main goal is to provide the reader with the information and tools needed to accommodate his or her specific application.

The opening chapter of the book is a review of the current capabilities of ICP-MS in its conventional configuration using pneumatic nebulization sample introduction. Instrumentation from existing manufacturers are described and compared to show their similarities and differences. The strengths, limitations and current applicability of ICP-MS, with and without mixed-gas plasmas, are assessed using a review of the literature.

The remaining chapters demonstrate how some of the current limitations of ICP-MS can be overcome by combining various other sample introduction techniques with ICP-MS. The techniques covered in this book all produce a transient or time-dependent analytical signal. These include flow injection analysis (Chapter 2), electrothermal vaporisation (Chapter 3), laser ablation (Chapter 4) and direct sample insertion (Chapter 5).

We note that chromatography and other separation techniques, which also generate transient signals, are absent from this book. The wealth of information on separation science is indeed such that these techniques will be thoroughly dealt with in a separate forthcoming book on ICP-MS based speciation methods.

Diane Beauchemin
D. Conrad Grégoire
Detlef Günther
Vassili Karanassios
Jean-Michel Mermet
Todd J. Wood

Chapter 1

Current status of ICP-MS

Diane Beauchemin

1.1 INTRODUCTION

The ever-increasing need for easier and faster ultratrace elemental analysis of environmental and high-tech materials [1] has forced instrument manufacturers to build reliable user-friendly instruments which do not require highly-trained personnel and can be operated unattended. In particular, instruments allowing ultratrace multielemental analysis of a variety of samples are more likely to be cost-effective and therefore worth the investment by analytical laboratories wanting to remain competitive. Up to the advent of inductively coupled plasma mass spectrometry (ICP-MS), three techniques were widely used: inductively coupled plasma atomic emission spectrometry (ICP-AES), a multielemental technique with detection limits at the parts-per-billion (ppb) level for most elements, flame atomic absorption spectrometry (FAAS), a monoelemental technique with similar or higher detection limits than ICP-AES, and graphite furnace atomic absorption spectrometry (GFAAS), a monoelemental technique with unrivalled detection limits (sub-ppb) for many elements [2]. All these techniques have the advantage of requiring samples in solution, which greatly facilitates control over homogeneity and allows quick calibrations. Highly desirable qualities for a new technique would therefore have to include the multielemental capability of ICP-AES and similar or lower detection limits to GFAAS, in addition to being designed for the analysis of samples in solution. This is what ICP-MS promised to be.

As its name implies, ICP-MS combines an ICP to a mass spectrometer and should, in theory, add the advantages of the ICP source to those of MS. When used as an emission source, the ICP features, in addition to a multielemental capability, good precision, a long linear dynamic range, and relative freedom from the chemical interferences which plague AAS [3]. On the other hand, MS provides simple spectra and a low background, hence resulting in detection limits similar to or better than GFAAS. In addition, MS has the ability to do rapid isotopic analysis which is not possible by GFAAS and difficult by ICP-AES (as it requires a very high resolution spectrometer).

The ICP has one additional feature which should ensure very simple spectra indeed: it first atomizes and then ionizes the sample. Ions extracted from the plasma into the mass spectrometer are therefore expected to produce a spectrum which is elemental, consisting of as many lines as there are stable isotopes for each element present. According to Fig. 1.1, this means that a sample containing all the elements from the periodic table would result in a total of a few hundred lines, which is quite in contrast to ICP-AES where a single element such as Fe or U can emit thousands of lines!

With such potential features, the introduction of ICP-MS was, needless to say, welcome. So, the technique quickly established itself

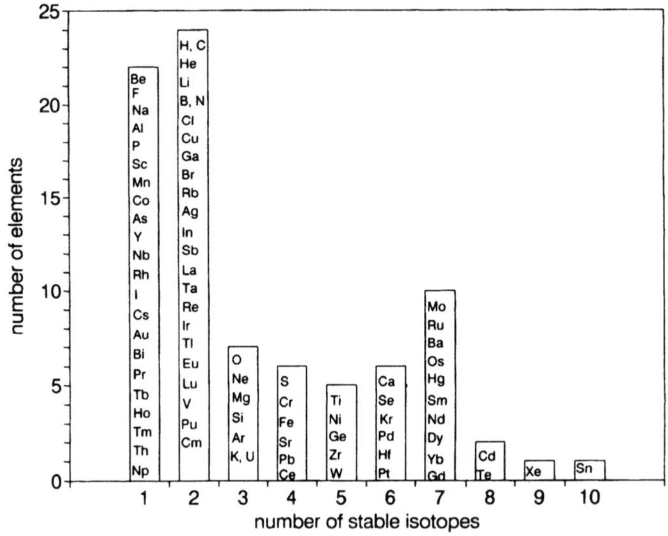

Fig. 1.1. Bar graph showing the number of stable isotopes available for elements of the Periodic Table.

and is now the focus of recommended procedures by governmental agencies, such as the U.S. Environmental Protection Agency [4]. It also has supplanted instrumental neutron activation analysis in geoanalysis [5]. This success occurred despite the fact that ICP-MS is still evolving, and does not quite feature all the advantages expected, as will be seen in the discussion of its present capabilities (Sections 1.3 to 1.5). However, the source of its limitations can only be found through a good understanding of the basic principles of the technique which are discussed below. Most of the book pays particular attention to the latest developments (i.e. which occurred in the 1990s). The reader may also consult other books [6–15], book chapters [16–20] and reviews [21–43]. For instance, the *Journal of Analytical Atomic Spectrometry* publishes an annual update that includes ICP-MS (for example, Refs. [44–50]).

1.2 DESCRIPTION OF THE TECHNIQUE

Since the following chapters deal with modifications to the sample introduction system, a detailed description of the ICP is first made, followed by a discussion of how it can be used as an ion source for MS. As well, a review is made of the common nebulization systems that are typically used to introduce aqueous samples into the ICP. The associated limitations of each MS or sample introduction system are also summarized.

1.2.1 Fundamentals of the ICP source

An Ar ICP is the standard ion source for MS on all commercial ICP-MS instruments. It is essentially identical to that which is currently used in commercial ICP-AES instruments as a result of the pioneering work of S. Greenfield and V.A. Fassel, as well as important contributions from P.W.J.M. Boumans, who demonstrated the excellent analytical characteristics of the ICP as a spectroscopic emission source [51]. A 1–2.5 kW plasma is typically generated by radiofrequency (rf) magnetic fields induced by a coil wound around the end of a quartz torch through which argon flows [52]. A high-voltage spark is used to seed the argon with electrons which collide with argon atoms, ultimately causing ionization and plasma ignition. This energy transfer is called inductive coupling, since it involves no electrodes. The motion of electrons and ions, which is continuously induced by the oscillating magnetic field,

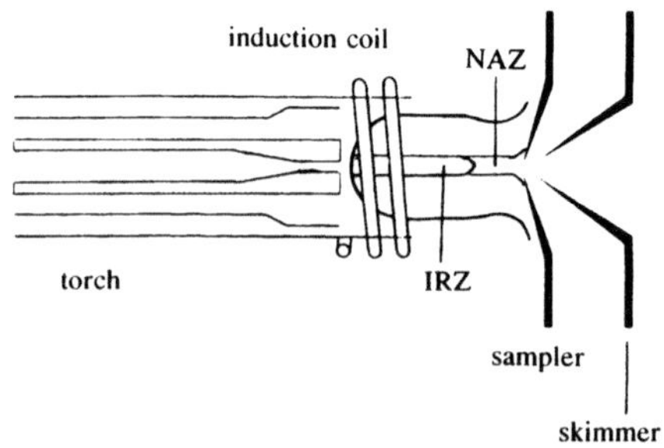

Fig. 1.2. Sketch of the plasma showing the quartz torch, the initial radiation zone (IRZ) and the normal analytical zone (NAZ), as well as the interface of the mass spectrometer.

causes collisions with Ar atoms i.e. ohmic heating of the gas [52], resulting in temperatures as high as 10,000 K [51]. The Ar ICP therefore consists of a high-temperature electrodeless discharge that is sustained in flowing argon at atmospheric pressure.

Although numerous torch designs are available [53,54], the standard torches used on commercial instruments are usually of the Fassel-type [53] and consist of three concentric quartz tubes (see Fig. 1.2): an outer tube for the plasma (outer) gas, an intermediate tube for the auxiliary gas, and a central tube (called injector tube) for the carrier gas (which can also be the nebulizer gas when a nebulizer is used for the introduction of solutions). Argon is commonly used as the operating gas for the three gas flows, each of which has a different function. The outer gas, which has the highest flow rate, is used to maintain the plasma (along with an rf power supply) and stabilize it, as well as to prevent the plasma from melting the outer tube. The auxiliary gas positions the plasma within the torch and also prevents the plasma from overheating the rim of the inner tube. Finally, the carrier gas is used to carry the sample and introduce it into the centre of the plasma. A quartz bonnet is often used to protect the end of the load coil (closest to the plasma) from plasma flashover [53].

At plasma ignition, only two gases are required: the outer gas and the auxiliary gas through which the spark propagates into the outer gas. Once the plasma is ignited, the carrier gas flow rate is increased and essentially punches a hole in the centre of the elliptical plasma, giving it an annular or toroidal shape [6]. Although the temperature in the induction region of the plasma (which is within and slightly above the load coil) may be as high as 10,000 K, the temperature is lower within the central channel created by the carrier gas, because of the

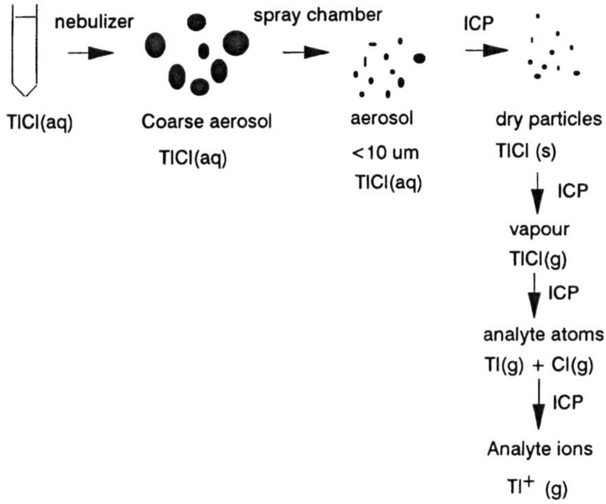

Fig. 1.3. Sketch of the various processes taking place in a conventional nebulization system and the plasma.

various processes taking place in this region (and thereby using energy).

In general, a sample is presented to the ICP in solution, and a nebulizer (see Section 1.2.3) is used to convert this solution into an aerosol (see Fig. 1.3). This aerosol is carried by the nebulizer gas (which typically also serves as carrier gas) through a spray chamber where large droplets condense and drain out. Only small droplets are allowed to reach the plasma, because large ones are (as will be seen in greater detail in Section 1.4.1) incompletely dissociated in the plasma and cause increased noise.

Once the wet aerosol penetrates the plasma, several processes take place sequentially in a few milliseconds [24] (Fig. 1.3): (1) desolvation whereby the solvent is evaporated off each droplet and results in an anhydrous salt; (2) vaporization which involves the physical transformation of the salt particles from solid to gas phase; (3) atomization of the gaseous molecules into free atoms; and (4) ionization of these atoms. Excitation of atoms and ions, where individual atoms and ions undergo spectroscopic transitions, also occurs, resulting in atomic and ionic emissions which are detected in ICP-AES. The introduction of a wet aerosol also causes shrinking of the plasma [55,56]. For instance, the diameter of the plasma was reported to be about 2 mm narrower with a wet than with a dry aerosol [55].

The first two steps require a relatively large amount of energy which translates into a residence time in the plasma of the order of 3 ms for 0.3 mg/s of injected aerosol in the plasma, depending on the size of the droplets [57], the thermal conductivity of the vapour, and the thermal conductivity of the plasma gas [58]. The desolvation process alone is typically carried out over a distance of about 10 mm in the plasma [56]. The atomization step also requires a long residence time (i.e. 2–7 ms) but, although most of the energy is spent on the production of free atoms, a large amount of energy is consumed by the desolvation process [59]. Both the excitation and the ionization occur much more rapidly in the plasma, requiring about a nanosecond. Hence, the important parameters to control for obtaining efficient desolvation, volatilization and dissociation are the residence time in the plasma and the thermal conductivity of the plasma gas. Since the aerosol carrier gas flow rate has a direct effect on the residence time, its control is critical. A mass-flow controller is recommended for this purpose.

The sequential aspect of the atomization and ionization processes can be clearly seen as two distinct regions in the plasma when a solution containing 1000 mg/l Y is nebulized. These two different zones are shown in Fig. 1.2. A slender inner red cone due to emission from YO^+ and neutral Y species is called "initial radiation zone" (IRZ), and a blue plume due to emission from Y^+ is called "normal analytical zone" (NAZ) since the latter is typically monitored in ICP-AES [60]. The temperature in the IRZ is about 5000 K and that in the NAZ about 7000 K. This increase in temperature in the central channel with increasing distance from the load coil has been tentatively attributed to the diffusion of hot argon gas toward and into the central channel [61]. In both the IRZ and NAZ regions, heating occurs mainly by radiation and conduction from the induction region. The relative positions of the IRZ and NAZ with respect to any chosen point depend on several operating parameters, especially the carrier gas flow and the rf power. Figure 1.4 shows the La^+ and LaO^+ signal observed while moving the IRZ away from the sampler, starting with the IRZ at the tip of the sampler. The two species clearly have maxima in different regions: La^+ exhibits a maximum when the IRZ is at 3.75 mm of the sampler orifice, i.e. well in the NAZ, whereas the best LaO^+ signal lies in the IRZ. The same observations were made by others [62,63] and can be extended to other species as well. For instance, the ion maxima for $BaOH^+$, BaO^+, Ba^+ and Ba^{2+} are seen in that order as one moves the plasma sampling position from within the IRZ to well within the NAZ [63].

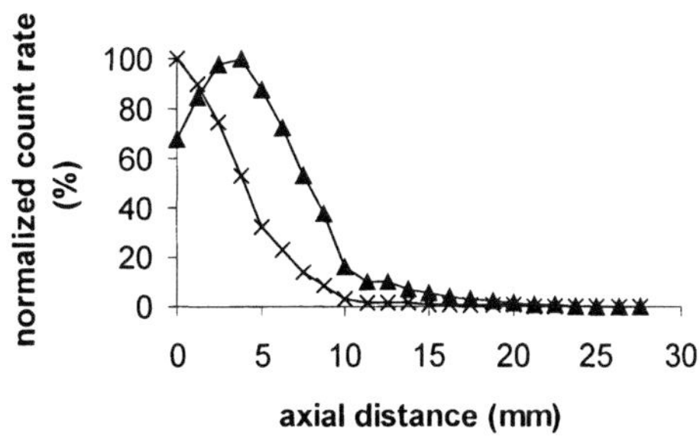

Fig. 1.4. Axial profiles of the ICP showing the normalized signal of $^{139}La^+$ (×) and $^{139}La^{16}O^+$ (triangles) obtained as the ICP is moved away from the interface. At 0 mm, the tip of the IRZ is at the sampler whereas at greater distances the sampler is within the NAZ of most of the elements.

The plasma sampling position, which is usually just downstream of the IRZ, is therefore important to maximize the density of analyte ions extracted from the ICP. This position is dependent on all ICP operating parameters. For instance, a change in the nebulizer flow rate (which affects both the residence time in the plasma and the amount of sample reaching the plasma) will require a change in either the sampling position [62] or the rf power [64] to keep the analyte signal at a maximum. It also depends on the droplet size distribution of the aerosol being introduced into the plasma, larger droplets requiring a longer residence time in the plasma than small ones to undergo the various processes. In fact, high-speed photographs have shown that large droplets and/or solid residues resulting from the desolvation of large droplets can pass directly into the sampler [65].

Three main ionization processes have been proposed: electron-impact ionization, charge transfer, and Penning ionization [66]. The former results from the collision of the analyte, M, with electrons if the kinetic energy of the colliding electron is greater than the threshold energy for ionization:

$$M + e^-_{fast} \rightarrow M^+ + 2e^-_{slow} \qquad (1)$$

Charge transfer between argon ions and analyte atoms can also occur:

$$Ar^+ + M \rightarrow Ar + M^{+*} \qquad (2)$$

In this case, the resulting analyte ion is generally in an excited electronic state. Although the detection of ions by MS is independent of their state of excitation, this process can only occur if the total energy needed for ionization and excitation of M is less than the first ionization potential of argon (15.76 eV). However, it is possible that analyte ions may be formed by charge transfer with other ions than Ar^+ (such as those derived from water) [67]. Finally, Penning ionization results from collision with metastable argon atoms, Ar^m:

$$Ar^m + M \rightarrow Ar + M^+ + e^- \tag{3}$$

The resulting analyte ion may be in ground or excited electronic state, depending on the energies involved in the collision, excess energy being carried away as kinetic energy by the electron. This mechanism is not as favoured as the other two because the population of Ar^m is low compared to that of electrons and Ar^+ in the ICP [68].

Thermal equilibrium would be established if the rate of each process occurring in the plasma was equal to that of the inverse process and if there was instantaneous response to any change in plasma conditions. This, however, is never achieved in laboratory plasmas because of the large temperature and density gradients resulting from photons escaping the plasma boundaries. However, thermal equilibrium can exist in localized regions of the ICP, where the equilibrium velocity, excitation, ionization products, and dissociation distributions are locally maintained. This is called a local thermal equilibrium (LTE) because temperatures can be specified only locally and instantaneously. Under these conditions, the microscopic thermodynamic state of the plasma can be described by macroscopic variables such as pressure, concentration, and temperature, without a need for a detailed knowledge of plasma particle interactions.

The population distribution of the ionization products is described by the Saha equation,

$$\frac{n_e n_1^+}{n_p} = 2\left(\frac{g_1^+}{g_p}\right)\left(\frac{2\pi m_e k T_{ion}}{h^2}\right)^{3/2} \exp\left(\frac{-E_{p\lambda}}{kT_{ion}}\right) \tag{4}$$

where n_1^+ and n_p are the number density of the ground state ion and excited atom level p, respectively; n_e is the free electron number density; the initial factor of two is the electron statistical weight; g_1^+ and g_p are the statistical weight of the two levels; m_e is the electron mass; h is Planck's constant; $E_{p\lambda}$ is the ionization energy from the level p; and T_{ion} is the ionization temperature of the species.

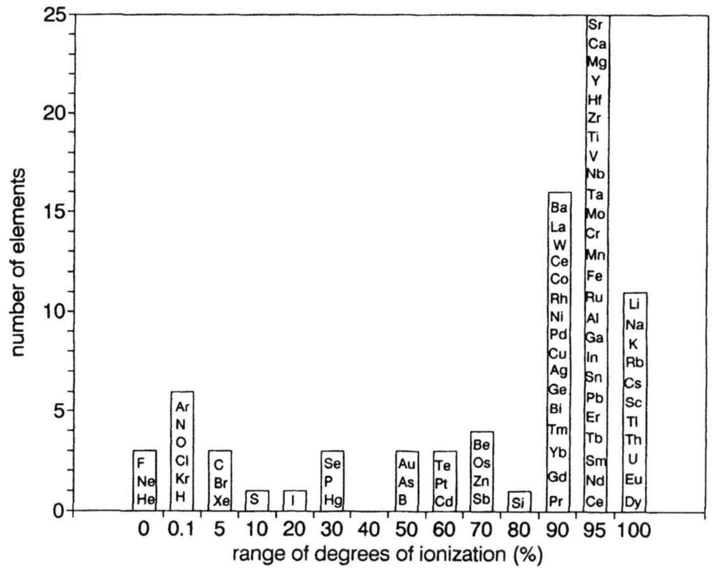

Fig. 1.5. Bar graph showing the number of elements possessing different degrees of ionization. Compiled from Ref. [24].

Calculations were made for 80 elements, assuming T_{ion} of 7500 K and n_e of 1×10^{-15} cm^{-3} [24]. Under these conditions, the degree of ionization, α, of 52 elements is expected to be ≥90% [24] (see Fig. 1.5), according to:

$$\alpha=\frac{n_i}{n_a+n_i}=\frac{\dfrac{(n_i n_e)}{n_a}}{n_e+\dfrac{(n_i n_e)}{n_a}}=\frac{K_m}{n_e+K_m} \tag{5}$$

where n_i, n_a and n_e are the number densities of ions (M^+), atoms (M) and electrons, respectively; and K_m is the Saha equilibrium constant which is defined as:

$$K_m=\frac{n_i n_e}{n_a}4.83\times10^{-5}T_{ion_{3/2}}\frac{Z_i}{Z_a}\exp\left(-\frac{V_i}{kT_{ion}}\right) \tag{6}$$

where k is the Boltzmann constant; T_{ion} is the ionization temperature; Z_i and Z_a are partition functions of the ion (M^+) and atom (M) respectively; and V_i is the ionization potential (IP) of the element.

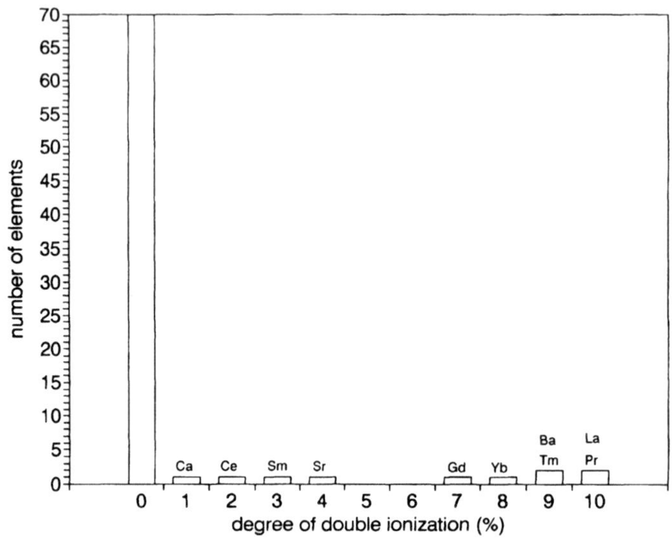

Fig. 1.6. Bar graph showing the number of elements possessing different degrees of double ionization. Compiled from Ref. [24].

All elements having a first IP ≤ 10 eV should have α > 50%. Only three elements (He, F and Ne) which possess a first IP greater than that of Ar would not be ionized and could therefore not be determined by ICP-MS with an argon ICP. Similar calculations of the levels of doubly-charged ions show that the highest degree of double ionization is 10% and occurs only for those elements with the lowest second IP (see Fig. 1.6) [24]. The ICP is therefore an efficient elemental ion source for MS since the majority of elements in the periodic table are singly ionized, with few doubly-charged ions being formed. This is greatly advantageous in terms of the simplicity of the resulting mass spectrum.

A portion of the plasma is sampled from the central channel of the plasma, extracted through a differentially pumped interface, which is maintained at approximately 1 Torr using a mechanical roughing pump, and then transmitted into the mass spectrometer where ions are separated from other species and analyzed according to their mass-to-charge ratios (*m/z*). The interface includes two water-cooled metal cones with ca. 1-mm orifice diameter, the sampler and the skimmer, which provide passage of the sample ions and allow a reduction in pressure from near atmospheric pressure (in the plasma) down to MS operating pressure. Because the plasma around the sampling orifice is in contact with the water-cooled sampler, it cools down to a temperature similar to that in the IRZ. This cooler region in contact with the sampler is called a boundary layer. Any analyte present in this region can therefore recombine with oxygen to form oxides.

According to calculations by Douglas and French [69], the region of the plasma which passes through the sampler orifice is ca. 8 orifice diameters wide. Because the pressure within the interface is much lower than in the plasma, a supersonic expansion or free jet results. The expanding gas forms shock waves, i.e. a barrel shock ending with a perpendicular Mach disk, against the background gases within the interface [69]. These waves surround the expanding region, which is also called the zone of silence since essentially no ion-molecule chemistry is expected to take place within this region [69]. (One study reported molecular ionization in the interface through charge exchange with Ar^+ [70]; but it was carried out with the skimmer positioned close to the Mach disk, i.e. under conditions where ions from the turbulent region outside the zone of silence could be entrained into the mass spectrometer.) The gas typically flows from the sampler to the skimmer in about 3 μs [67]. When the skimmer is located at about two thirds of the distance between the sampler and the Mach disk, only the centreline flow of the sampler is transmitted into the mass spectrometer. As a result, a representative sample of quasineutral plasma [71] from a ca. 0.8-mm diameter region [69] (slightly smaller than the cone orifice) enters a second vacuum region where the pressure is low enough and the mean free path long enough for the ions to be focused into the mass spectrometer [24]. Electrostatic ion lenses, which focus ions according to their kinetic energy, are in general used for this purpose (more details are provided in Section 1.2.2.1).

In any case, all of the instrumentation on the market uses a similar differentially pumped interface, with a sampler and a skimmer positioned as discussed above, to sample the plasma. Yet, a recent study showed that this usual skimmer position at 6–7 mm from the sampler results in substantial scattering and a beam profile that is less intense on the centreline and much wider than that which can be obtained by placing the skimmer 17–27 mm downstream of the sampler [72]. In fact, using a wider skimmer orifice of 2 mm instead of the usual 0.9 mm, in addition to positioning the skimmer at 17–27 mm of the sampler resulted in an even more intense beam, which was nonetheless still about half of that expected for an ideally skimmed beam [72].

Furthermore, the effective sampling diameter depends on the operating conditions, including the sampler–skimmer separation [73] and can be as high as 3 orifice diameters [20,24] which roughly corresponds to the central channel of the plasma, as can be seen in Fig. 1.7 where a transversal profile of the plasma was obtained while keeping the position of the IRZ constant (and close to the sampler orifice when on axis). All observed analyte-containing ions are within a 3-mm diameter region which roughly corresponds to three times the 1-mm diameter of the sampler orifice. The Ar_2^+ profile also shows that the plasma is about 14 mm wide. This means that, in the absence of turbulence, no surrounding air should be entrained when the sampler orifice is aligned with the centre of the plasma. Unfortunately, because of the non-passive plasma sampling process in ICP-MS, the sampler may disturb the gas flows, to an extent depending on the sampling position. Supporting evidence for this disturbance is the different noise amplitude spectra which are observed at different sampling positions in ICP-MS [74,75] but not in ICP-AES [75]. More evidence is also discussed at the end of Section 1.4.2.

Although Fig. 1.5 shows the expected degree of ionization of the elements, it does not indicate the position in the Ar plasma where this degree is reached. The dependence of the best sampling position on physical characteristics of the elements is illustrated in Fig. 1.8. It shows the longitudinal profile along the plasma axis that is observed for three elements, while moving the IRZ away from the sampler, starting with the IRZ at the tip of the sampler. Since the maxima for As^+ and La^+ are separated by 2.5 mm, a sampling position that is optimal for both of these elements cannot be selected with the present

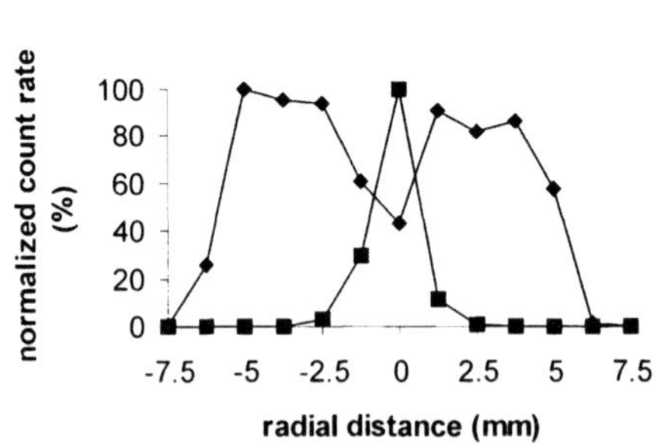

Fig. 1.7. Radial profile of the ICP showing the normalized signal of $^{76}Ar_2^+$ (lozenges) and 100 µg/l La^+ (squares) as the ICP is moved across the sampler while kept at a fixed sampling depth.

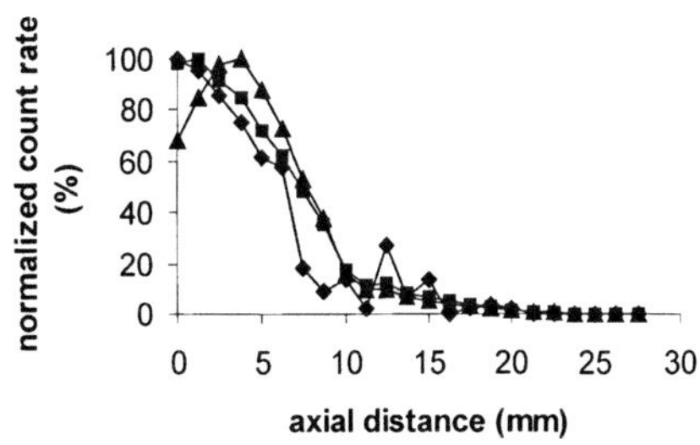

Fig. 1.8. Axial profiles of the ICP showing the normalized signal of $^{139}La^+$ (triangles), $^{75}As^+$ (squares) and $^{27}Al^+$ (lozenges) obtained as the ICP is moved away from the interface. At 0 mm, the tip of the IRZ is at the sampler whereas at greater distances the sampler is within the NAZ for most of the elements.

sampling system, which is restricted to two orifice diameters deep i.e. a 2-mm depth [20].

Furthermore, the order under which these ions appear in the plasma is not commensurate with their very different IPs: one would indeed expect the element with the lowest IP to exhibit its maximum ion population closest to the IRZ. Yet, the maximum for As (with an IP of 947.0 kJ/mol [7]) is at 1.25 mm of the IRZ whereas that of La (with an IP of 538.1 kJ/mol [7]) is at 3.75 mm. In fact, the IP does not seem to be that critical a parameter since Al (with an IP of 577.4 kJ/mol [7], similar to that of La) has a maximum at the IRZ. A difference of 369.6 kJ/mol in IP therefore translates into about 1.25 mm in the plasma, which implies that most of the singly-charged elemental ions should be sampled into the mass spectrometer. Others have also reported that elements with different IPs, i.e. As and Y, had the same optimal sampling position [76]. The reason why the maximum La^+ signal lies deeper into the NAZ is that it forms a refractory oxide: a longer residence time in the plasma is therefore required for its atomization and ionization [62].

The optimum sampling position can also be mass dependent: at a given rf power and sampling position, a higher aerosol carrier gas flow rate may be required to obtain maximum signal for lighter elements than for heavy ones [77,78] or for heavy analytes [79]. This mass dependence, which may or may not be observed, depending on the sample introduction system, has been attributed to mass transport effects [79]. Finally, the chemical form (i.e. speciation) of an element in solution may also influence its transport efficiency and where it undergoes the various processes (i.e. desolvation, vaporization, etc.) in the plasma, thereby determining where the maximum ion density will be in the plasma. This would explain why different sensitivities can be observed for different forms of a given element [80,81].

One of the most important operating conditions in ICP-MS is therefore the position of the IRZ with respect to the sampler since it determines which portion of the plasma is sampled. Since the NAZ widens as it progresses from the torch, the portion of the NAZ that will be sampled will decrease, along with the analyte signal, as the distance between the sampler and the IRZ is increased [82]. The IRZ position, defined as the distance between the tip of the sampler and the tip of the IRZ is a more informative plasma position parameter than the distance between the load coil and the tip of the sampler. The latter can indeed remain constant while a completely different region is being sampled as a result of a change in one or more of the plasma operating parameters. In particular, changes in aerosol carrier flow rate and/or rf power, which are two interdependent parameters, have the most determining effect on the IRZ position [83]. In the author's experience, changing the quartz torch can also change the IRZ position due to either small differences in torch design and/or changes in the position of the torch within the load coil. This is consistent with the observation that the optimal conditions found by simplex optimization were different if any part of the ICP system (i.e. torch, spray chamber, nebulizer, etc.) was changed [84]. Similarly, erosion of the sampler and/or skimmer, or the installation of new cones can have a significant effect on all operating parameters, including the optimal voltages to the ion lenses of the mass spectrometer. In any case, even specifying the sampling position relative to the IRZ is, at best, a time-averaged approximation since the position of the tip of the IRZ was observed to vary both axially and radially on a ms time scale [65].

In general, a univariate optimization of the plasma and MS operating parameters is carried out while aspirating continuously a solution containing analytes spanning the mass range to be covered during the analyses. These selected analytes and the background are monitored continuously while the position of the plasma with respect to the interface, the nebulizer flow rate and then the ion lenses' settings are optimized to maximize sensitivity. In general, optimum ion transmission is obtained when the applied lens voltage is comparable to the kinetic energy that an ion gains from supersonic expansion through the interface [85]. Although the rf power can also be included in the optimization, it is usually set at a pre-established value for aqueous solutions, and is only modified for significantly different matrices (i.e. which will affect the various processes occurring in the plasma), such as those containing organic solvents.

The presence of an organic solvent may indeed require a 500-W increase in power [86] for an efficient combustion of the solvent and/or a decrease in aerosol carrier flow rate to compensate for the increase in nebulization efficiency which results from the presence of solvents more volatile than water [87]. For instance, as little as 3% v/v methanol induced a decrease (by about 0.05 l/min) of the optimal aerosol carrier flow rate compared to that observed with all aqueous solutions [81]. On the other hand, an increase in power may not be required if the solvent load is reduced by decreasing the sample uptake rate and/or the temperature of the spray chamber, as demonstrated by a multivariate simplex optimization [87]. Nonetheless, oxygen-containing solvents behave completely differently than oxygen-free ones in the ICP [88,89]. For instance, methanol distributes itself between the central and outer regions of the plasma and induces a shrinkage of the plasma in both the axial and radial directions, whereas chloroform remains concentrated in the central channel [89]. In any case, an addition of O_2 to the plasma may be beneficial for the analysis of organic solutions, as discussed later in this chapter (see Section 1.6). More discussion of the optimization of instrumental settings is included at the end of Section 1.2.2.1.

1.2.2 Instrumentation

This section concentrates on instrumentation that was commercially available at the time of writing, and which can be sub-divided into three categories: quadrupole-based, sector-based and time-of-flight (TOF) based instruments. Quadrupole instruments provide low resolution (i.e. unit mass resolution) whereas TOF-type instruments allow either low (R = 450 at m/z 208, with Renaissance from LECO Corp.) or medium resolution (R = 1800 at m/z 238, with OptiMass 8000 from GBC Scientific Equipment) and sector-based instruments provide a choice or range of resolutions, from low to high resolution [90]. However, other instruments are being developed in research laboratories. In particular, those based on ion traps show great promise as elemental mass analyzers and may likely evolve into commercially available instrumentation [91]. Indeed, ion traps can act as gas-phase test tubes because of their storage capability, and allow beneficial ion-molecule reactions to decrease the extent of interferences [91]. For instance, Ar^+

could be selectively reduced through charge-transfer reaction with H_2 added to the trap [92]. They can also be used for collision-induced dissociation (CID) where gas-phase collisions eliminate interferences from polyatomic ions such as ClO^+, ArO^+ and $ArCl^+$ [91]. Furthermore, Fourier transform ion cyclotron resonance mass spectrometry provides resolution from 200,000 to 200,000,000 (full width at half height), which is much higher than that which can be achieved with sector-based instruments, and would basically circumvent any spectroscopic interference in ICP-MS [91]. At the time of writing, however, their quantitative ability was not yet competing with commercially available instruments [91].

1.2.2.1 Quadrupole-based ICP-MS

A schematic representation of selected quadrupole-based instruments is shown in Fig. 1.9 and their main characteristics and specifications are summarized in Tables 1.1 and 1.2, respectively. These instruments have several points in common. They have fairly similar dimensions, except for the benchtop HP 4500, the only one of this size, although the new ELAN 6000 is also substantially smaller than most of the other instruments. They all use a peristaltic pump to deliver the sample to the nebulizer in order to minimize physical interferences (due, for instance, to changes in viscosity from sample to sample or standard). It is interesting to note how "universal" this approach is, despite the fact that the pulsations induced by individual rollers of the peristaltic pump result in low frequency noises [75,93,94], i.e. fluctuations which can limit the precision attainable. Care should therefore always be taken to minimize this source of noise by adjusting the pressure of the rollers to the minimum required to provide a constant flow rate (while preventing a pneumatic nebulizer (PN) from freely aspirating the sample). Sample introduction is discussed in more detail in Section 1.2.3.

All the instruments use a similar interface with a sampler and a skimmer, although the shape of these cones, the material they are made of (patented NICONE (VG Elemental), Ni, Cu, with and without Pt insert, etc.), their orifice diameter and the distance separating them

Opposite: Fig. 1.9. Schematic diagrams of some of the commercially-available quadrupole-based ICP-MS instruments. (a) UltraMass. Copyright Varian Associates 1994. Used with permission. (b) SOLA. Courtesy of Finnigan MAT. (c) HP-4500. Courtesy of Agilent Technologies. (d) POEMS. Courtesy of Thermo Jarrell Ash. (e) ELAN 6100. Courtesy of Perkin-Elmer. (f) PQ Excel. Courtesy of VG Elemental.

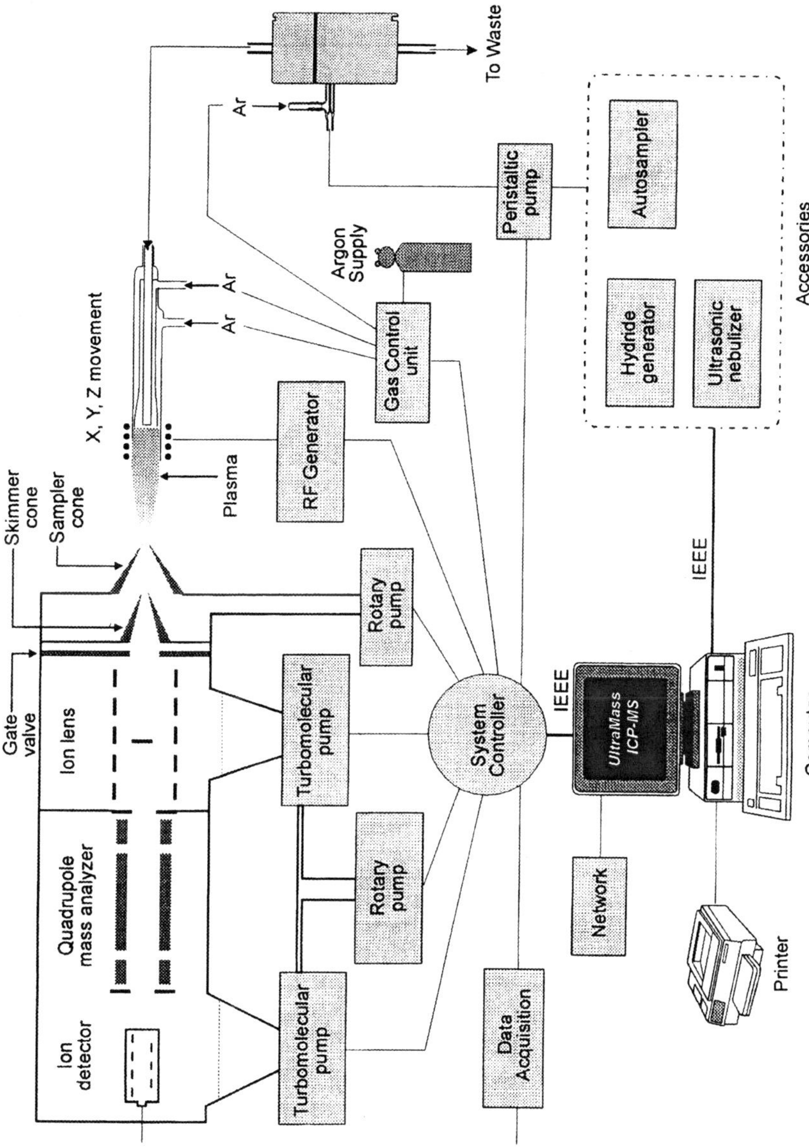

Gate valve
Ion lens
Quadrupole mass analyzer
Ion detector
Skimmer cone
Sampler cone
X, Y, Z movement
Plasma
Ar
Ar
Ar
Argon Supply
To Waste
Rotary pump
Turbomolecular pump
Rotary pump
Turbomolecular pump
RF Generator
Gas Control unit
Peristaltic pump
System Controller
Data Acquisition
Network
IEEE
IEEE
UltraMass ICP-MS
Computer
Printer
Hydride generator
Ultrasonic nebulizer
Autosampler
Accessories

TABLE 1.1

Main characteristics of selected quadrupole-based ICP-MS instrumentation

	Manufacturer							
	Agilent Technologies	Varian Inc.	Spectro	VG Elemental	Finnigan MAT (Turner)	Perkin-Elmer SCIEX	Seiko Instruments Inc.	Thermo Jarrell Ash
Model	HP 4500	UltraMass 700	Spectromass 2000	VG PQ ExCell	TS SOLA	ELAN 6100	SPQ 6100	Ion Scan/POE MS
Sample introduction system	peristaltic pump, Peltier-cooled spray chamber, V-groove nebulizer	computer-controlled 4-channel peristaltic pump; glass concentric nebulizer and inert Sturman-Masters spray chamber, external to torch box	water-cooled spray chamber, selection of nebulizers (Meinhard, crossflow, V-groove, etc.)	4-channel peristaltic pump; Peltier-cooled inert sample introduction system	4-channel peristaltic pump; glass concentric nebulizer and water-cooled spray chamber	2-channel peristaltic pump; Ryton Scott spray chamber (cyclonic spray chamber option) and Gem-Clean™ HF-resistant cross-flow nebulizer	peristaltic pump; pneumatic nebulizer and water-cooled Scott spray chamber with chamber gas	built-in peristaltic pump; inert cross-flow nebulizer with water-cooled glass or inert spray chamber
Torch		1-piece low flow (option: torch with demountable quartz or alumina injector)	partially demountable torch option with injectors of various i.d.	torch with demounta ble injector (sapphire, alumina or platinum)		torch with demountable alumina injector (available in various i.d.)		torch with demountable injector (quartz or sapphire)
RF generator; maximum power	27.12 MHz, solid state, crystal-controlled	40.68 MHz, solid state, crystal-controlled, water-cooled; 1.6 kW	27.12 MHz, water-cooled, free-running	27.12 MHz, solid state	27.12 MHz, crystal-controlled; 2 kW	40.68 MHz, free running; 0.7-1.8 kW; Ar-cooled coil		27.12 MHz, crystal-controlled; 2 kW

	Manufacturer							
	Agilent Technologies	Varian Inc.	Spectro	VG Elemental	Finnigan MAT (Turner)	Perkin-Elmer SCIEX	Seiko Instruments Inc.	Thermo Jarrell Ash
Plasma positioning	computer-controlled x,y,z translation stage	computer-controlled x,y,z translation stage		computer-controlled x,y,z adjustment	computer-controlled sliding plate mounted on 2 rods parallel to central axis for x-y adjustment			automatic axial positioning; manual y-z positioning
Plasma interface	ShieldTorch	Turner interlaced induction coils	water-cooled interface grounded coil	PlasmaScreen torch, fully automated		PlasmaLok	device to shield the ICP	
Ion lens feature	Omega lens off the quadrupole axis	1 photon stop	a photon stop	high efficiency ion guide; chicane deflector; off-axis quadrupole; collision cell option	accelerator cone; deflection towards quadrupole axis; phase-matching optics	shadow stop and AutoLens ion optic system	1 photon stop	offset ion lens system
Quadrupole	hyperbolic rods	cylindrical rods	30-cm long, 1.6 cm diameter Mo cylindrical rods	cylindrical rods	30-cm long, 1.6 cm diameter cylindric al rods	binary gold metallized ceramic, producing hyperbolic field	cylindrical rods	hyperbolic 250-mm quadrupole with 12-mm internal radius
Pumping	3 stages: 1 rotary pump, 2 turbomolecular pumps backed by 1 rotary pump	3 stages: 1 rotary pump, 2 turbomolecular pumps backed by 1 rotary pump	3 stages: 1 rotary pump, 2 turbomolecular pumps backed by 1 membrane pump	3 stages: 1 rotary, 2 turbomolecular pumps	4 stages: 1 rotary pump, 3 turbomolecular pumps backed by a rotary pump	3 stages: 1 rotary pump, 2 turbomolecular pumps backed by 1 rotary pump	3 stages	3 stages: 1 rotary pump, 2 turbomolecular pumps

TABLE 1.1 (continuation)

	Manufacturer							
	Agilent Technologies	Varian Inc.	Spectro	VG Elemental	Finnigan MAT (Turner)	Perkin-Elmer SCIEX	Seiko Instruments Inc.	Thermo Jarrell Ash
Detector	discrete dynode electron multiplier on axis	discrete dynode electron multiplier off axis	Faraday cup and secondary electron multiplier	discrete dynode electron multiplier	SEM and Faraday cup off axis	dual stage discrete dynode electron multiplier off axis	CEM off axis	CEM
Detection mode	pulse counting and analog	pulse counting		pulse counting and analog	pulse counting	simultaneous analog and pulse counting		analog and pulse counting
Dimensions (L × H × D)	1.1 × 0.58 × 0.6 m	1.64 × 1.21 × 0.8 m	1.69 × 1.31 × 0.74 m	1.55 × 1.25 × 0.9 m	1.76 × 1.327 × 0.75 m	0.99 × 1.17 × 0.73 m		
Software	MS-Windows	IBM OS/2		Windows NT	SIMDEQ	Windows NT		MS-Windows
Optional nebulizer		V-groove, DIN, MCN, USN	USN, MCN	low-flow inert nebulizers	USN	USN		USN, DIN
Other options	ETV	HG, LA	FIA, HPLC, ETV, LA, Spark ablation	LA, sample preparation system, glove-box enclosure	LA (UV and IR), GD, ETV, HG, FI	FI, LA, ETV, HG		SA

TABLE 1.2

Specifications for selected quadrupole-based ICP-MS instruments

Manufacturer	Agilent Technologies	Varian Inc.	Spectro	VG Elemental	Finnigan MAT (Turner)	Perkin-Elmer SCIEX	Thermo Jarrell Ash
Model	HP 4500	UltraMass 700	Spectro mass 2000	VG PQ ExCell	TS SOLA	ELAN 6100	Ion Scan/ POEMS
Resolution	0.65 amu (10% peak max.)	0.5-1.5 amu (5% peak max.)	0.4-1 amu (10% peak height)	0.75 amu (10% peak height)	1.0 amu (10% peak height)	0.3-4 amu (10% peak max.)	0.3-3 amu (10% peak max.)
Mass range	2-260	3-256 amu	1-240 amu	2-260 amu	0-250 amu	3-250 amu	1-300 amu
Abundance sensitivity	~10^7	10^7 at 1 amu higher; 10^6 at 1 amu lower	>10^7 on high mass side; >10^6 on low mass side		>10^6	>10^7 at 1 amu higher; >10^6 at 1 amu lower	1 x 10^6 and 2 x 10^6 at 23±1 amu
Background level across mass range	<2 cps	<20 cps	<10 cps at mass 8 and 230	<0.5 cps at mass 5 and 220	<50 cps	<10 cps	<20 cps
Scanning rate	2580 amu/s	2000 amu/s			2500 amu/s	2400 amu/s	

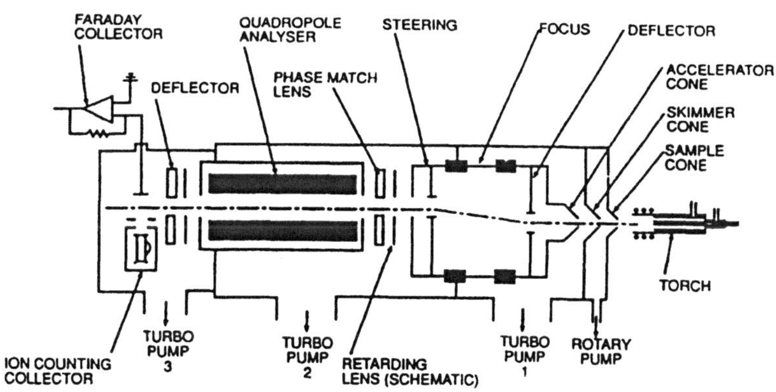

Fig. 1.9. (b) SOLA. Courtesy of Finnigan MAT.

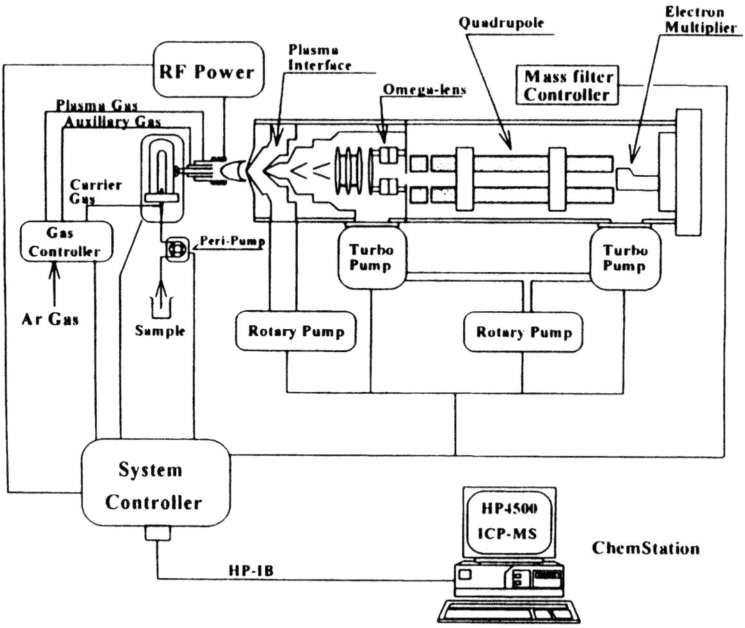

Fig. 1.9 (c) HP-4500. Courtesy of Agilent Technologies.

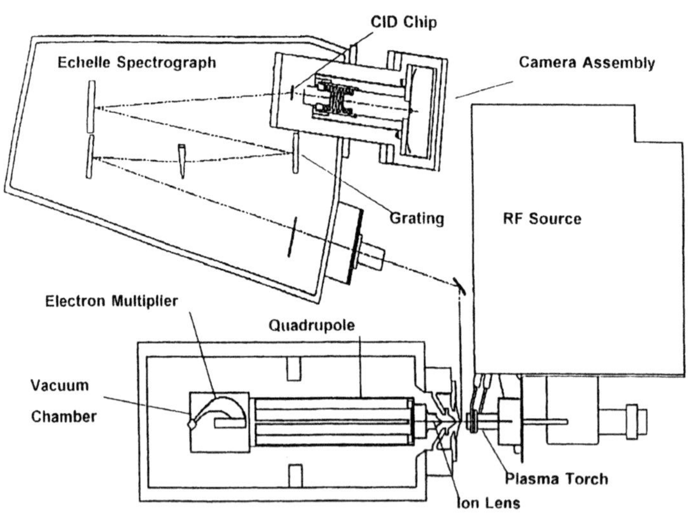

Fig. 1.9 (d) POEMS. Courtesy of Thermo Jarrell Ash.

Fig. 1.9 (e) ELAN 6100. Courtesy of Perkin-Elmer.

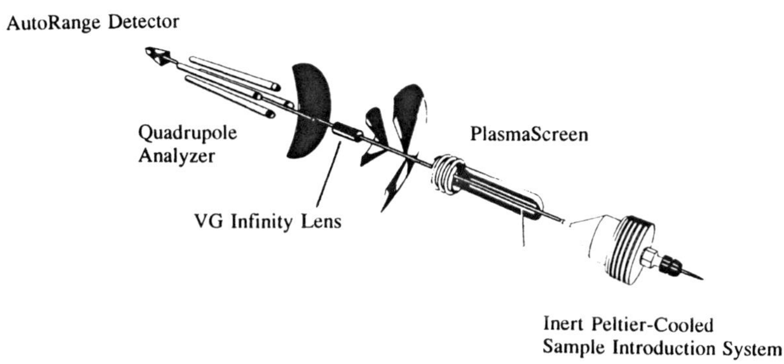

Fig. 1.9 (f) PQ Excel. Courtesy of VG Elemental.

vary between instruments. In particular, the Ion Scan (and the POEMS which consists in an Ion Scan with an optical detection system) uses 1.2-mm i.d. sampler and skimmer orifices (with appropriately larger pumping capacity) which makes these cones less prone to clogging [95]. All manufacturers provide a mass-flow controller for the aerosol carrier gas (and some of them also provide controllers for the other gases, for maximum stability). Except for the HP 4500 and the Ion Scan where the ion transmission, peak shape and better resolution at high mass are attributed to a hyperbolic quadrupole, the instruments use a cylindrical quadrupole.

Most of the manufacturers use a channel electron multiplier (CEM) as a detector. The current reaching the detector is continuously monitored by the software to ensure that no count rate exceeding 10^6 cps is measured as this would fall outside the linear response range of the detector and would also decrease its useful lifetime [7]. In the case of the TS SOLA, the ion beam is first deflected towards a Faraday cup and if the current is low enough, then deflected towards a secondary electron multiplier (SEM). Since the introduction of Active Film Multipliers by ETP Scientific Inc., several manufacturers have adopted this discrete dynode electron multiplier which features high ion signal count rate with a very low background count rate i.e. enhanced sensitivity compared to that achievable with a CEM [96]. For instance, installing this new type of detector on ELAN 5000 instruments resulted in sensitivity improvements of 30 to 100% without significant increase in the background count rate [96]. Furthermore, the lifetime of this detector is also 1.5–3 times that of a CEM. That is why it is now standard on the ELAN 6000 [97]. All detectors have a finite lifetime and are typically replaced when a significant loss in gain is observed. However, detectors near the end of their useful lifetime may not be adequate for the precise measurement of isotope ratios [98]. Significant changes in dead time were indeed observed before any change in gain was noticed [98]. Regular monitoring of the detector's dead time may therefore be a more objective method for deciding when it should be replaced [98].

In all cases, the software, which is mouse-driven, is multi-tasking and allows the user to customize routines for specific applications. Further user-friendliness is achieved by a substantial amount of automation i.e. computer control of a number of operating parameters in all instruments, some of them being totally computer-controlled (i.e. the only switch to activate is that of the computer). The first fully computer-controlled commercial ICP-MS instrument was the UltraMass from Varian which was released at the 1993 Pittsburgh Conference. A year later, at the 1994 Pittsburgh Conference, the first bench-top ICP-MS instrument (which is also fully computer-controlled), the HP 4500, was introduced by Hewlett-Packard (now Agilent Technologies).

In the original configuration which is still used by several of the manufacturers, the top of the load coil, i.e. the turn closest to the mass interface, is grounded (this is also called a front-grounded coil or an inverted-coil geometry). Under these conditions, the rf voltage anywhere on the coil is of the same polarity and is maximum at the other end of the coil. It also leads to a high rf voltage in the plasma (as high as 120 V [99]) as a result of stray capacitance between the coil and the plasma. This plasma potential with respect to the interface, if too high, results in electrical arcing between the plasma and the sampler ultimately producing a secondary discharge (i.e. a micro-plasma) into the sampler orifice [99]. This secondary discharge not only etches the cone orifice thereby producing a high background at the masses of cone-related elements, but it also constitutes a secondary ionization source which is not directly controlled by the operator. It therefore prevents a representative sampling of the ICP, unless a bias potential is applied to the extraction system to reduce its formation. Since the magnitude of the secondary discharge depends on the plasma operating conditions and sampling position, this bias potential then has to be readjusted after every change [63].

Douglas and French [99] were the first to find a simple solution to this problem: grounding the coil near its centre (which is then called a centre-grounded coil). Under these conditions, each half of the coil has a similar voltage to the other half but of opposite polarity, thereby resulting in a nearly zero voltage in the plasma. This approach was patented by SCIEX as the "PlasmaLok" and implemented on ELAN 250 and 500 instruments. With this modified impedance matching network, the plasma conditions can be optimized independently from the mass spectrometer parameters. In fact, constant ion energy population distributions are achieved for wet and dry plasmas (i.e. under completely different solvent loading conditions) which means that no re-optimization of ion focusing conditions is required [100]. This allows profiling of the plasma or modification of the plasma (for instance, the formation of mixed-gas plasmas) without having to re-adjust the ion lens settings after each change in the plasma operating and/or sampling conditions. Ions entering the mass spectrometer also have a narrower energy spread which means that a better resolution can be achieved at a given rf to dc ratio [99]. In addition, the level of doubly-charged species is reduced, as well as the background from ions originating from the sampler and/or skimmer material, and the lifetime of the sampler is extended [99]. The only drawback of the approach is a slightly higher analyte oxide ions level since the secondary discharge is no longer there to facilitate their breakdown [99].

A subsequent study performed by another group [101], which compared the centre-grounded and front-grounded coils found contradictory results. Although a narrower plasma was produced and the temperature of the interface was higher (30 vs 15°C) with the centre-grounded coil, the secondary discharge was not significantly reduced [101]. The load coil geometry also had only a minor influence on the analytical characteristics (sensitivity, detection limits, etc.) although it dictated the optimal operating conditions [101]. However, these two geometries were tested on a different instrument than that used by Douglas and French. A 3-turn coil grounded in the centre [101] was used instead of a 4-turn coil grounded at 1¾ turn from the exit end of the coil [99]. It is also unclear whether the sampler was grounded or not [101] whereas the sampler was grounded on the instrument used by Douglas and French [99].

The plasma interface is therefore critical to avoid a secondary discharge. Otherwise, a much lower n_e than expected is obtained behind the skimmer whenever the plasma potential is high, which, in turn, translates into degraded ion transmission [71]. On the other hand, the secondary discharge can be reduced by decreasing the solvent load in the plasma [7], which can be achieved by reducing the aerosol carrier gas flow rate or, even more efficiently, by desolvating the aerosol [102]. This is probably why several manufacturers are using the standard front-grounded geometry. Although similar analytical performance to that achievable on instruments free of a significant secondary discharge is possible (see Section 1.3), the mass spectrometer operating parameters (in particular, the potential applied to the lenses focusing the ion beam, and/or the bias potential to the interface that was discussed above) are not as independent of the plasma operating conditions [7]. More tuning may be required, especially if various analytes must be determined in a variety of sample matrices. In fact, even a change in the moisture content of the aerosol has been demonstrated to significantly affect the bias potential (as well as the optimum sampling position) [55].

Different modified tank circuit load coils have subsequently been used by Perkin-Elmer SCIEX [18] and others to reduce or eliminate the secondary discharge [103]. The improved PlasmaLok interface of the ELAN 5000 provides constant ion focus even when the plasma operating conditions are changed, thereby allowing a single nebulizer flow rate to be optimum for elements across the mass range, under wet or dry conditions [102,104]. (Although this peculiarity has been assigned solely to the PlasmaLok [104], other operational parameters must also have an effect since the observation made on instruments with a secondary discharge [77,78] that a higher nebulizer flow rate is required to maximize the response of lighter elements than for heavier elements, is also regularly reproduced on the ELAN 500 of the author.) Varian is using the "Turner" interlaced induction coils which were

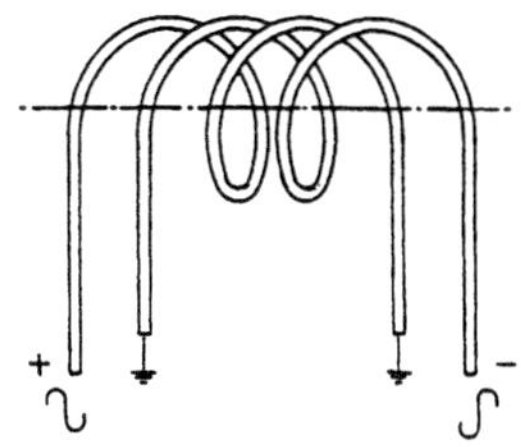

Fig. 1.10. Schematic of the Turner Interlaced Induction Coils. Copyright Varian Associates 1994. Used with permission.

specifically designed by Ian L. Turner (not to be confused with Patrick J. Turner, the originator of the TS SOLA instrument) to ensure an optimal coupling of rf power into the plasma [105]. The system essentially consists of a front-grounded coil which is interlaced with a rear-grounded coil (i.e. with the ground at the farthest turn from the interface), as illustrated in Fig. 1.10 [105]. With the appropriate impedance matching network, some components of the electromagnetic fields add up whereas others tend to cancel out. The net result of this approach is a measured coupling efficiency of 90% [105] with essentially no secondary discharge [106].

Both Seiko and Hewlett-Packard developed devices to reduce the secondary discharge, based on the shielded ICP described by Gray [107]. The SPQ-8000A uses a stainless steel tube inserted between the torch and the load coil. When the ICP is moved against the interface, the tail of the shielding tube touches a stainless steel pole which is itself connected to the grounded interface [108]. This decreases the average kinetic energy of analyte and polyatomic ions and narrows their kinetic energy distribution, which then improves their transmission efficiency and facilitates the optimization of ion lens settings [108]. The ShieldTorch interface from Hewlett-Packard consists in a grounded copper shield plate which is inserted between the torch and the load coil [77]. Using this device, no secondary discharge could be visually observed, and signals from C^+, N^+, Ar_2^+ and ArO^+ were completely suppressed [77].

Regarding the operating radio frequency of the generator, no consensus seems to exist. A study showed that 40.68 MHz lead to an enhanced secondary discharge with an associated increase in doubly-charged species, and a reduction in oxide species [109]. Detection limits were also slightly degraded but, on the other hand, non-spectroscopic interferences (see Section 1.4) were reduced [109]. It is likely that under discharge-free conditions, the advantages reported in ICP-AES of improved background stability and detection limits [53] would also be observed in ICP-MS. A study performed on a discharge-free ELAN-5000 instrument indeed showed that a lower aerosol carrier flow rate was required with a 40.68-MHz free-running rf generator than with a 27.12-MHz fixed-frequency rf generator to obtain maximum analyte sensitivity [100]. Since lower oxide levels are also achieved under these conditions, significant improvements in analyte sensitivity (at least a factor of six) resulted [100]. This generator, which was specifically designed for ICP-MS, has no moving parts for impedance matching and therefore allows a more rapid automatic tuning than is possible with other generators, thereby providing more robustness towards changes in sample matrix composition [100].

A variety of plasma torches is used by the various manufacturers. Some have a demountable injector tube to customize the torch to the type of samples analyzed or to the particular sample introduction system used. For instance, alumina should be used instead of quartz for corrosive samples, including those containing HF. In addition, the quartz injector should be replaced by one made from a different material (such as Pt) for the trace determination of Si, as this component was found to be the major source of Si background [110]. Both smaller i.e. 9-mm i.d. outer tube, working at 850 W and 8.7 l min^{-1} [111], and larger torches i.e. 23-mm i.d. outer tube working at 1.5 kW and 19.2 l min^{-1} [112] were tested. A water-cooled low-flow torch was also used to decrease the total argon flow rate to 2.3 l min^{-1} but resulted in increased ion energies [113]. This probably resulted from worsened secondary discharge formation as can be deduced from the huge doubly-charged ion level then observed [113]. The most encouraging results i.e. similar analytical characteristics to a conventional torch, were obtained with the 9-mm torch, but the optimal sampling position was then different for each element [111]. A type of torch which may yield interesting results is the linear-flow torch which can be relatively easily fabricated by adding an insert to a conventional tangential-flow torch [114,115]. This insert presumably makes the outer gas flow laminar instead of tangential as in regular torches, providing similar or improved detection limits in ICP-AES at a lower total gas flow rate because of reduced vortex instabilities and better long-term stability [114, 115]. It would, in fact, reduce or eliminate discrete frequency noises in the ICP [74]. In any case, the internal diameter of the injector tube is important. For instance, a 1.5-mm bore injector was observed to be less prone to clogging than a 1-mm one [116].

The positioning of the plasma with respect to the sampler is another important parameter which is entirely computer-controlled on some of the instruments. As was seen in Section 1.2.1, the diameter of the region of the plasma where analyte ions are present is fairly narrow, so the plasma should be well centred on the sampler orifice to maximize ion sampling and, therefore, transmission. Furthermore, any change in aerosol carrier gas flow rate, power and even moisture content of the aerosol [55] will modify the optimum sampling position. That is why several groups [117,118] have modified the alignment system of the ELAN (where the torch box was tilted on a single rod instead of being translated in three dimensions, parallel to the main axis) to facilitate this process. The new ELAN 6000 includes a three-dimensional translation system.

Once the ions exit the skimmer, they must be focused into the quadrupole. Electrostatic lenses are used for this purpose by all manufacturers, with a variety of configurations. Since, in most cases, the longitudinal axis of the plasma is identical to that of the quadrupole, the instruments include at least one photon stop on that axis to block as much plasma light as possible. Indeed, all the detectors used also respond to light. To further decrease this background, the transmitted ions are deflected into the detector which is then mounted off the quadrupole axis. (The Varian instrument has the detector housing on axis but the detector itself is off the quadrupole axis.) The size of the photon stop is fairly critical since it must be large enough to block light but not too large otherwise the efficiency with which ions can be deflected around it decreases [119].

The lower background count rate (see Table 1.2), despite a similar sensitivity (see Table 1.5), of the UltraMass compared to that observed on the other instruments employing photons stops, may be attributed to a photon stop of more optimal dimension and/or the low operating pressure of the second chamber (containing the ion lenses). Recent results have indeed demonstrated that part of the background originated from collisional processes in the ion beam [106]. The background was found to decrease as the pressure in the second chamber was lowered; and the background level was also affected by ion lens settings, two parameters which have no effect on the level of plasma photons reaching the detector [106]. In contrast, no significant change in background was noticed when the pressure of the quadrupole chamber was changed [106].

A completely different approach was used by Hewlett-Packard/Yokogawa Electric, where the longitudinal axis of the plasma, interface and ion lenses has a sufficient offset from that of the quadrupole to eliminate the need for a photon stop, and allows the detector to be mounted on the quadrupole axis. In particular, an Omega lens is used to slightly deflect the ion beam into the entrance of the quadrupole, whereas light is efficiently blocked by the wall of this third stage. As can be seen in Table 1.2, this approach seems the most efficient at reducing the background across the mass range. However, the low background may also partly result from a lower operating pressure (unspecified in the documentation received from most manufacturers) as discussed above.

The ion extraction system of the TS SOLA is also different from that of the other manufacturers. An accelerator cone, placed at about 10 mm downstream from the skimmer, and to which 2 kV is applied, generates an accelerating and convergent electrostatic field which partly counteracts space charge repulsions between ions [120]. In addition, a transfer lens system held at a relatively high potential helps to focus ions with a wide energy spread into the quadrupole of the TS SOLA, which is positioned parallel but with an offset to the longitudinal axis of the plasma [120]. A retarding lens is used to reduce the velocity of the ions to one compatible with the quadrupole mass filter. Hence, no photon stop is required. The accelerator cone also serves as a third sampling aperture, creating a second differentially-pumped region [120].

Alternatively, the Ion Scan uses offset ion optics. As shown in Fig. 1.11, although the quadrupole is on the longitudinal axis of the plasma and interface, the ion beam is deflected off axis and back on axis into the quadrupole by this offset ion lens thereby eliminating the need for a photon stop [95]. The position of the first ion lens element was also found to be critical in terms of sensitivity. A substantial increase in sensitivity was indeed observed when this lens was positioned as far as possible into the skimmer (too small a separation between the tip of the skimmer and the lens resulting in an electrical discharge) [95]. The improvement in sensitivity was attributed to a mitigation of space charge effects which are discussed in Section 1.4.3 [95].

In any case, all instruments provide a resolution of about 1 amu, defined as the peak width at 10% peak height. Whatever the lens system used, the potential applied to each component must be optimized for the analyte(s) determined, in addition to the plasma

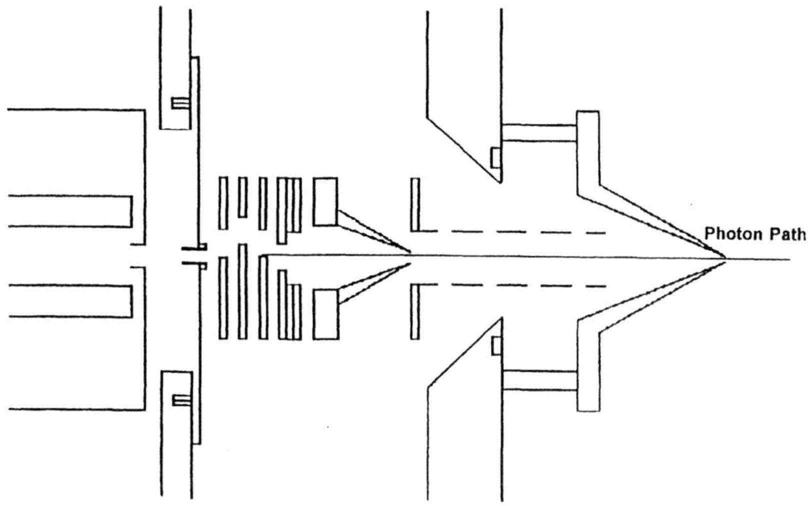

Fig. 1.11. Ion optic design of the Ion Scan. Courtesy of Thermo Jarrell Ash.

operating conditions. If widely different elements are to be detected, this optimization will, in most cases, involve the selection of compromise conditions since, except for the Sola and Ion Scan instruments (see Section 1.3), the MS response is not uniform across the mass range, and the optimum aerosol carrier flow rate generally depends on the mass of the analyte [83] (except on the ELAN 5000 and 6000, as discussed earlier). Alternatively, the elements may be subdivided into smaller mass ranges for which the lens settings can be better optimized [121]. If the whole mass range is covered, conditions are typically selected to maximize the response of an element in the middle of the range (Rh or In, for instance) while those of low and high mass elements are made equal to one another and as high as possible without degradation of the response of the middle range element.

A simplex optimization [122] may be carried out to facilitate this process. This approach has the advantage of varying all parameters concurrently to find the maximum of the response surface. However, depending on the efficiency of the simplex method used [123], it may yield a local maximum which is not the optimum, so univariate searches must also be carried out following the simplex optimization to verify that the optimum conditions have indeed been found. Furthermore, a suitable response function must be defined. The successful optimization of both the plasma operating conditions and ion lenses settings for the multielemental determination of a series of elements with various first IP and spanning the mass range was recently carried out [124]. The figure of merit used was $n/\sum(\mathrm{SBR})^{-0.7}$ where n is the number of elements and SBR is the signal-to-background ratio for individual elements [124]. This allowed optimization of the analyte signal without an increase in the corresponding background which could lead to a degradation of detection limit, and also prevented bias toward the most sensitive elements [124]. However, manual univariate optimization yielded conditions resulting in similar detection limits to those obtained after simplex optimization [124]. In any case, since the voltages applied to the various ion lenses are interdependent, it may be possible to find various settings resulting in a similar ion throughput. This was demonstrated with simulations of ion trajectories through the ion lens of an ELAN-250 [125], but is likely true for the various configurations used by all manufacturers.

Several manufacturers now offer instruments incorporating collision-cell technology where the ions pass through some sort of reaction cell prior to entering the quadrupole mass filter [90]. Selected interfering polyatomic ions can then be eliminated through ion-molecule reaction with a small amount of gas, such as H_2. Indeed, the admission of H_2 in the post-skimmer region was demonstrated to selectively reduce Ar^+ [92]. Various configurations are used by different manufacturers. For instance, Micromass uses a hexapole collision cell that is on at all times, even under standard operating conditions [90]. In contrast, the PQ ExCell comes with a collision cell as an option. The latter is used for selective attenuation of spectroscopic interferences from polyatomic ions as well as to lower the background to less than 0.1 counts/s [90]. The ELAN, which now comes as version 6100, is also available with a dynamic reaction cell (DRC) as an option. However, this device is claimed to be a much lower energy device than a collision cell, and hence to provide more selective ion chemistry [90]. Both the reaction gas and the bandpass can be selected to overcome spectroscopic interference. In general, the reaction gas should have a smaller mass than that of the analyte, to minimize scattering losses [126]. Furthermore, the bandpass, which changes the low-mass cut-off of the DRC, can be adjusted by changing the rf potential that is applied to the rods [126]. The collision-cell/DRC technology shows promise as a cheaper alternative to high resolution instruments, but it is still in its infancy [90].

1.2.2.2 High resolution

Double-focusing instruments were the second type of mass analyzers to be coupled to ICP-MS in commercially available instruments. Despite the facts that they are significantly more expensive than the other types of ICP-MS instruments and that they require more expertise on the part of the operator, the magnetic-sector ICP-MS market has steadily increased over the years since 1994 [90]. This is because the only general method for overcoming spectroscopic interferences from polyatomic species is to use high mass resolution, a feature that can be achieved with a double-focusing mass analyzer, i.e. combining a magnetic sector with an electric sector field analyzer [127,128]. The status of this instrumentation was recently reviewed [127, 128] and was the subject of an extensive bibliography [129].

Diagrams of selected high-resolution ICP-MS instruments are shown in Fig. 1.12 and their main characteristics and specifications are summarized in Tables 1.3 and 1.4. Since the ICP-MS market is in full expansion, several changes had been made as this chapter was being completed. For instance, newer versions (Element 2, PlasmaTrace 2, PLASMAX 2, etc.) have appeared, as well as instruments with multiple collectors for precise isotope ratio measurements (Neptune from Finnigan MAT, IsoProbe from Micromass, Nu Plasma from Nu Instruments) [90]. Nonetheless, the ICP source and sample introduction system are similar to those used for low-resolution ICP-MS (see Sections 1.2.2.1 and 1.2.3 for a detailed discussion of these components, respectively). However, owing to the greater complexity of the mass spectrometer, high-resolution instruments have larger dimensions than the quadrupole-based ones. All are double-focusing, i.e. use an electrostatic and a magnetic analyzer in series. As can be seen in Tables 1.3 and 1.4, instruments of reverse (i.e. with the magnet before the electric sector) or regular geometry can achieve similar resolution. The ELEMENT from Finnigan which was specifically designed for elemental analysis features a laminated magnet for rapid scan [130]. All instruments include a gate valve to allow a change of skimmer without losing vacuum in the chamber(s) containing the magnet and the electric sector.

The plasma interface is quite different from that for quadrupole-based instruments since, in addition to having to pass through several differentially pumped regions, the ions have to be accelerated to a kinetic energy of about 5–8 keV before entering the analyzer [131]. This was accomplished by floating the plasma and the whole interface of the

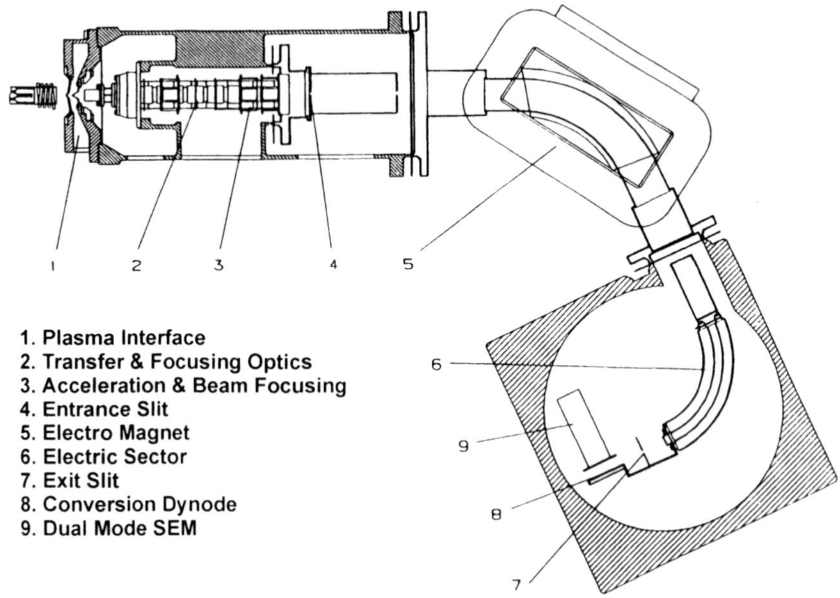

Fig. 1.12. Schematic diagrams of some of the commercially-available sector-based ICP-MS instruments. (a) ELEMENT. Courtesy of Finnigan MAT.

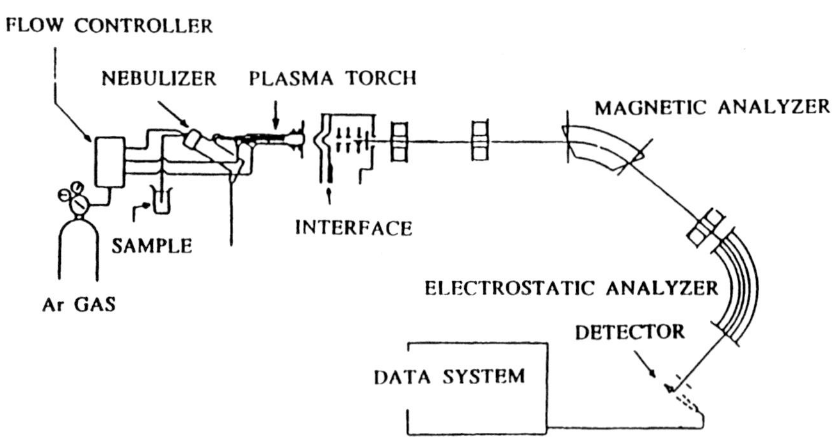

Fig. 1.12. (b) PLASMAX-1. Courtesy of JEOL.

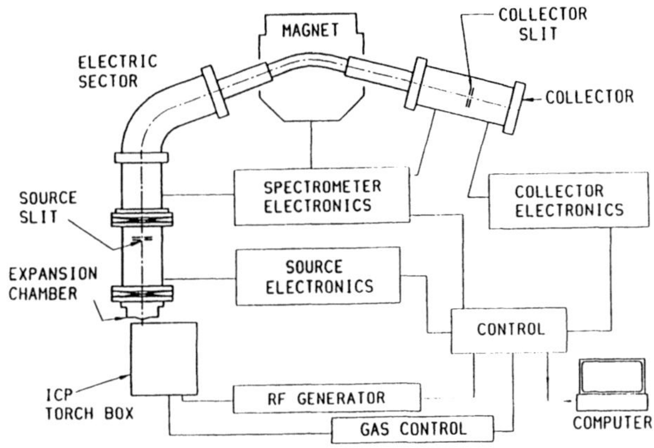

Fig. 1.12. (c) PlasmaTrace. Courtesy of Micromass.

TABLE 1.3

Main characteristics of selected double-focusing and TOF ICP-MS instruments

	Manufacturer					
	Finnigan MAT	JEOL	Micromass	VG Elemental	GBC Scientific Equipment	Leco
Model	ELEMENT	PLASMAX-1	Plasma-Trace	VG Axiom	OptiMass 8000	Renaissance
Sample introduction system	4-channel peristaltic pump, external water-cooled spray chamber and Meinhard nebulizer	pneumatic cross-flow nebulizer and Scott spray chamber		integral 4-channel peristaltic pump		4-channel peristaltic pump; concentric nebulizer; Wu-Hieftje F.A.S.T. spray chamber
Torch				fixed Pt M-Plasma-Screen between torch and load coil		
RF generator; maximum power	27.12 MHz, crystal-controlled (40 MHz option); 2 kW	40.68 MHz; 1.6 kW		27.12 MHz, solid state; 2.0 kW	27.12 MHz, solid state	dual 27.12/ 40.68 MHz

continued

	Manufacturer					
	Finnigan MAT	JEOL	Micromass	VG Elemental	GBC Scientific Equipment	Leco
Plasma positioning	x,y,z translation stage			computer-controlled x,y,z positioning		computer-controlled x,y,z positioning
Plasma interface	interface at ground potential (see text); gate valve to change cones while maintaining vacuum	interface at accelerating voltage (5 kV): require insulation		high-voltage interface; mass spectrometer at ground potential		
Mass spectrometer	double-focusing reverse Nier-Johnson geometry; laminated magnet	double-focusing reverse Nier-Johnson geometry	double-focusing	forward-geometry, double-focusing	orthogonal TOF	axial TOF
Pumping	5 stages: 1 rotary pump, 4 turbomolecular pumps	3 stages: 1 rotary pump, 2 turbomolecular pumps		4 stages	4 stages	3 stages
Detector	conversion dynode with 2-stage SEM	dynode electron multiplier	CEM	on-axis Faraday cup; off-axis electron multiplier (multicollector array detector option)	dual detection system with discrete dynode multiplier	electron multiplier
Detection mode	analog/ion counting		pulse counting or analogue			
Dimensions L × H × D (m)	2.2 × 1.39 × 0.88			2.46 × 1.48 × 0.72	1.2 × 0.6 × 0.82	1.68 × 1.35 × 0.79
Software	MS-Windows					MS-Windows based
Optional nebulizer	USN			USN, MCN, DIN	MCN, DIN, low-volume glass spray chamber	
Other options	ETV, HG, FI, LA, GD	GD	LA	LA, PrepLab	LA	LA, discrete solution analyzer

TABLE 1.4

Specifications for selected double-focusing and TOF ICP-MS instruments

	Manufacturer					
	Finnigan MAT	JEOL	Micromass	VG Elemental	GBC Scientific Equipment	Leco
Model	ELEMENT	PLASMAX -1	PlasmaTrace	VG Axiom	OptiMass 8000	Renaissance
Resolution	300, 3000, 7500 amu (10% valley)	500-10,000			1800 at *m/z* 238	450 at *m/z* 208
Mass range (amu)	2-260	1-500		2-260	3-260	1-275
Background level (cps) across mass range	<<0.1	<<0.1	<<0.1	<0.1		1-5 cps (per point)
Maximum scanning rate	full mass range <0.6 s				30,000 full range mass spectra per second	30,000 full range mass spectra per second

PlasmaTrace (including the sampler and skimmer) to a few kV, or floating the whole interface at a high voltage and using a front-grounded coil which is galvanically isolated from the plasma torch on the PLASMAX-1. In either of these configurations, where acceleration of the ions occurs in the second vacuum region, safety then dictates that all mechanical parts be isolated and the interface be fully shielded, unlike in quadrupole systems where the interface is readily accessible. In contrast, the ELEMENT from Finnigan MAT accelerates the ions in the third vacuum region [131]. As a result, the whole interface is grounded and a front-grounded load coil can be used without additional shielding, just as in some quadrupole-based systems. Also, because the interface is at ground potential, the gate valve is positioned directly behind the skimmer and the third vacuum region does not have to be ventilated and subsequently evacuated whenever the skimmer is changed, unlike in the other two systems [131]. In sector-field MS, it is important to minimize the energy distribution of the initial ion beam that is accelerated since a wide distribution would degrade the focusing abilities and sensitivity of the technique [132]. Waiting until the third vacuum stage (such as in the ELEMENT) for the acceleration process, ensures that the sampled beam consists principally of ions and hence minimizes energy distribution and loss in signal [132]. In contrast, in the second vacuum stage, the sampled beam probably has both plasma and ion-beam characteristics i.e. a relatively broad range of energy distribution, which must be reduced by some other means, such as the use of a torch shield [132].

Similarly to quadrupole-based instruments, several of the instruments use a single detector. A mass spectrum is obtained by varying either the electrostatic field (also called acceleration potential) or the magnetic field. An example of the mass spectrum resulting from a continuous scan of the electrostatic field at a resolution (R) of 300 (5% peak valley definition) is shown in Fig. 1.13. In contrast to quadrupole-based ICP-MS where the peak width is constant across the mass range (see Section 1.2.2.1), the peak width increases with m/z because R (defined as $M/\Delta M$, where M is the mass of the lightest of two adjacent peaks of similar intensity and ΔM is the mass difference between them) is constant. Furthermore, by suitable adjustment of the widths of the entrance and exit slits, peaks with a flat top can be obtained (as shown in Fig. 1.13c) which increases the reliability of peak-hopping. At low count rates, the effect of short-term fluctuations of the ICP (see Section 1.4.1) can degrade the peak shape, and peak area may be used for quantitation to minimize the associated fluctuations in peak height [133]. In any case, the background is negligible because of: (1) the curved path that the ion beam must follow, which efficiently prevents photons from reaching the detector; (2) the operating pressure which is considerably lower than that of quadrupole-based instruments, and as discussed in Section 1.2.2.1, reduces collisional processes in the ion beam [106].

An original feature of the ELEMENT is the SynchroScan where both the magnetic and the electrostatic fields are changed simultaneously to mimic the peak-hopping mode of quadrupole-based instruments [131]. The principle is illustrated in Fig. 1.14. Small changes in the electrostatic field are made to oppose the effect of the magnetic field on a small range of masses around a given mass such that only that given mass passes the exit slit during a small part of the magnetic field scan. The process is then repeated for the next mass, etc. The result is shown in Fig. 1.15 which compares the mass spectra obtained with a scan of the magnetic field only to that with SynchroScan.

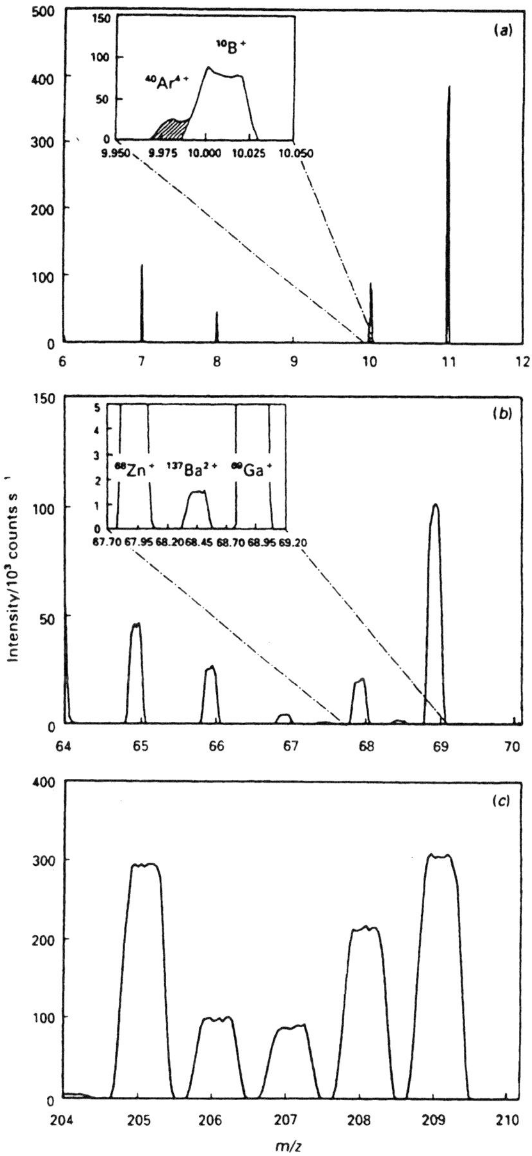

Fig. 1.13. Mass spectra obtained over (a) low, (b) medium and (c) high mass ranges on a high-resolution instrument operating in low resolution (i.e. R = 300) using 100 ms per data point and 20 data points per peak. Reproduced for Ref. [136] with permission.

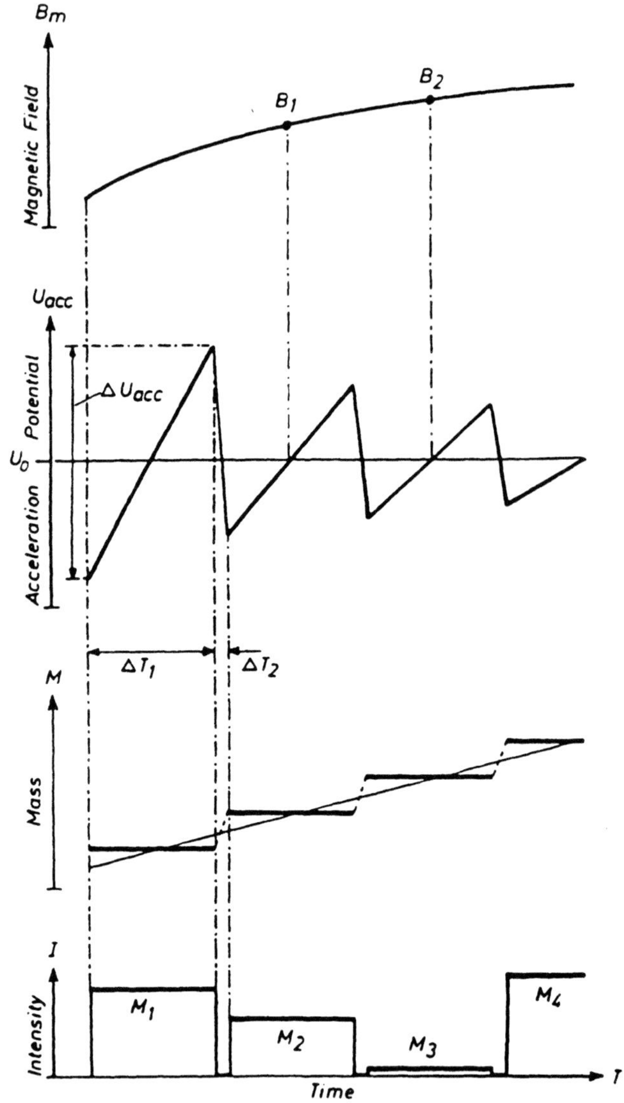

Fig. 1.14. Principle of SynchroScan where the acceleration potential and the magnetic field and changed simultaneously. B_m is approximately $t^{1/2}$, U_{acc} is approximately t/M_x, M_{Bm} is approximately t, M_x is the mass registered, Δt_1 is the measurement time and Δt_2 is the jump time.

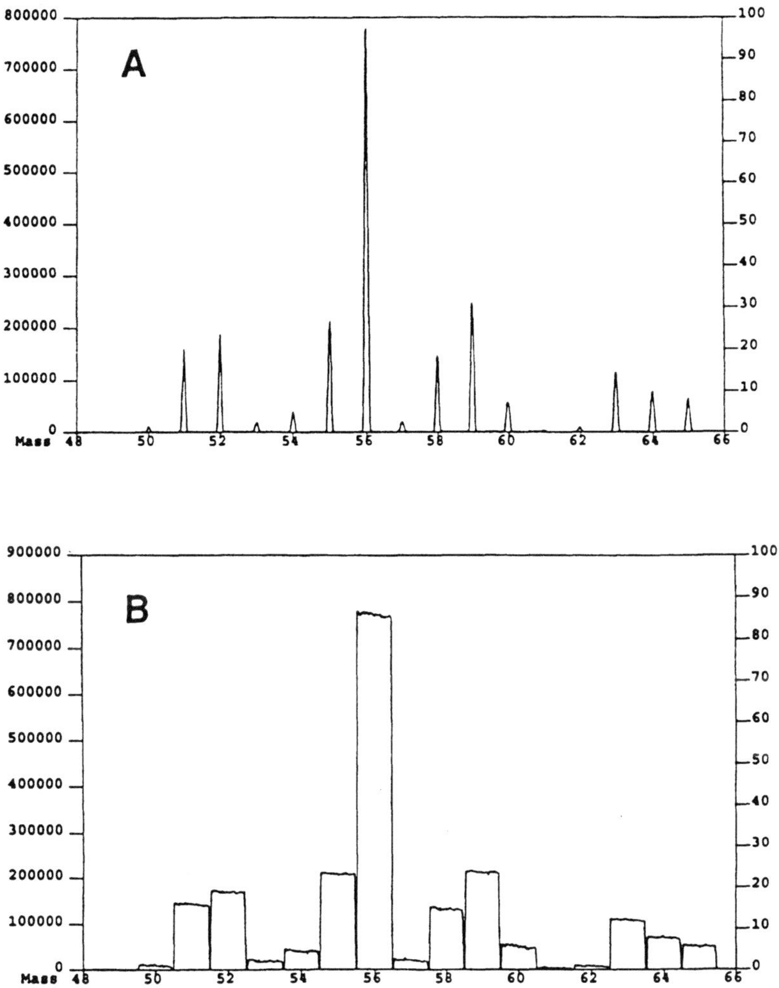

Fig. 1.15. Mass spectra obtained for a multielement standard using 10 ms/Da with (a) a scan of the magnetic field and (b) SynchroScan.

The Plasma 54 (from VG Elemental) is a modified version of the PlasmaTrace. It is operated with a resolution fixed at 400 (5% valley definition) and is equipped with seven moveable Faraday cups [134,135]. The mass spectrometer provides a 540-mm dispersion in the image plane perpendicular to the ion optical axis where the adjustable detectors are located [135]. It therefore allows a truly simultaneous detection over a narrow mass range. On the other hand, it can only be used to monitor the isotopes of one or two elements at a time (depending on the number of isotopes available per element and the mass range they cover), and is specifically suited to the measurement of isotopic ratios. Similarly, the other multi-collector instruments on the market cannot be operated in high resolution [128], the highest resolution available varying from 2500 to 4000 (10% valley definition) [90].

As far as optimization is concerned, precise mass calibration which is required for any mass spectrometer, is even more critical for high-resolution instruments. On the other hand, because the ions are acquiring high energy from the accelerating voltage applied at the interface and the energy focusing of the electrostatic analyzer, there is no mass dependence of the optimum aerosol carrier flow rate and the plasma rf power, unlike that observed with several quadrupole-based instruments [136]. In all cases, resolution is typically increased by decreasing the slit width of the entrance and exit slits. This, however, is commensurate with a reduction in ion transmission which translates into a loss of sensitivity (see Table 1.6). Furthermore, the slits have a finite lifetime. For example, the average lifetime of a high-resolution slit is about 300–400 operating hours [137]. Material sputtered off the low-resolution slit and deposited on the high-resolution slit caused an increase in resolution as a function of time with a concurrent decrease in sensitivity [137]. As a result, more frequent mass calibration is therefore required for a sector-based instrument than a quadrupole-based one.

1.2.2.3 TOF-ICP-MS

The most recent addition to the family of ICP-MS instruments has been the TOF-type. Although TOF mass analyzers traditionally provide a low resolution, and it is this type that, at the time of writing, uses an ICP as an ion source, recent developments in TOFMS have greatly improved mass resolution [138], so it is very likely that more developments will take place in ICP-MS in the near future. One of the commercially available TOF-ICP-MS instruments indeed provides a medium resolution of 1800 at m/z 238 [90]. In TOF-type instruments, ions are electrostatically accelerated to a uniform kinetic energy so that ions of different m/z travel at different velocities in the flight tube, lightest ions travelling faster than heavier ions. An ion mirror can be used to double the flight path so that a better separation of individual m/z species is achieved in time.

The two instruments that were available at the time of writing differed in configuration. One has the TOF analyzer orthogonal to the ICP-sampling interface whereas the other has it on axis so that the ions are extracted and accelerated along one axis. Whether orthogonal or axial acceleration is carried out, the ions that are extracted through an interface similar to that of other types of mass analyzers, must be guided by lenses to an acceleration region where a large voltage pulse must be applied to push all the ions into the field-free flight tube. The flight time of all the ions can be measured, in contrast to other mass analyzers where only a single mass or set of masses can be detected simultaneously. Furthermore, since full mass spectra can be obtained at the rate of 30,000 per second, this instrumentation should be particularly advantageous for applications where the speed of acquisition and/or the simultaneous full mass spectrum are needed, such as the accurate and precise measurement of isotope ratios, and hence ID analysis, calibration using internal standardization and the analysis of small-volume samples. This TOF technology may also become unrivalled when the ICP is coupled to sample introduction methodologies (such as those discussed in the following chapters) that generate a transient or time-dependent signal.

On the other hand, full mass spectra really means that all *m/z* are detected, including those of background ions such as Ar^+. Since this could drastically decrease the lifetime of the detector, means must be taken to get around this problem. On the Renaissance, a mechanism is used to repel bundles of matrix ions partially or totally out of the flight path before they reach the detector. Several *m/z* regions of various widths can be selected to significantly decrease not only background ions but also those originating from the matrix. Nonetheless, only about 10% of ions are accelerated to the TOF tube [139]. The duty cycle is therefore reduced compared to quadrupole-based instruments [139]. This translates into detection limits that are about an order of magnitude worse than those obtained by quadrupole ICP-MS [139]. As a result, TOF-ICP-MS is not expected to replace quadrupole ICP-MS but it is clearly advantageous for the detection of transient signal (which is the focus of the following chapters) [139].

1.2.3 Sample introduction

Since the main purpose of this book is to describe modifications to the sample introduction system to generate transient or time-dependent signals, a detailed discussion of common commercially available systems is warranted, including their main limitations. The following chapters will address at least some of these limitations. Most of the systems discussed were also used with ICP-AES.

1.2.3.1 Common systems with pneumatic nebulizers

As can be seen in Tables 1.1 and 1.3, the most common sample introduction system for solution nebulization consists of a pneumatic nebulizer (PN) (where a gas is used to transform the solution into an aerosol) and a spray chamber. They are generally made of glass, which is satisfactory for most applications. However, whenever corrosive samples (such as those containing HF) have to be analyzed, the PN and spray chamber (as well as the torch injector) should be made of a material which is resistant to that matrix.

A variety of PNs are commercially available. They can be subdivided into concentric, cross-flow and Babington-type nebulizers [140, 141] which include the conespray nebulizer [142] originally proposed by Sharp [141]. A schematic representation of each type of nebulizer can be seen in Fig. 1.16: an aerosol is respectively created when the nebulizer gas flows around a small capillary supplying the solution through a constricted orifice (a), under (at a 90° angle) the tip of a capillary (b) or as a gas jet under a film of solution (c). Pneumatic

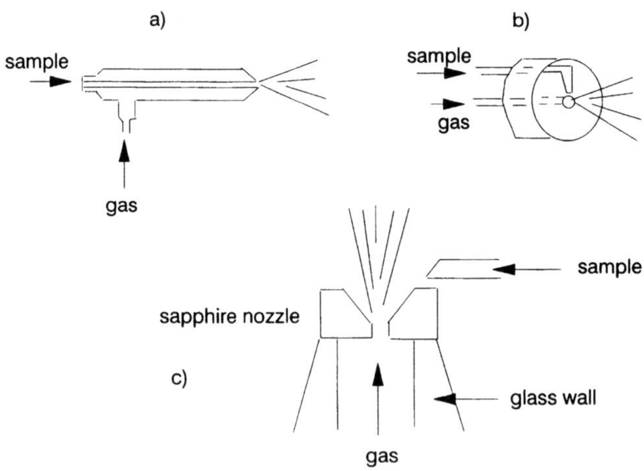

Fig. 1.16. Schematic diagram of pneumatic nebulizers: (a) concentric, (b) cross-flow, (c) conespray.

nebulizers are popular because they are relatively inexpensive and simple devices, which have been shown to provide good precision (typically 1% RSD and as good as 0.1% RSD [93]) and accuracy in ICP-AES [140].

The nebulizer gas serves two functions: formation of the aerosol and aerosol transport to the plasma. If no pump is used to supply the sample solution, it also serves as a pump. However, it is a rather poor pump since its aspiration rate is highly dependent on the viscosity of the solution being aspirated [143]. This is important whenever external calibrations are carried out: unless the matrix of the standard solutions is closely matched to that of the samples, determinate errors will result if a significant viscosity difference exists between these solutions. It is therefore not surprising that all manufacturers use a peristaltic pump to supply the sample solution to the nebulizer. The small sacrifice in precision which is commensurate with the pulsations of the rollers [75,93] on the flexible tube is a rather small price to pay in exchange for relative freedom from viscosity effects. Otherwise, a change in acid concentration and/or type could induce a change in the signal, as has been thoroughly discussed in a recent review paper [144].

Using a peristaltic pump also has the additional benefit of allowing a reduction in the uptake rate to the nebulizer independently from the nebulizer gas flow rate. In effect, the nebulizer can then be operated under 'starving' conditions i.e. at a lower uptake rate than the free aspiration one. For instance, the typical aspiration flow rate of a concentric nebulizer is around 3 ml/min. Yet, the sample uptake rate can be lowered down to about a third of this free uptake rate without a sacrifice in sensitivity [145]. Indeed, the amount of sample reaching the plasma barely changes as the sample uptake rate is changed from 0.4 to 2.0 ml/min [86]. Within this uptake rate range, transport efficiency into the ICP increases as the uptake rate is decreased, probably because less coagulation (as a result of fewer droplet collisions) then occurs within the spray chamber [86]. In any case, when starving the nebulizer, care should be taken to select a size of pump tubing small enough so that the pump can be operated at a relatively high speed to minimize the effect of pulsations. An improvement in precision can be gained by replacing the peristaltic pump with a pulseless pump, such as a dual-piston reciprocating pump with pulse damper or a gas displacement pump [93].

However, whether a pump is used or not, volatility differences between the analyte compounds in the sample and standards may lead to significant determinate error [146]. This problem arises when various forms of an analyte (such as different organometallic compounds) being introduced in a spray chamber have widely different boiling points: the vapour exiting the spray chamber may then be enriched in the more volatile compound [146]. This problem can be alleviated by measuring the analyte signal from several controls (i.e. containing compounds with significantly different volatility), and a standard solution (containing a non-volatile compound of the analyte) in addition to the sample at two different spray-chamber temperatures [146]. If all controls, standards and samples are prepared in a common solvent, then a volatility correction may be calculated even if the structure of the analyte compound in the sample is unknown [146].

The high-efficiency PN, which has recently become commercially available (from J.E. Meinhard Associates), provides 10–1200 µl/min uptake rates [147]. It yields similar detection limits as well as better short-term and long-term precisions than a conventional PN which operates at higher uptake rates (1–3 ml/min) [102]. In other words, this high-efficiency PN is at least 10 times more efficient at aerosol production than a conventional PN [102]. In addition, the droplets generated have a significantly narrower size distribution than that which is observed with a conventional PN, hence the better precisions obtained [147].

Nonetheless, in all cases, the aerosol generated is highly polydispersed i.e. droplets of a wide range of diameters, from submicrons to hundreds of microns, are produced [140, and references therein]. However, the biggest droplet size that can be desolvated by the plasma to yield analyte which can then be vaporized, atomized and ionized is restricted by the relatively short residence time of the droplets in the plasma which is dictated by the nebulizer gas flow rate (typically about 1 l min^{-1}) required to punch a hole in the plasma. For instance, a study performed with various nebulizers in ICP-AES reported that aerosol droplets ≤2.5 µm in diameter were more likely to yield analyte signal than larger droplets [148]. The acid concentration has been shown to affect the total aerosol volume that is contained in droplets smaller than 3 µm [144]. Not only can large droplets or particles "survive" the plasma as was clearly shown by Hobbs and Olesik [149,150], as well as Winge et al. [82], and can be inferred from stable isotope labelling experiments [151], but the vapour that can be generated from a droplet may locally cool the plasma [149,150] thereby degrading the local ionization efficiency of the latter and the sensitivity of the technique [149,150].

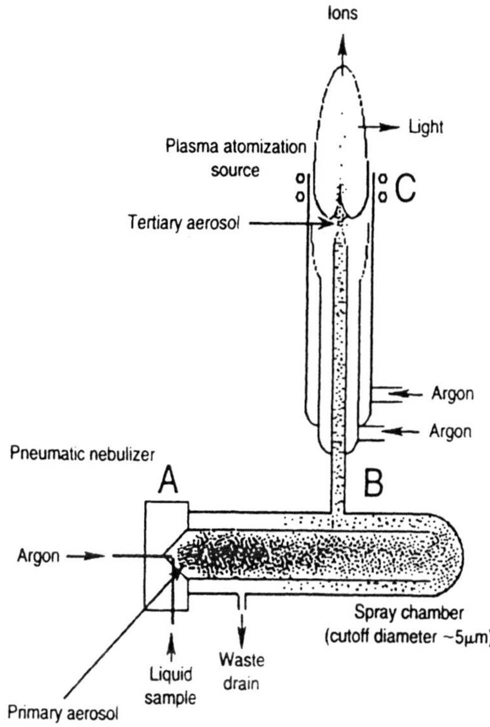

Fig. 1.17. Typical sample introduction system combining a pneumatic nebulizer to a Scott-type double-pass spray chamber. From Ref. [140] with permission.

A spray chamber is therefore typically used to 'filter out' large droplets. The most common is a Scott double-pass spray chamber, shown in Fig. 1.17 [152]. The aerosol has to flow through the inner tube of the chamber and then between its inner and outer tube before it can enter the plasma torch. Obviously, the smallest droplets can most efficiently follow this tortuous path without impacting and sticking somewhere on the walls of the spray chamber. Droplets exiting a Scott spray chamber typically have a diameter smaller than 5 μm although a small fraction of the bigger droplets can still remain [140]. The chamber has also been reported to dampen fluctuations due to the peristaltic pump [93]. It is usually made of glass although some manufacturers use other materials (like Ryton) to make it more corrosion-resistant and increase its wettability i.e. the ease with which solution sticking to its surface can drain down.

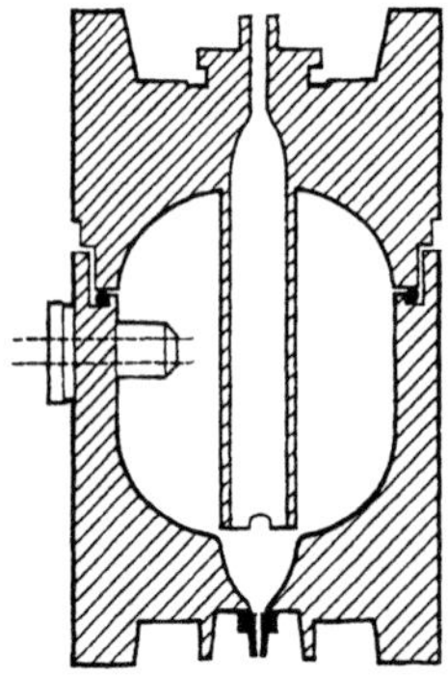

Fig. 1.18. Schematic diagram of the Sturman-Masters spray chamber. Copyright Varian Associates 1994. Used with permission.

The UltraMass instrument uses a completely different and original spray chamber design, the Sturman-Masters spray chamber, which is shown in Fig. 1.18. This chamber has a 95-ml internal volume, which is slightly smaller than the 110 ml of a typical Scott-type spray chamber [153]. The Renaissance TOF-instrument features an externally mounted, high-efficiency, rapid-rinse Wu-Hieftje F.A.S.T. spray chamber. Single-pass chambers with an impact bead (or an impact cup) [143] situated in front of the gas outlet are also available, especially in combination with Babington-type nebulizers. The impactor is then used to trap larger droplets and/or generate smaller ones. It has also been shown to reduce the noise induced by the peristaltic pump, especially when placed near (i.e. 3 mm) the capillary tip [154]. A cyclonic spray chamber has also been reported to greatly reduce the acid transient effects (where the analyte signal suddenly changes upon a change of acid concentration before stabilizing) that are observed with a double-pass spray chamber, presumably because the quicker drainage of the former reduces effects linked to solvent evaporation from the chamber walls [144].

Several cyclonic spray chambers are indeed available from various manufacturers. Those made of glass or polypropylene were reported to provide coarser tertiary aerosols, higher solution transport and solvent transport efficiencies, higher sensitivities and lower detection limits than a double-pass spray chamber in ICP-AES [155]. However, the signal was critically dependent on the position of the nebulizer within the cyclonic chamber [155]. Nonetheless, the glass cyclonic spray chamber provided the smallest memory effect when a Pd solution was nebulized [155]. Although a reduction in the size of the tertiary aerosol was noticed for all spray chambers upon an increase in acid concentration of the solution being nebulized, the cyclonic spray chamber made of glass provided the smallest change in the net emission intensity [155].

In any case, changes in the droplet size distribution can also occur through other mechanisms [140], including solvent evaporation which is faster for smaller droplets, particle coagulation where droplets combine to form bigger ones, and charging i.e. electrostatic interactions between particles or aerosol ionic redistribution [144]. The size distribution will therefore depend on a number of parameters: the nebulizer, the flow rate of the nebulizer gas, the spray chamber and the physical nature of the solution. For instance, a solution containing volatile solvents will lead to smaller droplets but also a greater solvent vapour load in the plasma.

Nonetheless, an improvement in detection limit may still result for several elements in the presence of organic solvents, despite a concurrent decrease in sensitivity in some cases, as illustrated in Table 1.9. Obviously, a degradation can be expected in low resolution for analytes at *m/z* corresponding to polyatomic ions (such as the interference of ArC on Cr). On the other hand, increasing the ethanol concentration was reported to increase the mean droplet size distribution compared to a similar aqueous solution [156].

The temperature of the spray chamber is therefore an additional parameter to control since it influences the plasma solvent loading (i.e. the amount of vapour reaching the plasma), and the aerosol characteristics (i.e. the droplet size distribution and the aerosol loading) [157, and references therein]. A decrease in spray chamber temperature will allow condensation of solvent vapour (for example, water) and may thereby reduce the formation of refractory oxides which, however, will not be completely eliminated since analyte-containing droplets will still reach the plasma. Nonetheless, a reduction of water vapour loading was reported to not only decrease the formation of polyatomic oxide species but also to improve analyte sensitivity [158] and reduce noise [154]. That is why several of the ICP-MS instruments maintain the spray chamber at a low temperature (typically 4°C) using a water-cooling jacket around it or a Peltier cooler. The latter has the advantage of being readily computer-controllable. However, reducing the temperature of a cyclonic spray chamber did not significantly reduce acid transient effects [144].

The effect of a change in water loading is, in fact, fairly complex. A decrease in moisture content, from that typical of a pneumatically generated wet aerosol down to a completely dry aerosol, induced a shift to shorter sampling position (i.e. the sampler had to be closer to the top of the load coil) as well as an increase in analyte sensitivity [55]. The latter presumably results from the increasing amount of plasma energy that is no longer being used for the dissociation of water molecules. The maximum analyte signal for a dry aerosol also lay within the load coil, and could therefore not be sampled unless the aerosol carrier flow rate was increased and/or the forward power decreased. The shift in optimal sampling position was visually observed to be commensurate with an axial shift of the whole plasma away from the injector [55].

In any case, other manufacturers (Varian and Finnigan) maintain the spray chamber at room temperature by installing the nebulization system outside the torch box. Although this approach can obviously not condense as much vapour as cooling the spray chamber, it prevents the gradual heating of the spray chamber by heat dissipated from the plasma, which typically results in signal drift. However, one might expect greater memory effects from the connecting tubing between the exit of the spray chamber and the injector of the torch. For instance, a 52-cm length of 6-mm i.d. silicone rubber tubing is used on the UltraMass, which essentially adds 14.7 ml to the internal volume of the spray chamber. On the other hand, losses of droplets due to settling in the tube are less likely to occur for small droplets, which are transported quickly [159]. Depending on its dimensions, the tubing can also act as an evaporation column and reduce the mean droplet size, without significantly modifying the total plasma solvent load [157]. With an appropriate set of conditions, the washout time may then be similar to that observed with a Scott spray chamber directly connected to the torch. That is observed with the UltraMass: 40 s is typically required, once a blank is nebulized following a 500 µg/l In solution, to reach 0.1% of the original signal. This clear-out rate could actually be modelled to a simple exponential dilution system with a total volume of 109.7 ml, as can be seen in Fig. 1.19. Like the washout, memory effects are similar to those observed with a Scott spray chamber.

Heating of the spray chamber could also be used. Under these conditions, the cutoff diameter of the spray chamber is larger, and a greater number of the large droplets from the aerosol can traverse the chamber and reach the plasma [157, and references therein]. An increase in spray chamber temperature from 5 to 15°C can increase,

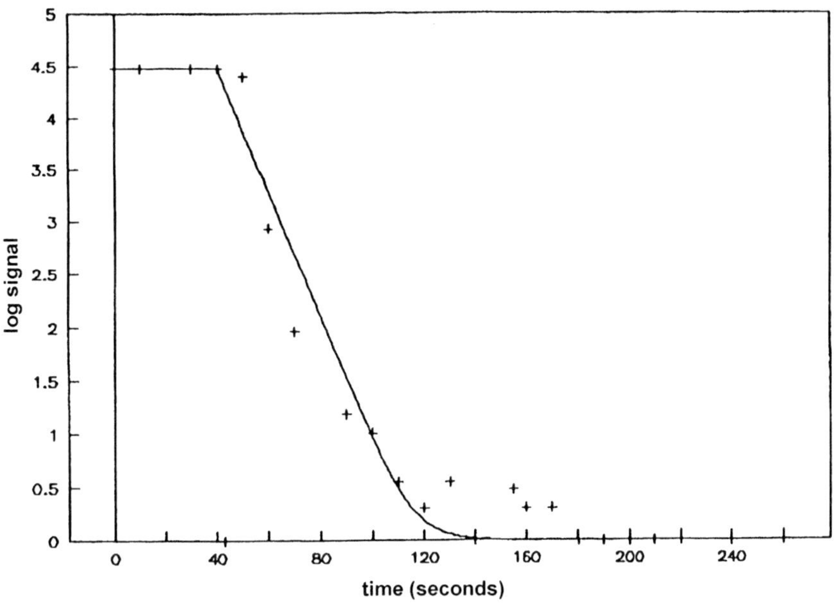

Fig. 1.19. Washout time with the Sturman-Masters spray chamber. The blank reaches the nebulizer at $t = 40$ s. The experimental data points (+) compare favourably with the calculated profile (solid line). Copyright Varian Associates 1994. Used with permission.

decrease or not affect sensitivity, depending on the analyte [157]. In ICP-AES, a three-fold improvement in signal was reported upon an increase in spray chamber temperature up to 45°C, although the improvement decreased when the solution nebulized contained salt [160]. In fact, the sample introduction system has a great effect on the analytical characteristics of ICP-MS. For instance, although a smaller mean droplet size was reported to increase the level of oxide ions, an increase in solvent vapour load resulting from a raise of spray chamber temperature could actually decrease this level [157].

All of these approaches decrease the instrument warm-up time by eliminating the waiting period which would otherwise be required for the spray chamber to reach a relatively stable temperature. They also eliminate a source of drift: that due to fluctuations in the temperature of the spray chamber concurrently with any change in the plasma temperature, which then vary the amount of solvent vapour and analyte aerosol reaching the plasma as a function of time. Another source of fluctuation arises from the small pressure changes that occur each time the U-shaped drain of the spray chamber overflows. This can be minimized by using a channel of the peristaltic pump to constantly and uniformly remove waste solution from the drain of the spray chamber while maintaining a sufficient back-pressure to force the nebulizer gas through the plasma. This simple approach is provided with the ELAN 5000 or 6000 and the UltraMass, but it can easily be implemented on all the other available ICP-MS instruments since they all use a peristaltic pump.

1.2.3.2 Limitations

Yet, with the exception of a high-efficiency version described earlier [102], pneumatic nebulization systems are rather wasteful of sample since their overall transport efficiency, i.e. the percentage of the sample reaching the plasma, is only about 2% [86], the rest of the sample going down the drain at typically 1–3 ml min^{-1}. This means that two orders of magnitude in response are potentially lost during the sample introduction process. This increase in signal would indeed result if 100% of the sample nebulized reached the plasma and, in addition, all the analyte was ionized and efficiently sampled by the interface.

Depending on the homogeneity and physical properties of the sample solution, the sub-sample reaching the plasma may not be representative of the original sample due to aerosol ionic re-distribution or fractionation. That sub-sample may also represent a different fraction than that resulting from a standard solution [143]. For instance, the presence of an increasing amount of NaCl in a solution has been shown to shift the particle size distribution upwards [161]. That is why, even when a peristaltic pump is used, matrix effects may be observed, depending on the complexity of the samples. In addition, the capillary of PNs (i.e. cross-flow and concentric) is prone to clogging by, for instance, salt deposition [143]. This problem, which inevitably leads to signal drift, is not very likely with Babington-type nebulizers where no capillary is used and the solution flows freely over the surface of the gas outlet. That explains why these nebulizers are provided as options with several of the instruments. An example is the Hildebrand grid nebulizer, which was used for the introduction of saline samples (synthetic sea water and urine) in ICP-MS [162]. No clogging of the nebulizer occurred, although blockage of the sampler and matrix effects were observed [162].

Memory effects are another potentially important limitation of pneumatic nebulization systems where left-overs from a sample in the nebulization system contaminates the next sample until all left-overs from the former have been washed out. Examples of elements which are notorious for sticking to glass and, as a result, are especially difficult to rinse out are B, Hg and Au [52]. Ironically, adding a small amount of Au to sample and standard solutions preserved Hg in even poly(ethylene terephthalate) containers [163]. (This approach obviously precludes the simultaneous determination of Au.) On the other hand, the addition of Au was not observed to reduce the washout time, in contrast to dicromate in HCl which significantly reduces the memory effect [164]. Usually, a rinsing solution must be used between samples, which decreases the sample throughput i.e. the number of samples that can be analyzed per hour. The memory effect from Hg can however be reduced by adding Au to the rinsing solution [165]. Memory effects are dependent on the nebulization process, the design of the spray chamber, the speciation of the analyte and the sample matrix. For instance, a persistent memory effect was observed for ^{127}I when it was in the form of NaI in 0.14 M HNO_3, whereas KIO_3 in 0.14 M HNO_3 or NaI in 0.5% v/v ammonia did not produce this effect [166]. Similarly, the larger the surface of the chamber that comes in contact with aerosol, the more time will be required to rinse it out. The peristaltic pump can, however, be programmed to speed up while it aspirates a washing solution between samples [164,167].

1.2.3.3 Alternative high-efficiency nebulization system

As can be seen in Tables 1.1 and 1.3, most manufacturers offer an ultrasonic nebulizer with desolvation (USND) as an option since it has been reported to improve detection limits by at least an order of magnitude [168,133]. This USND is commercially available from Cetac Technologies Inc. and is illustrated in Fig. 1.20. In an ultrasonic nebulizer, ultrasounds are applied to a piezoelectric transducer, resulting in high-frequency vibrations of its surface which if in contact with a film of solution will ultimately generate an aerosol. The latter can then be transported into the plasma by a carrier gas. This approach possesses some advantages. First, since the nebulization process is separate from the aerosol transport, each process can be optimized separately [169]. Furthermore, because of the larger surface used for aerosol production compared to a PN where the aerosol is generated from a small point, more aerosol can be produced. Typical analyte

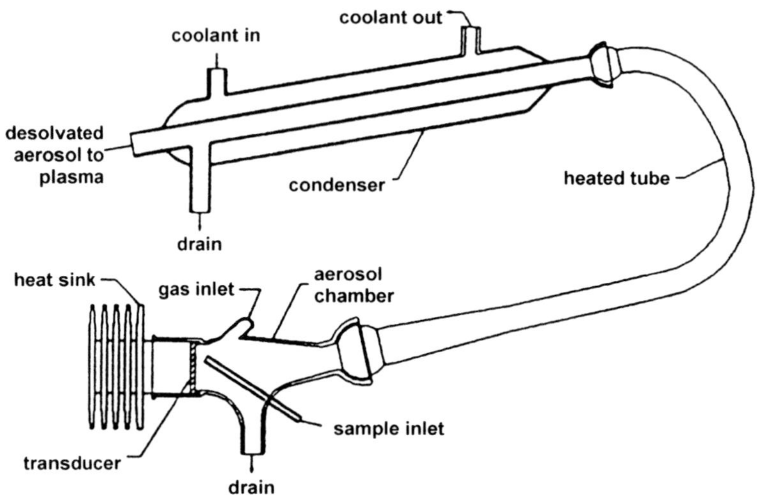

Fig. 1.20. Schematic diagram of an ultrasonic nebulizer with desolvation system (i.e. condenser). Courtesy of Cetac Technologies Inc.

transport efficiencies with the USN and a small spray chamber at 1–3 ml/min are 10–20% [140], clearly higher than with conventional PNs for a similar solution uptake rate.

However, because of the associated increased solvent load, a desolvation system must be used to prevent cooling of the plasma by the solvent. Otherwise, the beneficial effect of having more analyte reaching the plasma would be counteracted by either a degradation in the performance of the ICP or plasma extinction [169]. A typical desolvation system consists in a heated region to vaporize as much of the solvent as possible, followed by a condenser maintained at about 0°C to remove the vapour. Besides removing water, this approach is also efficient for organic solvents with boiling points near 100°C but not for more volatile solvents.

In addition to the aerosol carrier gas flow rate, several parameters must be optimized to obtain the best analytical performance from a USND [169]. The forward power to the piezoelectric crystal affects the capillary waves formed and, in turn, the size of the aerosol formed. For instance, a higher forward power increases the nebulization efficiency of samples with a relatively high level of total dissolved solids [169]. On the other hand, too high a forward power (i.e. >50 W) may shorten the lifetime of the crystal [169]. Furthermore, the sample delivery, in combination with the surface tension, density, heat capacity and heat of vaporization of the sample, influences the aerosol number density, the amount of analyte ultimately reaching the plasma and the wash-out time [169]. The temperatures of the heater and condenser tubes must be adjusted according to the hygroscopicity and level of total dissolved solids of the sample. For instance, as the dissolved solids content increases, the heat capacity of the sample solution also increases thereby requiring more heat for desolvation at a given residence time in the heater tube [169]. On the other hand, too high a heater temperature can overload the condenser.

Care should also be taken not to remove analytes along with the solvent. This is especially important for analytes that form volatile compounds, such as non-metals. In this case, a better transmission through the desolvation/condensation system can be achieved by increasing the aerosol carrier flow rate and adjusting the temperatures for the desolvator and condenser [170]. In any case, once the analytes exit the sample introduction system, they will enter into a dry and therefore hotter plasma, which should improve their degree of ionization and decrease spectroscopic interferences from polyatomic species (if the latter originate from within the plasma) such as oxides [170].

In general, all the above parameters should be optimized to achieve maximum S/N or best detection limit. Indeed, fluctuations induced by large particles, while not necessarily degrading sensitivity, will degrade the detection limit [169]. Matrix matching of the standards to the samples is also important to ensure that the aerosol generation process will be as similar as possible. Internal standardization can be used to compensate for changes in surface tension and density resulting from varying amounts of organic compounds [169]. The method of standard additions or the stable isotope dilution (ID) analysis, while being more time-consuming, will provide more accurate results.

In any case, when the USND system is combined to a membrane dryer which is illustrated in Fig. 1.21, even volatile solvents can be removed. When the aerosol from the desolvation system flows in the centre of a heated microporous PTFE tubular membrane dryer (held at 160°C), the solvent passes through the membrane and is carried out by a counterflow of Ar, whereas the analyte continues through the tube to the ICP [171]. This approach allows a significant reduction of the MO^+/M^+ ratios and MH^+ signal as well as a removal of HCl [172]. Under optimal operating conditions, it also provides a 25-fold improvement in sensitivity compared to PN with a spray chamber at 1°C [172]. As more

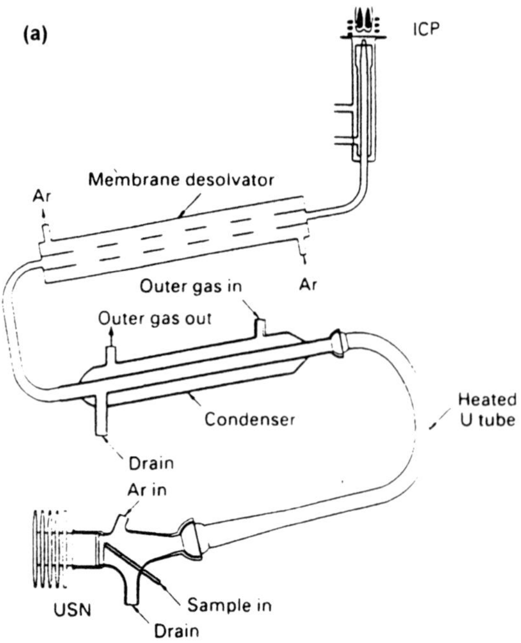

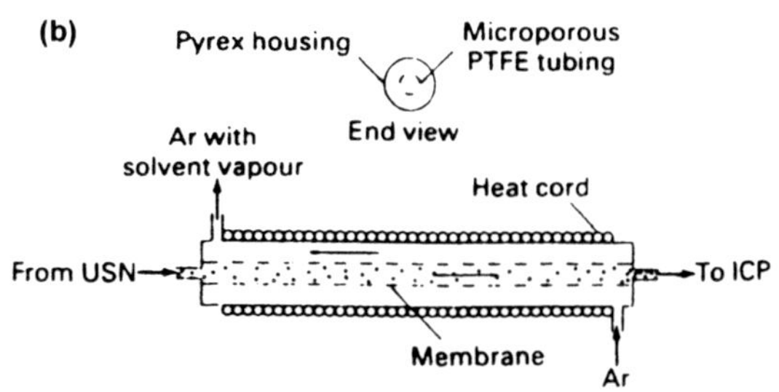

Fig. 1.21. Schematic diagram of (a) an ultrasonic nebulizer with desolvation system (i.e. condenser) coupled to a microporous membrane (b) for a second stage of desolvation. Courtesy of Cetac Technologies Inc.

and more water is removed, atomization and ionization occur sooner (i.e. further upstream in the plasma) and so the analyte signal maximizes at higher aerosol carrier gas flow rate [172]. This observation is true for any desolvation device.

Although a peristaltic pump cannot be used for a number of solvents which attack even Viton rubber tubing, it can push air into a sealed sample container thereby displacing the sample towards the nebulizer [171].

On the other hand, a membrane desolvation system can only be used for the analysis of samples with a high salt concentration with caution, as blockage of the membrane has been reported with a subsequent increase in oxide ratio as a result of the associated reduction in desolvation efficiency [173]. This system also requires longer washout times than a spray chamber [173].

An alternative system was also described which uses cryogenic desolvation to remove water [168,172,174] and HCl from aqueous solutions [172,174], and organic solvents [175] from the desolvated aerosol exiting the USND. Although the approach substantially reduced interferences from several polyatomic species by several orders of magnitude, it was more susceptible to matrix effects [174]. Furthermore, memory effects were also observed, which were dependent on the chemical form, i.e. the speciation and the physical properties of the analyte [175]. Nonetheless, compared to USND with and without membrane desolvation, this approach is the most efficient at reducing the MO^+/M^+ ratios and MH^+ signal as well as removing HCl [172]. Although it is more cumbersome, it also provides a similar improvement in sensitivity as the membrane desolvator compared to PN [172].

Nonetheless, the USND has been reported to offer improvements over concentric nebulization for the analysis of biological matrices by ICP-AES [169]. However, it still has several disadvantages in combination with ICP-MS, whether a membrane dryer is used or not. It is expensive, relatively cumbersome and is prone to memory effects because of the large surface with which the aerosol can come in contact. Furthermore, it makes the detector more susceptible to matrix effects since it concentrates both the analyte and the matrix [100], and can lead to injector and/or sampling orifice blockage if the concentrated aerosol contains a high concentration of salt. Although a sheathing device can be used to prevent blockage of the injector tip [176], it is not likely to prevent cone blockage as well. Furthermore, the increased matrix effects may cancel the above-mentioned improvement in detection limits. Nonetheless, it is a relatively easy way of improving the detection limits for samples with a relatively simple matrix, compared to those achievable with conventional nebulization systems coupled or not to a desolvation system. It also removes solvent and therefore drastically reduces the formation of refractory oxides.

In any case, water may not have to be removed completely. A study has indeed demonstrated that, when the solvent load in the plasma is kept low (and corrections are made for sample losses), the ICP-AES sensitivities observed for wet and dry aerosols were essentially identical [177]. Furthermore, water has been shown to play an important role in the plasma: by providing hydrogen (which has a higher thermal conductivity than Ar) to the plasma, it increases the axial temperature of the plasma [178]. Molecular emission from H_2 was indeed measured when water was introduced in the ICP, which demonstrates that H_2 can be formed from water [179]. The emission arose from the axial channel at 2 mm from the load coil or even closer to the torch injector, i.e. well below the NAZ [179]. In general, plasma temperatures and electron number densities increase upon introduction of water vapour into the ICP whereas a decrease is observed when a water aerosol is introduced [86]. That is why an improvement in sensitivities and especially detection limits was observed when the aerosol exiting the spray chamber was pre-vaporized prior to its entry in the ICP [180].

The desolvation temperature is therefore important: if it is too high, too much water vapour is removed and a reduction in analyte emission signal occurs, because less hydrogen is thereby produced by the dissociation of water [181]. The optimal desolvation temperature depends critically on the aerosol carrier gas flow rate. For instance, too low a flow rate promotes coagulation of droplets which, in turn, requires a higher desolvation temperature [181]. At a given aerosol carrier gas flow rate, an increase in desolvation temperature up to a certain optimal value (above which no significant change occurred), can increase the ionization temperature of the plasma [181]. Finally, as was mentioned in Section 1.2.3.1, an increase in plasma solvent loading can actually reduce the formation of interfering oxide ions under certain operating conditions [157].

One variation which shows great promise, although it is not yet commercially available, is the microflow USN (μ-USN) [182]. An HPLC pump with pulse damper was used to supply liquid to a 50-μm capillary placed in contact with the transducer (at an angle of 45–70° from the normal to the surface) [182]. As the liquid flows over the surface of the transducer, a very fine aerosol is formed and entrained into the ICP through a specially designed spray chamber [182]. At an uptake rate of 10–20 μl/min, the μ-USN provided detection limits similar to that observed with a pneumatic nebulizer which required, however, 100 times that uptake rate. Transport efficiencies of 100% at 5 and 80% at 10–20 μl/min were achieved since, because of the very small sample delivery rate, no desolvation was required [182].

1.2.3.4 Other sample introduction systems

Improvements in the analytical performance of ICP-MS can also be obtained by adding a desolvation system to a conventional PN/spray chamber system. For instance, by using a heater/condenser system of the type used in USND, an increase in sensitivity by a factor of 2–5 was observed [183], along with a reduction in the signal from some polyatomic species [158, 183]. The reduction in oxide and hydroxide ions was commensurate with a decrease in water vapour [158]. Similarly, the sensitivity improvement was due to a decrease in water vapour rather than to a change in the droplet size distribution which remained identical to that observed without desolvation [183]. However, in addition to the analyte, more matrix reached the plasma, which worsened matrix effects to a point where any gain in sensitivity was cancelled [183].

The addition of an efficient desolvation system is particularly important for the continuous nebulization of solutions containing organic solvents. For instance, two systems with 70% efficiency at removing solvent, were used for the analysis of diethyl ether [184]. One was a cooled spray chamber (at –40°C) whereas the other combined a heated spray chamber with a desolvation system involving Peltier coolers (at –40°C) [184]. The latter was so efficient at reducing CO^+ that it allowed the determination of Si using its most abundant isotope at *m/z* 28 [184].

Other desolvation systems (such as the jet separator) have also been described [53,185–187]. The reader is referred to these publications and the references therein for more information on these systems which are not commercially available. The membrane separator interface (or membrane drying tube) which was mentioned in the previous section, is commercially available from Perma Pure Products as well as Cetac Technologies. Another commercially available system (from Fisons Instruments) is the Mistral which consists in a radiatively heated spray chamber combined to a two-stage condenser [188]. The infrared radiation from a glow bar is used to heat the spray chamber to 120°C while the water vapour content is reduced by passage through a water-cooled region (at about 10°C) followed by a Peltier-cooled condenser maintained at about –2°C. This led to improvements by, on average, an order of magnitude in detection limits in ICP-AES, in addition to reducing the washout time [189]. In ICP-MS, the signal was increased by an order of magnitude without degradation of precision [188]. In addition, sample consumption was reduced by a factor of 10 compared to a conventional nebulization system because the sample uptake rate was 0.35 ml/min [188].

A recycling nebulization system with a disposable spray chamber was proposed to decrease sample waste with a concentric PN [190]. This closed system, which is illustrated in Fig. 1.22, contained 0.5 ml of sample and allowed 8 min of data acquisition per sample, with sensitivities similar to that obtained with a Scott spray chamber. However, the mini spray chamber had to be cleaned and dried before the next sample or washing solution (that was used between samples) could be loaded in the chamber, which could then be re-installed [190]. This approach obviously decreased sample throughput, and also completely eliminated memory effects. However, memory effects are not restricted to the sample introduction system: deposits may indeed be found in the injector of the torch in addition to the sampling orifices and even the ion optics of the mass spectrometer. ICP-MS, unlike ICP-AES, requires a regular maintenance of the whole instrument to achieve the specified performance.

A simple and low-cost alternative to the Scott spray chamber has also recently become available from Perkin-Elmer in the form of a cyclonic spray chamber [191], which is shown in Fig. 1.23. It is used with a conventional concentric nebulizer, and relies on centrifugal action to discriminate droplets according to their size. Using this system, sensitivities and detection limits were improved by a factor of two without any increase in either the level of oxides or that of doubly-charged species. Although short-term precision is slightly degraded, long-term precision over 100 min is similar to that observed with a conventional Scott spray chamber [191].

A powerful pneumatic alternative to the nebulizers described in Section 1.2.3.1, is the direct injection nebulizer (DIN) which was first used with ICP-MS by Wiederin et al. [192] and is illustrated in Fig. 1.24. It is a special concentric nebulizer that replaces the injector tube in a demountable torch, and therefore sits underneath the plasma. Since no spray chamber is used under these conditions, 100% of the sample enters the plasma, i.e. no sample is wasted. Analyte memory effects are also drastically reduced under these conditions, especially

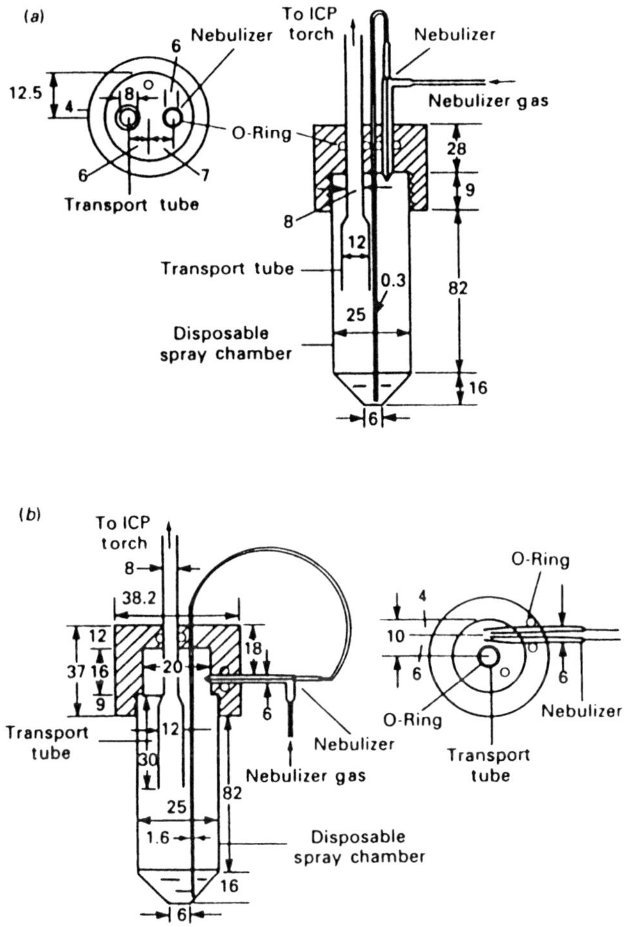

Fig. 1.22. Schematic diagram of two recycling nebulization systems: (a) parallel design, (b) cyclone design. Each system includes a concentric nebulizer inserted in a holder, a PTFE capillary tubing for the recycling process, a disposable screw-top spray chamber and a Pyrex transport tube that allows the aerosol to be carried into the torch. Reproduced from Ref. [190] with permission.

for volatile elements such as B, I and Hg [31 and references therein, 193]. However, a high-pressure pump and appropriate tubing must be used to adequately push solution through the very small bore of the DIN. Furthermore, 100% of the solvent penetrates the plasma and

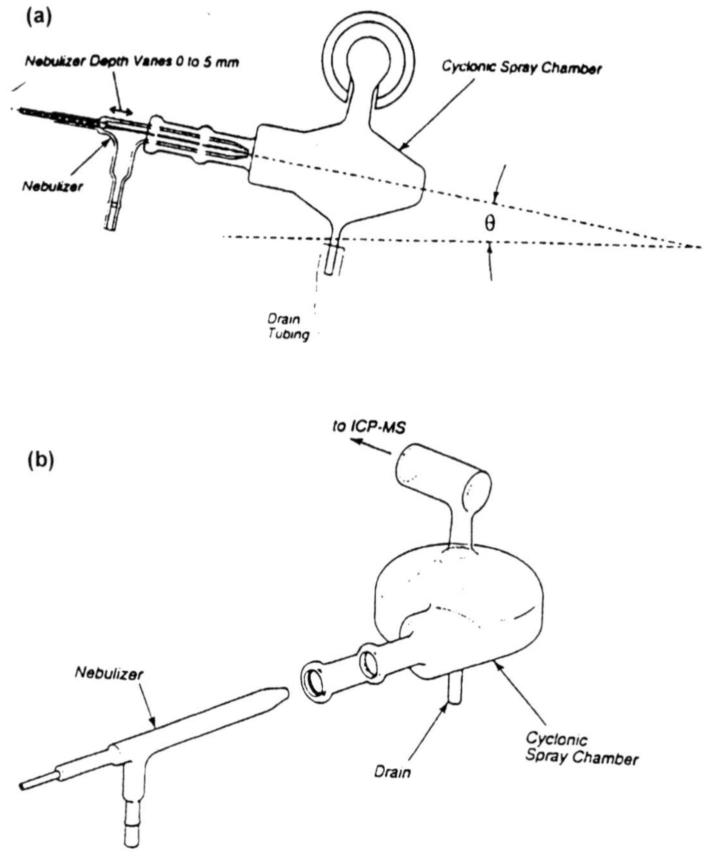

Fig. 1.23. Schematic diagram of a cyclonic spray chamber: (a) top view, (b) mounting configuration of a concentric nebulizer in the cyclonic spray chamber. From Ref. [191] with permission.

therefore limits the solution flow rate to about 100 µl min^{-1}. As a result, no significant improvement in detection limits based on concentration is achieved, although detection limits based on absolute mass are improved. Nonetheless, as will be seen in some of the following chapters, the extremely small dead volume of this nebulizer is a clear asset in combination with some of the approaches proposed to improve the analytical capabilities of ICP-MS. It is commercially available from CETAC Technologies Inc.

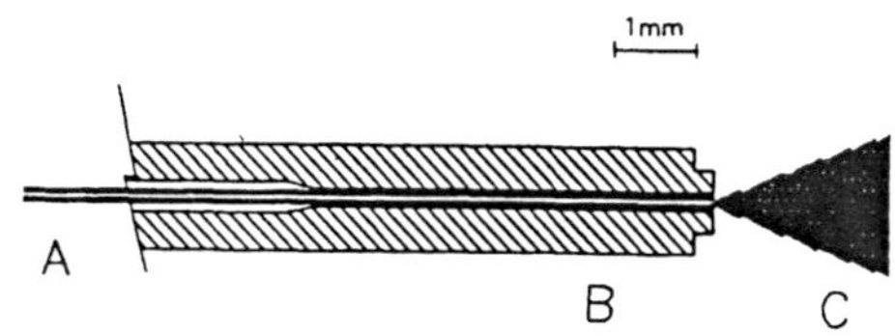

Fig. 1.24. Enlarged view of the direct injection nebulizer tip. (a) Sample delivery capillary, (b) ceramic nebulizer tip, (c) nebulized aerosol. Courtesy of Cetac Technologies Inc.

Another alternative, the direct injection high efficiency nebulizer (DIHEN), is available from J.E. Meinhard Associates, Inc. at a relatively low cost [194]. In contrast to the DIN, the DIHEN does not require a high-pressure pump for sample delivery, i.e. a peristaltic pump can be used [194]. Its dead volume is approximately 55 µl [194]. Like the DIN, because 100% of the sample is introduced in the plasma, it provides similar figures of merit (detection limit, sensitivity and precision) to those obtained with a cross-flow nebulizer and a Scott-type spray chamber but at a solution flow rate of 85 µl/min (instead of 1 ml/min) [194]. However, because the DIHEN produces smaller droplets than the DIN, detection limits with the DIHEN are lower by about a factor of two than those with the DIN [194]. On the other hand, since the primary aerosol is introduced directly into the plasma, a higher oxide level is observed, as with the DIN, by a factor of 2–3 higher compared to cross-flow and microconcentric nebulizers with spray chambers [195].

Other high-performance nebulization systems that have been used with ICP-MS include thermospray [78,168,196], hydraulic high-pressure nebulization [197], microconcentric nebulization [79,198–199] and monodisperse dried microparticulate injection (MDMI) [200]. The first, which is commercially available from Leeman Labs, consists in using a high-pressure pump to push the sample solution through a thin fused silica capillary with typically a 50-µm i.d. aperture, which is electrothermally heated, as illustrated in Fig. 1.25 [201]. With sufficient heat, a portion of the solution is vaporized and bursts out in a jet, entraining nebulization of the rest of the solution [196,201]. This process is very efficient and, as for the ultrasonic nebulizer, a desolvation system must be used to decrease the solvent load to the plasma. With a proper desolvation system, improved detection limits by at least an order of magnitude were reported using 75-µm and 180-µm capillaries,

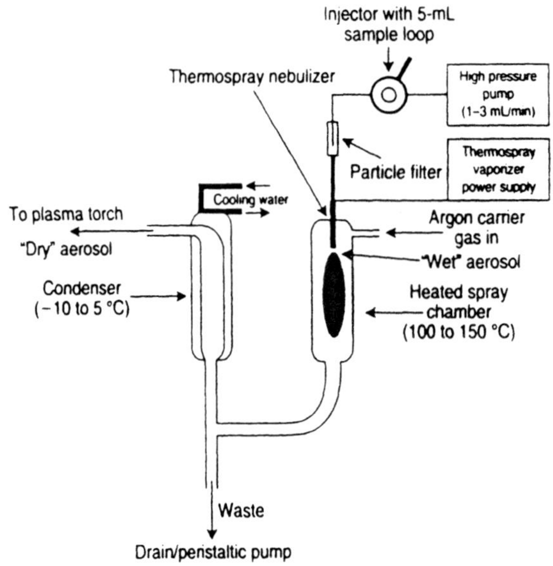

Fig. 1.25. Schematic diagram of a thermospray nebulizer sample introduction system with desolvation. Reproduced from Ref. [140] with permission.

compared to conventional PNs [168,202], even when the latter were combined to a desolvation apparatus [168]. In addition, the level of oxide and doubly charged ions were decreased by about a factor of 2.5 compared to a PN/spray chamber system without desolvation [202]. Furthermore, this nebulizer has potential for the non-chromatographic speciation analysis of some oxidation states of an element. Indeed, with a suitable adjustment of sample pH and thermospray operating parameters, Cr(IV) could be selectively determined in the presence of up to 10 ng/ml Cr(III) as the latter is then quantitatively precipitated as $Cr(OH)_3$ in the vaporizer [203,204]. Similarly, Se(IV) is reduced to Se(0) within the thermospray process and then trapped on the vaporizer, allowing the selective determination of Se(VI) [205]. However, like USND, thermospray nebulization concentrates the analyte along with the matrix. As a result, matrix effects should be worsened compared to those observed with pneumatic nebulization [140]. A modified version of this nebulizer, although not commercially available yet, was described, where microwave heating is performed instead of electrothermal heating [206].

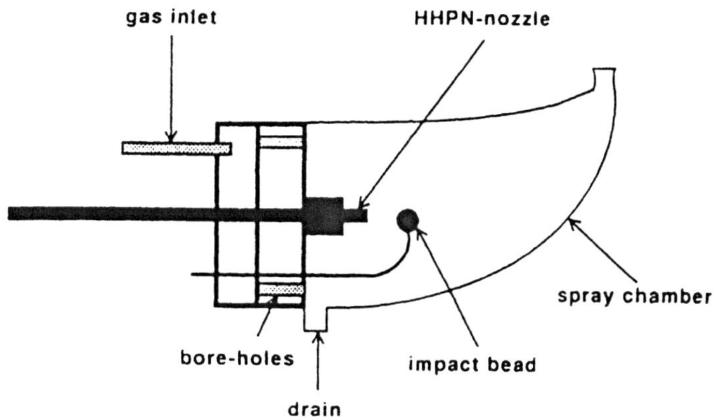

Fig. 1.26. Schematic diagram of a hydraulic high-pressure nebulizer. Reproduced from Ref. [197] with permission.

Hydraulic high-pressure nebulization, which is illustrated in Fig. 1.26, consists in pumping the solution at high pressure through a small (10-μm diameter) nozzle, aiming the resulting fine liquid stream directly on an impact bead within a small spray chamber [197]. The aerosol exiting the spray chamber represented 30% of the sample (in contrast to 1–2% with conventional PNs) [197]. As for USND and thermospray, the resulting aerosol has to be desolvated. A 30-fold increase in count rate was reported for a number of elements, as well as a decrease in water-related polyatomic species ($^{40}Ar^{16}O^+$ and $^{38}ArH^+$) [197]; this decrease is as expected from desolvation systems. Matrix effects are also expected to be worse than those observed with a PN/spray chamber system. This nebulizer is commercially available from Knauer (Berlin, Germany).

The microconcentric nebulizer (MCN), which is available from Cetac Technologies Inc., and other micro flow nebulizers (such as that from Agilent) are entirely made of HF-resistant materials (Chapter 3 of Ref. [11]). They therefore allow the direct analysis of samples requiring HF for their dissolution by acid digestion, hence eliminating the additional step that is typically needed to remove HF and a possible source of contamination and/or volatile analyte losses. The MCN can be used in conjunction with a spray chamber or a desolvation system. At 30 μl/min, it provides detection limits that are comparable to those obtained with a PN at 1 ml/min. It is therefore very advantageous for the analysis of volume-limited samples, such as several biological materials, which cannot be diluted without precluding accurate determination of several elements [79]. With the MCN, a multielemental analysis can be performed with only 100 µl of sample [79]. Furthermore, because a very small sample volume is required, waste is reduced, as well as the washout time [79]. Although it is not required, the MCN can also be combined to a heated spray chamber/membrane desolvation system and is commercially available in that version as well. In the latter case, a small amount of N_2 is typically added to the Ar aerosol carrier gas to assist in sample transport and reduce oxide ions [207]. Mixed-gas plasmas, which are discussed in Section 1.6, can indeed be used to decrease the level of spectroscopic interferences from polyatomic ions.

The reader should keep an eye out for other commercially available nebulizers. For instance, SCP SCIENCE now provides various Burgener parallel path nebulizers, including one, the Micro 3, which uses a high gas pressure to give a sample uptake rate ranging from 0.02 to 1 ml/min.

The MDMI, shown in Fig. 1.27, is a completely different and original nebulization system where discrete droplets of a very reproducible size are produced [200]. A droplet generator produces uniformly spaced 57-µm droplets [200]. This generator acts as a micro-pump by using a piezoceramic tube to apply pulsations to a glass tube. Since the glass tube has a sharp-edged 30-µm orifice at one end and a flow constriction near the other end, applying a pulse contracts the glass, thereby expelling a droplet. The uniform droplets, whose size can also be changed by varying the width and the height of the pulse applied to the piezoceramic, then pass through a 25-cm laminar flow oven maintained at 700–800°C where circulates the heated Ar carrier gas [200]. The solvent is evaporated (but not removed) in the laminar flow oven into the Ar carrier gas. The resulting aerosol consists of dried particulates which remain separated into a laminar flow of Ar that is partially saturated with solvent vapour. The actual size of the particulates is determined by the carrier gas flow rate and the temperature of the oven [208]. The position in the ICP where the droplet vaporized depends on the drop size, and can be situated below the load coil or more than 30 mm above it [208].

As with the DIN, a 100% sample introduction efficiency is achieved at a low flow rate (up to 35 µl/min). In addition, the oxide ratio at the optimum IRZ position for maximum analyte sensitivity is reduced compared to that observed with a PN [200]. The analyte signal is independent of the NaCl concentration of the sample (up to at least

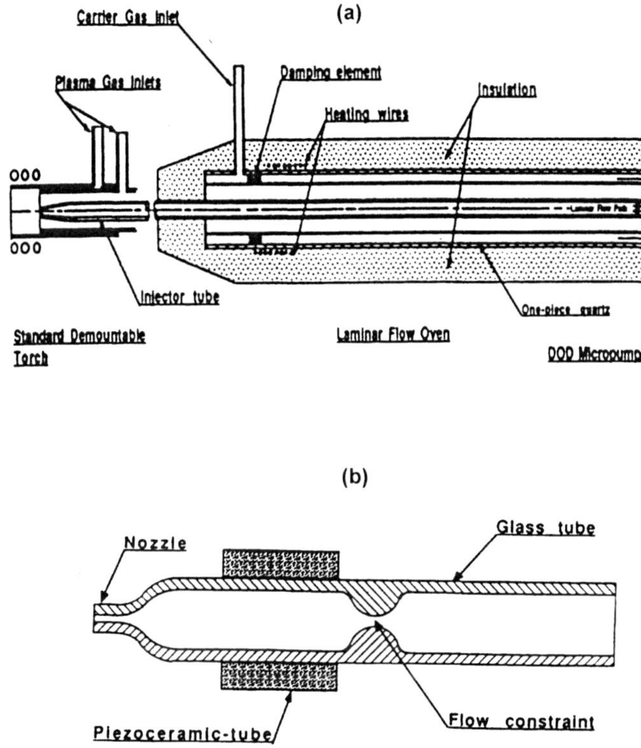

Fig. 1.27. Schematic diagram of a monodisperse dried microparticulate injection system. (a) Cross section of the MDMI apparatus. (b) Droplet generator design (of the micropump). Reproduced from Ref. [200] with permission.

10000 mg/l NaCl) [200]. Furthermore, the precision is improved down to the statistical level, which should allow the measurement of isotope ratios with a precision of 0.1% [200]. In fact, the MDMI should provide the best precision of all high-efficiency nebulizers discussed in this section since, in addition to eliminating fluctuations due to the desolvation process (see Section 1.4.1), fluctuations resulting from the polydispersity of the dry aerosol during the vaporization process are also completely eliminated as a result of the uniform size of the particulates. Although this system is not yet commercially available, it is very promising. Its main operational problem lies in the susceptibility of the system to air bubbles in the sample inlet which make the micropump lose its prime [200].

As forecasted, ICP-MS combines several features of ICP-AES with those of MS [6–39]. Tables 1.5 and 1.6 summarize the figures of merit of the available low and high resolution instruments, respectively. As can be seen, the linear dynamic range for a given element covers at least five orders of magnitude so that, in contrast to GFAAS, major, minor and trace elements may all be determined at once. Except for the UltraMass which uses a single detector in only one detection mode, pulse counting, and limits the maximum count rate to 3,000,000 counts/s to protect the detector and extend its useful lifetime, all manufacturers use hardware and/or software tricks to extend the linear dynamic range. In many cases, two modes of detection are used: pulse counting for low signals and standard analog detection with a lower gain to the detector for strong signals [210]. With this approach, a sample solution must be aspirated twice, once in the analog detection mode and once in the pulse counting mode to allow the determination of major and/or minor elements and trace elements, respectively. The TS SOLA uses two detectors, a channel electron multiplier for low signals and a Faraday cup for signals that could damage or drastically decrease the lifetime of the other detector. A completely different approach is used on the ELAN where the level of filtering is controlled through software (OmniRange) thereby allowing the determination of analytes at both low and high concentrations in a single scan [211]. The "filtering" is achieved by altering the voltage to the exit lenses between the quadrupole and the detector, thereby attenuating the ion beam by a known fraction that can be selected for each mass [212].

The multielemental capability of ICP-MS allows the quasi-simultaneous determination of, for instance, 70 elements in less than 2 min on sequential mass spectrometers, and requires less than 2 ml of solution at an uptake rate of 1 ml/min [52]. Obviously, the measurement time can be substantially less on TOF and multicollector instruments. Although some polyatomic species are detected, as will be discussed in detail in Section 1.4, the mass spectrum is much simpler than that observed in ICP-AES. The likelihood of spectroscopic interferences is therefore reduced, even more so with high-resolution instruments. However, when they occur, the number of alternative lines is small when existent.

Detection limits, i.e. the minimum concentration that will give rise to a detectable signal from background fluctuations [213], are typically

TABLE 1.5

Figures of merit for selected quadrupole-based ICP-MS instruments

	Manufacturer						
	Hewlett-Packard Yokogawa Electric	Varian	Spectro	VG Elemental	Finnigan MAT (Turner)	Perkin-Elmer SCIEX	Thermo Jarrell Ash
Model	HP 4500	UltraMass 700	Spectromass 2000	VG PQ ExCell	Sola	ELAN 6100	POEMS
Orders of dynamic range	8	5	8 with SEM; 12 plus Faraday cup	8	8 with Faraday; 10 plus SEM	8 in a single continuous scan	6 pulse counting; 4 analog extended range; 8 overlapping; 5 optical; 11 total overlap with optical
Short term precision (10 repeats) RSD	<1%	<3%	<2%	1%	<3% (<1.5 with internal standard)	<3%	<3%
Long term precision (4 hours) RSD	<2%	<3%	<2% (Rh), <3% (Mg, U)	1.5%	<5% (<3 with internal standard)	<4%	<4%
Sensitivity in Mcps/ppm	>8 Li; >12 Y, Tl	>1 Be, Pb; >5 Co, In	>8 ^{24}Mg, >30 Rh, >12 ^{205}Tl	>60	20 In	>30 In, >4 ^{24}Mg, >25 ^{238}U	10 Ho; 8.0-8.2 Li, Co, In, Bi
Detection limit in ng/l ($3\sigma_b$)	<1: 51 elements; 1-10: 18 elements (including Fe with ShieldTorch)	<5: 22 elements out of 28 quoted so far	<3: ^{9}Be; <2: Co, ^{138}Ba, ^{209}Bi; <1: Rh	0.01-0.1: 51 elements; 0.1–1: 8 elements; 1-10: 8 elements	<0.2: Li, Co; <0.4: In; <0.6: Be; <1: Pb, Bi; <1.5: U	<15: Be; <2: Co; <0.5: U and In	<2: Bi, In, Tb; <5: Cu, Pb; <10: Cd
Isotope ratio precision (RSD) for ^{107}Ag/^{109}Ag	<0.2%	^{206}Pb/^{208}Pb 0.3%; ^{86}Sr/^{88}Sr 0.7%;	≤ 0.2%	^{206}Pb/^{207}Pb 0.04%	0.2 %	<0.2 %	<0.3 %
Ratio of Ba doubly-charged ions	0.02 (^{70}Ce^{2+}/^{140}Ce^{+})	<0.03	1% (Ce^{2+}/Ce^{+})		<0.05	<0.03	
Ratio of Ce oxide ions	0.005	<0.03	<1%		<0.015	<0.03	0.011

**With single-element solutions.

TABLE 1.6

Figures of merit for selected double-focusing and TOF ICP-MS instruments

	Manufacturer					
	Finnigan MAT	JEOL	Micromass	VG Elemental	GBC Scientific Equipment	Leco
Model	ELEMENT	PLASMA X-1	PlasmaTrace	VG Axiom	OptiMass 8000	Renaissance
Orders of dynamic range	9			>9	7	7
Short term precision (10 repeats) RSD	2%			<2%		0.5% (with internal standard)
Long term precision (4 hours) RSD				<2%		2.0% (with internal standard)
Sensitivity in Mcps/ppm for In	20 at R=300; 1 at R=3000		2 GHz/ppm (R=400); 400 (R=3500)*			3
Detection limit ($3\sigma_b$) in ng/l	<0.1: In at R=300	1: Hg, In, U, Tl at R=500	0.0008-0.01: 35 elements; 0.02-0.2: 12 elements R=400*			
Isotope ratio precision (RSD)			0.014% for $^{235}U/^{238}U$**	<0.002%		
Ratio of Ba doubly-charged ions	< 0.03					<1%
Ratio of Ce oxide ions	< 0.01					<2%

*With USND [209].

**On a modified PlasmaTrace, the "Plasma 54", equipped with 7 Faraday cups [134].

≤1 ng/l for most elements with all quadrupole-based instruments, and slightly lower with high-resolution instruments. This difference arises from the lower background typically observed on high-resolution instruments [128]. It can also be attributed to the higher reliability in peak hopping, which results from the flat-top shape of the peaks observed in the low-resolution mode of high-resolution instruments [128,214]. This difference in peak shape is obtained by scanning the accelerating voltage while keeping the magnetic field constant, and is in contrast to the peaks observed with a regular magnetic scan, as illustrated in Fig. 1.15. Obviously, flat-top peaks are more tolerant of fluctuations and confer more stability to the measurements.

In any case, the background and sensitivity observed with high-resolution instruments is dependent on the resolution setting. Indeed, a background count rate of 0.2 count/s was obtained across the mass range (5–250) on a PlasmaTrace 2 at a nominal resolution of 400 [170]. This background decreased to 0.01 count/s with R = 3000 but the sensitivity was only 10% of that at R = 400. A further increase in resolution to 12000 decreased the background to 0.001 count/s and the sensitivity to 2% of that at R = 400 [170]. Using a cool plasma also decreases background levels with a concurrent sacrifice in sensitivity [170]. Nonetheless, high-resolution instruments may still be advantageous for determinations at low concentrations, because of their low background and high transmission. They are definitely advantageous for the determination of elements prone to spectroscopic interferences such as non-metals, without requiring a preliminary separation of the source of interferences [170].

Even though the HP-4500 features similar sensitivity to the other quadrupole-based instruments, it can nonetheless offer low detection limits for the greatest number of elements as a result of its low background count rate. Furthermore, the ShieldTorch interface and a similar device from Seiko allow distinctively low detection limits for K, Ca and Fe, as can be seen in Table 1.7. Impressively enough, these detection limits are comparable to those obtained using high-resolution instruments. They were in fact obtained by combining the shielded ICP with operating conditions resulting in a cooler plasma. Seiko added argon through a separate inlet of the spray chamber tangentially to the aerosol carrier gas [108]; whereas Hewlett-Packard operated the ICP at lower power and higher carrier gas flow rate than without the ShieldTorch [77]. Similar impressive results can be obtained on the ELAN 6000 without the attachment of shields, by simply operating the

TABLE 1.7

Detection limits ($3\sigma_b$, in ng/l) quoted for selected isotopes of P, S, K, Ca, Cr and Fe on some of the ICP-MS instruments

Instrument	^{31}P	^{39}K	^{44}Ca	^{52}Cr	^{56}Fe
HP 4500	100-1000	1-10*	10-100*	1-10	1-10*
PQ ExCell		0.5*	2 at ^{40}Ca*	0.1*	0.3*
ELAN 5000	100-1000	100-1000	1000-10000	10-100	100-1000
TS SOLA			10000		10000
UltraMass 700				<10	
SPQ 8000A*					3
PlasmaTrace (R=3000)	20 (R=1000-1500)***			0.09***	2***
ELEMENT (R= 3000)	100****		100****	20****	<100
ELAN 6000		15*****	50*****		5*****
PLASMAX-1 (R=3000)					20

*In cool plasma conditions (with ShieldTorch or PlasmaScreen Torch).
**From Ref. 108.
***From Ref. 209.
****Obtained on a prototype instrument; from Ref. 136.
*****From Ref. 97; the Ca detection limit was obtained at *m/z* 40.

plasma at lower power and higher carrier gas flow rate [97]. However, because of the cooler plasma, these approaches also result in lower sensitivity across the mass range, and are therefore only advisable for monitoring elements suffering from important interferences from the background, since the background is then even more reduced than the sensitivity at these masses [77]. In any case, the shield must be grounded. Indeed, a study has reported that although grounding the shield had no effect on the ion densities upstream of the sampler, a significant drop in ion densities occurred downstream of the skimmer when the shield was floated [215].

Sensitivity (expressed in terms of atomic concentration) was reported to be fairly independent of mass on the TS SOLA instrument [120]. A similar uniform mass response was also achieved on the Ion Scan. Since the ion beam is in essence deflected twice in the latter case [95], a slightly lower ion flux through the lenses is expected compared to the HP 4500 and the TS SOLA where the beam is only deflected once.

However, as can be seen in Table 1.5, sensitivity is still comparable because of the higher ion flux through the interface that results from the wider sampling orifices used for the sampler and skimmer [95]. Nonetheless, the uniform response [216] is in contrast to that which is typically observed on the other quadrupole-based instruments, where the response is fairly uniform only in the middle of the mass range whereas the response decreases at higher m/z and even more steeply at lower m/z, because of mass discrimination (see Section 1.4.3). In contrast, the response of a high-resolution instrument does not decrease at high m/z; in fact, the sensitivity normalized with respect to isotopic abundance and expressed in molar concentration, increases as a quadratic function of m/z. It is, however, possible to tune the focus offset parameter to maintain the focal depth of the low-resolution lenses at the plane of the entrance slit while varying the acceleration voltage during electric scans so that elements with similar first ionization potentials exhibit mass-independent sensitivity at each magnet setting [207].

Finally, since specific isotopes are detected, isotope ratios can be measured which in turn allows calibration by ID analysis [217,218]. The flat-top peaks which can be obtained in low resolution with high-resolution instruments, may allow the measurement of ratios with similar or better precision than that achievable with quadrupole-based instruments. However, because these mass spectrometers cannot be scanned as fast as quadrupoles, the resulting precision can be degraded compared to that obtained with quadrupole MS. For instance, the RSD for $^{240}Pu/^{239}Pu$ ratio was ≥2% with $R = 400$ whereas 0.2% RSD [206] and even 0.1% RSD (i.e. at counting statistics level) [219] was obtained with quadrupole MS. Furthermore, constant slit settings are recommended to obtain the best precision in isotope ratio measurements [170]. Indeed, changing the slits takes approximately 3 s so no compensation for short-term stability is possible during this period [170]. If ratios are measured at different slit settings, they must also be corrected for differences in sensitivity at the various settings [170], which complicates the procedure and may degrade precision through error propagation. In any case, the best precision is usually obtained with an electric scan at a fixed magnetic field, which allows a scan of the mass region from m_B to $(1 + 0.3)m_B$, where m_B is the mass selected by the magnet [220]. Under optimized conditions, Pb isotope ratios obtained in the low-resolution setting ($R = 300$) were more precise than those obtained by quadrupole-based instruments [220].

Notable exceptions are high-resolution instruments that are equipped with multiple detectors, and therefore allow a truly simultaneous determination at several *m/z* ratios. This translates into isotope ratios that are as accurate (after suitable corrections for dead time, spectroscopic interferences and mass discrimination as discussed in Sections 1.4.2, 1.4.3 and 1.5) and precise as those obtained by thermal ionization MS, but with a much higher sample throughput [164,165].

All these features explain why ICP-MS has attracted much attention for applications in a variety of areas, including environmental [221], clinical and geochemical studies, as will be seen in Section 1.5.

1.4 LIMITATIONS

The various types of interferences which plague ICP-MS were reviewed [222]. Although spectroscopic interferences (Section 1.4.2) are a bigger problem with low-resolution instruments, non-spectroscopic interferences (Section 1.4.3) are observed on all instruments. All are discussed in detail below.

1.4.1 Precision

The short-term precision of ICP-MS is limited by noise in the ICP, which arises from the plasma and its interaction with the interface [74,75,82,94,154,223] or aerosol processing (i.e. aerosol production, desolvation, and transport) [93,149,150,154]. Some of these sources of fluctuations were discussed in Section 1.2.3, i.e. pulsations from the peristaltic pump and pressure fluctuations in the spray chamber, which are physical effects that are common to ICP-AES. However, when a plasma is put in contact with the sampler, a whine can become clearly audible, depending on the instrument [82]. This whine is due to plasma fluctuations [223] that cause the NAZ to swell and contract periodically, on a millisecond time-scale [82]. It can be decreased by adding a sheathing bonnet between the end of the torch and the sampler [224]. Furthermore, the frequency of the plasma fluctuations decreases as the plasma is moved away from the sampler [82], consistent with a shift of discrete frequency noises to lower frequency [74,75], whereas no such change is observed in ICP-AES [75]. White noise is also higher in ICP-MS than ICP-AES [74]. Another significant source of fluctuations arises from the skimming process [225]. Because only a small fraction of the ions entering the interface actually traverse the skimmer, any spatial inhomogeneity in the ion distribution may result in fluctuations in the analyte signal [225]. Similarly, the analyte signal is very sensitive to fluctuations in carrier gas flow rate and changes in sampling position induced by the plasma fluctuations mentioned above, because it comes from a discrete spatial zone in the ICP [67].

However, the major source of noise (white, $1/f$ or discrete frequency) i.e. which is most responsible for degrading precision above the levels predicted by counting statistics, depends on the instrument [74]. For the ELAN in pulse counting mode, the noise (calculated as the standard deviation of the signal) is limited by counting statistics for any signal (S) below 10,000 counts/s [226]. In other words, the noise (N) can be estimated as $\sqrt{S}$ for $S \leq 10{,}000$ counts/s because N is mainly random shot noise which arises from the discrete nature in which ions are detected when they strike the detector. This means that the RSD will decrease as S increases up to 10,000 counts/s. Above that signal level, the RSD levels at about 1% because correlated flicker noise, which is mainly associated with the generation and transport of aerosol, then also becomes significant [226]. In fact, the total noise associated with any signal within the linear dynamic range of the instrument can be predicted by the following empirical formula [226]:

$$N = \sqrt{S} + 0.01 \cdot S \qquad (7)$$

In general, the precision can be improved by increasing the integration time, but only down to the levels predicted by counting statistics. Internal standardization, which is widely used to improve precision, will only be effective when significant proportional noise is present (i.e. for $S > 10{,}000$ counts/s), since the signals from different analytes are then correlated in time [226]. On the other hand, for signals that are limited mainly by random noise and would therefore be uncorrelated, internal standardization would not be effective and may even degrade precision [226]. In any case, the efficiency of internal standardization with sequential mass spectrometers depends on the noise frequency. Indeed, internal standardization will only be effective against noise components with periods that are longer than twice the time lapse between measurement of the individual masses [227]. This elapse time, t_e, is usually estimated as:

$$t_e = 2t_D n_p + t_s \qquad (8)$$

where t_D is the dwell time per channel, n_p is the number of points per peak and t_s is the settling time [227].

The short-term precision of ICP-MS is therefore expected and is reported to be degraded compared to ICP-AES because of the very different measurement process associated with ICP-MS. Fluctuations that may be averaged out in ICP-AES as a result of the relatively large plasma volume from which light is measured, will be more significant in ICP-MS since only a very small central portion of the plasma, which precisely coincides with the region where droplet processing occurs, is sampled [78,228]. As discussed earlier in this chapter (see Section 1.2.1), a droplet entering the plasma undergoes a series of sequential processes beginning with desolvation and ending with ionization. The position in the plasma where these will actually occur depends on the size of the droplet, when all other parameters are kept constant (such as the plasma operating conditions, the analyte solution, etc.). A small droplet may therefore be at the ionization stage while a nearby bigger droplet may just start to undergo desolvation or vaporization at the same time, thereby causing a local cooling of the plasma (that can be by more than 1500 K [86]), which in turn decreases the ionization efficiency. A decrease in analyte signal [149] and temperature [150] were indeed observed near incompletely desolvated droplets. If a number of assumptions are made regarding for instance the ion transmission of the mass spectrometer, the ionization efficiency of the plasma, etc., the standard deviation of the blank may be calculated theoretically, although that for real samples is not as easily predicted, because non-instrumental errors are then also involved [229].

In any case, the addition of a desolvation device as part of the sample introduction system would remove fluctuations in analyte density that are caused by passage of relatively large droplets [65]. It also makes the position of the IRZ and therefore the plasma sampling position more stable with time than when a wet aerosol is introduced in the plasma [65]. Furthermore, atomization and ionization of the analyte then occur in a narrower range of axial positions in the ICP, which may be advantageous for multielement determination [65]. A cheaper alternative was also described where a heated extension (i.e. 2-mm i.d. Pyrex tube wrapped with heating tape and maintained at 400°C) was inserted between the spray chamber and the torch [180]. This simple approach does not remove the solvent but rather pre-vaporizes the aerosol prior to its entry into the plasma [180]. Yet, numerous advantageous features were observed. Improved detection limits resulted from enhanced sensitivity and/or reduced noise since the pre-evaporation decreased the average size of the droplets entering the plasma [180]. The sensitivity of oxide-forming elements increased by a factor of 3–4 because the pre-evaporation shifted the atomic ion zone towards that of the other elements, making the compromise multielemental conditions optimal for a greater number of elements [180]. The analyte profile was also observed to become narrower with pre-evaporation [180]. With the sampling process being so highly localized spatially in ICP-MS, having the analyte in a narrower zone will allow fluctuations to occur in the IRZ position without degradation of sensitivity and/or precision since the analyte would then be moving within the portion of the plasma being sampled into the mass spectrometer.

The measuring conditions selected can affect the analytical characteristics of the ICP-MS instrument. For instance, two modes are typically available for quadrupole-based instruments: peak hopping where a single mass is selected at a time by setting the rf and dc potentials to the corresponding discrete values, and ramp scanning where the rf and dc potentials are continuously incremented across a mass range. Single-point peak hopping, from peak centre to peak centre, has been demonstrated to provide the best sensitivity, precision and detection limits compared to those obtained by measuring a full spectral profile using either peak hopping or ramp scanning [230]. However, instability in the mass calibration and/or mass discrimination characteristics of the instrument (this is especially true for older versions of the ELAN and PlasmaQuad) can limit precision, particularly for isotope ratios [94]. The shortest dwell time that still allows single-mass-at-a-time instruments to acquire signals properly should therefore be used to obtain isotope ratios with the best precision [94,223]. More time should also be spent on the least abundant isotope being measured in a given ratio, although simply using timings inversely proportional to the abundance will not necessarily provide the ratio with the best precision [231]. In fact, if the instrument provides a precision close to that predicted by counting statistics, then the best timing on each isotope can be found by mathematical simulation [231].

The acquisition time must also be optimized on multi-detector instruments and TOF systems, although in these cases, there is no worry about how simultaneous the measurements are. For example, an acquisition time of 30 s was found to be optimal for the simultaneous measurement of many isotope ratios by ICP-TOF-MS with a precision

≤0.05%, whereas degraded precision was observed at longer acquisition time per replicate [232]. Shorter acquisition times (i.e. <10 s) could be used to obtain similar or better precision than the best values reported with quadrupole-based instruments [232]. The fact that such precision could be obtained in a fast acquisition mode is an advantage of TOF over quadrupole-based instruments. Indeed, similar precision can be obtained for discrete sample introduction as for continuous nebulization [232]. This feature is important for all the sample introduction techniques that will be discussed in the following chapters.

Another drawback of ICP-MS, whatever its resolution, is that it cannot handle a solution having as high a concentration of dissolved salts as ICP-AES on a continuous basis. Because the plasma sampling process in ICP-MS is not as passive as the measurement of emitted light in ICP-AES, deposition of salt on the sampler and/or skimmer occurs, thereby gradually clogging the sampling orifices and causing a concurrent degradation in sensitivity, which limits the long-term stability [233]. This process is particularly important when the concentration of dissolved salt is higher than 0.2% [20]. Similarly, the analysis of sample solutions that contain organic solvents presents problems because of soot deposition [7]. One of the manufacturers (Thermo Jarrell Ash) uses a sampler and skimmer with 1.2-mm orifices (combined to an appropriate pumping system) to increase the tolerance of the instrument to clogging [216].

1.4.2 Spectroscopic interferences

One major problem, especially with quadrupole-based instruments, is spectroscopic interferences, even if they are less severe in ICP-MS than in ICP-AES because of the simpler background of the former. These interferences can result from isobaric overlap of isotopes of different elements, from doubly-charged ions, and from polyatomic ions arising from either the plasma [234] or the matrix [64]. They can be so numerous that a computerized data base is available to provide the ICP-MS (as well as the GD-MS) user with a comprehensive compilation of all the interferences identified up to 1992 [235]. Except for the most important ones i.e. those originating from the plasma (for example, that of $^{40}Ar^+$ on $^{40}Ca^+$), most isobaric interferences between elements (for example, that of $^{114}Sn^+$ on $^{114}Cd^+$) can be corrected for by the software, which measures the signal of an interference-free isotope of the interfering element (Sn in this example), applies a correction factor based on the natural abundances, and subtracts that contribution from the analyte signal, yielding the interference-corrected analyte signal. The same type of correction must be made on high-resolution instruments if the highest practical resolution setting is not sufficient to resolve the interference (as in this particular case of $^{114}Sn^+$ on $^{114}Cd^+$ (see Table 1.8).

TABLE 1.8

Resolution ($M/\Delta M$) required to separate peaks of similar intensity

Analyte ion	Interfering species	Resolution
$^{60}Ni^+$	$^{44}Ca^{16}O^+$	58
$^{28}Si^+$	$^{14}N_2^+$	958*
$^{32}S^+$	$^{16}O_2^+$	1795*
$^{64}Zn^+$	$^{48}Ti^{16}O^+$	1900
$^{64}Zn^+$	$^{32}S^{16}O_2^+$	1952*
$^{52}Cr^+$	$^{40}Ar^{12}C^+$	2375*
$^{56}Fe^+$	$^{40}Ar^{16}O^+$	2502*
$^{51}V^+$	$^{35}Cl^{16}O^+$	2572*
$^{63}Cu^+$	$^{47}Ti^{16}O^+$	3700
$^{39}K^+$	$^{38}ArH^+$	5689*
$^{75}As^+$ *	$^{40}Ar^{35}Cl^+$	7775*
$^{159}Tb^+$ *	$^{143}Nd^{16}O^+$	7800
$^{63}Cu^+$	$^{40}Ar^{23}Na^+$	7900
$^{165}Ho^+$ *	$^{149}Sm^{16}O^+$	8900
$^{169}Tm^+$ *	$^{153}Eu^{16}O^+$	9100
$^{80}Se^+$	$^{40}Ar_2^+$	9700
$^{58}Ni^+$	$^{58}Fe^+$	28000
$^{76}Se^+$	$^{76}Ge^+$	34570**
$^{114}Cd^+$	$^{98}Mo^{16}O^+$	36000
$^{150}Sm^+$	$^{150}Nd^+$	41471**
$^{148}Nd^+$	$^{148}Sm^+$	77000
$^{114}Cd^+$	$^{114}Sn^+$	190000
$^{40}Ca+$	$^{40}Ar^+$	>200000*
$^{87}Rb^+$	$^{87}Sr^+$	300000

*From Ref. 246 where exact masses to the fifth significant digit were measured.
**From Ref. 247.

As discussed in Section 1.2.1, most analyte ions formed in the plasma are singly charged and are therefore detected at the mass of their isotope(s). Doubly-charged ions, on the other hand, are detected at half the mass of the parent ion. Although only a few elements (most REEs have the second IP smaller than the first IP of Ar) will be partly doubly ionized, if they are present in relative excess compared to the analyte, they still can be a significant source of interference. The ratio of the doubly-charged ion to the singly-charged one was reported to decrease exponentially with the difference between the second and first ionization energies [78]. In any case, the extent of doubly-charged ionization can be decreased to 0.2–0.5% of the parent ion by changing the operating conditions, but this is usually accomplished at the expense of analyte sensitivity. For instance, to decrease the level of doubly-charged ions observed in the conditions corresponding to maximum sensitivity for singly-charged ions, a decrease in power and/or an increase in nebulizer flow rate is required, either of which degrade sensitivity [64]. The level of both doubly-charged and singly-charged ions also depends on the sample introduction system (i.e. droplet size distribution and plasma solvent loading) and the interface pressure [157]. A decrease in solvent loading reduced the level of doubly-charged and oxide ions [158,236]. In any case, conditions must be selected to reach the best compromise.

Spectroscopic interferences from polyatomic ions arising from elements present in the sample matrix [64] and/or the plasma [234] are more problematic. The most common elements in the plasma (Ar, H, O and N) can combine to form ions having the same nominal mass as an analyte ion. For example, $^{40}Ar^{16}O^+$ and $^{56}Fe^+$ both appear at the nominal mass-to-charge ratio (m/z) 56 and are not resolved by quadrupole MS. The ShieldTorch interface of the HP 4500, the shielded ICP of the SPQ 8100, and the ELAN 6000 with a cooler plasma [97] drastically reduce interferences from several polyatomic ions (such as $^{40}Ar^{16}O^+$, as clearly shown by Table 1.7). Very low levels of $^{40}Ar^{14}N^+$ and $^{40}Ar^{16}O^+$ were also reported on an instrument similar to the Ion Scan [237].

Other polyatomic ion interferences may arise from the matrix, including any reagent used during sample preparation. For instance, chlorine-containing acids result in the formation of $^{35}Cl^{16}O^+$ and $^{40}Ar^{35}Cl^+$, which interfere with the major isotope of V, ^{51}V, and monoisotopic ^{75}As, respectively. Similarly, carbon-containing samples will give rise to $^{40}Ar^{12}C^+$, which interfere with the major isotope of Cr, ^{52}Cr [238]. The formation of XCl^+, where X is an element from the matrix, can also occur and interfere with the determination of analytes at 35 and 37 amu above the mass of X [239]. This problem can be minimized by using nitric acid and hydrogen peroxide preferentially (or by evaporating undesirable acids off, followed by redissolution of the sample in HNO_3). Obviously, this approach cannot be successful if chlorine is part of the sample matrix (in the form of chloride salts, for example). Although mixed-gas plasma (through the addition of 2% N_2 to the aerosol carrier gas) or the addition of alcohol (up to 5% ethanol) to the sample solution can be used to reduce $ArCl^+$ interference, the level of chloride may require a separation of the analyte from the matrix using, for example, hydride generation with gas/liquid separation for the determination of As [137]. A notable exception may be the Ion Scan: extremely low levels of $^{35}Cl^{16}O^+$ and $^{40}Ar^{35}Cl^+$ were indeed measured on a prototype instrument while solutions containing either 1% HCl or 0.25% NaCl were being nebulized [237].

Nonetheless, for most low resolution instruments, polyatomic ions arising from argon, water, mineral acids and elements from the sample matrix can cause substantial spectral overlap for a number of elements, particularly those with m/z below 80 [64, 234]. Although the extent of these interferences can be significantly reduced by inserting a heated extension tube between the spray chamber and the torch so that pre-evaporation of the aerosol occurs prior to its entry into the plasma [180], if no alternative isotope is available, or is of low abundance, these interferences can drastically degrade the analytical capabilities of the instrument. In some cases, a correction can be made by using a standard solution to determine a correction factor. For instance, standard solutions of sodium carbonate were aspirated to find the relation between ^{13}C and $^{40}Ar^{12}C^+$. This relation was used to convert the ^{13}C signal measured for a sample into $^{40}Ar^{12}C^+$ which was then subtracted from the ^{52}Cr signal [238]. However, aside from leading to a degradation in precision, this type of correction will only be adequate when the correction factor (in this example, the ratio of mass 52 to mass 13) is independent of the matrix for all the standards and samples analyzed, and when all other polyatomic interferences (i.e. at mass 13 and 52) have been taken into account [240]. Spectral fitting, where the observed spectrum is simulated by a linear combination of spectra from individual components, including both atomic and polyatomic ions, was investigated with some success [241]. The approach requires the user to establish relationships between related polyatomic ions and to set lower and upper limits for a number of components (such as ratios, mass discrimination coefficients, etc.) [330]. A different approach based on Bayesian methods can also be used to infer the species at a given *m/z*, determine their contribution to the signal, and correct for spectroscopic interferences while calculating the errors associated with each sample [227].

The formation of refractory oxides is an important problem since, apart from causing isobaric interferences 16, 17 and/or 18 amu (atomic mass unit) higher than the parent ion (for instance, oxides from Ba or light REEs (rare earth elements) can interfere with heavier REEs) [52,239], the signal from the parent ion itself can be reduced significantly as ions are trapped in oxides. For example, a significant fraction of La and Ce form oxides (LaO and CeO) thereby reducing the available La and Ce analyte signals at the compromise sampling position for multielemental analysis. In any case, an exponential correlation was found between the oxide ratio and the analyte–oxygen (M–O) bond strength [78, and references therein], which can be used to predict the relative levels of oxides to be expected. Depending on the analyte and the operating conditions, the levels of MO_2^+ and MO_3^+ can also become significant [7]. A study measured the extent of interference from various polyatomic ions (including oxides) formed from matrix elements of biological materials, at concentrations up to 1 g/l, and reviewed means to reduce and/or correct for these interferences [242]. For instance, a decrease in solvent loading will reduce the level of oxide ions [158, 236]. For example, a micronebulizer with a heated Teflon desolvating membrane reduced oxide interferences to <1% of analyte signal [207].

The effects of doubly-charged ions, hydroxides and hydrides are generally less severe than those caused by the presence of oxides, but may be significant in some cases, such as the interference of ^{238}UH on ^{239}Pu [243]. In general, the formation of polyatomic ions can be minimized by changing the operating conditions, usually at the expense of sensitivity, but can hardly be eliminated (except where noted above). Although desolvation can be used to decrease the extent of these interferences [243,244], they must still be taken into account during mass selection and/or data processing. A spreadsheet template, created from the measured sensitivity of X and corresponding yield of polyatomic ions (obtained using single element standard solutions of X) can be used to predict the intensities of interfering polyatomic ions for a known concentration of X in a sample [239]. Corrections based on this approach yielded an accuracy of ±10% for rocks [239]. However, it is only valid for a given set of operating conditions on a given ICP-MS instrument. An alternative, which can be used for some elements, is to perform their determination by measuring their doubly-charged ions after optimization of the operating conditions to maximize the formation of the latter [245].

On sector-based instruments, the resolution can be increased to resolve several interferences. However, this is achieved at the expense of sensitivity (see Table 1.6) [246]. For example, the determination of As in drinking water necessitated a resolution of 7800 to separate the analyte signal from $ArCl^+$ interference but at the expense of sensitivity, resulting in a quantitation limit of 0.7 µg/l [137]. A better determination limit (of 0.5 ng/l) could be obtained in low resolution with a separation of the analyte from the matrix using hydride generation with gas/liquid separation [137]. Furthermore, Table 1.8 shows that not all interferences can be resolved, even at the highest resolution setting (typically 10,000) which still allows detection of the analyte ions [127]. Many spectroscopic interferences from oxides cannot be resolved by high mass resolution [173]. The highest resolution achievable on a PLASMAX-1 was increased to about 40,000 by modifying the interface (the i.d. of the sampler, skimmer and third cone were increased to 1.3, 1.3 and 3.0 mm instead of 1.0, 0.7 and 0.7 mm, respectively), the ion optic and the electronics [247]. Under these conditions, more interferences from polyatomic ions could be resolved, as well as some isobaric overlaps (for instance ^{76}Se and ^{76}Ge) but the associated detection limit was degraded to about 10 µg/l [247].

In any case, the resolutions listed in Table 1.8 and in Ref. [335] both assume that peaks of similar magnitude are to be separated. Since in real samples, the source of the interference will likely be in large excess compared to the analyte, even higher resolutions will be required [127]. Therefore, even if high-resolution instruments allow substantially more flexibility in the use of acid mixtures without fear of introducing a source of spectroscopic interference, they are not entirely free of such problems. Depending on the relative levels of analyte and interfering species, the same recommendations as for quadrupole-based instruments may have to be followed. The combination of desolvation with high resolution is particularly efficient at reducing polyatomic ion interferences from solutions containing HCl, leading to improved analytical performance compared to quadrupole-based instruments, because of the almost zero background then achieved [244]. In some cases of partial spectroscopic overlap, adequate results may also be obtained if the mass scanning range is limited to only a narrow segment in the centre of the peak, where the contribution from the interfering peak is negligible [248]. For example, by scanning only the central 10% of the Cu peaks, the Cu isotope ratio could be measured with a precision of ≤0.6% RSD (n = 5) in an Antarctic sediment digest despite partial spectroscopic overlaps on both Cu isotopes [248].

Whatever the instrument, multivariate optimization techniques can be used to find operating conditions with substantially reduced spectroscopic interferences. A simplex optimization using several criteria (such as the In signal and its RSD, the BaO^+/Ba^+ ratio, and the background signals at several m/z where interferences are expected from the plasma and/or matrix) could be carried out to find compromise conditions for multielemental analysis [84]. The plasma can also be operated under cool conditions i.e. at high aerosol carrier gas flow rate and low power to substantially decrease the background (Ar^+, ArH^+, etc.) originating from the ICP [67, and references therein, 85]. However, a cool plasma will only be useful if the plasma potential is low. Indeed, operating under cool conditions increases the plasma potential, which would exacerbate any secondary discharge and, in turn, maintain a significant level of Ar^+, etc. [67]. That is why a metal shield must be inserted between the torch and the load coil on many of the available instruments. Furthermore, this type of plasma is expected to only be useful for the analysis of clean samples [67]. Indeed, in contrast to the robust plasma conditions that allow a longer residence time in the plasma, the cool plasma conditions decrease the residence time, so complicated matrix may not be properly atomized. Care should also be taken to ensure that there are no spectroscopic interferences from oxide ions at the m/z of analyte ions, because the lower plasma temperature associated with cool plasma conditions favours the formation of oxide ions [85].

Several mechanisms for the formation of polyatomic ions have been proposed, but there is so far no consensus. The situation is complicated by the fact that at least two different mechanisms are involved. Indeed, under a given set of operating conditions, a signal maximum is clearly observed for Ar_2^+ while varying the aerosol carrier gas flow rate from 0.5 to 1.0 l/min [249, 250], but only a gradual decrease of $ArCl^+$, ClO^+ and ClO_2^+ concurrently occurs [250]. Similarly, a gradual decrease of $ArCl^+$ and ClO^+ was observed as the temperature of the water-cooled spray chamber was increased but a maximum was concurrently seen for Ar_2^+ under the same conditions [250].

Some have postulated that condensation reactions resulted in the formation of oxides in the expansion region downstream from the sampler [251,252], but Douglas and French [69] have later calculated that very few ion–molecule reactions are likely in this region. These calculations were supported by measurements performed on an ELAN (i.e. without a secondary discharge), where the ratio of analyte oxide signal over the sum of the analyte and analyte oxide signals was independent of the ion-extraction configuration [253]. On the other hand, the ratio changed with any modification to the plasma conditions, which indicated that the ratio was determined solely by plasma conditions [253]. In fact, modification to the sample introduction system can have a dramatic effect on the oxyde:analyte ion ratio [157].

The interface pressure was reported to be an important parameter, but since the results were obtained on an instrument with a secondary discharge, the change in pressure likely modified the intensity of the discharge, thereby altering the formation or destruction of oxide ions [157]. The latter assertion is further supported by results obtained using a shielded ICP, which visibly eliminated the secondary discharge while concurrently reducing substantially the formation of several weakly-bound polyatomic species (for instance, ArH, CO_2 and ArO) [77,108] but not that of refractory oxides [108]. This indicates that at least part of the polyatomic ions can be promoted by a secondary discharge. In fact, the signal of ArX^+ was not only dependent on the electron number density in the secondary discharge, but the measured rotational temperature of the discharge was consistent with a collision-induced reaction of Ar with X^+ [108].

Another possible source of polyatomic ions may be collisional reactions occurring in the boundary layer (formed by the deflection and cooling of the plasma against the interface) around the sampler orifice [21,254]. Although the flow of gas through the sampler is sufficient to puncture the boundary layer [7], this does not preclude sampling the layer by entrainment [254]. High-speed photographs have indeed shown the plasma plume (and therefore the boundary layer) passing downstream along the surface of the cone [82]. This is further supported by the very low levels of weakly-bound polyatomic species (such as $^{40}Ar^{16}O^+$ and $^{40}Ar^{35}Cl^+$) but normal levels of refractory oxide ions (for instance, LaO and CeO) that are observed with an offset ion lens [237]. Since the offset ion lens only accepts ions leaving the skimmer along its central axis, in contrast to instruments using photon stops where only ions leaving the skimmer off centre enter the ion lenses, this strongly suggests that a significant portion of the boundary layer is indeed sampled [237].

Finally, survival of polyatomic species, especially refractory metal oxide ions, through the plasma has also been postulated. Although oxide formation is unlikely at the calculated plasma temperatures [254], the inhomogeneous distribution of droplets (in terms of size) in the plasma can likely allow an analyte to react with oxygen from a nearby desolvating droplet [149-150]. This is supported by recent results which showed a decrease in excitation and Doppler temperatures under conditions favouring oxide formation i.e. when the analyte had a short residence time in the plasma [62]. Furthermore, Tanner reported that the kinetic energy of analyte oxide ions was smaller than that of the corresponding analyte ions which, in turn, indicated that the analyte oxide ions originated from a 761-K cooler plasma region than the analyte ions [255]. Finally, calculations have shown that, under steady state conditions, the plasma is not hot enough to fully dissociate refractory oxide ions [67].

1.4.3 Non-spectroscopic interferences

As a result of the rather intrusive sampling process, ICP-MS is also much more susceptible to effects of concomitant elements (also called matrix effects and non-spectroscopic interferences) than ICP-AES. In fact, analyte ion signal intensity in ICP-MS is affected by concentrations of other elements at levels which would usually have a negligible effect in ICP-AES. These interferences are usually observed as a suppression of analyte signals [21,116,240,256–267], although enhancements have also been reported [76,116,240,257–259,261,266, 268–272]. They depend on the analyte, the sample, the sample introduction system (for instance, a suppression was observed using a MCN with a double-pass spray chamber but an enhancement resulted when the same MCN was combined to a heated spray chamber/membrane desolvation system [173]), the interface (in particular, the skimmer orifice diameter [273]), the ICP operating conditions [261,274] and the lens configuration and settings [260,261,265,266,273], hence the range of observations reported and the unpredictability of these effects. They are caused by a relatively high concentration of matrix element(s). The change in signal is most often (but not always, see for example [264]) a function of the masses of the analytes and the concomitant elements, and can also be related to their ionization potentials [256,262,264,275], depending on the operating conditions and the ICP-MS instrument used. In general, the lower the atomic mass and, in some cases, the degree of ionization of an analyte, the greater will be the effect of a given concomitant element. Conversely, the greater the atomic mass and, in some cases, the degree of ionization of a concomitant element, the greater will be its effect on the signal of a given analyte. A variety of mechanisms have been proposed to explain these effects, although no consensus has been reached so far; one or more mechanisms may in fact predominate under certain experimental conditions [259].

It is possible to decrease the level of these effects by using robust ICP conditions (which have been most extensively described in ICP-AES by Mermet [276]) i.e. at low carrier gas flow rate, high rf power and using an injector with a large internal diameter. For instance, the absolute recovery (obtained by comparing the average intensity over the whole range of carrier gas flow rates studied in presence of the matrix to that with the reference matrix) observed while nebulizing five-fold diluted seawater was improved by using a 1.5-mm i.d. injector (instead of a 1-mm one) at high power (1.45 kW) and small sampling depth (i.e. 11 mm between top of the load coil and tip of the sampler) [275]. Similarly, a decrease in aerosol carrier gas flow rate was reported to significantly reduce the suppression observed in presence of 1000 mg/l Cs, albeit with a 50% sacrifice in sensitivity [261]. The relative changes in analyte signal with an increase in HNO_3 concentration were also reported to be less pronounced at lower aerosol carrier gas flow rate [228]. Both a higher forward power and a lower solution uptake rate were found to minimize the suppression observed with plant samples in axial ICP-AES [277]. (Because only the injector channel is monitored in axial ICP-AES, it is more similar, in terms of analytical characteristics, to ICP-MS than ICP-AES with lateral view of the plasma.) Modern instruments provide more than adequate sensitivity for many applications, so they could be operated under robust conditions that would not only provide still adequate sensitivity for these applications, but would also alleviate matrix effects substantially. In fact, it is really surprising to see that this simple approach has received so little attention [67]. On the other hand, the robustness of the ICP conditions is not expected to be as absolute as in ICP-AES, where the analyte signal can be made independent of the matrix, because the relatively small spatial region that is sampled in ICP-MS makes it much more sensitive to any change in aerosol properties and/or plasma temperature [228].

The presence of easily ionized concomitant elements (i.e. with a degree of ionization close to 100%) may interfere with the atom–ion equilibrium:

$$M \leftrightharpoons M^+ + e^- \tag{9}$$

by shifting it towards the atom side as a result of an increase in the number of electrons, therefore resulting in recombination and, consequently, signal suppression [256]. On the other hand, since the vast majority of the electrons in the plasma arise from the ionization of argon, which is in much greater excess than the concomitant element, the number of electrons is unlikely to be significantly increased by the ionization of an easily ionized element (EIE) [3,278]. However, it is possible that, because of the inhomogeneity of droplet size distribution in the plasma, the effect can be locally significant, similarly to the oxide-formation process. This hypothesis is consistent with ionic fluorescence measurements which showed a suppression of analyte signal in presence of some concomitant elements in the central channel of the ICP but little change in analyte intensity off axis [279].

Additional evidence for a change in the atom-to-ion equilibrium was also obtained, where an increasing suppression of both atomic and ionic emissions was caused by an increasing concentration of acid (HNO_3, HCl or H_2SO_4) under certain plasma operating conditions with a concurrent decrease in electron number density (without a change in the excitation temperature) of the plasma [280]. However, this does not explain why the concentration and type of acid present in the matrix can affect some elements and not others. For instance, no significant change in the ICP-MS signals of Co, Y and In was observed with 0.04–0.96 M of HNO_3, whereas that of As was increasingly suppressed [76]. On the other hand, a small suppression occurred for Co, Y and In with 0.06–0.35 M H_2SO_4 while the signal of As was suppressed with 0.006–0.06 M H_2SO_4 but partly recovered at higher H_2SO_4 concentrations [76].

A change in the ionization equilibrium over a limited range of first ionization energy was proposed by Allain et al. [269] to explain the enhancement observed in the presence of carbon-containing compounds (glycerol and methane) for elements with a first IP between 9 and 11 eV (such as Hg, As, Au, Se and Te) whereas only minor changes occurred for other elements under identical conditions. Similarly, addition of up to 20% methanol, ethanol or isopropanol increased As sensitivity [137]. However, these two studies were made using a constant forward power and aerosol carrier flow rate, despite the fact that the presence of an organic solvent changes the optimal sampling position, as discussed at the end of Section 1.2.1. Nonetheless, an enhancement of As [76,81] and Se [81] signals was reported after re-optimization of the aerosol carrier flow rate, in the presence of 5% v/v Triton X-100 [76] and 3% v/v methanol [81], which was greater than what could be ascribed to the increase in nebulization efficiency resulting from the addition of these chemicals to the matrix. Similar enhancements were also observed with an equivalent amount of ammonium carbonate [81]. The carbon content of samples or the residual carbon present after their digestion can also induce a significant enhancement of analyte signal [116,281], which may be compensated by internal standardization [281] or not [116]. Alternatively, matching the matrix of the standard solutions (such as in ethanol content, in the case of the analysis of wines [282]) may also be required for accurate determinations.

A more recent study looked at the enhancing effect of methanol, ethanol, propanol, butanol, glycerol, acetonitrile and acetic acid on the Se signal [267]. A lower aerosol carrier gas flow rate was required to maximize Se sensitivity in presence of all these matrices, except acetic acid, which optimized at the same flow rate as aqueous solutions [267]. An enhancement factor of 5–6 was obtained with all these matrices, with no correlation to the molar carbon concentration, and no dependence on the Se concentration [267]. Furthermore, the sensitivities were equal for various Se species (selenite, selenate, selenomethionine and trimethyselenonium iodide) [267]. It is possible that the degree of ionization of elements (which are not 100% ionized in an all-argon plasma), could be increased through collisions with the significant population of carbon-containing ions created in the central channel of the plasma by introducing a carbon-containing chemical [76,81]. This would, for instance, explain why the argon dimers almost double in presence of 5% acetic acid [267]. Therefore, only elements with a first IP lower than that of C (11.26 eV or 1086.2 kJ/mol) and higher than about 9 eV or 868 kJ/mol (i.e. where the degree of ionization is significantly lower than 100%) [7] could benefit.

Whatever the ionization process taking place in the plasma in the presence of carbon-containing compounds, its efficiency will likely be dependent on the operating conditions and the amount of carbon-containing compound. For instance, Table 1.9 shows the effect of 30% v/v methanol on the figures of merit for selected elements covering a range of IP and mass. These results (unpublished so far) were obtained in the author's laboratory. According to the above discussion, an increase in sensitivity would be expected for Zn, As and Cd. Yet, only those of Zn and As were enhanced. A suppression actually occurred for the heavier elements, whereas sensitivity was unaffected for elements of lower mass. However, a three-fold improvement in detection limit still resulted in most cases, even despite a concurrent decrease in sensitivity for Cd and Pb. This is likely due to the reduction of noise which would result from smaller droplets entering the plasma when the matrix is made more volatile [283] by the addition of some methanol. An improvement in signal-to-noise ratio by a factor of 3–4 was also reported in presence of 3% v/v methanol [81]. The different physical characteristics of carbon-containing compounds also usually result in a memory effect of the enhancement effect. For instance, it was observed to last about 20 min after withdrawal from 5% acetic acid and 50 min with 3% methanol [267].

Grégoire [260] and Hobbs and Olesik [279] considered ambipolar diffusion to explain effects of concomitant elements, whereby analyte ions would diffuse out of the central channel of the ICP and decrease the analyte population available for sampling in the centre of the plasma. However, Hobbs and Olesik [279] pointed out that electrons possess the highest diffusion rates and, provided that the electron numbers are sufficiently high (which should be the case with an EIE), diffusion of the analyte ions should not depend on the masses of other positive ions present, and, therefore, mass-related non-spectroscopic

TABLE 1.9

Change in sensitivity and detection limit observed when switching from a 1%-HNO_3 (at 1.2 kW; 29 PSI nebulizer pressure) to a 30%-methanol matrix (at 1.4 kW, 21 PSI nebulizer pressure)

Analyte	First IP* (kJ/mol)	Sensitivity ratio (MeOH/aqueous)	Detection limit ratio (aqueous/MeOH)
^{51}V	650	1.1	2.9
^{55}Mn	717.4	1.0	1.0
^{60}Ni	736.7	1.0	0.6
^{64}Zn	906.4	1.6	3.0
^{75}As	947.0	2.6	5.0
^{114}Cd	867.6	0.8	3.6
^{208}Pb	715.5	0.4	3.3

*From Ref. 7.

interferences should not arise in the plasma. This was supported by the fact that the Sr ion fluorescence suppression caused by 0.005 M Ba was identical to that caused by 0.005 M Li, thereby demonstrating that matrix-related mass-dependent effects did not occur in the plasma [279]. Similarly, another study looking at fluorescence at the tip of the sampler showed that neither 118 ppm Mg nor 1000 ppm Pb had any significant effect on the ionic fluorescence of Ba or Sc in front of the sampler [284].

Beauchemin and co-workers [259] proposed a combination of mechanisms to explain the variety of effects that they observed: suppression, enhancement, analyte mass dependence (the effect being worst for light analytes) and the fact that the concomitant element effects depended on the mass of that element. They explained the enhancement by suggesting that an increase in the number of atom–electron collisions as the gas flows through the sampler orifice would lead to enhancement of the analyte signal if the number of electrons increased (as in the presence of EIEs) since it would favour collisional ionization (Eq. 1) over recombination with electrons. The latter process would however prevail if the analyte element forms refractory oxides that can deposit on the interface, thereby decreasing the collisional rate and leading to analyte signal suppression. All mass-related effects were attributed to space charge effects inside the mass spectrometer. Similarly, Kawaguchi et al. [262] could not explain the non-spectroscopic interferences that they observed by a single mechanism. They attributed the mass dependence of the matrix effect to collisions within the interface and space charge effects in the ion lens system. A shift in the ion-to-atom equilibrium could also explain the dependence of the suppression on the ionization potential of the concomitant element [262].

Space charge effects were discussed by Gillson et al. [278] and in greater detail by Tanner [285,286]. Tanner et al. have demonstrated that, at least for instruments free of secondary discharge, the plasma flows through the interface (i.e. the sampler and the skimmer) without charge separation i.e. a neutral beam emerges from the skimmer [287,288]. Once the beam enters the MS region immediately downstream of the skimmer, the ion lenses repel electrons and collect the positive ions into a confined beam. The total beam current and the repulsion of like charges limits the narrowness of this beam. The greater the imbalance of charge, the greater the magnitude of the space charge field i.e. repulsion effects. In particular, light ions are more readily deflected away from the axis than heavy ions which stay closer to the centre of the ion beam. This has been demonstrated by inserting a graphite disk behind the skimmer and then looking at the ion-beam profile that was deposited from a 0.01 mM multielemental solution by spatially resolved laser ablation ICP-TOF-MS [289]. The results clearly showed that, at the low acceleration potential that is typical of quadrupole-based instruments, the ion beam width was larger for low mass elements. Specifically, that for ^{7}Li was three times that for ^{208}Pb [289]. Similarly, scanning electron microscopy studies of the distribution of elements deposited on a target placed along the central axis past the ion lens region, showed that the spatial distribution of lighter ions was slightly broader than that of heavier ions [290]. This widening of the ion beam was also confirmed by ion fluorescence measurements 22–45 mm downstream of the skimmer that reported a larger FWHM (full width at half maximum) for Sc than for Ba [291]. In fact, the latter measurements showed that the ion beam was even broader than had been determined by the above deposition studies [291]. Ion fluorescence measurements also showed that Mg had a focusing effect on the Ba and Sc ion beams without suppressing them, whereas Pb dramatically suppressed them [291]. In accordance with the space charge theory, the Sc^+ signal was more heavily suppressed than that of Ba^+; in contrast however, the FWHM did not change for Sc^+ whereas it increased for Ba^+ [291].

Similarly, both scanning electron microscopy and energy dispersive X-ray analysis did not reveal any broadening in the Sc ion beam in presence of 1000 mg/l Cs, only a suppression of the Sc signal [292]. In fact, a more recent study cautioned the use of deposition experiments to study the behaviour of the ion beam, since the distribution depended on the composition of the analyte stream. As well, codeposition of neutral analyte atoms and charged or neutral microparticulates cannot be discriminated against [293]. An increase in acceleration potential to 4000 V produced an insignificant correlation between beam width and analyte mass, which may explain why deleterious space charge effects are less significant in sector-field instruments [289]. Nonetheless, whether these effects result in beam broadening or not, they are also observed on sector-field ICP-MS, although to a lesser extent than in quadrupole-based ICP-MS, as witnessed during the nebulization of five-fold diluted seawater [275]. Furthermore, some effects do arise from the interface since installing an additional external rotary vacuum pump to decrease the vacuum in the first stage from 1.8 mbar to 0.8 mbar significantly decreased matrix effects [294].

Hence, on instruments using a low acceleration potential, a matrix composed of heavy elements will suppress the analyte signal more extensively than one composed of light elements; and heavy analyte ions will be suppressed less severely than light ones. Similarly, the greater the concentration of the concomitant element, the higher will be the ion beam current, and the more extensive will be the effect.

Space charge effects within the ion optics are further supported by the fact that changes to the ion lens voltages [237,265] or configuration [295] can drastically mitigate the observed effects of concomitant elements. In instruments with a secondary discharge, the interface pressure was reported to change the effects from concomitant elements [157], which would indicate that charge separation occurs within the interface and depends on this discharge. Furthermore, a study using MDMI for sample introduction and a twin quadrupole ICP-MS instrument suggested that space charge effects could occur between the skimmer and the first ion lens or between the first and second ion lenses [296]. Indeed, the simultaneous detection of Li and Pb from individual droplets allowed the observation of the appearance of a shoulder on the leading edge of the Li^+ signal in presence of a Pb matrix, a shoulder that was not observed on the Pb channel, was absent when Pb was not present, and could be modified by changing the voltages applied to the first two ion lenses [296]. In any case, i.e. whether charge separation occurs within the interface and/or in the ion optics, the space charge effect is a likely explanation for mass-dependent analyte signal suppressions.

Space charge effects can be minimized by reducing the charge density of the beam. This can be achieved by accelerating the ions as they emerge from the skimmer, before dispersion takes place, as was carried out on the TS SOLA [120] and in high-resolution instruments. An alternative is to reduce the flux of the ion beam by passing it through a third orifice of smaller i.d. than the skimmer [287,288]. Either approach drastically reduced effects from concomitant elements [120,287]. In the case of instruments where a uniform response was also observed as a function of mass, a single internal standard could be used to correct for these effects across the mass range [120,287]. In theory, cool plasma conditions can also be used. Indeed, because the degree of ionization is reduced compared to normal plasma conditions, the total ion current is reduced, which in turn should reduce the contribution of space charge effects to the observed matrix effects [85]. On the other hand, since these operating conditions are the complete opposite of the robust plasma conditions that have been demonstrated to reduce matrix effects that are not space-charge related, worse matrix effects have been reported than in normal plasma conditions [85]. However, internal standardization with NO^+ or another sample-derived ion could be used to largely compensate for these effects [85].

However, space charge effects do not explain the signal enhancement reported by several groups, nor the matrix effects observed in high-resolution instruments where space charge effects should be reduced [116,128,136]. In fact, in a comparison of quadrupole-based ICP-MS with double-focusing ICP-MS, more important matrix effects were observed with the latter during the multielemental trace analysis of human serum, making the use of multiple internal standards mandatory to cover the mass range [116]. An explanation for the latter probably lies with the sample introduction system and the sample solution, which both determine the aerosol droplet size and plasma solvent loading. Since the ratio of major to minor cations in a given droplet is dependent on the size of that droplet [297], a variety of ratios can therefore be obtained with various sample introduction systems operated under different conditions, which likely yield widely different effects from concomitant elements. For instance, by increasing the spray chamber temperature to 15°C (which increased the average droplet size as well as the plasma solvent loading) and the interface pressure to 3 Torr, the suppressing effect of 10 mM of selected concomitant elements on several analytes could be eliminated or transformed into an enhancement [157]. The effect of a given concomitant element on a given analyte element was also reported to vary with pH and/or the complexing ability of the matrix, which determined the species formed by the analyte and, ultimately, how efficiently the plasma could atomize and ionize these species [157].

Using a conespray nebulizer combined with a glass cyclonic spray chamber reduced matrix effects compared to those observed in ICP-AES with a cross-flow nebulizer combined with a double-pass spray chamber [298]. However, the position of the nebulizer in this spray chamber is critical and should be optimized since it affects the characteristics of the tertiary aerosol [155]. In any case, a decrease in the solvent load in the plasma was observed with both sample introduction systems when Na was present in the solution [298]. The presence of Na also resulted in a notably finer tertiary aerosol [298]. Finally, some phenomenon similar to aerosol ionic redistribution is also possible since an enrichment of analyte was observed in the solution being drained [298]. The sample introduction system is therefore a key component in matrix effects and should be selected carefully. At the moment, this selection can, however, only be made empirically since the spray chamber design parameters that would minimize aerosol changes still have to be investigated [298].

Furthermore, some matrix effects have been demonstrated to occur in the first vacuum stage [284]. Indeed, a study looking at fluorescence at the tip of the sampler and inside the skimmer showed that, although neither 118 ppm Mg nor 1000 ppm Pb had any significant effect on the ionic fluorescence of Ba or Sc in front of the sampler, both (Mg marginally more than Pb) suppressed the Ba and Sc signals inside the skimmer [284]. Furthermore, the rate at which the Sc^+ beam intensity dropped was greater than that of Ba^+ [284]. These effects could perhaps be linked to the flow disturbance that was recently measured at or slightly upstream of the skimmer tip [299]. In any case, since the sampling interface is fairly common to all instruments, matrix effects in this region could very well correspond to the matrix effects observed in high-resolution instruments.

Changes in transport efficiency can also be involved. For instance, an increase in acid concentration can change aerosol formation and decrease aerosol transport efficiency [280]. The spray chamber has been shown to have a key role in this source of matrix effect since it affects both aerosol transport and filtering [276]. This is also true for elements forming volatile species: if the matrix reduces or prevents the formation of these species, the resulting signal will likely be smaller than that obtained for the same analyte concentration in a matrix favouring the volatile form of the analyte, since the latter will be more efficiently transported to the plasma than the former who may be drained out [166]. Even if the use of a spray chamber is avoided as recommended in a study [276], a change in sample solution composition can change the average droplet size as well as the width of the droplet size distribution. For instance, the enhancement of sensitivity observed with a DIN when 20% MeOH is present in solution, is likely the result of the smaller droplets which are produced with a narrower size distribution than in the absence of MeOH [300]. The source of the various effects from concomitant elements is therefore quite complex since it is affected by a combination of interdependent parameters.

Mass discrimination is a final limitation of ICP-MS, which arises from the more efficient transport of the heavier particles through the interface [3], as well as space-charge effects [278,285] and the inherent transmission efficiency of the mass spectrometer. In instruments where space charge effects exist, the extent of defocusing of analyte ions in an ion beam composed largely of Ar^+ will depend on their mass as explained above. Heavy ions will be transmitted much more efficiently than lighter ones. So, an equimolar solution of analytes of various masses will give rise to various signals, the lowest ones being observed for light analytes. Even with minimal (if existent) space charge effects, a mass-dependent response has been observed [136]. This limitation seems to have been alleviated on the SOLA and Ion Scan for which the manufacturers claim a uniform response over the entire mass range (see Section 1.3). The accuracy of isotope ratios measured with the latter two instruments is therefore expected to be higher (i.e. without bias towards the lower mass) than that on the other systems, which require correction for mass discrimination. Even the high-resolution Plasma 54, which is equipped with multiple detectors for simultaneous detection, is not free of mass discrimination and requires a correction [134,135]. The same is true of axial TOF-based instruments where a mass discrimination of 1% per mass unit was observed to be similar to that on quadrupole-based and sector field instruments [232].

This correction involves the measurement of isotope ratios for a standard of known isotopic composition to determine the mass bias. The corrected ratio, R_{true}, is then given by:

$$R_{true} = R_{meas}(1 + C)^{\delta m} \tag{10}$$

where R_{meas} is the measured ratio, C is the mass bias factor or mass discrimination per mass unit and δm is the mass difference [135]. Alternatively, mass bias correction can be based on linear or exponential relationships [231]. The mass bias factor is usually constant over a limited mass range. The correction can be determined using either an external standard or an internal standard (such as an element with an isotopic ratio that is constant in nature) [214]. There are advantages and disadvantages to both approaches. Obviously, an external standard allows the measurement of the same isotope ratio and should therefore provide the best correction if there is negligible drift in the mass bias. On the other hand, an internal standard drastically reduces errors arising from drift in the mass bias but measuring the isotopes of the internal standard requires time, which may reduce the number of elements that can be determined, degrade the precision of isotope ratios if less time is available for measurement of the isotopes of interest [214], and/or not provide the best correction [231].

If only an isotopic standard from a different element than the analyte is available, then the bias factor obtained using that element may need to be corrected by mass as well [301]. The latter correction requires the measurement of the bias factor for a series of elements over the mass range to find the relationship between mass bias and mass [301]. If the sample matrix induces additional mass discrimination to that inherent to the instrument and operating conditions used, then a matrix-matched standard of known isotopic composition is required for the correction to be appropriate [302]. Obviously, this requirement will likely be impossible to satisfy for samples with a complex matrix, in which case a separation of the analyte from the matrix is then recommended [302]. For light elements (such as Li), the isotopic ratio may also vary with the analyte concentration itself thereby making an additional requirement that the isotopic standard be concentration-matched to the sample analyzed [302]. Finally, in cases where the background signal is not negligible compared to the analyte signal, a background correction procedure is also required, where a background signal is used as an internal reference to correct for variations in sensitivity [232].

1.5 PRESENT CAPABILITIES

Despite the limitations, ICP-MS has been applied to the analysis of a variety of samples. This is illustrated in Table 1.10 where a summary is presented of selected applications that have appeared in peer-reviewed journals. Also, some books include chapters reviewing applications of ICP-MS [6,7,19]; while others consist in proceedings of conferences. These publications include environmental [437–446], geological [445, 447–451], hydrological [452-458], nuclear [459], metallurgical [456, 460–464], industrial [465–467], biological [468–473], food-related [472, 474] and archaeological [475] applications, as well as the use of ICP-MS in the production of standard reference materials (SRMs) [451,461,477, 478]. In addition, a review by Crews [479] describes the application of ICP-MS in food science, whereas Barnes summarizes how ICP-MS was used to study human nutrition and heavy metal toxicology [480]. Because of its low detection limits, care must be exercised in minimizing all sources of contaminations. Even gloves (vinyl and especially latex) have indeed been reported to be a significant source of some biologically important elements (such as Fe and Zn) [481].

TABLE 1.10

Selected applications of ICP-MS

Element	Sample	Sample preparation	Calibration strategy	Comments	Instrument; sample uptake rate	Ref.
Water analysis						
Ba	seawater	1 ml pH 2 seawater + 1 ml ^{135}Ba enriched standard with same specific gravity as seawater; dilute to 100 ml with 0.2 M HNO_3	isotopic dilution correction for mass discrimination	each sample equilibrated with plasma ~1 min before data collection; <4% accuracy if 0.5<^{138}Ba/^{135}Ba <5.0	PQ1; 0.8 ml/min uptake rate	303
Te	industrial waste water	organo-Te converted into Te(IV) and Te(VI) by adding NaOH to pH ≥ 12.5; hydroxide precipitation prevented by adding EDTA	on-line internal standardization by pumping ^{125}Te and sample through a mixing coil	calibration graphs usng second-order polynomial since the sample already contained some ^{125}Te	ELAN 500; each line at 0.55 ml/min uptake rate; spray chamber at 1°C	304
U	pore water of marine sediments, river water, seawater	0.25-ml sample spiked with 0.1 ml ^{236}U and diluted to 5 ml (except for fresh waters)	ID using ^{236}U as spike	^{236}U not present in natural waters; correction for mass discrimination	PQ1; 0.4 ml/min uptake rate; water-cooled spray chamber (unspecified temperature)	305
U	standard solutions with 2.5 g/l $NaNO_3$		external calibration with internal standardization using Th	lower detection limit with USND instead of pneumatic nebulization with water-cooled spray chamber; adequate compensation of 60% suppression; applicable to industrial wastewater	PQ1; 1.5 ml/min uptake rate; USND with heater at 140°C and chiller at 5°C	306
Al, Ca, Cl, Fe, Mg, Mn, Na	fresh water		external calibration with internal standardization using Co for Al, Mn, Fe, and Sc for Na, Mg, Ca	rinsing with 0.14 M HNO_3 between samples and standards to avoid memory effect; non-optimal nebulizer flow rate for too concentrated elements to expand normal working range	PQ; 0.9 ml/min uptake rate; spray chamber at 10°C	307

Element	Sample	Sample preparation	Calibration strategy	Comments	Instrument; sample uptake rate	Ref.
Hg and 13 elements	potable water		external calibration with matrix-matched standards and internal standardization using Rh	presence of Au preserve Hg in PET containers, allowing both Hg and other metals to be determined simultaneously	ELAN 6000; wash with 5 mg/l Au in 1% HNO_3	163
Al, As, Cd, Cr, Cu, Mn, Ni, Pb, Zn	groundwater	50 ml sample acidified with 0.5 ml concentrated HNO_3 and filtered	external calibration with internal standardization using Rh	separate determination of C, Cl and Na for expected interferences on ^{52}Cr (ArC, ClOH), ^{53}Cr (ClO) and ^{63}Cu (ArNa); ^{53}Cr used for low chloride samples otherwise ^{52}Cr with correction as in [174]; similar correction for CaO on ^{60}Ni using ^{46}Ca	ELAN 5000; unspecified uptake rate	308
Ni, Zn, V, Ba, Pb	groundwater SRM		external calibration with internal standardization using In	a high-efficiency Meinhard nebulizer was used with desolvation	ELAN 5000; 85 µl/min uptake rate	102
Ti	seawater	4-l acidified sample spiked with ^{49}Ti, equilibrated, adjusted to pH 4.0 with NaOH and NH_4OAc, pumped through 8-HOQ resin; rinse with 0.02 M KHP and water to remove Ca^{2+}, Mg^{2+} and SO_4^{2-}; elution with 10 ml 2.4 M HNO_3	estimate of Ti concentration obtained by external calibration prior to ^{49}Ti spiking; ID analysis	400-fold preconcentration; removal of Ca, P and S required to prevent interferences from $^{48}Ca^+$, POH^+ and SO^+ on $^{48}Ti^+$ and $^{49}Ti^+$; background due to NO_2^+ must be monitored; detection limit of 0.2-0.4 ppt	PQ1 and PQ2 Turbo Plus; unspecified uptake rate; spray chamber at 15 and 8°C, respectively	309
Mo	seawater	10-fold dilution with 0.14 M HNO_3	external calibration with internal standardization using In (or Rh, after anion exchange separation)	correction for Br-containing polyatomic species by adding $NaBrO_3$ to blank and standards to match Br concentration of sample, or anion exchange separation of Mo from Br, or using experimentally-determined correction factors	PQ; 0.9 ml/min, spray chamber at 10°C	310

continued

TABLE 1.10 (*continuation*)

Element	Sample	Sample preparation	Calibration strategy	Comments	Instrument; sample uptake rate	Ref.
Al, Cd, Co, Cu, Fe, Pb, Mn, Mo, Ni, Pb	fresh water	filtered sample split into fractions which are pumped through columns of Chelex (Ca form) which collects free metal ions and weak complexes, DEAE resin which collects humic complexes, and Sep-Pak C-18 which traps non-polar organic compounds	unspecified	fractions determined by difference i.e. sample analyzed before and after passage through each column rather than after elution; two-channel sample introduction system to quickly switch between sample and rinsing solution (1% HNO_3)	PQ1; 0.9 ml/min uptake rate; spray chamber at 10°C	311
40 elements	river and lake water, rainwater	filtered through 0.45-μm cellulose acetate filter just before analysis	external calibration with internal standardization using In	sample diluted as needed to keep analyte concentration within working range of instrument	PlasmaTrace (R= 400-3000); 2-4 ml/min uptake rate USND with heater at 120°C and chiller at 1°C	209
REEs	groundwater, seawater	50 μL of 65% HNO_3 added to 10-ml sample; seawater diluted 5-fold with DDW	external calibration with internal standardization using In	operating conditions optimized to minimize BaO^+/Ba^+ ratio; 30% signal suppression in diluted seawater	ELAN-5000; unspecified sample uptake	312
Au	groundwater, spring water	2-L filtered and acidified sample extracted into MIBK; evaporated; treated with aqua regia; evaporated and diluted to 5 ml	external calibration with internal standardization using Pt	400-fold preconcentration; memory effects necessitated 5-min wash with 4% *aqua regia* between samples	PQ; 0.8 ml/min uptake rate	313
B	groundwater	20-30 fold preconcentration by evaporation; passage through Dowex AG50W-X8 cation-exchange	measurement of $^{11}B/^{10}B$ with correction for mass discrimination by bracketing each sample between isotopic SRMs	quartz spray chamber used to minimize retention of B on glass; no fractionation induced by preconcentration steps	PQ STE with enhanced sensitivity interface; no peristaltic pump	314

Element	Sample	Sample preparation	Calibration strategy	Comments	Instrument; sample uptake rate	Ref.
Cd, Co, Cu, Ni, Pb, Zn, Mo, U, REEs	estuarine and seawater SRMs	separation of matrix and preconcentration on cellulose-immobilized EDTA at pH 4; elution with 1 M HNO_3	external calibration	contaminations observed for Zn and Pb, as well as for Sm	PQ2 STE; 0.5 ml/min uptake rate	315
Cu, Cr, Cd, Ni, Pb, Zn, U, Co, Mn	seawater	sample (adjusted at pH 8) passed through column of silica-immobilized 8-HOQ; wash with DDW; elute with 10 ml of 1 M HCl/ 0.1 M HNO_3	ID analysis; standard additions for monoisotopic elements and Cr	$^{37}Cl^{16}O^+$ interference on $^{53}Cr^+$ precluded the determination of Cr by ID analysis	ELAN 500; 0.95 ml/min uptake rate	316
REEs, Y	seawater	sample at pH 1.5 spiked with In; extracted with HDEHP/H_2MEHP in heptane; back-extracted in octyl alcohol and evaporated	external calibration with internal standardization using Cd	100-200 fold preconcentration required to bring concentration at detectable level	PQ; 0.4 ml/ min sample uptake	317
28 trace elements	coastal seawater	Al coprecipitation at pH 9.5; dissolution of precipitate in HNO_3	external calibration with internal standardization using Ge, In and Bi	suppression induced by Mg (partly precipitated at this pH) is compensated by internal standardization	SPQ-8000; 0.7 ml/min sample uptake	318
Al, Cu, Cd, In, Sn, Sb, Ba, Pb, Bi	rainwater	30-ml sample evaporated to 5 ml or analyzed directly	external calibration (assumed since unspecified)	preconcentration necessary for the determination of some elements; no significant matrix effects observed	PQ; 0.5 ml/ min sample uptake; spray chamber at 5°C	319
Pb	rainwater		isotope ratios with a dead time correction of 30 ns; ^{187}Re used for normalization of blank subtraction	memory effect avoided with a 3-min wash time; 2σ precision from 0.3% for $^{207}Pb/^{206}Pb$ to 0.7% for $^{208}Pb/^{204}Pb$	PQ2+; spray chamber at 10°C; unspecified sample uptake	320

continued

TABLE 1.10 (*continuation*)

Element	Sample	Sample preparation	Calibration strategy	Comments	Instrument; sample uptake rate	Ref.
Cr	river and ground waters, sewage and industrial effluents	sample acidified to 1% (v/v) HNO_3 and heated to 80°C in a microwave oven	external calibration (assumed since unspecified)	measurement at ^{13}C used to find $^{40}Ar^{12}C$ interference; correction factor established with Na_2CO_3 standards	ELAN-5000; 1 ml/min sample uptake	238
REEs	seawater	11.5 L sample acidified, filtered and spiked wit Lu; passed on Chelex-100 (NH_4^+ form); elution with 30 ml 3.5 M HNO_3	external calibration with matrix-matched standard and internal standardization using Lu	330-fold preconcentration required to bring concentration at detectable level	ELAN 500 with USND; 0.5 ml/min sample uptake	321
Al, Zn	river water, snow	sample placed in a container along with Chelex-100 resin (with and without fulvic acid) and stirred; subsamples pumped out through a filter	unspecified	kinetic study of analyte transfer from naturally-formed complexes to a ligand (Chelex-100) by monitoring the total analyte concentration vs time after ligand addition; no actual species determined	ELAN 5000; 1 ml/min uptake rate	322
REEs + 9 other elements	river and estuarine waters SRMs		external calibration with internal standardization where analyte mass referenced to linearly interpolated amu value of Be, In and Bi signals	direct analysis possible with enhanced ion transmission interface (S mode), which provided smaller matrix effects and enhanced sensitivity	PQ3 in S mode; 1.5 ml/min uptake rate; spray chamber at –5°C	294
Si	semiconductor grade ultra-high-purity water	sample evaporated down by a factor of 10 in a class 100 clean room	external calibration	preconcentration required to bring Si concentration higher than background level from instrument	HR-ICP-MS PlasmaTrace; 0.5 ml/min uptake rate	110

Element	Sample	Sample preparation	Calibration strategy	Comments	Instrument; sample uptake rate	Ref.
Rb	natural water	sample evaporated to dryness and dissolve in DDW; then passed through Bio-Rad AG 50W-X8 resin; Rb eluted with 0.5 M HCl	ID analysis	cation-exchange separation used to remove isobaric interference of ^{87}Sr on ^{87}Rb	PQ2+; 1 ml/min sample uptake	323
22 elements	ultrafiltrates of ground waters	anaerobic ultrafiltration (99% Ar, 1% CO_2) or passage through series of flat bed membrane filters	external calibration or method of standard additions	reliable results only obtained by the method of standard additions	ELAN-5000; 1.1 ml/min sample uptake	324
Pb, Cu, Cd	seawater	1.3 ml acidified seawater spiked with enriched isotopes; $Mg(OH)_2$ precipitated with high-purity ammonia; after centrifugation, precipitate washed and redissolved in 100 µl 5% HNO_3	ID analysis using ^{204}Pb (^{208}Pb as reference), ^{110}Cd (^{114}Cd as reference) and ^{65}Cu (^{63}Cu as reference)	washing step necessary for Cu only (to remove residual Na); recoveries were 90% for Pb, 32% for Cd and 57% for Cu; double precipitation required for Pb to reach appropriate detection limit	PQ II+; MCN free uptake rate; water-cooled spray chamber	325
Environmental analysis						
major (Na, Mg, A., Ca, Ti, Mn, Fe) and 40 trace and ultra-trace elements	soils	1 g ground sample digested with 10 ml $HClO_4/HNO_3$ and then 15 ml $HClO_4$/HF in Teflon beaker; residue redissolved in HNO_3	external calibration with internal standardization using Sc for major elements and In for all other elements	higher dilution factor used for the determination of major elements	PQ and Plasma-Trace (low resolution); unspecified uptake rate	326
REEs	soil and sediment SRMs	digestion (unspecified)	internal standardization with In and partial least square regression	30 standards required for the model, with a matrix similar to that of the sample (or standard additions required)	PQ+; unspecified uptake rate	327
Co, Cu, Zn, Cd, V, Cr, As, Ni	soil and sediment SRMs	pressurized digestion with HNO_3 or various acid mixtures; dilution with DDW	external calibration with external drift monitor and correction for interferences from polyatomic species	digestion with HNO_3 only suitable for As, Cd, Co, Cu and Zn; but HF also required for Cr and Pb	ELAN 5000; unspecified operating conditions	328

continued

TABLE 1.10 (*continuation*)

Element	Sample	Sample preparation	Calibration strategy	Comments	Instrument; sample uptake rate	Ref.
Pb	blood, dust, paint, sediment, soil	blood digested with HNO_3 under high pressure or diluted with dilute HNO_3 and Triton X-100; other samples digested in closed vessels under microwave heating	isotopic ratios measured to identify likely sources of lead poisoning	various dissolution procedures compared; total dissolution of soil unnecessary for isotope ratio measurements	ELAN 250; 0.3 ml/min uptake rate	329
REEs, U, Th	sediment and soil SRMs, aquatic plant, mussel tissue	microwave digestion with different acid mixtures, depending on the sample type	external calibration with internal standardization using In (for REEs) and Ru, Rh or Re (for U)	negligible oxide interferences with desolvation; memory effect for Th required 4-min washout with 10% HNO_3 at high pump speed	ELEMENT; MCN; heated spray chamber with membrane desolvation	173
Sc, Cr, Co, Ni, Cu, Zn, Cd, Sn, Cs, Pb, Th, U	estuarine sediment SRM	repeated digestion with HNO_3/HF and H_2O_2; down to dryness; comparison with 4-step sequential extraction	external calibration with internal standardization using In (for Sc-Cs) and Tl (for Pb-U)	In not best internal standard for Sc-Ni; incomplete dissolution for Sc and Th	ELAN 5000; 1 ml/min uptake rate	330
Pb, Ni	crude oils	1 g oil + 5 ml MIBK + standard addition + 25 ml 40% v/v HNO_3; shaked 1 min; filtered; aqueous layer diluted to <5% HNO_3 and analyzed	standard additions (5-point for each sample) with solutions of oil-soluble salts (i.e. Ni and Pb cyclohexane butyrate)	MIBK used to reduce oil viscosity; xylene better than MIBK for Pb determination	ELAN 250; 1 ml/min uptake rate	331
Cd	coal fly ash	microwave digestion with $HCl/HNO_3/HF$; dilution	multivariate external calibration with matrix-matched standards and internal standardization using Rh	no interference-free Cd isotope; multivariate calibration finds Cd concentration in presence of spectroscopic interferences	PQ2+; 1.0 ml/min uptake rate; spray chamber at 6°C	332

Element	Sample	Sample preparation	Calibration strategy	Comments	Instrument; sample uptake rate	Ref.
Pb	atmospheric aerosols; suspended particulate matter (SPM)	digestion of 100-150 mg aerosol with $HF/HNO_3/HClO_4$; SPM collected by continuous flow centrifugation; digested with 1:3 HNO_3/H_2O_2	$^{206/207}Pb$ isotopic ratio measured to identify the pathways of anthropogenic lead in the environment	ratios corrected for mass discrimination using mass bias factors obtained using a Pb isotope standard; 0.2-0.5% precision	ELAN 500; 1.0 ml/min uptake rate PQ1; 0.9 ml/min uptake rate; spray chamber at 6°C	333
16 elements	aerosol from waste incinerator and traffic	aerosol digested with $HF/HNO_3/HClO_4$ and diluted with DDW	5-point external calibration with internal standardization using Sc, Pr and Lu; $^{206/207}Pb$ isotopic ratio	weighed regression lines; isotopic ratios corrected for mass discrimination using a Pb isotope standard	ELAN 500; 1.0 ml/min uptake rate	334
53 elements	tobacco smoke condensate (TSC)	TSC collected from 100 cigarettes using an automatic smoking machine; TSC digested with HNO_3 in microwave oven; diluted with DDW	external calibration with internal standardization using In or semiquantitative determination using Al, Co, In and Pb as response elements (In as internal standard)	2-min rinse with 4% v/v HNO_3 between solutions	PQ1; 2.0 ml/min uptake rate; spray chamber at 10°C	335
V, Cr, Cu, Zn, As, Se	sewage, percolate water, light sandy soil, human serum	soil refluxed with aqua regia and filtered; serum digested with $HNO_3/HClO_4$ and taken to near dryness; filtrate or digest diluted with 0.14 M HNO_3; anion-exchange separation	external calibration with internal standardization using Ge	separation with Dowex-1 resin which strongly binds Cl- and S-containing anions thereby preventing the formation of polyatomic species which interfere with the analytes	PQ; 0.9 ml/min sample uptake; unspecified spray chamber temperature	336
Pb	gasoline	1 g sample + 75 µL 3% m/v iodine in toluene; mixed and allowed to react 5 min; add 50 g 10% v/m HNO_3; shake 5 min; HNO_3 layer analyzed	external calibration with internal standardization using Bi; $^{206/207}Pb$ isotopic ratio	100±2% recoveries for Pb level> 0.1 µg/g; ratios corrected for mass discrimination using appropriate standards; relative precision of ± 0.3% for $^{206}Pb/^{207}Pb$	PQ2 Turbo+; 1.0 ml/min uptake rate; spray chamber at 10°C	337

continued

TABLE 1.10 (*continuation*)

Element	Sample	Sample preparation	Calibration strategy	Comments	Instrument; sample uptake rate	Ref.
Ni, Mo, Cd, Ag	leaf and soil SRMs	sample digested with $HCl/HNO_3/HF$; electrochemical separation for Ag; ion-exchange separation for other analytes with additional separation for Ni and Cd	external calibration with internal standardization using In; or ID analysis using ^{62}Ni, ^{97}Mo, ^{109}Ag and ^{111}Cd	separations removed potential interferences from polyatomic ions; isotope ratios corrected for mass discrimination, assuming linear drift with time	PQ; 0.3 ml/min sample uptake; spray chamber at 2°C	338
Mo	domestic sludge SRM	sample digested with $HNO_3/HClO_4/HF$ (with a cycle to pH 8 with NH_4OH); separation of Mo from matrix by chelation with Chelex 100 (NH_4^+ form)	ID analysis using ^{97}Mo	separation required to remove interference from polyatomic ions; correction for mass discrimination as above	PQ2 Turbo+; 0.3 ml/min sample uptake; spray chamber at 2°C	339
Cd, Pb	Antarctic sediment SRM	sample treated with HCl overnight; microwave digestion of precipitate with HF/HNO_3; recombination with supernatant; dissolution with 0.14 M HNO_3	ID analysis with correction for mass discrimination and for isobaric interference from Sn on Cd	detailed uncertainty calculation made for each analyte	PQ2+; 1 ml/min sample uptake	340
Al, Cr, Fe, Mn, Ni, Pb, Zn	Antarctic sediment	microwave digestion with $HNO_3/H_3PO_4/HClO_4/HBF_4$	unspecified calibration procedure	corrections for ArN, ArN_2, ArO and CaO polyatomic interferences	ELAN-5000; 1 ml/min sample uptake	341
Sn	sediment SRM	open-vessel digestion with $HNO_3/HClO_4/HF$ or $LiBO_2$ fusion; extraction with TOPO/IBMK (or DIBK) and back-extraction with 5 M HNO_3	ID analysis with ^{118}Sn (^{120}Sn as reference) or method of standard additions	isotope ratio of back-extracted sample sometimes >1% RSD whereas 0.3% RSD with aqueous solutions; suppression from IBMK in back-extracted sample	HP-4500; 0.2 ml/min sample uptake	342
Cd	lichens SRM, marine sediment SRM	microwave digestion with HNO_3/HF (sediment) plus $HClO_4$ (lichens) then with H_3BO_3 (to remove excess HF); dilution	ID analysis (on- or off-line) with ^{111}Cd (^{114}Cd as reference); correction for differences in density when on-line	similar accurate results whether ID is performed on-line or not but simpler procedure when on-line	HP4500; 0.4 ml/min sample uptake; spray chamber at 2°C	343

Element	Sample	Sample preparation	Calibration strategy	Comments	Instrument; sample uptake rate	Ref.
^{230}Th, ^{234}U and ^{235}U	marine sediment	open-vessel digestion with HNO_3 then $HClO_4$/HF; evaporation; dissolution in HNO_3; separation using anion-exchange resin AG1-X8; elution with 9 M and 1 M HCl	ID analysis using ^{229}Th and ^{236}U	medium resolution required because of strong tailing of ^{232}Th; separation resin could be re-used several times without significant Th and U contamination	ELEMENT (R=4430); micronebulizer; 0.03 ml/min sample uptake	344
Pu	soil SRM, sediment SRM, fish flesh SRM	repeated leaching of ashed samples with 8 M HNO_3; $NaNO_2$ used to convert all Pu to Pu(IV); separation of Pu(IV) using Dowex 1X8; elution with NH_4I (5%)/10 M HCl	ID analysis using ^{242}Pu (^{239}Pu and ^{240}Pu as reference)	Both Dowex 1X8 and TEVA resins (but Dowex most effective) selectively separated Pu(IV) from matrix and U	PMS-2000; 0.2 ml/min sample uptake	345
Cu, Ni, Mn, Cr, Cd, Ba, As, V, Sb, Pb, Zn	airborne particulate matter	microwave digestion of PTFE filters loaded with sample using HNO_3/$HClO_4$/HF; dilution with DDW	external calibration	recovery >90% except for Cr and V	ELAN-5000; no operating conditions provided	346
16-18 elements (major, minor, trace)	airborne particulate matter SRM	microwave digestion of PTFE filters loaded with sample using HNO_3/$HClO_4$/HF; evaporation of HF; dilution with DDW	external calibration	determination of As, V and Cr jeopardized by use of $HClO_4$; low recovery for Cr in NIST SRM 1648	ELAN-5000; 1.0 ml/min sample uptake	347, 348
13 trace elements	coal SRM, coal fly ash SRM	open-vessel digestion with H_2SO_4 then treated with $HClO_4$/HF at 200°C, evaporated to dryness and dissolved in HNO_3; or microwave extraction with HNO_3	4-point external calibration; drift monitoring using 1 standard solution every 3 samples; linear drift correction applied to raw signal	HF essential for complete dissolution; loss of volatile elements with open-vessel digestion; As and Se associated with non-silicate minerals	PQ2+STE; de Galan V-groove nebulizer; water-cooled spray chamber (5°C); 0.5 ml/min sample uptake	349

continued

TABLE 1.10 (*continuation*)

Element	Sample	Sample preparation	Calibration strategy	Comments	Instrument; sample uptake rate	Ref.
As, V	incineration ash SRM, soil SRM, coal fly ash SRM, urban particulate matter SRM	open-vessel digestion with HNO_3/H_2O_2/HF (5:3:2 v/v), evaporation to dryness; dilution	method of standard addition	HF necessary to destroy silicate matrix; chemical loss of V from chlorinated samples due to the formation of volatile V compound	ELAN-5000 and Plasma-Trace 2; unspecified uptake rate	350
Pt	dust	digestion with *aqua regia* compared to pressure decomposition with HNO_3	external calibration for dust with internal standardization using Rh	pressure decomposition with HNO_3 gave low Pt recoveries	ELAN-5000; 1 ml/min uptake rate	351
Mn	washing powders	sample grinded; HNO_3 leach; or $LiBO_2$ fusion with 7500-fold dilution	external calibration with correction for drift using external drift monitors; or standard additions	7500-fold was minimum dilution factor to prevent suppression of analyte signal by Li and B in matrix	prototype; no operating conditions provided	352
Cu, Cr, As	chromated copper arsenate treated wood and its pyrolysis residues	refluxing with HNO_3 for total dissolution or sequential extraction using NH_4OAc, NaOAc/HOAc, NH_2OH, H_2O_2 and HF/$HClO_4$	external calibration	when sequential extraction was used, the matrix of the standards was the leaching solution to minimize non-spectroscopic interferences	PQ2+; 1 ml/min uptake rate	353
Au	plant materials	dry ashing; digestion with *aqua regia*/HF; down to dryness; 5% *aqua regia* added; centrifugation; supernatant dilute with 5% *aqua regia*	external calibration with internal standardization using Ir	wash with 8% *aqua regia* between samples required to minimize memory effects	SOLA; 2 ml/min uptake rate	354
As, Pb, Cd, Se, Hg	brine shrimp *Artemia*, salt ponds water, biological & seawater SRMs	microwave digestion of solids with HNO_3/HF/H_2O_2s; evaporation; water diluted 1:10	unspecified calibration procedure	ArCl interference on As reduced by filtering the samples on On-Guard Ag cartridges prior to analysis	ELAN-5000; 1 ml/min uptake rate	355

Element	Sample	Sample preparation	Calibration strategy	Comments	Instrument; sample uptake rate	Ref.
Pb	several SRMs: leaves, air filters, urine, sludge, dust, paint	continuous-flow microwave digestion of sample slurries (1% solids) using $HNO_3/HClO_4/HF$	ID analysis using ^{206}Pb (^{208}Pb as reference isotope)	5-min wash (3 M HNO_3/0.3 M HCl/1.5 M HF) required to minimize carryover observed with organically rich samples	PQ II Turbo Plus	356
U	soil	leaching with HNO_3; extraction of U with MIBK; back-extraction	isotope ratio of U or U oxide ions (the latter can be used for verifying results or in case of interferences) with correction for mass bias	separation of U required to prevent nebulizer clogging, matrix effects and spectroscopic interferences	ELAN 6000; DIHEN; 85 µl/min using syringe pump	195
Forensic science						
48 elements	glass	crushed sample washed with 20% v/v HCl; air-dried; digested with HF/HCl (1+2 v/v) in ultrasonic bath for 1 h; heated uncovered after adding $HClO_4$ to remove fluorides; diluted to 5 ml with 5% v/v HCl	external calibration; correction for doubly-charged ions & minor polyatomic interferences from analyte oxide and hydroxide species using in-house databank of measured ratios for standard solutions	special thin-walls PTFE digestion vessels were designed for maximum heat transfer; 2-min wash with DDW between samples; all sample solution used for one scan; interferences from TiO on both Cu isotopes precluded its determination	ELAN 250; 1.0 ml/min uptake rate	357
U, Th	glass	sample grinded in agate mill; dissolved in $HF/HClO_4$; evaporated to dryness (a second time with $HClO_4$); residue dissolved in HNO_3 and diluted	external calibration with internal standardization using Bi (to compensate for up to 60% suppression)	removal of Si and B matrix constituents sufficient for ICP-MS analysis; no need for lengthy extraction and separation procedure	PQ; unspecified uptake rate	358

continued

TABLE 1.10 (*continuation*)

Element	Sample	Sample preparation	Calibration strategy	Comments	Instrument; sample uptake rate	Ref.
36 trace elements	glass	cut sample cleaned with cold HCl in ultrasonic bath; digestion with HF/HNO_3 overnight; drying and digestion with HNO_3	1-point external calibration using basalt SRM with internal standardization using Be, As, Rh, In, Tm, Re and Bi	drift corrections by interpolating between internal standards closest in mass to the analyte, except for Li (normalized to Be), Th and U (normalized to Bi)	ELAN-6000; cross-flow nebulizer; unspecified uptake rate	359
Al, Ba, Mn, Pt, Rb, Sr	drug ((-)-ephedrine hydrochloride)	addition of HNO_3 and dilution with DDW	external calibration with internal standardization using In	unexplained variation in day-to-day reproducibility; no assessment of accuracy	PQ2+; 1 ml/min uptake rate	360
Mg, Al, Ti, Mn, Ni, Cu, Cd, Sr, Ba, Pb, Fe, Cr	polyethylene film, coloured garbage bags	microwave digestion with HNO_3/H_2O_2 and dilution	semi-quantitative external calibration with internal standardization using In; $^{208}Pb/^{206}Pb$ and $^{207}Pb/^{206}Pb$ isotope ratios	internal standard required to compensate for signal enhancement caused by residual carbon; lower detection limits required for analysis of cling film	ELAN 5000; cyclonic spray chamber; unspecified uptake rate	281
Nuclear field						
Fe, Mn, Ni, Co	primary water of pressurized water reactors	preconcentration/separation by passage through Waters Guard Pak or Dionex MetPac CC-1 and elution with 0.1 or 1M HNO_3 to remove boric acid	external calibration with standard solutions in HNO_3 (concentration matching the eluent)	Al sampler to decrease Ni background caused by Ni sampler	PQ2+; 1 ml/min uptake rate	361
18 elements	spent nuclear fuel [202, 203]; leachates of spent fuel and high-activity waste [203]	sample dissolved in 3-7 M HNO_3; residue filtered, dried, weighed and dissolved in HCl/HNO_3 (9+1); diluted in glove-box	qualitative & semiquantitative analysis of spent nuclear fuel; ID analysis of residues (stand. additions with internal standardization for monoisotopic elements)	isotopic ratios corrected for mass discrimination using natural standards; either Zr, Mo, Ru or Pd can be used as internal standard for determination of Tc and Rh	ELAN-250 modified to handle radioactive samples in glove-box; 1 ml/min uptake rate	362 363

Element	Sample	Sample preparation	Calibration strategy	Comments	Instrument; sample uptake rate	Ref.
28 elements	uranium oxide	1-g sample digested with 3.2 ml HNO_3 and diluted to 50 ml with DDW	standard additions with internal standardization using Ga, Tm and Tl	on-line extraction of U in 0.2 M TOPO in cyclohexane; aqueous phase with analytes fed to nebulizer	ELAN 250; 0.7 or 1.0 ml/min sample uptake	364
^{237}Np	isotopically enriched U solutions	hydroxylamine chlorohydrate used to reduce Pu and Np to valences III and IV; Np extracted in 1 M HNO_3 with 0.5 M thenoyltrifluoroacetone in xylene; Np re-extracted in 10 M HNO_3; evaporated; diluted with 2% HNO_3	external calibration with internal standardization using ^{165}Ho	quartz torch with alumina injector used; tailing of $^{238}U^+$ (10-100 mg/L U in final sample solution) interferes with $^{237}Np^+$: Omnirange used to reduce it to negligible level; but 100-fold loss in detection limit	ELAN 5000; 0.8 ml/min uptake rate	365
^{99}Tc	radioactive wastes	fusion with NaOH; digestion with HNO_3 and then NH_4OH; separation by passage through extraction column; purification by extraction and back-extraction	external calibration	separation required to eliminate spectroscopic interferences from ^{99}Ru and ^{98}MoH	PQ2+; 1.2 ml/min uptake rate	366
^{99}Tc	high-purity $^{99}TcO_2$	dissolution in 10% v/v HNO_3/30% H_2O_2; dilution with DDW	external calibration	Tc memory required washes with 10% HNO_3; TcO and Tc^{2+} formed to lesser extent than with lanthanides and actinides	PQ2 in glove-box; 1.5 ml/min sample uptake; spray chamber at 10°C	367
U, Th, Pu	radioactive waste		isotope ratio of analyte or analyte oxide ions (the latter can be used for verifying results or in case of interferences)	precision of isotope ratio dependent on elemental concentration and magnitude of the ratio	Elan 6000; DIHEN; 85 µl/min using syringe pump	195

continued

TABLE 1.10 (*continuation*)

Element	Sample	Sample preparation	Calibration strategy	Comments	Instrument; sample uptake rate	Ref.
^{226}Ra, ^{230}Th, ^{237}Np, ^{238}U, ^{241}Am	radioactive waste solutions		external standard calibration; isotope ratios with a correction for mass bias using isotope SRM	better sensitivity with USN but smaller sample size preferable for radioactive sample	ELAN 6000; micronebulizer with minicyclonic spray chamber; 0.085 ml/min	272
Metallurgical analysis						
23 elements	liquid-phase WF_6 residue	residue digested with concentrated ammonia and diluted with DDW	4-point external calibration using matrix-matched standards with internal standardization using Be, In, Bi	internal standardization using an interpolation procedure instead of assigning one internal standard to each element; HNO_3 level must be matched in all standards for maximum stability	PQ2 Turbo+; 1 ml/min sample uptake; USND with heater at 140°C and chiller at 4°C	368
Ga	Al	0.2 g Al dissolved in 5N HNO_3, spiked and diluted to 1% HNO_3; or Ga extracted in isopropyl ether and back-extracted into DDW	ID analysis with correction for mass discrimination using a Ga standard solution	Ga first determined semiquantitatively; ID then made with spike such that peak area ^{71}Ga/^{69}Ga ratio>2; preconcentration recommended for 99.999% Al	PQ; unspecified sample uptake	369
Ag, Sb	Cu	Cu dissolved in 50% HNO3, diluted and analyzed; or preconcentrated by coprecipitation with Te (for Ag) or La (for Sb)	ID analysis with correction for mass discrimination using Ag and Sb standard solutions	ID made with spike such that ^{107}Ag/^{109}Ag and ^{123}Sb/^{121}Sb ratio was about 4 for maximum precision; preconcentration used for 99.9999% Cu	PQ; unspecified sample uptake	370
Pb	Pb isotopic SRMs	digestion in 0.5 M HNO_3; dilution with DDW	measurement of Pb isotopic ratios, and those of Sr, Nd, Hf, Re, Os and Tl to establish bias factor *vs* mass curve	correction for mass discrimination using bias factor found for Tl, but corrected by mass	Plasma-54; 0.6 ml/min (no peristaltic pump); spray chamber at 4°C	301

Element	Sample	Sample preparation	Calibration strategy	Comments	Instrument; sample uptake rate	Ref.
21 elements	Al_2O_3 and SiC ceramic materials	Al_2O_3 digested in HCl/H_2SO_4 and diluted 1+2500; SiC grinded, digested in $HF/HNO_3/H_2SO_4$ and diluted	4-point external calibration using matrix-matched standards with internal standardization using Rh	dilution to <200 mg/l Al otherwise clogging and drift due to ablation of sampler and glassy deposits on skimmer	PQ2 Turbo+; 0.75-1.1 ml/min sample uptake; cooled quartz spray chamber at 10°C	371
36 elements	SiC powders	digestion with $HNO_3/H_2SO_4/HF$; down to near dryness; dilution with 2% HNO_3	external calibration with internal standardization using Rh	evaporation step removes Si and C from matrix; without it, severe interferences on Mg, Ca, Sc and Ti	PQ2 Turbo+; 1.1 ml/min sample uptake; cooled spray chamber at 10°C	372
Mg, Mn, Ni, Cu, Zn, Ga, Zr, Sn, Ti	Al alloys	0.1 g sample dissolved in 10 ml 6M HCl and diluted to 100 or 1000 ml	external calibration using matrix-matched standards or with internal standardization using Co, Y and Rh	method applicable to low alloy and not high alloy Al whose high Si content makes it difficult to dissolve	ELAN-250; unspecified sample uptake	373
Cd	Fe- and Ni-based alloys SRMs	microwave digestion with $HCl/HNO_3/HF$; dilution	multivariate external calibration with matrix-matched standards and internal standardization using Rh	no interference-free Cd isotope; multivariate calibration finds Cd concentration in presence of spectroscopic interferences	PQ2+; 1.0 ml/min uptake rate; spray chamber at 6°C	332
B	Ti	80-mg Ti cube dissolved in 5 ml DDW with several 700-µl portions of HF; diluted to 50 ml with 0.14 M HNO_3	external calibration with internal standardization using Be	minimal blank with inert nebulizer/spray chamber/connector to torch and alumina injector; blank varied from day to day	PQ; 1.8 ml/min sample uptake; PTFE spray chamber at 10°C	374
Sc	high-purity Ti, Ti SRM	0.1-g Ti pieces dissolved in 8 M HCl by refluxing 25 h; Ti separated from Sc using Bio-Rad AG 50W-X12 cation-exchange resin	external calibration with internal standardization using Co	matrix separation required to eliminate spectroscopic interferences from Ti; recovery of Sc was 92 ± 12% (1s)	PQ2+; 1 ml/min sample uptake	375

continued

TABLE 1.10 (*continuation*)

Element	Sample	Sample preparation	Calibration strategy	Comments	Instrument; sample uptake rate	Ref.
Li, Mn, Zn, As, Ag, Cd, Te, Pb	high-purity Se	digestion with HNO_3; down to dryness; dissolution in DDW; passage through Dowex 1-X8 resin; elution with 0.5M HNO_3	external calibration	chloride removed by evaporation to dryness for Ag recovery; evaporation to dryness also required following elution to eliminate non-spectroscopic interferences from acetic acid (column conditioner)	SPQ8000A; 1 ml/min sample uptake (no peristaltic pump)	376
25 elements	Ta	digestion in HF/ HNO_3; dilution with HCl; extraction with MIBK; Rh added to aqueous phase then evaporated for analysis	external calibration with internal standardization using Rh	low recoveries for Fe, Ga and Tl which could be determined if HNO_3 was used instead of HCl	ELAN-500; unspecified sample uptake rate	377
REEs	high-purity CeO_2	digestion with HNO_3/H_2O_2; addition of $KMnO_4$; extraction with EHEHP in cyclohexane; back-extraction into dilute HNO_3	external calibration with internal standardization using Cs	$KMnO_4$ oxidized Ce to Ce(IV) which is not back-extracted into acidic solution; quantitative separation of REEs from Ce achieved	PQ; 1.5 ml/min sample uptake	378
REEs	high-purity Eu_2O_3	purified Eu_2O_3 dissolved in 1% HNO_3	external calibration with internal standardization using In (for light REEs) and Tl (for heavy REEs)	similar results to those by standard additions; determination of Tm hampered by EuO interference	PMS 2000: 0.24 ml/min sample uptake; cooled spray chamber at 0°C	379
Al, Mn, Co, Zr, Nb, Mo, In, Sn, Sb, Ta, Pb, Bi	$SrTiO_3$ (ceramic)	pressure decomposition with HCl/HF; evaporated; residue dissolved in HCl and diluted	external calibration with internal standardization using Au (required to compensate for suppression induced by Sr and Ti)	method unsuitable to determination of Mg, V, Cr, Fe, Cu and Zn because of spectroscopic interferences from matrix	SPQ-6500; no operating conditions provided	380
trace and REEs (34 elements)	poly-metallic nodule SRMs	digestion with HCl/ $HNO_3/HClO_4/HF$; residue dissolved with H_2O_2; dilution with DDW	external calibration with internal standardization using Rh	results from standard calibration negatively biased unless corrected using data from SRMs	PQ1; 0.8 ml/min sample uptake; spray chamber at 9°C	381

Element	Sample	Sample preparation	Calibration strategy	Comments	Instrument; sample uptake rate	Ref.
Geological analysis						
Cd, Cu	zinc ore SRM	digestion with $HNO_3/HClO_4/HF$	ID analysis using ^{65}Cu and ^{111}Cd	correction for mass discrimination using a spike calibration standard (with the nominal isotopic ratio of the spiked samples) every two samples	PQ2 Turbo+; 0.3 ml/min sample uptake; spray chamber at 2°C	339
U, Th	silicate rocks	0.1 g finely powdered sample spiked with enriched isotope and digested with HF, HNO_3 and $HClO_4$; residue dissolved in 7M HNO_3	ID (with correction for mass discrimination) or external calibration with internal standardization using Bi	separation of U and Th from matrix on AG1-X8 resin unnecessary with ID; Bi not adequate as internal standard for determination of Th	PQ2+; 0.8 ml/min uptake rate; spray chamber at 0°C	382
Ru, Rh, Pd, Os, Ir, Pt, Au	rocks	5 g each ground rock, Ni and quartz + 3 g S + 10 g Na_2CO_3 + 30 g borax, fused at 1000°C; NiS button ground and digested with HCl; Au coprecipitated with Te; dissolved in HNO_3/HCl	external calibration with internal standardization using Tl	quantitative recovery for all elements except Au and Os (about 85% recovery); Pb fire assay gave higher recoveries	ELAN 5000; 1 ml/min sample uptake	383
U, Th	basaltic rocks	sample digested with HF/HNO_3; Si volatilized as SiF_4 by evaporation to dryness; dissolved in HNO_3 and diluted	external calibration with internal standardization using Bi	2-min rinse with 5% HNO_3 between samples reduced Th memory effects; less adsorption of Th on HDPE than on PE, PP and borosilicate glass vessels	PQ2; 0.8 ml/min sample uptake	384
14 REEs, Sc, Y	rocks	sample fused with $K_2B_4O_7$ (or K_2CO_3); dissolved; $KClO_4$ precipitated and centrifuged out; HF added; evaporated to dryness to expel B and Si; dissolved in HNO_3 and diluted	2-point external calibration with internal standardization using Rh (for Sc, Y, La, Ce, Pr and Nd) and Re for other analytes	separation of K from fusion cakes eliminates extra 500-fold dilution which would otherwise be required	ELAN-5000; 1.0 ml/min	385

continued

TABLE 10.1 (*continuation*)

Element	Sample	Sample preparation	Calibration strategy	Comments	Instrument; sample uptake rate	Ref.
Re, Os	massive sulfide and mineralized peridotite	3 g powdered sample treated with 10 ml HCl at 50°C for 3 h; ^{185}Re, ^{190}Os spike added; $HClO_4$ added to oxidize Os; OsO_4 distilled and trapped in DDW	ID analysis; correction for mass discrimination ; Re/Os isotope ratio also measured	Os trap (i.e. DDW) is unstable at room temperature and must be analyzed immediately	ELAN-250; conditions in reference therein	386
Re	rocks, mineral samples	sample digested with HNO3 and filtered; filtrate evaporated to dryness; residue dissolved in 1M HCl; anion exchange separation from matrix	$^{185}Re/^{187}Re$ isotopic ratio measured	anion-exchange separates Re (eluted with 8M HCl) from Ni, Cu and Fe (eluted with 1M HCl) and Os and W (retained on resin)	ELAN-250; unspecified sample uptake; nebulizer & spray chamber outside the torch box	387
Pt, Pd, Rh, Ir, Ru	rocks	sample treated by fire assay; NiS button dissolved, filtered and filtrate analyzed; or button submitted to Te coprecipitation during dissolution	5-point external calibration with internal standardization using Tl and Tm	Te coprecipitation technique increased recovery by 10-15%	ELAN-250; 1 ml/min sample uptake	388
Pt, Pd, Rh, Ru, Ir, Os, Au	rocks	rock pulp placed with admixed powdered NaCl in chlorination chamber; heated 3.5 h to 580°C under a continuous flow of chlorine; product dissolved in 10% HCl	2-point external calibration or method of standard additions	admixed NaCl essential for good Pt recovery and repeatability but also leads to salt build-up on cones and drift downwards	ELAN-500; 0.9 ml/min sample uptake	389
14 REEs, Ba	basalts	.25 g rock powder treated with HF and $HClO_4$ overnight at 120°C and taken to dryness; residue dissolved in HNO_3 and diluted	5- or 6-point external calibration; drift correcting standard analyzed every 4 or 5 samples; polynomial fit used to correct for curvilinear drift vs mass and time	all standards are matrix-matched to samples; n-1 (n= number of drift correction standards) order polynomial curve is fitted to each analyzed isotope	PQ2+; sample uptake specified as tubing i.d. and pump setting; spray chamber at 12°C	390

Element	Sample	Sample preparation	Calibration strategy	Comments	Instrument; sample uptake rate	Ref.
47 major, minor, trace elements	rock standards	powdered sample reacted with $HClO_4$/HF and evaporated to dryness; residues dissolved in HNO_3 and diluted to 100 ml	7-point external calibration with internal standardization using Rh	poor agreement for some elements at concentrations ≤ 1 µg/g; additional internal standard of lower mass would improve results for lower mass elements	ELAN-500; unspecified sample uptake	391
W, Mo	soil, sediment and rock SRMs	fusion with Na_2CO_3/$NaNO_3$; leach with warm DDW; adjust pH to 1.5-2.0 and add oxine and activated charcoal; ash; heat with 4M HCl; decant; dilute	ID or external calibration with internal standardization using Re (for W) and Ru (for Mo)	procedure for selective separation of W and Mo (other elements are not quantitatively adsorbed); precision with internal standardization slightly degraded compared to ID	ELAN-250; 1 ml/min sample uptake	271
Zr, Nb, Hf, Ta	various geological materials	sample fused with $LiBO_2$ at 950°C for 30 min; melt dissolved in HCl/HF and filtered; analytes precipitated with 6% cupferron solution at 10°C; precipitate dissolved with HNO_3/H_2O_2 and diluted	external calibration with internal standardization using Ru (for Zr and Nb) and Re (for Hf and Ta)	standard solutions prepared by spiking procedural blanks; method may also be applicable to the determination of Mo and V	ELAN-250; 1.5 ml/min sample uptake	392
REEs	bauxites	powdered sample fused with $NaKCO_3$/$Na_2B_4O_7$ at 1100°C for 45 min; melt dissolved in HNO_3/HF at 70°C, transferred and diluted	6-point external calibration with internal standardization using Rh; qualitative & semiquantitative analyses also performed	long-term drift due to deposition of Al_2O_3 on sampler; calibration checked by measuring some standard solutions and corrected for drift with time	ELAN-250; 1.0 ml/min sample uptake	393
Zr, Hf	rocks	ground sample spiked with enriched isotopes and digested with HNO_3/HF and then $HF/HClO_4$ for 4 days; HF evaporated; final solution in 0.2 M HNO_3	ID analysis using ^{91}Zr (^{90}Zr as ref.) and ^{179}Hf (^{178}Hf as ref.) with correction for mass discrimination & blank subtraction	ID provided similar accuracy to external calibration but with improved precision	ELAN-5000; 1.0 ml/min sample uptake	394

continued

TABLE 1.10 (*continuation*)

Element	Sample	Sample preparation	Calibration strategy	Comments	Instrument; sample uptake rate	Ref.
Pt, Pd, Ru, Ir, Rh, Au	rock SRMs	fusion with sodium peroxide and dissolution in DDW; coprecipitation of analytes with Te; final solution in dilute *aqua regia*	ID using ^{101}Ru (^{99}Ru as reference), ^{105}Pd (^{108}Pd as ref.), ^{193}Ir (^{191}Ir as ref.) and ^{194}Pt (^{195}Pt as ref.) with correction for mass discrimination & blank subtraction or external calibration (Rh and Au)	good accuracy if isotopic spikes added after dissolution of sodium peroxide cake such that the resulting isotopic ratios were between 1 and the geometric mean of their value in the spike and sample	prototype; 0.5 ml/min sample uptake; spray chamber at 5°C	395
Gd	rock SRMs	digestion with HF and then HCl/ HNO_3; dilution with DDW	partial least square regression using 27 standard solutions with varying levels of Ce, La, Pr, Nd and Gd	method corrects for spectroscopic interferences from Ce, La, Pr and Nd on Ga isotopes	PQ; 0.8 ml/ min sample uptake	396
REEs	rock SRMs	microwave digestion with HCl/HNO_3/HF; down to dryness; final solution in 1% HNO_3	external calibration with internal standardization using Rh	accuracy varies, depending on analyte and matrix	ELAN-500; 0.2 ml/min sample uptake	397
34 elements (trace and REEs)	rock SRMs	open acid digestion with HF/HNO_3/ $HClO_4$ or pressure decomposition with $HClO_4$/HNO_3/HF/ HCl or fusion with $LiBO_2$	external calibration with internal standardization using Rh and Bi	closed-vessel digestion gave the lowest blanks; fusion induced loss of some elements (Zn, Cs)	PQ1; operating conditions in another paper	398
40 trace elements	rock SRMs	digestion with HF/ HNO_3; down to dryness; reflux with 6M HNO_3; down to dryness; dissolution in HNO_3 and dilution	external calibration using one SRM with internal standardization using ^{6}Li, ^{84}Sr, Rh, In, ^{147}Sm, Tm, Re, Bi, ^{235}U	corrections for sensitivity change using internal standards and external calibration, with linear interpolation *versus* mass	PQ2+ STE; 1 ml/min uptake rate; spray chamber at 1°C	399

Element	Sample	Sample preparation	Calibration strategy	Comments	Instrument; sample uptake rate	Ref.
REEs	rock and coal fly ash SRMs	sintering with sodium peroxide in Zr crucible at 480°C for 45 min; dissolution in 6 N HCl; dilution to 0.15% total dissolved solids	external calibration with internal standardization where analyte mass referenced to linearly interpolated amu value of In and Bi signals	enhanced ion transmission interface (S mode) provided smaller matrix effects together with enhanced sensitivity and stability	PQ3 in S mode; 1.5 ml/min uptake rate; spray chamber at -5°C	294
Ru, Pd, Ir, Pt	meteorite and rock SRMs	alkali fusion followed by microwave digestion; separation on anion exchange column (AG 1X-8); elution with 5 M HCl/5 M $HClO_4$	ID analysis using ^{99}Ru, ^{105}Pd, ^{191}Ir and ^{196}Pt (^{101}Ru, ^{108}Pd, ^{193}Ir and ^{195}Pt as reference) (mass discrimination <1%)	anion exchange separation removes potential sources of spectroscopic interferences; acid digestion not recommended (loss of Ru)	PQ; water-cooled spray chamber; unspecified uptake rate	400
REEs	chondritic meteorite, basalt SRMs	powdered sample digested with HNO_3/HF in many steps; evaporation; residue dissolved in 1:1 HNO_3 and diluted with 2% HNO_3	external calibration with internal standardization using Rh	corrections for oxide and hydroxides of Ba, Ce, Pr and Nd made by establishing the proportion of these species in pure solutions of Ba, Ce, etc.	ELAN-6000; unspecified uptake rate	401
Li	Li_2CO_3 isotopic SRM	open digestion with HF/HCl; separation of Li from matrix using 50W-X8 cation exchanger; elution with 0.23 M HCl	measurement of ^{7}Li/^{6}Li with correction for mass discrimination using concentration matched standard	matrix separation required to eliminate matrix-induced mass discrimination	ELAN-5000; 1 ml/min sample uptake	302
Pb	rare earth minerals and compounds	dissolution in HNO_3; dilution with DDW to 0.6% HNO_3	measurement of ^{207}Pb/^{206}Pb and ^{208}Pb/^{206}Pb with blank subtraction on individual isotopes	ratios used to determine initial composition and origin of the samples	PQ1 and ELAN-500; no operating conditions provided	402

continued

TABLE 1.10 (*continuation*)

Element	Sample	Sample preparation	Calibration strategy	Comments	Instrument; sample uptake rate	Ref.
Ru, Rh, Au, Pd, Ir, Pt	rocks, sand, ores, Ni-Cu matte, sulfide flotation concentrate SRMs	microwave digestion with *aqua regia*/HF; down to near dryness; digestion with 0.5 M HCl; fusion of insoluble residues with Na_2O_2/Na_2CO_3 or Na_2O_2; dissolution in 0.5 M HCl	external calibration with acid-matched standards	method restricted to mineralized samples with limits of determination between 0.2 and 1 μg/g	PQ2+; 0.5 ml/min sample uptake; spray chamber at 13°C	403
Ru, Rh, Au, Pd, Ir, Pt	Ni-Cu matte SRM, sulfide flotation concentrate SRM, Pt ore SRM	microwave digestion and fusion as in ref. 366; anion-exchange separation on Dowex 1-X8; elution with 0.3 M thiourea/ 0.1 M HCl and 12 M HCl	external calibration with acid-matched standards	determination of low levels hampered by dilution required for the analysis of each fraction, especially the thiourea one	PQ2+; 0.5 ml/min sample uptake; spray chamber at 4°C	404
Re, Au, Ru, Rh, Pd, Ir, Pt, Fe, Ni, Cu	inclusions of Fe or Ni-rich sulphides in diamonds	digestion with *aqua regia* and ethanol; dilution with DDW	external calibration with internal standardization using In	precision of Fe, Ni and Cu concentrations poorer than that obtained with energy dispersive X-ray spectrometry	PQ2+; 0.5 ml/min sample uptake; spray chamber at 5°C	405
Ru, Rh, Au, Pd, Ir, Pt	sulfide and magnetite minerals, NiS bead SRM	repeated dissolution in HCl/evaporation to dryness; cation-exchange separation with Amberlite CG-120; elution with 0.2 M HCl	external calibration with internal standardization using Cd (for Ru, Rh and Pd) and Tl (for Ir, Pt and Au)	backgrounds for Pd and Au were interpolated from two bracketing blanks to reduce memory effects	ELAN 250; 0.4 ml/min sample uptake: merged with 0.2 ml/min internal standard solution	406
Bbiological analysis						
15 elements	prawn parts (eyes, abdomen muscles, hepatopancreas, abdominal exoskeleton)	dry and ground sample digested (by closed-vessel microwave digestion) with HNO_3, taken to near dryness, digested with HNO_3/ H_2O_2, taken to near dryness and diluted	not indicated; (canonical variate analyses using the elemental concentrations found)	last digestion step used to remove volatile interfering matrix elements and ensure complete digestion	ELAN-5000; 1 ml/min sample uptake	407

Element	Sample	Sample preparation	Calibration strategy	Comments	Instrument; sample uptake rate	Ref.
Th, U	faecal material	sample dried at 150°C, ashed at 650°C, digested in HNO_3/HCl/$HClO_4$ and diluted	unspecified (external calibration assumed)	dilution used to eliminate matrix-induced signal suppression; Th memory effect minimized with long wash times and selected operating conditions	PQ; 0.7 ml/min sample uptake; spray chamber at 7°C	408
Cu, Zn	pig faecal matter	sample dried, ashed at 500°C for 24 h, dissolved in HNO_3 and diluted	isotope ratios with 0.1% RSD using 5-10 min measurements (timing factors for less abundant isotopes)	ratios corrected for dead time; drift compensated by frequent calibration	ELAN 5000; unspecified sample uptake	219
Zn	urine, milk	digested urine and ashed milk solutions buffered with CH_3COONH_4 at pH 5.5, passed through CPPI chelating resin; elution with 2 M HNO_3	$^{70}Zn/^{68}Zn$ isotope ratio	Na, Ca and Cl concentrations decreased by separation procedure thereby reducing interference on Zn isotopes	ELAN-250; 0.6 ml/min sample uptake	409
Se	urine SRM, human urine	dilution 5-fold to final concentration of 5% CH_3COOH/0.14 M HNO_3	method of standard addition with internal standardization using either Ga, Y or In	only ^{82}Se is suitable for quantitative analysis; acetic acid enhanced Se sensitivity 6-fold	ELAN-6000; 1.2 ml/min sample uptake	267
Cd	urine SRM, horse kidney SRM	dilution of urine; microwave digestion of kidney with HNO_3 and dilution	ID analysis (on- or off-line) with ^{111}Cd (^{114}Cd as reference); correction for differences in density when on-line	similar accurate results whether ID is performed on-line or not but simpler procedure when on-line	HP4500; 0.4 ml/min sample uptake; spray chamber at 2°C	343

continued

TABLE 1.10 (*continuation*)

Element	Sample	Sample preparation	Calibration strategy	Comments	Instrument; sample uptake rate	Ref.
Zn	plasma, red blood cells, urine, faeces, food	sample either digested (urine, red blood cells, plasma) or ashed (faeces, food); Zn extracted from dry sample with APDC into CCl_4; extract evaporated to dryness; residue dissolved in 1.2 M HNO_3	semi-quantitative analysis; $^{70}Zn/^{68}Zn$ isotope ratio determined	extraction procedure separates Zn from elements leading to potential polyatomic or doubly charged ion interferences; it also preconcentrates Zn	ELAN-250; 0.6 ml/min sample uptake	410
Fe	whole blood	blood diluted 25-fold with DDW after addition of 0.14 M ammonia, 0.003 M EDTA, 0.029 M ammonium dihydrogen phosphate and 50 g/L Triton X-100	measurement of $^{57}Fe/^{56}Fe$ and $^{58}Fe/^{56}Fe$ isotope ratios	operating conditions chosen to minimize $^{40}Ar^{16}O^+$; isotope ratios corrected for mass discrimination and dead time; best precision and accuracy with >5 mg/L Fe	PQ2+; 0.8 ml/min sample uptake	411 412
Cr, Ni	human blood	microwave digestion with HNO_3/H_2O_2; down to dryness	external calibration using internal standardization with Rh	correction for ArC on ^{52}Cr and CaO on ^{60}Ni made based on ^{13}C and ^{44}Ca signals	ELAN-5000; 1.0 ml/min sample uptake	413
Zn	biological SRMs, human breast milk	Zn extracted from digested sample into CCl_4 with DDDC, washed with DDW, back-extracted into 1.2 M HNO_3; dilution to 200-400 µg/l Zn	ID analysis using ^{67}Zn to find the concentration of natural ^{66}Zn, as well as ^{68}Zn and ^{70}Zn tracers	separation removes interferences from the matrix; isotope ratios corrected for mass discrimination and dead time	ELAN-5000; 0.5 ml/min sample uptake	414
As	biological SRMs	sample digested with HNO_3/H_2SO_4 and H_2O_2	external calibration using internal standardization with Y			
Pt	serum, cerebrospinal liquors	sample diluted 20-fold and acidified to 1% (v/v) HNO_3	external calibration or method of standard additions	Au unstable in sample solutions: could not be used as an internal standard to improve accuracy of external calibration	ELAN-5000; 0.75 ml/min sample uptake	415

Element	Sample	Sample preparation	Calibration strategy	Comments	Instrument: sample uptake rate	Ref.
Li, B, Mo, Cd, Sn, Sb, Cs, Ba, Hg, Pb, Bi	human serum	serum diluted 5- or 50-fold with 0.14 M HNO_3	4-point external calibration using single-element standard solutions, with internal standardization using Be, In and Tl	memory effects for Li, B, Mo, Sn, Hg and Bi were minimized with 2-min wash with 0.14M HNO_3 between solutions	PQ; 0.9 ml/min sample uptake; spray chamber at 10°C	416
V, Fe, Cu, Zn and Ag	human serum	serum diluted 4- (for Ag and V) or 8-fold with 0.14 M HNO_3	external calibration with internal standardization using In (for Ag) and Co (for other elements)	R=300 for Ag, and 3000 for other elements, to resolve spectroscopic interferences	ELEMENT; 0.9 ml/min sample uptake; spray chamber at 5-6°C	417
Au	human serum	serum diluted 2000-fold with 0.1% Triton X-100 and 0.3 M HCl	external calibration with matrix-matched standards or method of standard additions	dilution with HNO_3 not recommended for Au determination	SPQ-8000NP; unspecified sample uptake	418
14 trace elements, Ca	human serum, bovine serum SRM	1+4 dilution	external calibration with internal standardization using Sc (for Al), Ga (for m/z=44-66, Y (for m/z= 82-111) and Tl (for Pb and U)	R=3000 to resolve interferences on m/z<82; low results for Cu and Mn; no suitable internal standard for Se (too high) and Cd (too low)	ELEMENT; unspecified sample uptake; spray chamber at room temperature	116
I	milk powder, human serum	milk powder digested with 5% ammonia in microwave oven; serum diluted (5 or 10-fold) with 0.14 M HNO_3	external calibration with internal standardization using Te (for 5% ammonia) or In (for 0.14 M HNO_3); or method of standard additions	2-min rinse with solvent between solutions; 0.14 M HNO_3 lead to low recovery from milk powder	PQ; 0.9 ml/min sample uptake; spray chamber at 10°C	166

continued

TABLE 1.10 (*continuation*)

Element	Sample	Sample preparation	Calibration strategy	Comments	Instrument; sample uptake rate	Ref.
Al	serum, haemodialysis, haemofiltration and peritoneal dialysis solutions, tap water	serum diluted 1+4 and haemodialysis solution diluted 1+35 with DDW; HNO_3 added to all solutions (final concentration of 0.16% m/v)	external calibration with internal standardization using Sc (50 µg/l); or method of standard additions	standard additions (or matrix-matched external calibration) must be used for diluted serum and dialysis solutions	PQ2+; unspecified sample uptake and spray chamber temperature	419
Mo	human blood plasma, bovine serum SRM	microwave digestion with HNO_3/H_2O_2; evaporation to dryness; dissolution in 0.01 M HNO_3; sample passed through alumina column; Mo eluted with 3.75 M NH_4OH; evaporation to dryness; dissolution in 0.15 M HNO_3	ID analysis with correction for mass discrimination	anion-exchange alumina column removes cations (e.g. Na^+, K^+) and part or all of some anions (Cl^{-1}, PO_4^{3-}, SO_4^{2-}) so that spectroscopic interferences are reduced to negligible levels	TS Sola; 1.04 ml/min sample uptake; USND with heater at 140°C and condenser at 0°C	420
Cu, Mo	12 new biological reference materials	sample digested with HNO_3 and HCl; insoluble siliceous residue removed from solution prior to analysis	ID analysis	significant interferences on Cu (by ArNa) and Mo was observed with some biological materials	ELAN-250; unspecified operating conditions	421
As	biological SRM	microwave digestion of sample with HNO_3, H_2SO_4 and H_2O_2 in 4 stages	external calibration with internal standardization using Y	Triton X-100 added to all samples and standards to make them 5% v/v Triton, so there is no difference in response due to varying carbon content of samples	PQ2 STE; unspecified sample uptake; spray chamber at 0°C	76
Cr	biological SRM	microwave digestion with HNO_3; down to dryness; dissolution in dilute HNO_3	standard additions, monitoring ^{53}Cr because of ArC interference on ^{52}Cr from residual organic carbon	ID with ^{53}Cr (^{52}Cr as reference) is unsuitable because of spectroscopic interference on ^{52}Cr	ELAN 5000; 0.3 ml/min sample uptake	422

Element	Sample	Sample preparation	Calibration strategy	Comments	Instrument; sample uptake rate	Ref.
Pb	whole blood	digestion with $HNO_3/HClO_4$ in covered Teflon beaker on hot plate; down to near dryness; diluted with 2% HNO_3	ID with ^{206}Pb (^{208}Pb as reference isotope) with correction for mass discrimination and dead time using Pb isotopic standards	absence of spectroscopic interference verified by comparing the results with those obtained after matrix removal	PQ; 0.3 ml/min sample uptake	423
Pb, As, Cd, Tl	whole blood, urine	sample diluted 1:10 with solution containing Ga, Rh, Bi at 100 µg/L in 1% HNO_3; supernatant of diluted blood sample analyzed	external calibration with internal standardization using Ga for As, Rh for Cd, and Bi for Pb and Tl	correction for $^{40}Ar^{35}Cl$ interference on As made based on the count rate for $^{35}Cl^{16}O$	ELAN-5000; unspecified sample uptake rate	424
Hg	whole blood, urine	urine diluted 1:10 and whole blood diluted 1:20 with 2% HCl/57 mg/L $K_2Cr_2O_7$	external calibration with internal standardization using Bi; sum of signals at m/z 198-202 used for analysis	Au enhances Hg signal but no effect on washout; fast pump mode more effective for washout; dichromate with HCl minimizes memory effect	ELAN-5000; 1 ml/min with 20-s increase to 2.5 ml/min upon a change of solution	164
Pt	biomedical silicone oil and gel	digestion with *aqua regia*; rinse with diethyl ether; organic layer back-extracted with dilute *aqua regia* and DDW; combined aqueous layers dried and dissolved in dilute HNO_3	external calibration with internal standardization using Rh	sample preparation allows 100% recovery of Pt from silicone gel-filled breast implant	ELAN-250; 0.6 ml/min sample uptake	425
Se	bovine serum, whole blood, erythrocytes	addition of 1% Triton-X-100 and modifier (1% ammonia, 1.16 g/L EDTA in 0.33% ammonium phosphate) and dilution with 1% butan-1-ol	external calibration with matrix-matched standards and internal standardization using In	1% butanol gave the largest enhancement of Se signal; amount of internal standard had to be increased in whole blood because of greater suppression experienced	ELAN-5000; 0.8 ml/min sample uptake	426

continued

TABLE 1.10 (*continuation*)

Element	Sample	Sample preparation	Calibration strategy	Comments	Instrument; sample uptake rate	Ref.
Sr	prehistoric human bone	cleaned sample homogenized in agate ball mill; then digested with HNO_3 in Teflon bomb for 6 h at 160°C; diluted with DDW	isotope ratios corrected for deadtime and mass bias using SRMs of known isotopic composition	mass bias factor had to be determined prior to each measurement because it drifted; Sr isotope ratios obtained with a precision of 0.05% RSD	ELEMENT (R=300); microconcentric nebulizer; spray chamber cooled to 4°C; 40-50 µl/min	304
Pt	cells exposed to cisplatin	digestion with 0.5 M NaOH and then HNO_3/HCl; or sonication; digestion in hot HNO_3 and dilution with DDW	method of standard additions	NaOH sample preparation induced severe non-spectroscopic suppression of analyte signal	ELAN-500; 0.9 ml/min sample uptake	427
Mg, Cr, Mn, Fe, Ni, Co, Cu, Zn	hair SRM	microwave digestion of 0.05 g hair with 1 ml HNO_3; dilution to 5 ml	external calibration with internal standardization using In	free aspiration used to reduce noise; only 100 µl required for multielemental analysis	PQII+; 30µl/min sample uptake with MCN	79
REEs	human hair	dry ashing method or microwave digestion with HNO_3/H_2O_2; addition of $HClO_4$ and evaporation; dissolution in HNO_3 and dilution	external calibration with internal standardization using Rh (correction for BaO^+ interference on Eu using the fraction of BaO in a standard Ba solution)	dry ashing, microwave digestion or open vessel digestion gave similar results but lower blanks obtained with first two methods	POEMS; 1.8 ml/min sample uptake; cyclonic spray chamber	428
Food analysis						
B	hay, wheat flour, beef kidney	sample digested in microwave oven with HNO_3/H_2O_2 and diluted with 1.5 M HNO_3 to < 100 ng B/g solution	external calibration with internal standardization using Be; or standard additions; or ID analysis using $^{11}B/^{10}B$ ratio after correction for mass discrimination	B memory effect eliminated by 60-s rinse with 0.02 M HNO3/2 mg/g NaF solution between samples; standards acid-matched to samples; external calibration preferred for its simplicity and sample throughput	ELAN-5000; 1.0 ml/min sample uptake	429

Element	Sample	Sample preparation	Calibration strategy	Comments	Instrument; sample uptake rate	Ref.
REEs	wheat flour, tea, bush twigs	dry ashing method or microwave digestion with HNO_3/H_2O_2; addition of $HClO_4$ and evaporation; dissolution in HNO_3 and dilution	external calibration with internal standardization using Rh (correction for BaO^+ interference on Eu using the fraction of BaO in a standard Ba solution)	dry ashing, microwave digestion or open vessel digestion gave similar results but lower blanks obtained with first two methods	POEMS; 1.8 ml/min sample uptake; cyclonic spray chamber	428
43 elements	spinach, cabbage, domestic sludge	sample treated with HNO_3 (and HF for sludge) using pressure digestion, and diluted	external calibration with internal standardization using Rh	acid-matched standard solutions used for calibration; some low results for the sludge because of incomplete sample dissolution	ELAN-5000; 1.2 ml/min sample uptake	430
Pb	wine	sample diluted 10 times with 0.14 M HNO_3	external calibration with internal standardization using Tl (accuracy and precision similar to standard additions and ID analysis)	dilution required to reduce signal suppression (largely induced by ethanol) as well as soot deposition and clogging	PQ; 0.9 ml/min sample uptake; unspecified spray chamber cooling temperature	431
many elements	wine	dilution 1:1 with distilled water	semiquantitative analysis using Be, Ge, In and Bi with internal standardization using Rh	quantitative analysis required standards to contain 6% ethanol and internal standard	ELAN-5000; 1 ml/min sample uptake	282
I	milk powder and bovine liver SRMs, infant food	microwave digestion with HNO_3/$HClO_4$ or extraction with TMAH at high temperature in closed vessel	ID analysis using $^{129}IO_3^-$ spike and measuring $^{127}I/^{129}I$	TMAH method less expensive but requires 3 hours instead of 45 min for digestion, and its extraction efficiency may not be 100%	ELAN-5000; 1 ml/min sample uptake; quartz Scott spray chamber	432

continued)

TABLE 1.10 (*continuation*)

Element	Sample	Sample preparation	Calibration strategy	Comments	Instrument; sample uptake rate	Ref.
I, Mn, Co, Ni, Cu, Zn, Rb, Cd, Pb	food SRMs (oyster, egg & milk powders, tomato and peach leaves)	sample pressed into pellet; incinerated in a flow of O_2; volatile I condensed above the combustion chamber and recovered in CFA-C amine solution	method of standard additions	sample holder modified to slow the combustion of samples with high fat content and prevent leaks; 5% amine solution in all solutions minimized I memory effect	ELAN-5000; unspecified sample uptake	433
30 elements	food SRMs (wheat and rice flours, apple, peach and tomato leaves)	closed-vessel microwave digestion with HNO_3/HF	external calibration with internal standardization using In	0.1 ml HF per 0.5 g sample is so low that no need for post-digestion with H_3BO_3; more complete dissolution with HNO_3/HF than HNO_3 alone	ELAN 5000; GemTip cross-flow nebulizer; Ryton spray chamber; 1 ml/min sample uptake	272
15-20 elements	spinach, mussel SRMs	0.3-g sample digested with 5 ml HNO_3 at 160°C for 3 h, and diluted to 10 ml	semi-quantitative analysis with internal standardization using Rh	large deviations observed for some elements (Al, Cr, Ni, Pb, Zn and Fe)	ELAN-500; 1 ml/min sample uptake	434
Zn	infant formula, milk powder	dry ashing at 375°C overnight; wet digestion with HNO_3/H_2O_2; down to dryness; dissolution in 0.1 M HNO_3	ID with ^{67}Zn (^{68}Zn as reference isotope)	^{67}Zn spike added following total dissolution of the sample	PQ2; 0.7 ml/min sample uptake	435
High-purity process chemicals						
Si, P, S, Cl, Se, Br	H_3PO_4, H_2SO_4, $H_2S_2O_8$, HF, HCl, TMAH, HNO_3, H_2O_2	up to 25-fold dilution or evaporation to dryness and dissolution in 2% HNO_3	isotope ratios	With R=2500-4000 for Si, P and S, and R=8000-12000 for Cl, Se and Br, most polyatomic interferences resolved	Plasma-Trace 2; 1.0 ml/min sample uptake	170
Cr, Ni, Cu, Zn, Ag, Cd, Tl, Pb, Th, U, Fe	HF	sample evaporated to dryness and redissolved in 10% HNO_3	5-point external calibration with internal standardization using In and Bi; or ID analysis	evaporation step necessary to improve detection limits and eliminate HF matrix; external calibration results comparable to those by ID analysis	ELAN-5000; 0.8 ml/min sample uptake	436

One prominent measurement in Table 1.10 is that of stable isotope ratios which, when they are free from spectroscopic interferences and are corrected, if needed, for mass discrimination and dead time, will usually be less affected by effects of concomitant elements than the absolute signal of either isotope. The correction for mass discrimination (induced or not by the sample matrix) was described at the end of Section 1.4.3. The dead time, τ, which is the time period during which the detector is unable to count incoming ions (i.e. the time required to process a single count), is more significant at high count rates. It can be determined for a single pair of isotopes whose isotopic ratio is >3 but <10 [423] using a series of standard solutions covering the linear dynamic range for the analyte, and the following expression [219]:

$$S_c = S_m/(1 - S_m\tau) \tag{11}$$

where S_m and S_c are respectively the measured and corrected signals (i.e. count rates). The averaged ratio of corrected signals over the range of analyte concentrations is computed for various values of τ. The value of τ yielding the averaged ratio with the smallest RSD is then selected and used for all measurements [219]. Alternatively, the measured isotope ratio is normalized to the true value and plotted as a function of different values of dead time for each of the standard solutions of known isotopic composition [214,482]. The dead time then corresponds to the intersection of all the curves [214,482]. (Note that in this latter case, the magnitude of the ordinate at the intersection i.e. deviation from unity may also be used to estimate the mass bias factor [214,482].) However, these two methods are based on trial and error.

More recently, a method was described to calculate the dead time. It involves the measurement of different concentrations of a poly-isotopic element having at least three isotopes [98]. Ideally, two of the isotopes should offer a unity ratio to allow for mass bias correction (since the dead time has a negligible effect on unity ratios) and two have a ratio <0.05 for the dead time calculation after mass bias correction [98]. The dead time can indeed be obtained from the slope of the ratio of true ratio over measured ratio (corrected for mass bias) as a function of analyte signal [98]. Its value depends on the instrument. For the ELAN 5000, the dead time is typically around 4.0×10^{-8} s [219]. Obviously, it is unlikely to be significant for trace analysis [219]. This correction is only required to obtain ratios with the highest precision [423]. It must always be applied to the raw data instead of peak area to compensate for changes in count rate while the peak is being swept [475]. If no correction is applied, a significant bias may result which will be more important for larger ratios or, in the case of a given ratio, for higher analyte concentrations [219,475]. This may explain, at least partly, why one study reported a dependence of the detector dead time on analyte mass with a quadrupole-based instrument but not with a sector-based instrument [482]. Indeed, the study does not mention any correction for mass discrimination, which is expected to be more important in quadrupole-based than sector-based instruments. Another study, which involved mass bias correction, did not observe any correlation of the dead time with mass on a different quadrupole-based instrument [98].

Other pre-requisites must be fulfilled in order to obtain ratios with a precision that is limited only by counting statistics. Enough counts must be accumulated so that the statistical uncertainty will be small [219,425]. Furthermore, any change in mass discrimination as a function of time must be compensated, as it would induce a drift in the measured isotope ratio. Three different approaches can be used to correct for drift. If available and free of spectroscopic interferences, a third isotope of the analyte can be monitored [219]: the relative drift in a known ratio involving this isotope can then be used to correct the ratio of interest (through Eq. 10). If not, two internal standards, with masses bracketing the region of interest, could be added, or frequent recalibration could be carried out, using a standard solution of known isotopic abundances [219].

Isotope ratios can be used for accurate analysis by the quasi-absolute method of ID analysis. This method essentially consists in spiking the sample with a known amount of an enriched isotope of the analyte, and relating the change in isotopic ratio measured before and after spiking, to the original analyte concentration [217,218]. Because of mass discrimination, this normally standardless technique requires a prior determination of mass bias using standards of known isotopic composition in order to correct the measured isotope ratios for mass discrimination [339], hence the qualification of “quasi-absolute”. Equilibration between the spike and the analyte is also required. This means that, if a specific species is being determined after a separation (rather than a total element concentration), the spike will be added prior to the separation only if it is in the same chemical form as the analyte species [483]. Otherwise the addition of the spike may shift complex equilibrations. The spike should be added following the separation if it is not available in the same form as the analyte [483]. In any case, because of its high accuracy, ID analysis is invaluable for the production of certified reference materials [323,339,340,423]. Isotope ratios can also be used for stable isotope tracer studies [480,484], kinetic modelling [480] and to identify the likely source(s) of toxic elements such as Pb [329,480].

Table 1.10 demonstrates that various approaches (other than the techniques discussed in the following chapters) can be used to overcome spectroscopic interferences without making major modifications to the ICP-MS instrument, such as adding one [485] or two quadrupoles [28], and operating one quadrupole in an rf-only mode so that it can serve as a cell where gas-phase collisions are performed to break polyatomic ions, although some of these options are available commercially. As mentioned in Section 1.4, the operating conditions, particularly the carrier gas flow rate and the rf power, can be varied in order to minimize element oxides, hydrides, and doubly charged ions [83,121]. Multivariate calibration, such as partial least square regression, can also be used to correct for spectroscopic interferences [327,396,471,486,487]. However, this method requires a large number of standards i.e. a series of standards for each analyte and source of interference, and is therefore specific to individual samples, although a method has been described to decrease the number of standards required [488]. This is even more critical if non-spectroscopic interferences are also present: the instrumental sensitivity for the model data set must indeed be similar to that for the samples [327].

Alternatively, multicomponent analysis of a full mass spectrum of the sample can be carried out, where the contribution from interferents and analytes is modelled from their isotopic pattern in the mass spectrum of the actual sample [489]. By adding or deleting interferents from the model, optimal interference correction is achieved, and the corrected analyte signal is obtained from its full isotopic pattern (rather than from a single interference-free isotope which may be of low abundance) [489]. However, this powerful approach requires that only elements of known natural abundances be present, and that there not be too many species and/or strong background lines [489].

That is why many of the approaches summarized in Table 1.10 consist in separating the problematic element(s) from the analytes during sample preparation (by extraction, chelation, selective precipitation, etc.). Although they can be quite effective at reducing both spectroscopic and non-spectroscopic interferences (if the source of the latter is also removed during sample processing), they can be time-consuming and increase the likelihood of contaminations. Chapter 2 will describe how these separations can be carried out on-line with ICP-MS, in a closed system, thereby reducing sources of contamination, and often sample consumption as well. Alternative separation procedures that selectively isolate the analyte and transform it into a volatile species are also discussed in Chapter 2.

As discussed in Section 1.4.3, variations in the signal and effects from concomitant elements can be compensated using internal standardization [222]. Careful selection of internal standard is paramount [144]. Ideally, the internal standard must have the same physical properties (especially mass and ionization potential) as the analyte, so that it can behave similarly to the analyte in both the plasma and the detection system [259]. However, depending on the instrument and operating conditions used, an internal standard that is only close in mass to the analyte may be adequate [490]. This similarity in mass is important since, at least in quadrupole-based ICP-MS, the time required for ions to travel from the sampler to the detector was reported to exhibit a square root dependence with ion mass [491]. In any case, because of the additional problem of mass discrimination, several internal standards are typically required for multielemental analysis, in order to cover the entire mass range, which complicates the analytical procedure [259,266,490]. The range of internal standards can be extended through the use of enriched isotopes [399]. In any case, if an analyte lies between internal standards separated by a relatively large mass difference, then sensitivity changes (whether instrumental or matrix-induced) may not be adequately corrected, even if a linear interpolation of the internal standard correction factors (obtained by ratioing the signals of the internal standards measured for the initial solution of the run to those of each solution) as a function of mass is carried out [399]. An additional correction based on an external calibration may also be needed, where a linear or polynomial interpolation versus time is then carried out using standards bracketing the sample [399].

Nonetheless, the widespread use of internal standardization, which is evident in Table 1.10 demonstrates the efficiency of this approach. It can, in fact, be easily implemented on-line with ICP-MS by simply merging an internal standard solution with the sample using a multi-channel peristaltic pump [241,406]. An alternative is to use an external drift monitor which has a matrix similar to that of the sample [352]. Although this approach requires the additional and repeated analysis of the drift monitor, it has the advantages of not involving any addition to the samples (such as when the internal standard is added off-line), hence eliminating a potential source of contamination, and avoids possible systematic bias since the behaviour of each analyte is then monitored and corrected accordingly [352]. A change in the plasma operating conditions, in particular, a change in power or a decrease in aerosol carrier flow rate, can also reduce effects of concomitant elements but at the expense of sensitivity [261,274]. In some cases, optimization of the lens settings can also be carried out in presence of the matrix to counteract the effect of the latter [265,266]. The conditions thereby found can then only be used in combination with matrix-matched standard solutions, the method of standard additions or ID analysis [266]. Chen and Houk indicated that changing the ion lens settings so that the ions are brought into the lens region as a relatively broad beam would decrease space-charge related matrix effects [292]. When possible, a simple dilution of the sample can be carried out to minimize the effects of concomitant elements, since the extent of the effect only depends on the absolute amount of concomitant element [260,261].

Table 1.10 demonstrates that ICP-MS can be used for the determination of one or more elements, when a relatively fast analysis with low detection limits is required. Note that although detection limits are usually reported, a more important figure-of-merit for the analysis of real samples is the limit of quantitation, which is typically defined at 10 times the detection limit [213]. However, on some instruments the relative standard deviation (RSD) of the net signal does not vary with 1/concentration over the whole linear dynamic range [213]. It would therefore be more advisable to make a log–log plot of the %RSD as a function of concentration to determine the limit of quantitation i.e. the concentration corresponding to 5% RSD [213].

In any case, depending on the sample, direct analysis [307,319] or after minimal treatment (such as filtration or evaporation) of the sample [110,209,308] can be performed. Fast semi-quantitative analysis can also be carried out [492]. External calibration with matrix-matched standards and corrections for spectroscopic interferences often provide adequate results. However, the multielemental analysis of samples with complex matrices may be quite complicated, requiring frequent recalibration and drift correction procedures if an external calibration is used [390]. This may be simplified by also using internal standardization (as discussed above), although a few internal standard elements can hardly be ideal for the determination of numerous elements.

Better accuracy requires either the method of standard additions with or without internal standardization, or the stable ID analysis technique. Since the latter uses an isotope of the analyte as an internal standard, it provides the best precision of all calibration strategies. Monte Carlo simulation can determine which isotope and what amount of it should be used as an internal standard in order to achieve the best precision [493]. In general, however, a relatively wide range of ratios will correspond to a small propagated uncertainty [423]. As a rule of thumb, a good compromise between acceptable ratios (in terms of count rates) and uncertainty propagation is obtained by adding 1 µg spike for each µg of analyte in the sample [423]. Alternatively, the uncertainty resulting from error propagation can be minimized by adding enough isotopic spike so that the resulting ratio is close to the geometric mean of its value in the spike and sample [395]. In any case, the approximate concentration of the analyte is required to an order of magnitude [271]. If corrections are made for both mass discrimination and dead time, the precision achieved on concentrations determined by ID may be similar to that achieved on isotope ratio measurements [423]. Furthermore, once the isotopic spike has been added, analyte losses do not affect the accuracy, as long as not all the analyte is lost and a good chemical equilibrium is established between the spike and the sample. Isotope dilution has also been demonstrated to be independent of the solvent [420] and fairly insensitive to instrumental drift [420,494]. It therefore also provides the results of best accuracy. However, it can only be used if there are two isotopes free of spectroscopic interferences for each analyte. Chapter 2 will show how this method can be made as simple as an external calibration, including the requirement for only one isotope free of spectroscopic interferences for elements whose isotopic composition does not vary in nature, but with the added accuracy and precision typical of ID.

Needless to say, the calibration strategies providing better accuracy are also more time-consuming since they involve spiking each sample with known amounts of each analyte or analyte isotope. However, this time-consuming step can be essentially eliminated if it is performed on-line. For instance, a simple approach, which was described for FAAS, should be applicable to ICP-MS [495]. Basically, a standard solution is pumped (using a peristaltic pump) to the nebulizer through a tee whose other tip is immersed in the sample solution [495]. As the pump flow rate is increased, from full stop to a rate matching the free nebulizer uptake rate, the nebulizer aspirates a decreasing volume of sample which is being mixed with an increasing amount of standard [495]. This approach would be useful only if the sample can be nebulized continuously without inducing problems (such as clogging) and it does not cause non-spectroscopic interferences. Otherwise, the matrix of the standard solution would have to match that of the sample.

The same set-up can also be used to generate calibration curves: the sample is then simply replaced by a blank [495]. An alternative approach uses a gradient HPLC pump to vary the proportions of standard and diluent being mixed to yield a calibration curve, all under computer control [496]. The addition of the isotopic spike for ID analysis can also be made on-line, by continuously merging the spike solution with the sample [343,494]. One relies on a non-linear calibration curve obtained by plotting isotopic ratios as a function of the concentration of standard solutions [494]. Similarly to conventional ID, where a preliminary analysis is required to adjust the amount of isotopic spike required, a suitable standard concentration range must first be selected, the median of which is then used to calculate the isotopic spike concentration [494]. A different approach requires a knowledge of the densities of the sample and natural standard solutions which are sequentially pumped and mixed with the isotopic spike solution [343]. Chapter 2 will show various ways of carrying out any type of calibration on-line. In particular, several examples will be given of how the method of standard additions can be performed automatically on-line, without having to worry about matrix-matching.

In all cases, solid samples must first be put in solution by digestion, fusion, dry-chlorination, etc., prior to analysis, which introduces potential sources of contamination as well as spectroscopic and non-spectroscopic interferences. Sample preparation steps were reviewed [497] as well as microwave-assisted procedures [498]. One relatively simple way of avoiding the use of acids is to introduce the sample as a slurry [144]. Indeed, slurry analysis is similar enough to solution analysis that conventional instruments can be used with a high-solid nebulizer (many types of which, such as the V-groove one, are commercially available), and calibration may be accomplished with aqueous standards [499]. The approach was recently thoroughly reviewed by Ebdon and co-workers [499]. Slurry preparation can be as simple as the "bottle-and-bead" approach, where the sample is placed in a bottle along with a dispersant and beads of a harder material, and is shaken for several hours [500]. It has the advantage of involving fewer chemicals than total dissolution approaches, thereby reducing potential contaminations. Furthermore, slurry nebulization can avoid analyte loss, incomplete dissolution and the hazardous use of $HClO_4$ and/or HF [501].

However, in order for an external calibration with standard solutions to be successful for the analysis of slurries, several parameters are important. The particle size of the slurry is obviously very important since 3-μm solid particles have been seen to traverse the central channel intact [65]. In addition, the dispersion i.e. colloidal stability of the slurry, and the ICP system (i.e. nebulizer, spray chamber and torch injector), will also determine the particle size that will ultimately reach the plasma [499]. So, in order to minimize transport interferences, a single particle should ideally be present within each aerosol droplet [65], and the size of that particle should be such that the overall density of the droplet is not significantly different from that of a particle-free droplet [502]. Under these conditions, the slurry aerosol would then behave like a solution aerosol in the sample introduction system and provide accurate results using an external calibration with aqueous solutions. A wider-bore torch injector is also recommended to increase the residence time in the plasma and hence reduce atomization interferences.

Because of the heterogeneous nature of several samples, such as soils, the slurry particles may have varying density, and segregation of different analyte-containing phases may occur during aerosol transport to the plasma. Such transport effects are in fact a most important source of non-spectroscopic interference in slurry nebulization [503]. The smallest possible solid particle size is therefore recommended to minimize variations in densities amongst the resulting aerosol droplets [502]. ICP-MS analysis of 0.05% w/v slurries of reference soils, with particle sizes <3 μm, was carried out using continuous nebulization without sampler cone blockage [500]. Of the six trace elements determined using an external calibration with aqueous standards, the results for V, Cr and Ni were in agreement with the certified values whereas those for Cu, Co and Zn were higher than expected. The analysis of refractory materials can also be facilitated by using mixed-gas plasmas (see Section 1.6), which have been demonstrated to reduce atomization interferences, as exemplified by the ICP-AES analysis of highly refractory firebrick particles [504] and improve the evaporation of slurry particles from ceramic powder [503]. In any case, internal standardization using a matrix element as an intrinsic internal standard is recommended to correct for transport and atomization interferences [503].

Chapter 3 (on electrothermal vaporization (ETV)) and especially Chapter 4 (on laser ablation) and Chapter 5 (on direct sample insertion (DSI)) describe alternative solid sample introduction techniques that eliminate the time-consuming sample dissolution step. These techniques are very promising and/or valuable, especially for samples which are difficult to dissolve. Furthermore, two of them (ETV and DSI) can also be readily used for the analysis of solutions.

1.6 EXTENDED CAPABILITIES WITH MIXED-GAS PLASMAS

As discussed in Section 1.2.1, the default gas that is used to sustain an ICP ion source for MS is argon. It is an obvious good choice because it is monoatomic, chemically inert, has the fourth highest first IP of the Periodic Table (after He, F and Ne), and is the least expensive of the rare gases. An argon ICP is also easier to ignite and requires a lower rf power for operation than discharges in molecular gases (i.e. air, nitrogen, oxygen and hydrogen), because, among other factors, of the lower thermal conductivity, heat capacity and electrical resistivity of argon compared to these other gases [385].

However, the use of argon imposes several limitations. Since the non-metals and the halides have a relatively high first IP, their degree of ionization is ≤50% (see Section 1.2.1), which degrades their detection limits by at least two orders of magnitude compared to the values quoted in Tables 1.5 and 1.6. Argon-containing polyatomic species, arising from either combination with other elements from the plasma itself or combination with elements from the matrix, can interfere with the determination of several analytes, thereby degrading their detection limits. As discussed in Section 1.4.2, many interferences observed with low-resolution instruments (for instance, $^{40}Ar^+$, $^{40}Ar^{16}O^+$ and $^{40}Ar_2^+$, on respectively $^{40}Ca^+$, $^{56}Fe^+$, and $^{80}Se^+$) are resolved with high-resolution instruments, although a sacrifice in sensitivity results from the required increase in resolution. Nonetheless, some interferences (for instance $^{40}Ar^{35}Cl^+$) are still not resolved, even with R = 10000 (see Table 1.8) and, hence, also degrade the analytical capabilities of high-resolution ICP-MS. As discussed in Section 1.4.2, instrumental conditions can be adjusted in some cases to decrease or even eliminate these interferences, but only to the detriment of sensitivity.

Furthermore, the low thermal conductivity of argon limits the efficiency of atomization and, hence, ionization [58]. As discussed in

Section 1.2.1, the efficiency of desolvation, volatilization and dissociation is determined by the size of the droplets, their residence time in the plasma and the thermal conductivity of both their vapour and the plasma gas [57]. In particular, the thermal conductivity controls the rate of heat-transfer between the plasma and the aerosol particle surface. A higher thermal conductivity should therefore result in more efficient desolvation, volatilization, etc. A theoretical comparison of all-Ar and all-N_2 ICP discharges indeed concluded that, although the plasma temperature was higher in the Ar plasma, the N_2 plasma should be more effective in decomposing Al_2O_3 particles because of the higher thermal conductivity of N_2 [505].

That is why mixed-gas plasmas, where a foreign gas (which usually possess a higher thermal conductivity than Ar) is added to one or more of the argon gas flows [506], have been considered in ICP-AES and as alternative ion sources for MS, in an attempt to overcome the above-mentioned limitations of the Ar plasma. Molecular-gas plasmas, where a molecular gas completely replaces one or more of the argon gas flows, have also been studied in ICP-AES [506]. As witnessed by book chapters [507,508] and reviews [506,509,510], the number of publications on mixed-gas (Ar–N_2, Ar–O_2 and Ar–H_2) and molecular gas (N_2) ICPs has increased substantially over the last decade. Some publications [506–510] also discuss in detail the generation and characteristics of these plasmas which, in particular, can better decompose refractory analyte aerosol particles, and can handle larger amounts of solvent and analyte than an Ar ICP, thereby extending the range of samples that can be analyzed and reducing sample preparation [506]. This section will demonstrate how these plasmas can be used to expand the capabilities of ICP-MS, even for samples in simple aqueous matrices. As will be seen, enhancement of sensitivity, reduction and/or elimination of spectroscopic and non-spectroscopic interferences, and reduction of mass discrimination are distinctive features of these plasmas.

1.6.1 Modifications required for the addition of a foreign gas to the plasma

In general, addition of a foreign gas to the plasma will require re-optimization of the operating conditions of both the ICP and MS. One notable exception has been reported for the ELAN 5000 (and 6000) instruments which use a 40-MHz free-running rf generator and the patented PlasmaLok interface which essentially decouples the ICP from the MS (see Section 1.2.2.1) [100]. The generator can handle changes in both sample and plasma compositions very efficiently (i.e. without requiring manual re-tuning of the impedance network); whereas the special interface alleviates the need for re-optimizing the ion lenses settings since the instrument response function shape is constant [100].

In any case, all instruments with a secondary discharge can be expected to require more tuning than discharge-free instruments since the secondary discharge may also be modified by the addition of a foreign gas to the plasma, thereby complicating the overall result (in terms of optimal, stable operating conditions). For instance, the ion lenses settings had to be tuned after each change in aerosol carrier flow rate, with mixed-gas plasmas on a PQ [511] whereas no such change was required on an ELAN 500 [512]. Whether the strength of the electric field of the plasma is changed upon addition of a foreign gas, depends on the properties of the latter. For instance, computer simulations have shown that O_2 has no effect on the electric field whereas N_2 enhances it and would therefore worsen any secondary discharge [513]. A torch shield may be used to alleviate the secondary discharge, although it would then restrict the maximum power that can be applied. Indeed, with over 1.8 kW, the metal shield plate that is inserted between the quartz torch and the load coil was reported to melt [514]. The presence of a secondary discharge will also restrict the sampling positions that can be selected, as well as the lifetime of the sampler and skimmer [514]. For instance, even if Cu sampler and skimmer with a thick top were used in combination with an air–Ar ICP, a maximum power of 2 kW had to be applied to minimize damage to the sampler [514].

In general, mixed-gas plasmas require higher rf power than an all-argon plasma. Furthermore, those involving the addition of a molecular gas (such as N_2 or O_2) will result in narrower and shorter plasmas than an Ar ICP, hence requiring sampling at shorter sampling positions [513]. In any case, simplex optimization can be carried out to find the optimal operating conditions [84,124,515–519]. Various figures of merit can be used for this optimization. Plasma operating conditions which maximized the ratio of Sb^+ or Co^+ to an interference signal such as $ArCl^+$ [513–515], ArO^+ [516,517], ClO^+ [516,517], as well as $ArNa^+$, SO_2^+ and S_2^+, and PO_2^+ [517], minimized these interferences without sacrificing the analyte signal and facilitated the determination of As [515–517], Fe, V and Se [516,517], Cu [517,519] and Zn [517], although the accuracy of the results depended on the matrix (see Table 1.11). Similarly, plasma operating conditions maximizing Ce^+/CeO^+ minimized metal oxide formation and allowed the determination of Eu [516], Gd [516–518] and Sm [516] in the presence of Ba, Ce and La. Alternatively, a simplex optimization using several criteria, i.e. the In signal and its RSD, the BaO^+/Ba^+ ratio, and the background signals at several *m/z* where spectroscopic interferences are expected from the plasma and/or matrix, could be carried out to find compromise conditions for multielemental analysis [84].

A simplex optimization of both the plasma operating conditions and the ion lenses settings was also carried out for the multielemental determination of elements with various first IP and spanning the mass range [124]. Since all the analytes selected were free of spectroscopic interferences, the figure of merit used was $n/\Sigma(\mathrm{SBR})^{-0.7}$ where n was the number of elements and SBR the signal-to-background ratio for individual elements [124]. The analyte signal could therefore be maximized without an undue increase in background (which could degrade detection limit), and without bias toward the most sensitive elements [124]. However, similar SBR could be obtained by univariate optimization (i.e. manual tuning) [124]. In any case, although this has not been reported yet for ICP-MS, it should also be possible to perform a multielemental optimization of SBR and minimize non-spectroscopic interferences as was carried out in ICP-AES for a molecular gas ICP with N_2 as the outer gas [520].

Depending on the instrument (this is especially true for older versions), a more efficient cooling of the interface may be required to offset the increase in plasma temperature that may result from the addition of a foreign gas to the plasma [513]. (While the interface is in contact with a mixed-gas plasma, the temperature of the water coming out of the interface can simply be monitored to ensure that it does not unduly increase.) Otherwise, the lifetime of the sampler and skimmer may be significantly reduced. On an ELAN 250 or 500, an improvement in the cooling efficiency can simply be accomplished by modifying the piping system. In the original configuration, the water that cools the interface has indeed passed through the cryopump and therefore has a higher temperature than when it was supplied from the refrigerated recirculator. The piping can therefore be modified such that part of the input of cooling water to the ICP-MS instrument goes directly to the interface.

On some instruments (for instance, an ELAN 500), the addition of O_2 to the plasma and the analysis of organic solutions required modification of the roughing pump system to avoid shut down of the instrument every 2–3 weeks to clean the solenoid valve to the interface and change its O-ring [521]. Modifications consisted in using one roughing pump solely for the interface (thereby removing the susceptible solenoid valve) and a second roughing pump to back up the cryogenic pump [521].

Although torches with the injector tip made of boron nitride are generally required for the generation of molecular-gas plasmas, mixed-gas plasmas resulting from the addition of a molecular gas to the argon flow can be sustained in conventional ICP torches [507]. A simple modification to the gas delivery system may therefore be the only instrumental requirement for the successful generation of these mixed-gas plasmas. In fact, many ICP-MS instruments are supplied with an extra flow meter or mass-flow controller for the addition of O_2 that is often required for the analysis of organic samples. On the other hand, the formation of mixed Ar–He plasmas [522] and all-He plasmas [523,524] is not as straightforward, requiring special torches [523–525] and modifications to the interface [523,525], the vacuum pumping system [522,525] and/or the load coil geometry [522,523].

Obviously, the handling of reactive and/or explosive gases has to be done with care. For instance, in the case of H_2, the installation of a back-flow-prevention valve is warranted, in addition to severing the H_2 flow prior to shutting off the plasma [526].

1.6.1.1 Gas proportioner or T-piece on a gas line

A gas proportioner is most widely used to add various percentages of a foreign gas to one or more of the plasma gas flows. The principle of this device is simple: two flowmeters determine the flow rate of the foreign and the argon gases which both flow through one common outlet, hence being mixed. This device is equivalent to using two flowmeters to separately control the gas flows into a T-piece, the outcoming mixture being fed to the appropriate torch outlet(s). If the gas proportioner or T-piece approach are used instead of the regular flowmeter which would normally control the argon flow if no other gas was added, a constant total flow rate is achieved by decreasing the argon flow rate proportionally to the increase in foreign gas flow rate. However, an easier control of the total flow rate can be achieved by simply connecting the outlet of the gas proportioner or T-piece to the inlet of the regular flowmeter (which would otherwise no longer be used). This latter flowmeter can then be used to keep the total flow rate of argon constant without having to readjust the argon flow rate each time the foreign gas flow rate is changed. In other words, the latter configuration allows the gas proportioner to select the fraction of foreign gas in argon whereas the flowmeter determines the total flow rate reaching the torch.

Alternatively, a T-piece can be inserted between the spray chamber and the torch to add a foreign gas to the aerosol carrier gas flow. For instance, the organics sampling system which is optional with the ELAN uses a T-piece between the outlet of the spray chamber and inlet of the torch to add small amounts of O_2 (typically 10–80 ml/min) into the torch injector tube [527]. A mass flow controller is used to precisely control the amount of O_2 [527].

A mixed-gas plasma is usually not ignited directly. Instead, an all-argon plasma is first initiated (as described in Section 1.2.1) and allowed to warm up while water is being nebulized. The foreign gas is then slowly introduced into the selected gas stream of this already ignited Ar plasma until the desired percentage of foreign gas is reached. The initial addition of foreign gas must be carried out very carefully to prevent the plasma from extinguishing as a result of either a sudden shrinkage of the plasma if the foreign gas is added to the outer and/or the intermediate gas flows, or as a result of the sudden widening of the central channel when the foreign gas is added to the aerosol carrier gas flow. An example of a mixed-gas plasma with only 2% N_2 in the outer argon flow is shown in Fig. 1.28 in comparison to an Ar ICP at the same power and aerosol carrier flow rate. In contrast, the appearance of the plasma would be similar to that shown in Fig. 1.30 upon addition of a foreign gas to the aerosol carrier gas flow.

A greater percentage of foreign gas can usually be added to the outer gas than to the aerosol carrier gas while still sustaining a stable plasma. When the foreign gas is added to the outer and/or intermediate gas flows, the outer gas flow rate is usually increased compared to that normally used for an Ar plasma, in order to provide additional cooling to the torch wall, especially when the foreign gas has a higher thermal conductivity than Ar, since this will result in a greater heat dissipation towards the outer tube of the torch. This increase in flow rate is not necessary to the generation of a mixed-gas plasma which, depending on the foreign gas and its percentage in Ar, can be quite stable under conventional conditions. Nonetheless, it does substantially reduce the likelihood of torch melting while the plasma operating conditions are varied (i.e. during the optimization process), according to the author's experience. In the case of Ar–He ICP, He also had to be introduced into the auxiliary gas flow in addition to the outer flow in order to stabilize

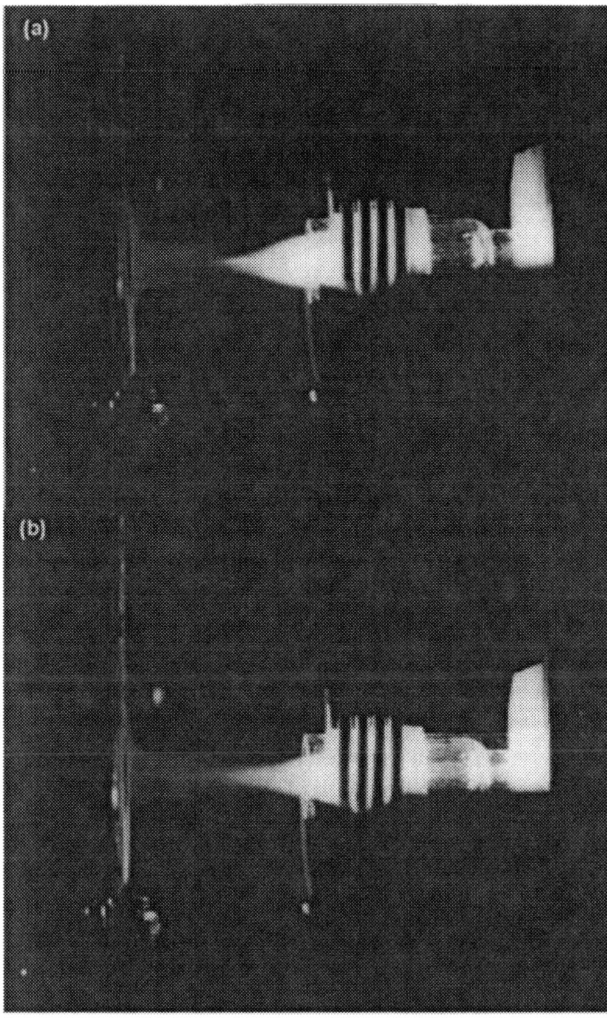

Fig. 1.28. Photographs of the ICP with a 500 mg/l Y solution being nebulized. (a) All-Ar plasma at 1.2 kW; (b) plasma observed after the addition of 2% N_2 in the outer plasma gas, under otherwise identical conditions.

the discharge [522]. In any case, if the ICP-MS instrument does not automatically re-tune itself for the mixed-gas plasma, the reflected power must then be minimized by manual fine-tuning. Otherwise, the plasma could extinguish and/or melt the torch at the first disturbance.

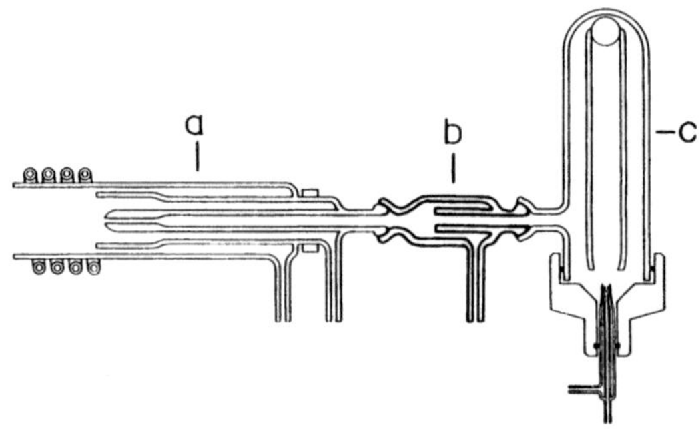

Fig. 1.29. Schematic diagram of the sample introduction system showing the insertion of a sheathing device (b) between the torch (a) and spray chamber (b).

1.6.1.2 Sheathing device

An alternative to using a gas proportioner or T-piece to add a foreign gas to the aerosol carrier gas, involves the insertion of a sheathing device between the spray chamber and the plasma torch as illustrated in Fig. 1.29. An additional flowmeter is then used to control the amount of foreign gas being introduced using the device, as a sheath around the aerosol carrier gas flow. Since this essentially results in the addition of foreign gas to the aerosol carrier flow, the addition must be made as described in the previous section i.e. into an already ignited plasma, and slowly to prevent the sudden widening of the central channel from extinguishing the plasma altogether. An example of the resulting mixed-gas plasma is shown in Fig. 1.30, in comparison to an Ar ICP under otherwise the same operating conditions. The main difference between the sheathing device and the gas proportioner or T-piece for the addition of a foreign gas to the aerosol carrier gas resides in the extent of mixing with the carrier gas: it is thorough with the gas proportioner or T-piece, whereas the sheathing device keeps at least some of the foreign gas relatively unmixed at the periphery of the central channel of the plasma.

1.6.2 Physical properties of the resulting plasmas

1.6.2.1 Addition to the outer gas

The addition of a foreign gas to the argon plasma has been reported to change the fundamental characteristics of the plasma [268,528] since it significantly alters the electron density distribution in the plasma [22]. The shrinkage of the plasma which occurs when certain gases are

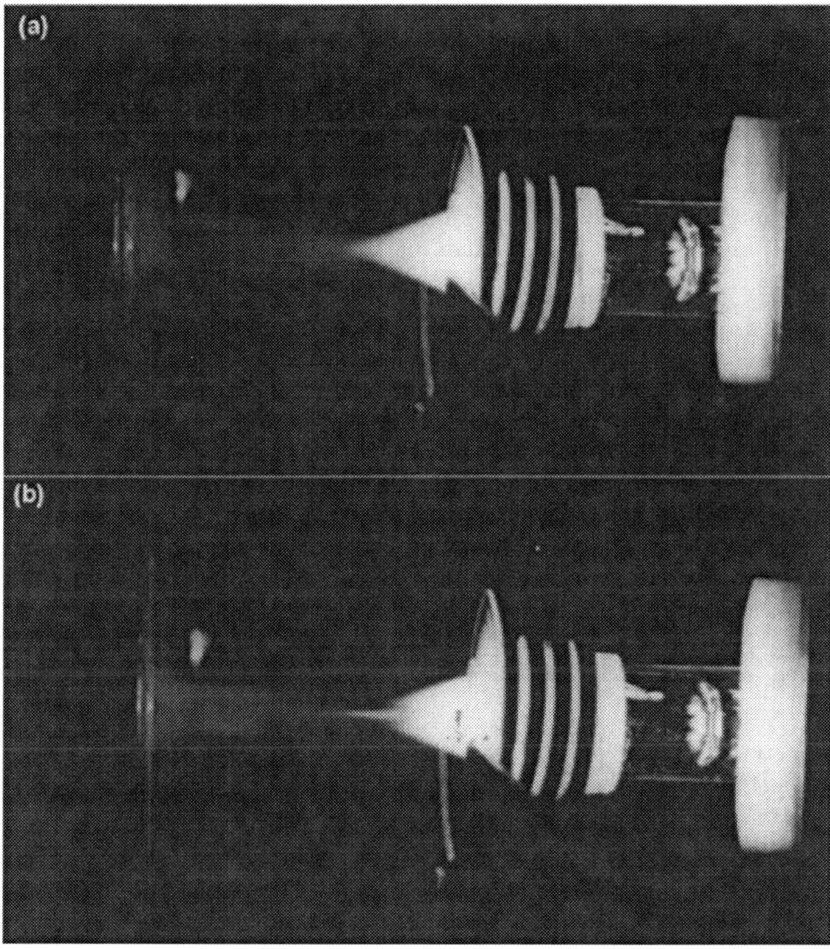

Fig. 1.30. Photographs of the ICP with a 500 mg/l Y solution being nebulized. (a) All-Ar plasma at 1.2 kW; (b) plasma observed after the addition of a trace of H_2 as a sheathing gas, under otherwise identical conditions.

introduced into either the outer and/or the intermediate plasma gas flows results from either the extra energy needed to dissociate a molecular gas and/or the "thermal pinch" effect [506]. The latter arises from the difference in thermal conductivity and electrical resistivity between the foreign gas and argon [529]. If the foreign gas, such as molecular gases, has a higher thermal conductivity than that of argon, the outer gas flow is cooler than if pure argon is used, leading to a reduction in electron density in the outer part of the plasma. When the electron density is too low to dissipate power i.e. the discharge can no longer be sustained in that region, the plasma becomes constricted to a smaller volume. In general, the larger the amount of foreign gas added, the smaller the volume of the resulting plasma discharge. The power is therefore concentrated into a smaller volume, resulting in a plasma with a higher power density than an Ar ICP. This is supported by the results of Choot and Horlick [530] who reported a significant increase (about 30%) in electron density with 10% N_2 or O_2 in the outer gas. These mixed-gas plasmas are therefore more energetic than a conventional Ar ICP and may therefore be better ionization sources for MS. In addition, the energy transfer within the plasma is greater, which facilitates the various processes (such as desolvation, vaporization, atomization and ionization) taking place in the plasma.

Not only is the overall size of the plasma reduced, but the diameter of the central channel also changes as illustrated in Fig. 1.28 where it is clearly narrower. This is as expected from an increased power density, and is in agreement with a previous report in ICP-AES [531]. However, this narrower channel (which was observed in ICP-MS with Perkin-Elmer instruments) contradicts the observation of a wider and fainter central channel that was made using a different (PQ2 from Fisons) ICP-MS instrument [515]. The tuning characteristics of the two brands of instruments are obviously quite different. This is further supported by the fact that a maximum of 2% N_2 could be tolerated in the outer gas flow with the Fisons instrument, whereas 20% N_2 could easily be added with a Perkin-Elmer ELAN 250 [532] and up to 100% N_2 was used on a prototype instrument [533]. In all fairness, however, more recent and advanced versions of the instruments are now available, and the above differences may no longer exist. Nonetheless, differences may be expected between mixed-gas plasmas generated on all current instruments with a secondary discharge compared to those on discharge-free instruments since the secondary discharge will also be modified by the addition of a foreign gas to the plasma, thereby affecting both the analyte signal and optimal operating conditions. In fact, an addition of N_2 is expected to exacerbate any secondary discharge [513].

Compared to an all-Ar plasma, the best sensitivity with the resulting mixed-gas plasma is usually observed at a higher forward power, higher aerosol carrier gas flow rate and a smaller sampling position in ICP-MS [532]. The plasma temperatures have been reported to be dependent on the percentage of foreign gas in the plasma. For instance, both T_{ion} [533] and T_{exc} [534] were significantly lower in a 100% N_2-cooled plasma (and 100% O_2-cooled plasma [535]) than in an all-argon plasma. Houk et al. [533] were the first to make use of Ar–N_2 plasma discharges in ICP-MS, adding up to 100% N_2 in the outer gas. Normalized intensities for nitrogen-containing ions generally increased along with the N_2 content of the outer gas flow, indicating either gas mixing or transport of nitrogen species from the outer gas region into the axial channel of the ICP. Furthermore, T_{ion} was approximately 1000 K lower with 100% N_2 in the outer gas than without N_2, which should, in turn suppress formation of Ar^+ in the axial channel. Yet, Ar^+ remained the major ion extracted from the central channel whether or not Ar was present in the outer gas, which lends further credibility to the transport of nitrogen species into the axial channel of the ICP, where a charge-transfer reaction (depicted by Eq. 2) between N_2^+ and Ar would then become possible.

However, with only 5–20% N_2 [534] or 5–50% O_2 [535] in the outer plasma gas, the axial T_{exc} was approximately 500–2000 K [534] or 500–1800 K [535] higher than in an Ar plasma at the same power. In general, with ≤20% foreign gas in the outer Ar gas flow, the axial T_{exc} is higher with N_2 than O_2, that with air being intermediate [535]. Vertical spatial emission profiles on Ar–N_2 plasmas also demonstrated that the observation zone for maximum intensity moved toward the load coil as the percentage of N_2 was increased, and that ionic emission increased to a larger extent than atomic emission [536], again suggesting that mixed-gas plasmas would be better ionization sources for MS. The highest sensitivities were obtained with 5–10% N_2, whereas more than 20% N_2 reduced sensitivities significantly [536], in agreement with the decrease in n_e which was observed as the percentage of N_2 was increased beyond 15% [533].

Mixed Ar–He ICP and all-He ICP are potentially more attractive than Ar ICP as ion sources for MS because of several unique characteristics of He. It has the highest first IP of the whole Periodic Table which implies that all elements should have a high degree of ionization in a He-containing ICP. Furthermore, the fact that ^4He is essentially monoisotopic (the only other isotope, ^3He, being 0.0001% abundant) and of low mass, should also result in a much simpler plasma background than that observed with Ar, which should, in turn, reduce potential interferences from polyatomic species [523]. The first study to report the introduction of aqueous solutions, using USND, into

He ICP-MS, confirmed the great simplicity of the background spectrum, since the highest *m/z* at which a major background ion occurred was 32 [525].

Like many molecular gases, He has a higher thermal conductivity than Ar [507,508]. However, He also has a higher electrical resistivity which makes power transfer into a He plasma more difficult and explains why stable pure He plasmas [507,523,525], and to a lesser extent, Ar-He plasmas [522] are more difficult to form than an Ar ICP. Using a specially designed MACOR-based low-gas-flow torch, a He ICP could be self-ignited away from the MS interface when approximately 500 W was applied, but it could not be sustained over an extended period at more than 800 W when the He ICP was coupled to MS [523]. So far, these plasmas have only been generated on Fisons and Delsi-Nermag (the latter company no longer supplies ICP-MS) instruments because earlier models of Perkin-Elmer instruments used a He cryopump for the main vacuum chamber (which could therefore not have successfully evacuated He gas). Unfortunately, these instruments (at least the models used for these studies) suffer from a residual secondary discharge that worsened with He [523] as a result of the stronger magnetic and electrical fields (by approximately an order of magnitude) generated by the He ICP in comparison to the Ar ICP [537]. This, for instance, resulted in the presence of orifice ions in the mass spectrum and increased levels of doubly-charged ions [523]. The intensity of the secondary discharge could be significantly reduced by using an anodized Al sampler [525] or a boron nitride sampler [524]. (The effect of a secondary discharge and some ways to circumvent it are discussed in greater detail in Section 1.2.2.1.)

1.6.2.2 Addition to the aerosol carrier gas

In contrast, when a molecular gas is added to the aerosol carrier gas, the diameter of the central channel increases with the percentage of foreign gas, presumably because energy in the vicinity of the channel is used to dissociate the molecular gas. This, in turn, leads to a decrease in temperature in that region, which translates into an apparent increase of the diameter of the central channel [506]. Furthermore, molecular gas with a higher thermal conductivity than Ar favours energy transport within the plasma thereby facilitating the various processes, in particular desolvation and vaporization, which then occur sooner. As a result, the IRZ and NAZ move towards the load coil [503]. A small decrease in aerosol carrier gas flow rate is therefore generally required to maximize analyte sensitivity compared to that in an all-Ar plasma [124,511]. A thermal pinch effect or reduction in plasma volume [503] is also observed upon an addition of molecular gas to the aerosol carrier gas, i.e. the plasma shrinks in height (but not in diameter) [528]. Furthermore, as the amount of molecular gas is increased, the plasma starts to pull away from the torch injector [528]. In any case, the reduction in volume of the plasma combined to the same incident power translates into a higher power density [528].

However, an addition of up to 16.7% H_2 decreased n_e in the central channel of the plasma [528]. (Note that plasma instabilities increased with a higher percentage of H_2, and plasma extinction might occur with more than 20% H_2 [503].) On the other hand, both the electron and the gas-kinetic temperatures increased by about 1500 and 2000 K respectively [528]. This increase in gas-kinetic temperature is commensurate with the higher thermal conductivity of H_2 than Ar. Adding up to 4% H_2 to the optimal aerosol carrier flow rate for the nebulization conditions used, with or without desolvation, increased the ionization temperature of the plasma [181]. Although a larger increase can be induced by desolvation at the appropriate temperature, a significant further increase resulted from the addition of H_2 [181] which, because of its higher thermal conductivity than Ar, increases the axial temperature of the plasma (see Section 1.2.3.3). In fact, the maximum ionization temperature and n_e were obtained with 4% H_2 at a desolvation temperature of 100°C for USND [181].

An addition of H_2 to a dry aerosol has also been shown to have a similar effect to adding an increasing amount of moisture [55]. It shifted the optimum sampling position to higher values (i.e. the sampler had to be moved away from the top of the load coil) and concurrently decreased sensitivity, above 5 ml/min H_2 in the aerosol carrier gas flow [55]. However, the initial addition of 2.5–5% H_2 increased analyte sensitivity by more than 50% [55]. The decrease in sensitivity observed at higher percentages of H_2 presumably results from the greater amount of plasma energy which is required to dissociate H_2.

An increase in n_e was also measured in the ICP as well as the interface plasma between the sampler and skimmer, upon an addition of 2–10% H_2 to the aerosol carrier gas flow of a dry plasma [538]. This increase in n_e may favour electron impact ionization (see Eq. 1) thereby explaining the increase in analyte sensitivity reported under these conditions during the introduction of a dry aerosol (from an electrothermal vaporizer) [538]. In addition or alternatively, charge transfer between background ions (especially Ar^+ and ArH^+ which are in greater population) and the analyte (see Eq. 2) may also be promoted by the increase in background ions which also results from H_2 addition [538]. As for an Ar ICP (which was thoroughly discussed in Section 1.2.1), the operating conditions are critical. For instance, at a smaller sampling depth, a decrease in analyte signal resulted as the percentage of H_2 in the aerosol carrier gas flow was increased from 5 to 15%, under otherwise fairly similar conditions [539].

A totally different behaviour was reported with an addition of up to 16.7% N_2 to the aerosol carrier gas [528]. Indeed, n_e and both the electron and gas-kinetic temperatures substantially decrease in the central channel and upper regions of the plasma as a result of the higher heat capacity of N_2 than Ar [528]. Nitrogen then acts as an energy sink which induces a drop by more than 5000 K in the central channel [528]. A suppression of sensitivity may therefore result.

In contrast to an addition of molecular gas, introducing up to 16.7% He in the aerosol carrier gas does not significantly change the shape or volume of the plasma [503], nor the temperature of the electrons, but increases n_e in the central channel and the gas-kinetic temperature throughout the plasma [528]. The latter temperature increase (by about 1500 K) is again likely due to the higher thermal conductivity of He than Ar which improves the energy transport within the plasma [528]. This, in turn, means that the desolvation and vaporization processes will occur sooner in the plasma, thereby shifting the IRZ and NAZ towards the load coil.

Addition of a gas to the central channel can also be made using a sheathing device. For instance, it was used in ICP-AES to introduce various gases into the central channel of an Ar ICP in an attempt to modify the energy transfer between the annular region of the plasma and the central channel [58,540,541]. This device was originally developed to prevent deposits on the inner wall of the torch injector by restricting the aerosol to the inner region of the central channel [58, and references therein]. It can also be used to modify the layer, and therefore the energy exchange, between the outer annular region of the plasma and the central channel, through the introduction of a sheathing gas which more or less results in the formation of an intermediate plasma between the plasma ring and the central channel [540]. A study with various sheathing gases (N_2, H_2, O_2, Ar and He) showed that N_2 and H_2 significantly modified the charge-transfer process with Ar [540]. In fact, a small amount of H_2 in an Ar sheathing gas increased

T_{exc} and resulted in a shift in the position of maximum emission intensity toward the load coil. The latter was combined to an enhancement in intensity when the emission lines had excitation energies greater than 4 eV [58]. These observations confirm an improved heat transfer from the annular region to the central channel as a result of the higher thermal conductivity of H_2 as compared to that of Ar.

Adding a sheathing gas to the aerosol carrier gas also changes the residence time of the analyte in the plasma, and therefore the IRZ position (see Section 1.2.1). If the distance between the plasma and the sampler is fixed (as is the case on several ICP-MS instruments which do not allow an easy change in plasma position while the plasma is ignited), then adding a sheathing gas will move the IRZ away from the load coil, and a change in plasma operating conditions will be required in order to bring it back to its original position, so as to sample the same region of the plasma. In general, the aerosol carrier gas flow rate is decreased as the sheathing gas flow rate is increased [542]. This, however, means that the nebulization efficiency (i.e. the percentage of sample solution actually exiting the spray chamber) will also likely decrease.

1.6.2.3 Addition to other or all gas flows

Similarly to an addition to the aerosol carrier gas, adding a molecular gas to the auxiliary gas flow increases the diameter of the central channel and induces a thermal pinch effect, except that in this case, the plasma shrinks both in height and in diameter [528]. Again, a molecular gas with a higher thermal conductivity than Ar favours energy transport within the plasma thereby increasing the rate of desolvation and vaporization, which, in turn, moves the IRZ and NAZ towards the load coil. Hence the optimal sampling depth above the load coil is smaller than that with an Ar ICP [514].

An addition of up to 16.7% H_2 to the auxiliary gas flow increased n_e. As well, both the electron and the gas-kinetic temperatures increased by 1500 and over 2000 K, respectively [528]. Similarly, introduction of 0.3% H_2 into all three argon gas flows increased T_{exc} and n_e compared to an Ar plasma [543]. It also changed the radial distribution of n_e which became higher near the central channel as a result of the "thermal pinch" effect [543].

A maximum of 3% v/v N_2 could be added to the auxiliary gas flow because the plasma shrank away from the sampler and the reflected power increased [511,515]. As the amount of N_2 was increased above

3%, the plasma started to fluctuate and eventually extinguished itself [511]. With 1–3% N_2, a smaller plasma than usual resulted, with a wide central channel similar to that resulting from an addition of N_2 to the aerosol carrier gas flow [515]. However, this plasma was of little analytical utility, as will be seen in Section 1.6.3.3.

Yet, addition of up to 16.7% N_2 to the auxiliary gas flow was accomplished on an ICP-AES instrument [528]. It decreased n_e in the central channel and upper regions of the plasma [528]. On the other hand, an overall increase in electron and gas-kinetic temperatures also resulted throughout the plasma (i.e. in both the central channel and the toroidal zones) [528]. On the other hand, when both the plasma gas and the aerosol carrier gas were replaced with air, the resulting plasma was observed to be less brilliant than an Ar ICP, and the kinetic energy distribution of each ion was found to be wider [514].

In contrast, an addition of up to 16.7% He to the auxiliary gas flow does not significantly change the shape or volume of the plasma nor the temperature of the electrons, but decreases n_e [528]. On the other hand, because of its higher thermal conductivity than Ar, it increases the gas-kinetic temperature by about 1500 K throughout the plasma [528] and thereby shifts the IRZ and NAZ towards the load coil.

An addition of He to both the outer and the auxiliary gas flows (as the latter was required to sustain a stable discharge) was carried out in a modified Greenfield torch with a modified load coil [522]. This resulted in a radially narrower and axially shorter plasma, with a central channel less clearly defined, than an Ar ICP, which therefore required a decrease in the sampling depth in order to sample a given region of the plasma [522]. It was not, at least under the conditions used, possible to also have a mixed He-Ar gas for the aerosol carrier gas, since the central channel became even less clearly defined i.e. the carrier gas could hardly punch a hole in the plasma [522].

1.6.3 Features and limitations of each approach

Mixed-gas plasmas have been successfully used to attenuate the formation of polyatomic species in ICP-MS. In particular, efforts have concentrated on the reduction of argon-, chlorine- and oxygen-containing polyatomic species. Some of them have also been used to reduce or eliminate non-spectroscopic interferences. These features and some of the associated limitations are thoroughly discussed in this section.

1.6.3.1 Addition to the outer gas

In general, a higher power and aerosol carrier gas flow rate are required compared to the typical plasma operating conditions used with an Ar ICP. However, the optimal plasma conditions vary with a number of parameters, including the foreign gas added as well as the instrument, torch, sample introduction system and which analytical capability is to be improved.

The hotter plasma resulting from the addition of 5–10% N_2 to the outer gas reduced analyte–oxide species as well as some other species (especially, ArO^+, $ArOH^+$, ClO^+ and $ArCl^+$, and to a lesser extent Ar_2^+) by either facilitating their decomposition, or allowing competitive reactions between argon and nitrogen species for the available oxygen, chloride, etc. [532]. The addition of 5% N_2 also induced an enhancement in signal by up to a factor of four [532], provided the aerosol carrier flow rate was increased and the distance between the tip of the sampler and the top of the load coil was reduced to compensate for the reduced size of the plasma which, as discussed before, shifted the IRZ toward the load coil. The extent of signal enhancement was dependent on either the analyte-oxide bond strength (for elements forming oxides), or the IP of the analyte [532]. The enhancement observed for elements with a high IP might arise from charge-transfer reactions with NO^+ which could induce additional ionization of the analyte [532]. On the other hand, the background originating from nitrogen-containing polyatomic species increased. Addition of air gave similar results to those with N_2 whereas O_2 had no effect on analyte sensitivity and caused only a small shift in IRZ position [532].

The above results for N_2, which were obtained on an ELAN 250, were reproduced by others with ELAN 500 who also observed moderate analyte signal enhancements and significant reduction in oxide species and other ions (such as ArO^+, $ArOH^+$, Ar^+, ClO^+ and $ArCl^+$) upon addition of 5–8% N_2 to the outer gas [544,545]. In fact, 8% N_2 was more effective at reducing the level of oxide species than a decrease in water loading achieved using either a water-cooled spray chamber or a desolvation apparatus consisting of a heater and a condenser [544]. However, an even more extensive reduction of oxygen-containing ions was obtained by using an Ar–N_2 plasma in combination with a desolvation apparatus [544,545]. This reduced $^{40}Ar^{16}O^+$, $^{47}Ti^{16}O^+$, $^{49}Ti^{16}O^+$, $^{94}Zr^{16}OH^+$ as well as $^{91}Zr^{16}O_2^+$ and/or $^{88}Sr^{35}Cl^+$ to such an extent as to allow the accurate determination of Fe, using the most abundant $^{56}Fe^+$, by external calibration, as well as that of Cu, Cd and Sb by isotope dilution analysis using the following reference/spike isotope pairs: $^{63}Cu/^{65}Cu$, $^{114}Cd/^{111}Cd$ and $^{121}Sb/^{123}Sb$ [544].

In contrast, no significant reduction of analyte–oxide levels could be achieved on PQ [511] and PQ2 [515] instruments with Ar–N_2 plasmas and water-cooled spray chambers. Furthermore, only a maximum of 2 [515] or 3% [511] N_2 could be added to the outer gas flow before an unstable plasma (with high reflected power) was reached. However, addition of about 2% N_2 completely removed ClO^+ and $ArCl^+$ in the presence of up to 1000 and 10,000 mg/l chloride, respectively [515]. Whereas 1% N_2 was sufficient to eliminate $ArCl^+$ and dramatically reduce ClO^+ from a 3% HCl matrix [511]. A slightly less efficient reduction of ClO^+ and $ArCl^+$ caused by 3% HCl could also be achieved with 1% CH_4, but this foreign gas led to carbon deposition on the sampler and could therefore not be used for extended periods of time without resulting in complete clogging of the sampler [511].

The fact that some polyatomic species can be reduced on both types of instruments whereas others can only be reduced on Perkin-Elmer instruments, is likely linked to the presence of a secondary discharge on VG Elemental instruments. Indeed, shielding an Ar ICP, which visibly eliminated the secondary discharge, was reported to reduce substantially the formation of several weakly-bound polyatomic species (for instance, ArH, CO_2 and ArO) [77,108] but not that of refractory oxides [108]. This similarity with the above results suggests that the addition of a foreign gas to the outer gas flow may reduce the formation of the secondary discharge. The addition of a gas with a different thermal conductivity and electrical resistivity than Ar may indeed change the potential of the plasma with respect to the interface and, therefore, the extent of the secondary discharge. This has certainly been predicted by computer simulations [513].

In addition to reducing interferences from polyatomic species and improving sensitivity for some elements, adding N_2 to the plasma outer gas also improved the stability of the plasma and reduced the effects of concomitant elements and mass discrimination. For instance, the accuracy and precision of $^{57}Fe^+$:$^{56}Fe^+$ and $^{76}Se^+$:$^{78}Se^+$ isotopic ratios was improved, even in the presence of up to 0.1 M Na, although a reduction in analyte signal intensity was then observed [268]. A 10% addition of N_2 to the argon outer gas of a 1.2-kW ICP eliminated the enhancing effect of 0.01 mol dm^{-3} K over the *m/z* range 51–114 while reducing mass discrimination over the *m/z* 51–78, but with a concurrent decrease in sensitivity [546]. The reduction of sensitivity in these two studies was a result of the constant aerosol carrier flow rate and solution uptake rate which were used [268,546]. In any case, the reduction/elimination of matrix effects by an addition of N_2 is not unexpected. Indeed, the addition of up to 100% N_2 or O_2 to the outer flow in ICP-AES completely eliminated the suppressing effect of phosphate on Ca ionic emission as well as the enhancing or suppressing effect (depending on the observation height) of easily ionized elements (such as K) on Ca ionic emission [547].

Several studies have reported a significantly increased S/N and S/B for some elements, such as Fe and Se, with 2–10% N_2, compared to those in an Ar plasma at the same powers [268,546]. This further substantiates the fact that adding N_2 decreases the background signals from argon-containing polyatomic species ($^{40}Ar^{16}O^+$, $^{40}Ar^{16}OH^+$, $^{40}Ar^{36}Ar^+$ and $^{40}Ar^{38}Ar^+$) [249,515,532,544] thus resulting in improved S/B and S/N. Furthermore, adding N_2 improved mass discrimination [546].

Experiments carried out at an optimal sampling position (i.e. adjusting the nebulizer flow rate for maximum sensitivity after each change in power and/or N_2 addition) compensate for any shift in the IRZ position but also include an increased nebulization efficiency (from an increase in the aerosol carrier gas flow rate) [512]. In any case, a 10% N_2 mixed-gas plasma at 1.3 kW only slightly degraded sensitivities and detection limits compared to those obtained in an Ar plasma (in fact, improvements were observed for a few elements) while eliminating the effect of 0.01 mol dm^{-3} Na across the mass range, and reducing that of 0.1 mol dm^{-3} Na to a uniform level across the mass range, thereby allowing the use of a single internal standard [512]. Under these conditions, mass discrimination was either similar or slightly degraded compared to an all Ar ICP, depending on *m/z* [512]. Mixed-gas plasmas can therefore be used to eliminate the effect of concomitant elements, with little or no sacrifice in sensitivity and detection limit. They clearly have the potential of allowing the analysis of samples with a variety of matrices using a simple calibration with aqueous standard solutions and internal standardization. This is, as will be seen later, demonstrated in Table 1.11.

1.6.3.2 Addition to the aerosol carrier gas

Because the aerosol carrier gas flow rate is typically less than 10% of the outer gas flow rate, a smaller absolute amount of foreign gas is needed to obtain a given gas mixture. A greater variety of gases (including some which are more expensive than argon, such as Xe, or are flammable, such as CH_4) can therefore be used without unduly increasing operating expenses. Furthermore, since the foreign gas is added to the central channel of the plasma which is being sampled into the mass spectrometer, a more drastic change to the mass spectrum is expected than from an addition to the outer gas which can only indirectly affect the central channel. These are some of the reasons why more publications have appeared using this approach than the previous one (and, in fact, any other approach). Another reason is that, as mentioned earlier, some instruments do not easily allow the formation of stable mixed-gas plasmas where the foreign gas is added to the outer gas. As for the other mixed-gas plasmas, the optimal conditions depend on several parameters.

Evans and Ebdon [548] were the first to report a reduction of interferences from $ArCl^+$ (from 3% HCl) and Ar_2^+ by adding 3% N_2 or O_2 to the aerosol carrier gas flow. Spiking the sample and standard solutions with 10% propan-2-ol in combination with 3% O_2 in the aerosol carrier gas flow reduced the interferences but not as much as the mixed-gas plasma alone [548]. They also reported a reduction of a number of other polyatomic ions (Cl_2^+, CCl^+, NCl^+, OCl^+, ArC^+, ArN^+, ArO^+ and $ArOH^+$) relative to an analyte signal (In^+) under the same conditions [249]. Overall, 3% N_2 led to the greatest reduction (except for nitrogen-containing polyatomic species) and was preferred over 3% O_2 because of the latter's detrimental effect on the lifetime of the sampler [548]. Their data suggest that the competitive formation of certain polyatomic ions (such as ArN^+ with 3% N_2 plasma) reduced the amount of others (such as Ar_2^+ and $ArCl^+$). Even in a dry plasma, an addition of up to 1% N_2 to the aerosol carrier gas flow decreased the level of several background ions (ArH^+, Ar^+, OH^+ and N^+) [538].

In any case, with continuous nebulization, adding N_2 eliminated the $ArCl^+$ interference from up to 1000 mg/l NaCl on 100 μg/l As [515, 549, 550] and that of ClO^+ from up to 10,000 mg/l NaCl on 100 μg/l V [515]. In general, for a given chloride concentration, the percentage of nitrogen required to eliminate the interference increased as the level of As or V decreased [515,550]. On the other hand, the interference from $^{14}N^{37}Cl^+$ on $^{51}V^+$ became more significant, at lower concentrations of V, as the percentage of N_2 and the chloride level increased [515]. However, the success of this approach depends on the extent of the spectroscopic interference. Indeed, an addition of N_2 to the aerosol carrier gas was found less efficient at reducing the ArCl interference observed in high-resolution with drinking water samples than separation of As from the matrix through hydride generation [137].

In any case, the elimination of the above-mentioned interferences is, in general, to the detriment of sensitivity which gradually decreases as the amount of N_2 increases [511]. However, a decrease in background often occurs concurrently thereby yielding similar or slightly degraded detection limits to those obtained with an all-argon plasma, except for elements whose *m/z* correspond to a nitrogen-containing polyatomic species where a more important degradation of detection limit then results [250]. This is particularly true if the background arises from polyatomic ions originating from the matrix. For instance, a 200-fold improvement of the detection limit for Cu (at *m/z* 63) in a 1% v/v artificial seawater matrix resulted from the addition of 0.8% N_2 to the aerosol carrier gas flow, after simplex optimization of all plasma operating conditions, because this addition drastically suppressed $ArNa^+$ [519]. The detection limits of Mn, Co and Zn in the same matrix were also improved by at least an order of magnitude [519]. However, the resulting detection limits were not low enough to allow the analysis of seawater after a 100-fold dilution. This dilution is required to prevent the clogging problems which would otherwise result from continuous nebulization of undiluted seawater.

Nonetheless, operating conditions (i.e. outer, auxiliary and aerosol carrier gas flow rates, amount of N_2, power and sampling depth) can be found which provide a compromise in terms of sensitivity, precision, oxide levels, and background levels (at *m/z* 52, 60 and 78 in HNO_3 matrices as well as that at *m/z* 51 and 75, in HCl matrices) [84]. Although a simplex optimization of all these variables may take up to 2 days, the resulting conditions, which are suited to multielemental analysis, not only substantially reduced spectroscopic interferences compared to what could be achieved with an all-argon plasma even if optimized using the simplex approach, but also significantly decreased non-spectroscopic interferences [84].

In any case, the optimum aerosol carrier gas flow rate also decreases as the percentage of N_2 increases [511]. Nonetheless, N_2–Ar mixed-gas plasmas allow determinations which would otherwise require a matrix separation from the sample solutions. This was achieved with, at the most, 8% N_2 in the aerosol carrier gas flow [250]. At this level of N_2, a reduction of oxide levels was also obtained [250]. On the other hand, addition of 20% N_2 in the aerosol carrier gas flow was reported to destabilize the plasma and slightly suppress the analyte signal, except for Te (an element with a high IP) which was slightly enhanced [532]. However, these conditions essentially eliminated the formation of oxides such as LaO^+ [532].

Contrary to N_2, an addition of 5% H_2 to the aerosol carrier gas flow enhanced argon-containing polyatomic ions ArO^+, ArN^+ and $ArCl^+$ with nebulization [518]. A drastic increase in the level of several background ions (ArH^+, Ar^+, OH^+, O^+ and N^+) as well as that of selected analytes (Tl, Pb, Ag, Bi and Te) was also reported upon introduction of 2–10% H_2 in the aerosol carrier gas of a dry plasma, using electrothermal vaporization as a means of sample introduction (see Chapter 3) [538]. The enhancement was correlated to the IP of the analyte, i.e. elements with higher IP were enhanced the most [538]. On the other hand, an addition of H_2, like N_2, suppressed metal oxides down to 0.5%, i.e. an order of magnitude smaller than the levels obtained with an all-Ar plasma under typical operating conditions [518]. Interestingly, addition of 2% H_2 in the aerosol carrier gas flow rate degraded sensitivity and increased oxide ratio under otherwise conventional conditions (i.e. when the solvent load was not reduced), but beneficial effects (i.e. enhanced sensitivity and further reduction in oxide ratio) were observed in combination with cryogenic desolvation [526].

In contrast, addition of N_2, H_2 or CH_4 to the aerosol carrier gas flow of an Ar ICP with nebulization, and optimization of both the plasma and MS operating conditions do not result in any significant improvement of either SBR or detection limits in the absence of interferences [124]. However, in the case of a dry plasma, where mercury vapour was introduced in the plasma, addition of 0.5% N_2 to the aerosol carrier gas flow increased Hg sensitivity four-fold whereas 1% H_2 increased it six-fold [551]. The enhancement in sensitivity which occurred with 0.5% N_2 was, in fact, similar to that which could be obtained by humidifying the aerosol carrier gas by bubbling it through 0.1 g/l $KMnO_4$ solution [551]. Similarly, the enhancement caused by H_2 is consistent with the increase in ionization temperature and n_e which were measured upon an addition of H_2 to the aerosol carrier flow rate [181]. This increase suggests an improved energy transfer from the plasma into the central channel, and can be used to facilitate the evaporation of slurry particles [503].

Hutton [552] was the first to introduce 5–7% O_2 into the aerosol carrier gas flow to facilitate the analysis of organic-based samples by ICP-MS. A water-cooled spray chamber and a reduced uptake rate were also used to reduce the solvent load which would normally be greater than that with aqueous solutions since organic solvents are often more volatile than water. This approach prevented the deposition of soot from the burnt solvent onto the sampler (as well as the injector and skimmer) through the stoichiometric conversion of carbon into CO_2, and thereby eliminated the orifice blockage which would otherwise occur [552,553]. Otherwise, the greater the ratio of carbon to oxygen in a solvent molecule, the more important would be carbon deposition and C_2 emission [175]. This approach also resulted in detection limits which were similar to those observed for an aqueous matrix for a range of elements [552,553].

However, the increased carbon loading of the plasma and the addition of O_2 increases the formation of carbon-containing and oxygen-containing polyatomic species which degrade the detection limit of several elements [552,553]. Examples of these include $^{12}C_2^+$ on $^{24}Mg^+$ [552,553], $^{40}Ar^{12}C^+$ on $^{52}Cr^+$ [552,553], $^{12}C^{16}O_2^+$ on $^{44}Ca^+$ [552] and $^{40}Ar^{12}C_2^+$ on $^{64}Zn^+$ [553]. Therefore, the amount of O_2 has to be carefully controlled since too little will result in orifice blockage whereas too much may widen the orifices of the cones and unduly reduce their lifetime [527,553]. For this reason, an all-O_2 ICP would not be advisable as an ion source for MS.

The percentage of O_2 may be optimized while monitoring an oxide-forming element and its oxide as well as the sampler material element and corresponding oxide (for instance La, LaO, Pt and PtO could be monitored with a Pt interface), such as to minimize Pt, PtO (which indicates erosion of the Pt cone) and LaO (which may indicate a reduction of analyte sensitivity) while observing no carbon deposition on the sampler (which would be seen as a red glow on the sampler orifice) [553] or green C_2 emission [175].

The sample introduction of organic-based samples deserves some attention. Although a peristaltic pump minimizes physical interferences as discussed in Section 1.2.2.1, the flexible tubing used must be resistant to the sample matrix. In some cases, for instance when the solvent has a greater viscosity than water and a pneumatic nebulizer is used, allowing the nebulizer to also act as a pump and freely aspirate the sample naturally results in a lower uptake rate than that of water. This avoids the use of a peristaltic pump which eliminates any elastomer deterioration and will also greatly reduce memory effects since the contact surface area in the path of the sample is thereby drastically decreased [554]. Even if cryogenic desolvation is used, which is most efficient at reducing the solvent load in the plasma, it also results in a long sample path and important memory effects may occur, particularly if the analyte is in the form of neutral or volatile complexes [175].

The use of a 40-MHz, free-running rf generator has been reported to allow the change-over from aqueous solutions to organic solvents without any need for manual impedance network readjustment [527]. In addition, no change to the ion lenses settings are required when switching from aqueous to organic solutions with the patented Plasma-Lok interface which efficiently decouples the ICP from the mass spectrometer [527].

Since O_2 can react with carbon to prevent its deposition, the opposite may also be possible i.e. use a carbon-containing gas to reduce the formation of oxides. This approach was successfully used in both ICP-AAS (atomic absorption spectrometry) and ICP-AFS (atomic fluorescence spectrometry) [555 and references therein]. In ICP-MS, on the other hand, the addition of 2–6% CH_4 (or, alternatively, a glycerol or glucose matrix) only enhanced the signals of elements with an IP between 9 and 11 eV (with the exception of I) [269]. In fact, no significant change in signal could be observed for La, Eu and Ho, all elements forming refractory oxides, and therefore no correlation could be made with the chemical properties of the elements [269]. However, these results were obtained for a fixed set of plasma operating conditions, with continuous nebulization. In contrast, when 0.1% acetylene was added to the aerosol carrier gas flow of a relatively dry plasma (because cryogenic desolvation was used), a further reduction of the UO^+/U^+ ratio was achieved over cryogenic desolvation alone [174].

Using simplex optimization, a set of plasma conditions was found to eliminate $ArCl^+$, whereas other sets had to be used to reduce ArO^+, ClO^+ or CeO^+ [516]. In all cases, the elimination or reduction of interference was achieved without significantly affecting the analyte signal compared to what it would be in an Ar plasma, by reducing the applied power and increasing the aerosol carrier gas flow rate (except for the reduction of CeO^+ where a decrease was required) [516]. However, some conditions seemed very specific to a given interference; for instance, the plasma conditions reducing CeO^+ did not result in a significant reduction of BaO^+ and LaO^+ compared to what could be achieved with an all Ar ICP [516]. Nonetheless, adding CH_4 to the aerosol carrier gas flow was more efficient than an addition of N_2 to any of the plasma gas flows at improving the detection limits of As, Se and V [516]. On the other hand, an addition of CH_4 will increase the level of carbon-containing polyatomic ions (such as ArC^+, CO_2^+ and CO_2H^+) thereby degrading the detection limits for some analytes (such as Cr and Ti) [511]. Furthermore, it will also lead to carbon deposition on the sampler and can therefore not be used for long periods of operation [511].

In contrast, the addition of 2 ml/min CHF_3 to the aerosol carrier gas flow increased the sensitivity of ^{78}Se by an order of magnitude while significantly decreasing the corresponding background level, thereby resulting in a 30-fold improvement in detection limit compared to that achievable in an Ar plasma [556]. However, since no improvement was observed for other elements with a high IP, including Hg, no correlation could be found between signal enhancement and IP of the analyte. A slight improvement was obtained for As and, in addition, its detection limit was then unaffected by up to 0.04 M NaCl and 0.01 M $Ca(NO_3)_2$ [556]. Similarly, the $^{40}Ar^{23}Na^+$ interference on $^{63}Cu^+$ originating from 0.02 M Na_2SO_4 could be eliminated [556]. Other interferences (such as $^{31}P^{16}O_2^+$ on $^{63}Cu^+$, or SO_2^+ on Zn^+) were unaffected by the addition of CHF_3 [556].

However, as with the addition of CH_4, an increase in the level of carbon-containing polyatomic species (such as ArC^+) also occurred, as well as that of fluorine-containing species (such as F_2O^+ and $^{40}Ar^{19}F^+$), thereby degrading the detectability of several elements (such as ^{48}Ti, ^{51}V, ^{52}Cr, ^{54}Fe and Co). Elements from CHF_4 can also react with elements from the matrix to increase the background and deteriorate the detection limit of certain analytes in some matrices. An example is the formation of $COCl^+$ from an HCl matrix which completely precludes the determination of Cu (using either isotope) [556].

The most efficient removal of interferences from polyatomic species was achieved with the addition of 0.16–0.41% ethene to the aerosol carrier gas flow [517]. Conditions were found by simplex optimization to reduce specific interferences (ArO^+, $ArCl^+$, Ar_2^+, ClO^+, CeO^+, $ArNa^+$ and SO_2^+ and/or S_2^+) [517]. These single-element optimizations resulted in the elimination of the $ArCl^+$ and ClO^+ interferences from up to 33,000 and 1000 mg/l Cl respectively, as well as a reduction of the detection limit for $^{78}Se^+$ by up to two orders of magnitude compared to that achievable with an Ar plasma [517]. As well, a set of compromise conditions could be established which still significantly decreased the levels of all these interferences (except CeO^+) compared to those observed in an Ar plasma, thereby improving the multielement analysis of a variety of matrices [517].

The addition of rare gases to the aerosol carrier Ar flow can also be used to reduce the background levels of argon- and chlorine-containing polyatomic species. An addition of Xe resulted in a suppression of both the background and analyte count rates, probably because the heavier Xe (compared to Ar) induced a non-spectroscopic interference [557]. The degradation of analyte sensitivity was especially important for elements with high IP (>8 eV), possibly because the lower IP of Xe compared to that of Ar moderates the ionization conditions in the plasma [557]. Nonetheless, an even greater reduction of the signals from various polyatomic ions such as N_2^+, H_2^+, NO^+, ArH^+, ClO^+, ArC^+, $ClOH^+$, ArN^+ and ArO^+, was also achieved thereby facilitating the determination of Si, K, V, Cr and Fe, as well as isotopic ratios: $^{28}Si/^{29}Si$, $^{39}K/^{41}K$, $^{54}Cr/^{52}Cr$, $^{53}Cr/^{52}Cr$, $^{56}Fe/^{54}Fe$ [557]. Furthermore, except for the total or partial isobaric overlap of Xe^+ peaks with some of Te, Cs and Ba, no Xe-containing polyatomic ions were observed [557]. In contrast, the addition of up to 10% He to the aerosol carrier gas flow of a dry plasma, where sample introduction was carried out by electrothermal vaporization, had little effect on the signals of Ar^+, ArH^+, OH^+, N^+ and O^+ [538]. On the other hand, with a wet plasma, addition of up to 15% He to the aerosol carrier gas improved the evaporation of slurry particles, but to a lesser extent than a similar percentage of H_2 [503].

As mentioned earlier, the addition of a sheathing gas to the aerosol carrier gas was originally carried out to prevent diffusion of the aerosol carrier gas toward the inner wall of the injector, thereby reducing the deposition of salts as well as memory effects in the injector [540 and references therein]. When a gas other than Ar is used as sheathing gas, a change in the charge-transfer process occurs between the annular plasma and the central channel. A significant improvement of the process was observed with either N_2 or H_2 as the sheathing gas, He and O_2 inducing little change [540].

Therefore, the first study to look at the effect of adding a sheathing gas around the aerosol coming out of the spray chamber in ICP-MS, compared N_2 and H_2 to Ar as a sheathing gas [542]. As expected from the reduced nebulization efficiency resulting from the lower aerosol carrier flow rate which had to be used when any sheathing gas was added, a decrease in sensitivity and degradation of detection limits was generally observed compared to those obtained with a conventional plasma i.e. without a sheathing gas [542]. The detection limits of only a few elements were similar or slightly improved compared to those with a conventional plasma, as a result of a decrease in the corresponding background count rates and/or improvement in the stability of the background signal [542]. Since the background RSD was not improved for the majority of elements, the plasma stability did not increase significantly. (More plasma stability might have been expected from the smaller amount of aerosol reaching the plasma, which would thereby decrease fluctuations due to the vaporization of large droplets (see Section 1.4.1).)

However, a significant decrease in mass discrimination resulted from the addition of a sheathing gas of either N_2 or H_2 [542]. This decrease is advantageous for the determination of isotopic ratios and isotope dilution analysis, as it may eliminate differences in response between various isotopes of a given element and thereby simplify the procedure by avoiding the measurement of standards of known isotopic composition to find correction factors for any such differences. The decrease in mass discrimination suggests that, by constricting the aerosol inside the injector, a sheathing gas may have a focusing effect on the ion beam reaching the MS.

The addition of a small amount of N_2 decreased the $^{40}Ar^{16}OH^+$ background much more drastically than that of $^{40}Ar^{16}O^+$; but it also significantly increased the background due to nitrogen-containing polyatomic ions [542]. In contrast to the addition of N_2 to the aerosol carrier gas flow, which quenched Ar_2^+ thereby facilitating the determination of Se, no such improvement resulted from the use of a N_2 sheathing gas. This supports the assertion that the mixing of the sheathing gas with the aerosol carrier gas is not as thorough as when the foreign gas is added directly to aerosol carrier gas flow, before its introduction into the nebulizer. A small amount of H_2 as a sheathing gas did not degrade sensitivities and detection limits as much as N_2 because, as discussed earlier, it does not increase interferences from polyatomic ions [542].

1.6.3.3 Addition to other or all gas flows

In general, an addition of a foreign gas to only the auxiliary gas flow has not been as fruitful as addition to the other gas flows. For instance, with 3% N_2 (the maximum that could be withstood by the plasma), the resulting plasma was of little practical value since, even with a simplex optimization, no conditions could be found which reduced interferences from polyatomic ions without also substantially suppressing the analyte signal [511,515].

Addition of up to 200 ml/min N_2 to all plasma gases drastically reduced ArO^+ such that the detection limit for $^{56}Fe^+$ improved by a factor of six [58]. The CeO^+/Ce^+ oxide level was concomitantly reduced by over a factor of eight while the level of doubly-charged ions (Ba^{2+}) only increased two-fold [58]. Similarly, an addition of 30 ml/min (or 0.2%) N_2 to the main Ar inlet maximized Rh^+, Ce^+, Ba^+, Mg^+, As^+ and Pb^+ while keeping CeO^+/Ce^+, Ce_2^+/Ce^+ and Ba_2^+/Ba^+ below 3% [558]. The plasma behaved similarly to that resulting from an addition of N_2 to the outer gas flow since a higher aerosol carrier flow rate was required to maximize sensitivity [558]. However, a concurrent increase in the level of nitrogen-containing polyatomic ions degraded the detection limit for ^{31}P, ^{51}V, ^{52}Cr, ^{54}Fe and ^{55}Mn [558]. On the other hand, only 15 ml/min H_2 was required to keep the above-mentioned ratios below 3% without sacrificing sensitivity and without any significant increase in background levels compared to an Ar plasma [558]. Although either N_2 or H_2 improved the signal-to-background ratio (S/B) for a wide range of elements, H_2 is therefore more attractive since it does not enhance background interferences.

Replacing both the outer and the aerosol carrier gas flows with air gave a plasma with a background mass spectrum similar to that of a N_2 ICP, i.e. where the signals of Ar-containing ions are greatly reduced whereas those of N^+, O^+, NO^+, N_2^+ and O_2^+ are more important [514]. Although the relative distribution of ions in the air–Ar plasma was similar to that in an Ar ICP, maximum signals were obtained at higher aerosol carrier gas flow rate [514]. Furthermore, sensitivity was dependent on the first IP, being higher than in Ar ICP for elements with a low IP but degraded compared to Ar ICP for elements with high IP, especially Pt, Au and Hg [514]. The ratio of doubly to singly-charged ions was also observed to decrease compared to that in an Ar ICP for elements with a relatively low second IP (such as Ba and Y), whereas it was enhanced for elements with a high second IP (such as Pb) [514]. In addition, the ratio of monoxide ion to singly charged ion was constant along the plasma axis and with increasing rf power. In comparison to those observed in Ar ICP, the oxide ratio in air–Ar ICP was either similar, for elements with high oxide dissociation energy, or higher for elements, such as Pb, with a relatively low oxide bond strength [514].

In contrast, He could only be added to both the outer and the auxiliary gas flows (at least with the modified Greenfield torch and modified load coil used) [522]. Attempts made at also adding He to the aerosol carrier gas flow, only resulted in a loss of sensitivity for all analytes studied [522]. Nonetheless, with an Ar aerosol carrier flow, the degree of ionization of elements with a high IP improved [522]. The best signal-to-noise ratio (S/N) was obtained at a power of 1.65 kW (higher powers seriously degrading the sampler within seconds) with 20% He in the auxiliary gas flow and either 20% He for metals or 30% He in the outer gas flow for halides [522]. In all cases, the resulting detection limits were lower than those obtained with an Ar ICP, the most dramatic improvements being observed for elements with high IP [522]. In contrast to Ar–N_2 plasmas (with N_2 added to the outer gas), no reduction of effects from concomitant elements was observed [559]. However, less clogging of the sampler by up to 1% dissolved solids was observed with the Ar–He ICP than an Ar one [559]. With an all-He ICP, lower detection limits were reported for non-metals than for metals [525].

1.6.4 Applications

The determination of As and Se in urine was successfully carried out by the method of standard additions and internal standardization (using In) in He ICP-MS with USND [524]. However, the precision of these results was relatively large [524] and certainly does not compete with what can be achieved with mixed-gas Ar ICP-MS. Further development of He ICP-MS is therefore required before it can compete with either Ar or mixed-gas Ar ICP-MS.

Table 1.11 summarizes applications of mixed-gas plasmas as ion sources for MS. As explained at the beginning of Section 1.6.3.2, a larger number of applications involved the use of mixed-gas plasmas produced by adding a foreign gas to the aerosol carrier gas flow. All the applications of Table 1.11 would be impossible with a regular Ar ICP, unless additional sample pretreatment was carried out to eliminate the interfering matrix. However, this additional sample processing would not only reduce the sample throughput but also increase sources of contamination. Mixed-gas plasmas therefore extend the range of applications of ICP-MS by: (1) eliminating or dramatically reducing spectroscopic interferences from polyatomic ions, especially argon-, and chlorine- containing ones as well as oxides; (2) eliminating or dramatically reducing non-spectroscopic interferences; (3) allowing the analysis of organic-based samples; and (4) facilitating the analysis of solids.

Although mixed-gas plasmas produced by adding a foreign gas to the outer gas flow have not yet been as widely applied in ICP-MS, it is only a question of time since their usefulness is regularly being demon-

TABLE 1.11

Selected applications of mixed-gas ICPs in ICP-MS

Analytes	Sample	Sample preparation	Foreign gas	Comments	Instrument; sample introduction	Ref.
13 elements	fuel oil SRM	sample diluted 10-fold with white spirit	5-7% O_2 to aerosol carrier gas flow via T-piece on gas line	external calibration with Conostan standards diluted with white spirit; results in good agreement with certified values	PQ; spray chamber cooled to 0°C	552
V, Ni, Co, As	naphtha	emulsification with 2.5% solution of Triton X-100; stirring until introduction into ICP	8% O_2 to aerosol carrier gas	external calibration with matrix-matched organometallic standards; internal standardization using Rh; no need for organic diluent (such as xylene); recoveries ≥86%	PQ 3 with Pt sampler and Ni skimmer; spray chamber cooled to -5°C; 0.5 ml/min sample uptake	560
22 elements	soya oil	sample diluted 5-fold with xylene	O_2 to aerosol carrier gas flow via T-piece at the base of the torch	external calibration with Conostan organic solvent standards spiked with xylene; 80-120% recoveries for 19 elements	ELAN 5000; spray chamber cooled with water/glycol mixture; low sample uptake rate Meinhard nebulizer	527
Cu, Cd, Sb	river water and marine sediment SRMs	water analyzed directly; sediments digested	8% N_2 to outer plasma gas via gas proportioner	external calibration for river water; isotope dilution analysis for sediments; results in good agreement with certified values	ELAN 500; heating chamber (at 40°C) and cold-finger condenser inserted between spray chamber and torch	544
V, Ni, As	river and estuarine waters, seawater and urine SRMs	sample (except river water) diluted 10-100 fold; acidified to 1% HNO_3	H_2 to aerosol carrier gas flow via T-piece at the base of the torch (2% H_2 in gas flow)	5-point method of standard additions; H_2 enhanced analyte signal 2-3 fold and decreased oxide ratio with cryogenic desolvation; excess HNO_3 required to remove Cl from NaCl as volatile HCl during desolvation process	ELAN 250; USN with cryogenic desolvation; flow injection	526

Analytes	Sample	Sample preparation	Foreign gas	Comments	Instrument; sample introduction	Ref.
Fe, Cu, Mo	seawater SRM	400-900 ml acidified sample adjusted to pH 3 with ammonia and passed through a column of silica-immobilized 8-hydroxyquinoline; column washed with DDW; elution with 10 ml 1 M HCl/0.1 M HNO_3	8% N_2 to outer plasma gas via gas proportioner	isotope dilution analysis; preconcentration procedure required to separate the salts from the analytes and increase the concentration of the latter; correction for procedural blanks necessary; results in good agreement with certified values	ELAN 500; partial desolvation (as in Ref. 544)	316
V, As, Se	urine SRM	sample diluted with DDW and spiked with In	3% N_2 to aerosol carrier gas flow via T-piece on the gas line	external calibration with internal standardization using In; results in good agreement with certified values using ^{51}V, ^{75}As and ^{82}Se	PQ; spray chamber cooled to 6°C; flow injection	511
34 elements	biological SRMs	0.5-g sample digested overnight with 5 ml HNO_3 at 150°C, cooled and diluted to 10 ml with DDW	8% N_2 to aerosol carrier gas flow via T-piece on gas line	external calibration with internal standardization using In and Bi; results in good agreement with certified values for most elements including ^{51}V, ^{53}Cr, ^{67}Zn, ^{75}As and ^{77}Se	PQ1; spray chamber cooled to 10°C	250
As	standard with 760 mg/L NaCl		2 ml/min CHF_3 to aerosol carrier gas flow via T-piece at the base of the torch	result obtained by external calibration in agreement with expected concentration	PQ2; spray chamber cooled to 0–10°C	556
As, Se, Fe, V	seawater and biological SRMs	seawater and urine diluted 2- and 50-fold; other samples digested with HNO_3 or HNO_3/H_2O_2 in microwave oven	1% (0.45% for V) CH_4 to aerosol carrier gas flow via gas proportioner	external calibration with internal standardization using Sb, Co or In; optimal plasma conditions different for each analyte; good agreement except for As in seawater (too high) and Fe (too low)	PQ2; spray chamber cooled to 1°C	518

continued

TABLE 1.11 (*continuation*)

Analytes	Sample	Sample preparation	Foreign gas	Comments	Instrument; sample introduction	Ref.
As	biological SRMs	urine diluted 10-fold; other samples digested with HNO_3 in microwave oven	30 ml/min N_2 to aerosol carrier gas flow via T-piece on gas line	external calibration with internal standardization using In; good agreement with certified values	PQ2; V-groove high-solid nebulizer	549
Fe, As, Se, V	water and biological SRMs	seawater and freshwater diluted; other samples digested with HNO_3/H_2O_2 in microwave oven	0.16-0.41% ethene to aerosol carrier gas flow via gas proportioner	external calibration with internal standardization using Ho, Co or In; optimal plasma conditions different for each analyte; good agreement for V and As in lobster pancreas; low or high values in other cases	PQ2; V-groove high-solid nebulizer; spray chamber cooled to 1°C	517
REEs	marine particulate samples; basalt rock SRM	digestion with HNO_3/HF; dilution	0.014 l/min N_2 in aerosol carrier gas	ID analysis using ^{145}Nd and ^{171}Yb; the same two isotopes are also used as internal standards for the determination of the other REEs; oxide interferences <1% of analyte signal	ELEMENT; MCN-6000 i.e. MCN with heated spray chamber (75°C) and membrane desolvation	207
Pr, Nd, Sm, Gd, Dy	Bastnasite mill feed ore SRM	slurries prepared by shaking sample with zirconia beads and dispersing agent in a plastic bottle	5.75% H_2 to aerosol carrier gas flow via gas proportioner	external calibration with aqueous standard solutions; good agreement with certified values, including for ^{152}Sm despite BaO interference	PQ2; V-groove high-solid nebulizer; spray chamber cooled to 1°C; slurry nebulization	519
Nb, Mo, Ta, W	silicate rock SRMs	sample digested with HNO_3 and HF, acidified with HCl; pF adjusted to 3-3.5; analytes extracted into $CHCl_3$/BPHA; residue (after solvent evaporation) dissolved in 80°C n-pentanol	55 ml/min O_2 to aerosol carrier gas flow via T-piece on gas line	external calibration with internal standardization using Mo (for Nb) and W (for Ta); isotope dilution analysis for Mo and W; fair or good agreement with certified values; extraction required for successful determination of Mo	PQ; spray chamber cooled to 10°C	554

strated in ICP-AES. For instance, air–Ar ICPs were reported to facilitate the analysis of organic solutions by eliminating carbon deposition on the injector and improving the detection limits compared to an Ar plasma [561]. Similarly, the ICP-AES analysis of solid samples, which were introduced either in the form of slurries [562] or using a direct sample insertion device (see Chapter 5) [563], was facilitated with Ar–N_2 and Ar–O_2 mixed-gas plasmas. The maximum signal-to-background ratio (S/B) was observed lower in mixed-gas plasmas than in an Ar one, thereby demonstrating their higher efficiency at vaporizing and atomizing a sample [562]. In fact, a 20% O_2 plasma was more capable of vaporizing the sample into its elemental components than a 20% N_2 plasma [563]. The fact that they reduce or eliminate non-spectroscopic interferences without sacrificing sensitivity, should also increase the interest in these plasmas.

Although their operation is empirical and depends on the ICP-MS instrument used, all types of mixed-gas plasmas can enhance the features of various sample introduction approaches. Specific examples are given in the following chapters, which deal with sample introduction systems that generate transient signals. Note that chromatographic and electrophoretic techniques (i.e. separation techniques), which also result in transient signals, will be the focus of a forthcoming book, as there is currently an exponential growth of publications in this area. Whenever a transient signal is acquired, the speed of data acquisition is of paramount importance since it determines how many elements can be monitored simultaneously as well as how many data points can be measured during the transient peak. The processing of these signals is obviously different from that of a steady-state signal where integration can be easily carried out. In the case of a transient signal, either the peak height or the peak area can be used for quantitation, depending on the sample introduction technique and/or methods used. In any case, some data smoothing will often be required prior to the determination of peak height or area. Several techniques are available and the most efficient should be determined on a case-by-case basis [564].

ACKNOWLEDGEMENT

D.B. is grateful to the Natural Sciences Engineering Research Council for financial assitance.

REFERENCES

1 H.M. Ortner, Ultratrace analysis—Facts and fiction, *Fresenius J. Anal. Chem.* 343 (1992) 695.
2 W. Slavin, A comparison of atomic spectroscopic analytical techniques, *Spectroscopy*, 6 (1991) 16.
3 J.W. Olesik, Elemental analysis using ICP-OES and ICP/MS. An evaluation and assessment of remaining problems, *Anal. Chem.*, 63 (1991) 12A.
4 E. Pruszkowski, R.D. Ediger and E. Denoyer, ICP-MS: an alternative method for the USEPA Contract Laboratory Program, *Am. Lab.*, 21(8) (1989) 36.
5 P.J. Potts, Geoanalysis: Past, present and future, *Analyst*, 122 (1997) 1179.
6 A.R. Date and A.L. Gray (Eds), *Applications of Inductively Coupled Plasma Mass Spectrometry*, Blackie and Son Ltd., Glasgow, 1989.
7 K.E. Jarvis, A.L. Gray and R.S. Houk, *Handbook of Inductively Coupled Plasma Mass Spectrometry*, Blackie and Son Ltd., Glasgow, 1992.
8 G. Holland and A.N. Eaton (Eds), *Applications of Plasma Source Mass Spectrometry*, The Royal Society of Chemistry, Cambridge, 1991.
9 G. Holland and A.N. Eaton (Eds), *Applications of Plasma Source Mass Spectrometry II*, The Royal Society of Chemistry, Special Publication No. 124, Cambridge, 1993.
10 G. Holland and S.D. Tanner (Eds), *Plasma Source Mass Spectrometry: Developments and Applications*, The Royal Society of Chemistry, Cambridge, 1997.
11 A. Montaser (Ed.), *Inductively Coupled Plasma Mass Spectrometry*, Wiley-VCH, New York, 1998.
12 S.J. Hill (Ed.), *Inductively Coupled Plasma Spectrometry and its Applications*, Sheffield Academic Press, Sheffield, England (CRC Press LLC, Boca Raton, Florida for North America), 1999.
13 E.H. Evans, J.J. Giglio, T.M. Castillano and J.A. Caruso, *Inductively Coupled and Microwave Induced Plasma Sources for Mass Spectrometry*, The Royal Society of Chemistry, Cambridge, 1995.
14 H.E. Taylor, *ICP-MS Practices and Techniques*, Academic Press, London, 1999.
15 C.M. Barshick, D.C. Duckworth and D.H. Smith, *Inorganic Mass Spectrometry Fundamentals and Applications*. Practical Spectroscopy Series, Vol. 23. Marcel Dekker, Inc., New York, 2000.
16 A.L. Gray, Inductively coupled plasma source mass spectrometry, In: F. Adams, R. Gijbels and R. Van Grieken (Eds), *Inorganic Mass Spectrometry*. Chemical Analysis Series, Vol. 95. John Wiley & Sons, Inc., London, 1988.
17 G. Horlick and Y. Shao, Inductively coupled plasma-mass spectrometry

for elemental analysis, In: A. Montaser and D.W. Golightly (Eds), *Inductively Coupled Plasmas in Analytical Atomic Spectrometry*, 2nd edn., VCH Publishers Inc., New York, 1992.

18 D. Douglas, Fundamental aspects of inductively coupled plasma-mass spectrometry, In: A. Montaser and D.W. Golightly (Eds), *Inductively Coupled Plasmas in Analytical Atomic Spectrometry*, 2nd edn., VCH Publishers Inc., New York, 1992.

19 H.E. Taylor and J.R. Garbarino, Analytical applications of inductively coupled plasma-mass spectrometry, In: A. Montaser and D.W. Golightly (Eds), *Inductively Coupled Plasmas in Analytical Atomic Spectrometry*, 2nd ed., VCH Publishers Inc., New York, 1992.

20 G.E.M. Hall, Inductively coupled plasma mass spectrometry, In: M. Thompson and J.N. Walsh (Eds), *Handbook of Inductively Coupled Plasma Spectrometry*, 2nd edn., Blackie and Son Ltd, Glasgow, 1989.

21 A.L. Gray and A.R. Date, Plasma source mass spectrometry of inorganic samples—recent developments of the technique, *Int. J. Mass Spectrom. Ion Phys.*, 46 (1983) 7.

22 D.J. Douglas and R.S. Houk, Inductively coupled plasma-mass spectrometry, *Prog. Anal. At. Spectrosc.*, 8 (1985) 1.

23 A.L. Gray, The ICP as an ion source—origins, achievements and prospects, *Spectrochim. Acta*, 40B (1985) 1525.

24 R.S. Houk, Mass spectrometry of inductively coupled plasmas, *Anal. Chem.*, 58 (1986) 97A.

25 A.L. Gray, Ions or photons—an assessment of the relationship between emission and mass spectrometry with the ICP, *Fresenius Z. Anal. Chem.*, 324 (1986) 561.

26 H.E. Taylor, Inductively coupled plasma–mass spectrometry: An introduction, *Spectroscopy*, 1 (1986) 20.

27 A.L. Gray, The evolution of the ICP as an ion source for mass spectrometry, *J. Anal. At. Spectrom.*, 1 (1986) 20.

28 J. Douglas, Some current perspectives on ICP-MS, *Can. J. Spectrosc.*, 34 (1989) 38.

29 G.M. Hieftje and G. H. Vickers, Developments in plasma source/mass spectrometry, *Anal. Chim. Acta*, 216 (1989) 1.

30 R.S. Houk, Elemental and isotopic analysis by inductively coupled plasma mass spectrometry, *Acc. Chem. Res.*, 27 (1994) 333.

31 R.S. Houk, S.C.K. Shum and D.R. Wiederin, Frontiers in elemental analysis by mass spectrometry, *Anal. Chim. Acta*, 250 (1991) 61.

32 D. Beauchemin, Inductively coupled plasma mass spectrometry in hyphenation: a multielemental analysis technique with almost unlimited potential, *Trends Anal. Chem.*, 10 (1991) 71.

33 G.M. Hieftje, Towards the next generation of plasma source mass spectrometers, *J. Anal. At. Spectrom.*, 7 (1992) 783.

34 I.B. Brenner and H.E. Taylor, A critical review of inductively coupled plasma mass spectrometry for geoanalysis, geochemistry, and hydrology. Part I. Analytical performance, *CRC Crit. Rev. Anal. Chem.*, 23 (1992) 355.
35 G.M. Hieftje, P.J. Galley, M. Glick and D.S. Hanselman, New developments and final frontiers in inductively coupled plasma spectrometry, *J. Anal. At. Spectrom.*, 7 (1992) 69.
36 H. Falk, Methods for the detection of single atoms using optical and mass spectrometry, *J. Anal. At. Spectrom.*, 7 (1992) 255.
37 M.A. Marabini, B. Passariello and M. Barbaro, Inductively coupled plasma–mass spectrometry: capabilities and applications, *Microchem. J.*, 46 (1992) 302.
38 G. Horlick, Plasma-source mass spectrometry for elemental analysis, *Spectroscopy*, 7 (1992) 22.
39 D. Beauchemin, A comparison of ICP atomic spectrometry techniques, *Spectroscopy*, 7 (1992) 12.
40 D.W. Koppenaal, Atomic mass spectrometry, *Anal. Chem.*, 64 (1992) 320R.
41 N. Jakubowski and D. Stuewer, Recent progress in plasma mass spectrometry, In: Special Publication No. 154, Royal Society of Chemistry, 1994, pp. 121–144.
42 D. Colodner, V. Salters and D.C. Duckworth, Ion sources for analysis of inorganic solids and liquids by MS, *Anal. Chem.*, 66 (1994) 1079A.
43 J.S. Becker and H.-J. Dietze, Inorganic trace analysis by mass spectrometry, *Spectrochim. Acta*, 53B (1998) 1475.
44 J.R. Bacon, A.T. Ellis and J.G. Williams, Atomic spectrometry update-inorganic mass spectrometry and x-ray fluorescence spectrometry, *J. Anal. At. Spectrom.*, 6 (1991) 229R.
45 J.R. Bacon, A.T. Ellis, A.W. McMahon, P.J. Potts and J.G. Williams, Atomic spectrometry update-atomic mass spectrometry and x-ray fluorescence spectrometry, *J. Anal. At. Spectrom.*, 7 (1992) 279R.
46 J.R. Bacon, A.T. Ellis, A.W. McMahon, P.J. Potts and J.G. Williams, Atomic spectrometry update-atomic mass spectrometry and x-ray fluorescence spectrometry, *J. Anal. At. Spectrom.*, 8 (1993) 261R.
47 J.R. Bacon, A.T. Ellis, A.W. McMahon, P.J. Potts and J.G. Williams, Atomic spectrometry update-atomic mass spectrometry and x-ray fluorescence spectrometry, *J. Anal. At. Spectrom.*, 9 (1994) 267R.
48 J.R. Bacon, J.S. Crain, A.W. McMahon and J.G. Williams, Atomic spectrometry update-atomic mass spectrometry, *J. Anal. At. Spectrom.*, 12 (1997) 407R.
49 J.R. Dean, O. Butler, A. Fisher, L.M. Garden, M.S. Cresser, P. Watkins and M. Cave, Atomic spectrometry update—environmental analysis, *J. Anal. At. Spectrom.*, 13 (1998) 1R.
50 M.R. Cave, O. Butler, J.M. Cook, M.S. Cresser, L.M. Garden, A.J. Holden and D.L. Miles, Atomic spectrometry update—environmental analysis, *J.*

Anal. At. Spectrom., 14 (1999) 279.
51 P. Tschöpel, In: G. Svehla (Ed.), *Wilson and Wilson's Comprehensive Analytical Chemistry*, Vol. IX, Elsevier, Amsterdam, 1979, p. 179.
52 I. Jarvis and K.E. Jarvis, Plasma spectrometry in the earth sciences: techniques, applications and future trends, *Chem. Geol.*, 95 (1992) 1.
53 S. Greenfield and A. Montaser, Common RF Generators, Torches, and Sample Introduction Systems, In: A. Montaser and D.W. Golightly (Eds), *Inductively Coupled Plasmas in Analytical Atomic Spectrometry*, 2nd edn., VCH Publishers Inc., New York, 1992.
54 L. de Galan and M.T.C. de Loos-Vollebregt, Low-Gas-Flow Torches for ICP Spectrometry, In: A. Montaser and D.W. Golightly (Eds), *Inductively Coupled Plasmas in Analytical Atomic Spectrometry*, 2nd Edition, VCH Publishers Inc., New York, 1992.
55 N. Jakubowski, I. Feldmann and D. Stuewer, Diagnostic investigations of aerosols with varying water content in inductively coupled plasma mass spectrometry, *J. Anal. At. Spectrom.*, 8 (1993) 969.
56 D.C. Schram, J.A.M. van der Mullen, J.M. de Regt, D.A. Benoy, F.H.A.G. Fey, F. de Grootte and J. Jonkers, Fundamental description of spectrochemical inductively coupled plasmas, *J. Anal. At. Spectrom.*, 11 (1996) 623.
57 G.M. Hieftje, R.M. Miller, Y. Pak and E.P. Wittig, Theoretical examination of solute particle vaporization in analytical atomic spectrometry, *Anal. Chem.*, 59 (1987) 2861.
58 M. Murillo and J.-M. Mermet, Improvement of the energy transfer with added-hydrogen in inductively coupled plasma atomic emission spectrometry, *Spectrochim. Acta*, 44B (1989) 359.
59 Y.Q. Tang and C. Trassy, Inductively coupled plasma: the role of water on axial excitation temperatures, *Spectrochim. Acta*, 41B (1986) 143.
60 S.R. Koirtyohann, J.S. Jones and D.A. Yates, Nomenclature system for the low-power argon inductively coupled plasma, *Anal. Chem.*, 52 (1980) 1965.
61 M.W. Blades and G. Horlick, Interference from easily ionizable element matrices in inductively coupled plasma emission spectrometry—a spatial study, *Spectrochim. Acta*, 36B (1981) 881.
62 E. Poussel, J.-M. Mermet and D. Deruaz, Dissociation of analyte oxide ions in inductively coupled plasma mass spectrometry, *J. Anal. At. Spectrom.*, 9 (1994) 61.
63 G.H. Vickers, D.A. Wilson and G.M. Hieftje, Spatial dependence of ion concentrations in inductively coupled plasma mass spectrometry, *Spectrochim. Acta*, 45B (1990) 499.
64 M.A. Vaughan and G. Horlick, Oxide, hydroxide, and doubly charged analyte species in inductively coupled plasma/mass spectrometry. *Appl. Spectrosc.*, 40 (1986) 434.
65 R.S. Houk, R.K. Winge and X. Chen, High speed photographic study of wet

droplets and solid particles in the inductively coupled plasma, *J. Anal. At. Spectrom.*, 12 (1997) 1139.
66 M.W. Blades, B.L. Caughlin, Z.H. Walker and L.L. Burton, Excitation, ionization, and spectral line emission in the inductively coupled plasma, *Prog. Anal. Spectrosc.*, 10 (1987) 57.
67 H. Niu and R.S. Houk, Fundamental aspects of ion extraction in inductively coupled plasma mass spectrometry, *Spectrochim. Acta*, 51B (1996) 779.
68 T. Hasegawa, M. Umemoto, H. Haraguchi, C. Hsiech and A. Montaser, Fundamental Properties of Inductively Coupled Plasmas in A. Montaser and D.W. Golightly (Eds), *Inductively Coupled Plasmas in Analytical Atomic Spectrometry*, 2nd Edition, VCH Publishers Inc., New York, 1992.
69 D.J. Douglas and J.B. French, Gas dynamics of the inductively coupled plasma mass spectrometry interface, *J. Anal. At. Spectrom.*, 3 (1988) 743.
70 H. Togashi, A. Hashizume and Y. Niwa, Molecular ionization in the interface of an inductively coupled plasma mass spectrometer, *Spectrochim. Acta*, 47B (1992) 561.
71 H. Niu and R.S. Houk, Langmuir probe measurements of the ion extraction process in inductively coupled plasma mass spectrometry: I. Spatially resolved determination of electron density and electron temperature, *Spectrochim. Acta*, 49B (1994) 1283.
72 T.N. Loney, W. Chen and D.J. Douglas, Gas dynamics of the ICP-MS interface: impact pressure probe measurements of gas flow profiles, *J. Anal. At. Spectrom.*, 14 (1999) 9.
73 J.W.H. Lam and G. Horlick, Effects of sampler-skimmer separation in inductively coupled plasma-mass spectrometry, *Spectrochim. Acta*, 45B (1990) 1327.
74 J.S. Crain, R.S. Houk and D.E. Eckels, Noise power spectral characteristics of an inductively coupled plasma mass spectrometer, *Anal. Chem.*, 61 (1989) 606.
75 N. Furuta, C.A. Monnig, P. Yang and G.M. Hieftje, Noise characteristics of an inductively coupled plasma-mass spectrometer, *Spectrochim. Acta*, 44B (1989) 649.
76 M.J. Campbell, C. Demesmay and M. Ollé, Determination of total arsenic concentrations in biological matrices by inductively coupled plasma mass spectrometry, *J. Anal. At. Spectrom.*, 9 (1994) 1379.
77 K. Sakata and K. Kawabata, Reduction of fundamental polyatomic ions in inductively coupled plasma mass spectrometry, *Spectrochim. Acta*, 49B (1994) 1027.
78 F. Vanhaecke, C. Vandecasteele, H. Vanhoe and R. Dams, Study of the Intensity of M^+, M^{2+} and MO^+ Signals in ICP-MS as a Function of Instrumental Parameters, *Mikrochim. Acta*, 108 (1992) 41.
79 S.D. Lofthouse, G.M. Greenway and S.C. Stephen, Microconcentric

nebulizer for the analysis of small sample volumes by inductively coupled plasma mass spectrometry, *J. Anal. At. Spectrom.*, 12 (1997) 1373.
80 M.J. Campbell, G. Vermeir, R. Dams and P. Quevauviller, Influence of chemical species on the determination of mercury in a biological matrix (cod muscle) using inductively coupled plasma mass spectrometry, *J. Anal. At. Spectrom.*, 7 (1992) 617.
81 E.H. Larsen and S. Stürup, Carbon-enhanced inductively coupled plasma mass spectrometric detection of arsenic and selenium and its application to arsenic speciation, *J. Anal. At. Spectrom.*, 9 (1994) 1099.
82 R.K. Winge, J.S. Crain and R.S. Houk, High speed photographic study of plasma fluctuations and intact aerosol particles or droplets in inductively coupled plasma mass spectrometry, *J. Anal. At. Spectrom.*, 6 (1991) 601.
83 G. Horlick, S.H. Tan, M.A. Vaughan and C.A. Rose, The effect of plasma operating parameters on analyte signals in inductively coupled plasma-mass spectrometry, *Spectrochim. Acta*, 40B (1985) 1555.
84 T. van der Velde-Koerts and J.L.M. de Boer, Minimization of spectral interferences in inductively coupled plasma mass spectrometry by simplex optimization and nitrogen addition to the aerosol carrier gas for multi-element environmental analysis, *J. Anal. At. Spectrom.*, 9 (1994) 1093.
85 S.D. Tanner, Characterization of ionization and matrix suppression in inductively coupled 'cold' plasma mass spectrometry, *J. Anal. At. Spectrom.*, 10 (1995) 905.
86 J.W. Olesik, Fundamental research in ICP-OES and ICP-MS, *Anal. Chem.*, 68 (1996) 469A.
87 L. Ebdon, E.H. Evans and N.W. Barnett, Simplex optimisation of experimental conditions in inductively coupled plasma atomic emission spectrometry with organic solvent introduction, *J. Anal. At. Spectrom.*, 4 (1989) 505.
88 D.G. Weir and M.W. Blades, Characteristics of an inductively coupled argon plasma operating with organic aerosols Part 1. Spectral and spatial observations, *J. Anal. At. Spectrom.*, 9 (1994) 1311.
89 D.G.J. Weir and M.W. Blades, The response of the inductively coupled argon plasma to solvent plasma load: spatially resolved maps of electron density obtained from the intensity of one argon line, *Spectrochim. Acta*, 49B (1994) 1231.
90 B.E. Erickson, ICPMS—Beyond quadrupole, *Anal. Chem.*, 71 (1999) 811A.
91 D.C. Duckworth and C.M. Barshick, Ion traps: what do they hold for elemental mass analysis?, *Anal. Chem.*, 70 (1998) 709A.
92 G.C. Eiden, C.J. Barinaga and D.W. Koppenaal, Selective removal of plasma matrix ions in plasma source mass spectrometry, *J. Anal. At. Spectrom.*, 11 (1996) 317.
93 S. Luan, H. Pang, S.C.K. Shum and R.S. Houk, Noise characteristics of aerosols produced by inductively coupled plasma nebulizers, *J. Anal. At. Spectrom.*, 7 (1992) 799.

94 I.S. Begley and B.L. Sharp, Occurrence and reduction of noise in inductively coupled plasma mass spectrometry for enhanced precision in isotope ratio measurement, *J. Anal. At. Spectrom.*, 9 (1994) 171.
95 K. Hu, P.S. Clemons and R.S. Houk, Inductively coupled plasma mass spectrometry with an enlarged sampling orifice and offset ion lens. I. Ion trajectories and detector performance, *J. Am. Soc. Mass Spectrom.*, 4 (1993) 16.
96 K. Hunter, Ion detection in ICP–mass spectrometry using active film multipliers, *At. Spectrosc.*, 15 (1994) 17.
97 E.R. Denoyer, An advanced ICP-MS instrument, *Am. Lab.*, 27(4) (1995) 48D.
98 A. Held and P.D.P. Taylor, A calculation method based on isotope ratios for the determination of dead time and its uncertainty in ICP-MS and application of the method to investigating some features of a continuous dynode multiplier, *J. Anal. At. Spectrom.*, 14 (1999) 1075.
99 D.J. Douglas and J.B. French, An improved interface for inductively coupled plasma mass spectrometry, *Spectrochim. Acta*, 41B (1986) 197.
100 E.R. Denoyer, Current Trends in ICP-Mass Spectrometry, *At. Spectrosc.*, 12 (1991) 215.
101 B.S. Ross, P. Yang, D.M. Chambers and G.M. Hieftje, Comparison of center-tapped and inverted load-coil geometries for inductively coupled plasma mass spectrometry, *Spectrochim. Acta*, 46B (1991) 1667.
102 S.-H. Nam, J.-S. Lim and A. Montaser, High-efficiency nebulizer for argon inductively coupled plasma mass spectrometry, *J. Anal. At. Spectrom.*, 9 (1994) 1357.
103 A.L. Gray, R.S. Houk and J.G. Williams, Langmuir probe potential measurements in the plasma and their correlation with mass spectral characteristics in inductively coupled plasma mass spectrometry, *J. Anal. At. Spectrom.*, 2 (1987) 13.
104 E.R. Denoyer, Perkin Elmer Technical Summary No. TSMS-24.
105 I.L. Turner, Varian UltraMass Technical Bulletin No.1 (1994).
106 M. Plantz, J. Shkolnik and S. Anderson, Origin and control of continuum background correction in ICP-MS, Paper F6, 1994 Winter Conference on Plasma Spectrochemistry, San Diego, January 1994.
107 A.L. Gray, Influence of load coil geometry on oxide and double charged ion response in ICP-MS, *J. Anal. At. Spectrom.*, 1 (1986) 247.
108 N.S. Nonose, N. Matsuda, N. Fudagawa and M. Kubota, Some characteristics of polyatomic ion spectra in inductively coupled plasma mass spectrometry, *Spectrochim. Acta*, 49B (1994) 955.
109 G.H. Vickers, D.A. Wilson and G.M. Hieftje, Comparison of 27- and 40-MHz Plasmas in Inductively Coupled Plasma Mass Spectrometry, *J. Anal. At. Spectrom.*, 4 (1989) 749.
110 Y. Takaku, K. Masuda, T. Takahashi and T. Shimamura, Determination

of trace silicon in ultra-high purity water by inductively coupled plasma mass spectrometry, *J. Anal. At. Spectrom.*, 9 (1994) 1385.
111 B.S. Ross, D.M. Chambers, G.H. Vickers, P. Yang and G.M. Hieftje, Characterisation of a 9-mm torch for inductively coupled plasma mass spectrometry, *J. Anal. At. Spectrom.*, 5 (1990) 351.
112 L.A. Norman, M. Muñoz, D.P. Myers, B.S. Ross, G.M. Hieftje, Characterization of a 23-mm torch for use in inductively coupled plasma mass spectrometry, *Appl. Spectrosc.*, 46 (1992) 448.
113 J.S. Gordon, P.S.C. van der Plas and L. de Galan, Use of a water-cooled low-flow torch in inductively coupled plasma mass spectrometry, *Anal. Chem.*, 60 (1988) 372.
114 G.D. Rayson and D.Y. Shen, Description and characterization of a linear-flow outer gas flow torch for inductively coupled plasma atomic emission spectrometry, *Appl. Spectrosc.*, 46 (1992) 1245.
115 N.N. Sesi, P.J. Galley and G.M. Hieftje, Evaluation of a linear-flow torch for inductively coupled plasma atomic emission spectrometry, *J. Anal. At. Spectrom.*, 8 (1993) 65.
116 C.S. Muñiz, J.M. Marchante-Gayón, J.I. Garcia Alonso and A. Sanz-Medel, Multi-elemental trace analysis of human serum by double-focusing ICP-MS, *J. Anal. At. Spectrom.*, 14 (1999) 193.
117 D. Beauchemin, J.W. McLaren and S.S. Berman, Use of External calibration for the determination of trace metals in biological materials by inductively coupled plasma mass spectrometry, *J. Anal. At. Spectrom.*, 3 (1988) 775.
118 H. Klinkenberg, T. Beeren, W. Van Borm, B. Mevissen and C. van Dongen, Instrumental improvement of the XYZ-translation system for a Perkin-Elmer Sciex Elan 500 ICP-MS, *Spectrochim Acta*, 48B (1993) 475.
119 H. Kawaguchi, T. Tanaka and A. Mizuike, Continuum background in inductively coupled plasma mass spectrometry, *Spectrochim. Acta*, 43B (1988) 955.
120 P.J. Turner, Some Observations on Mass Bias Effects Occurring in ICP-MS Systems, In: G. Holland and A. N. Eaton (Eds), *Applications of Plasma Source Mass Spectrometry*, The Royal Society of Chemistry, Cambridge, 1991, pp. 71–78.
121 M.A. Vaughan, G. Horlick and S. H. Tan, Effect of operating parameters on analyte signals in inductively coupled plasma mass spectrometry, *J. Anal. At. Spectrom.*, 2 (1987) 765.
122 J.-P. Schmit and A. Chauvette, Simplex approach to the optimisation of the ion optics bias potentials of an inductively coupled plasma mass spectrometer, *J. Anal. At. Spectrom.*, 4 (1989) 755.
123 P. Hedlund and A. Gustavsson, Design and evaluation of an effective modified simplex method, *Anal. Chim. Acta*, 391 (1999) 257.
124 M.J. Ford, L. Ebdon, R.C. Hutton and S.J. Hill, Simplex optimization of

the plasma parameters and ion optics of an inductively coupled mass spectrometer with pure argon and doped argon plasmas, using a multi-element figure of merit, *Anal. Chim. Acta*, 285 (1994) 23.
125 M.A. Vaughan and G. Horlick, Ion trajectories through the input ion optics of an inductively coupled plasma-mass spectrometer, *Spectrochim. Acta*, 45B (1990) 1301.
126 J. Olesik, D. Rago, C. Hensman and S. Olesik, Developing ion-molecule reaction methods to overcome spectral overlaps in ICP/MS, paper #58, 26th Annual Conference of the Federation of Analytical Chemistry and Spectroscopy Societies and 45th International Conference on Analytical Sciences and Spectroscopy, October 1999, Vancouver, Canada.
127 L. Moens and N. Jakubowski, Double-focusing mass spectrometers in ICPMS, *Anal. Chem.*, 70 (1998) 251A.
128 N. Jakubowski, L. Moens and F. Vanhaecke, Sector field mass spectrometers in ICP-MS, *Spectrochim. Acta*, 53B (1998) 1739.
129 C.B. Douthitt, The complete sector field ICP-MS bibliography, *ICP Inf. Newslett.*, 25 (1999) 90.
130 U. Giessmann, A new technology in high performance ICP-MS, Paper F5, 1994 Winter Conference on Plasma Spectrochemistry, San Diego, January 1994.
131 U. Giessmann and U. Greb, High resolution ICP-MS - a new concept for elemental mass spectrometry, *Fresenius J. Anal. Chem.*, 350 (1994) 186.
132 T.W. Burgoyne, G.M. Hieftje and R.A. Hites, Reducing the energy distribution in a plasma-source sector-field mass spectrometer interface, *J. Anal. At. Spectrom.*, 12 (1997) 1149.
133 C.-K. Kim, R. Seki, S. Morita, S. Yamasaki, A. Tsumura, Y. Takaku, Y. Igarashi and M. Yamamoto, Application of a high resolution inductively coupled plasma mass spectrometer to the measurement of long-lived radionuclides, *J. Anal. At. Spectrom.*, 6 (1991) 205.
134 A.J. Walder and P.A. Freedman, Isotopic ratio measurement using a double focusing magnetic sector mass analyser with an inductively coupled plasma as an ion source, *J. Anal. At. Spectrom.*, 7 (1992) 571.
135 A.J. Walder, I. Platzner and P.A. Freedman, Isotope ratio measurement of lead, neodymium and neodymium–samarium mixtures, hafnium and hafnium–lutetium mixtures with a double focusing multiple collector inductively coupled plasma mass spectrometer, *J. Anal. At. Spectrom.*, 8 (1993) 19.
136 I. Feldmann, W. Tittes, N. Jakubowski, D. Stuewer and U. Giessmann, Performance characteristics of inductively coupled plasma mass spectrometry with high mass resolution, *J. Anal. At. Spectrom.*, 9 (1994) 1007.
137 B. Klaue and J.D. Blum, Trace analyses of arsenic in drinking water by inductively coupled plasma mass spectrometry: high resolution versus hydride generation, *Anal. Chem.*, 71 (1999) 1408.

138 R.J. Cotter, The new time-of-flight mass spectrometry, *Anal. Chem.*, 71 (1999) 445A.
139 F.C. Adams, X. Tian and M. Heisterkamp, ICP time-of-flight mass spectrometry for elemental and speciation analysis of environmental samples, EnviroAnalysis 2000, Ottawa, Canada.
140 J.A. Koropchak, Liquid sample introduction to ICP spectrometries, *Spectroscopy*, 8 (1993) 20.
141 B.L. Sharp, Pneumatic nebulisers and spray chambers for inductively coupled plasma spectrometry, *J. Anal. At. Spectrom.*, 3 (1988) 613.
142 J.C. Ivaldi, J. Vollmer and W. Slavin, The conespray nebulizer for inductively coupled plasma atomic emission spectrometry, *Spectrochim. Acta*, 46B (1991) 1063.
143 I.L. Garcia, C. O'Grady and M. Cresser, Pneumatic nebulisers-poor pumps and inferior sub-samplers, *J. Anal. At. Spectrom.*, 2 (1987) 221.
144 J.-L. Todolí and J.-M. Mermet, Acid interferences in atomic spectrometry: analyte signal effects and subsequent reduction, *Spectrochim. Acta*, 54B (1999) 895.
145 M. Thompson and J.N. Walsh, *Handbook of Inductively Coupled Plasma Spectrometry*, 2nd edn., Blackie and Son Ltd, Glasgow, 1989, pp. 70–71.
146 A.S. Al-Ammar, R.K. Gupta and R.M. Barnes, Correction for volatility differences between organic sample analytes and standards in organic solutions analyzed by inductively coupled plasma-atomic emission and mass spectrometry, *J. Anal. At. Spectrom.*, 14 (1999) 973.
147 H. Liu and A. Montaser, Phase-doppler diagnostic studies of primary and tertiary aerosols produced by a high-efficiency nebulizer, *Anal. Chem.*, 66 (1994) 3233.
148 J.A. Koropchak and H. Aryamanya-Mugisha, Aerosol particle size effects on plasma spectrometric analysis, *J. Anal. At. Spectrom.*, 4 (1989) 291.
149 S.E. Hobbs and J.W. Olesik, The effect of desolvating droplets and vaporizing particles on ionization and excitation in Ar inductively coupled plasmas, *Spectrochim. Acta*, 48B (1993) 817.
150 S.E. Hobbs and J.W. Olesik, Inductively coupled plasma mass spectrometry signal fluctuations due to individual aerosol droplets and vaporizing particles, *Anal. Chem.*, 64 (1992) 274.
151 A.A. van Heuzen and N.M.M. Nibbering, Elemental composition and origin of (polyatomic) ions in inductively coupled plasma mass spectrometry disclosed by means of isotope exchange experiments, *Spectrochim. Acta*, 48B (1993) 1013.
152 R.H. Scott, V.A. Fassel, R.N. Kniseley and D.E. Nixon, Inductively coupled plasma optical emission analytical spectrometry, a compact facility of trace analysis of solutions, *Anal. Chem.*, 46 (1974) 75.
153 T.J. Brotherton, P.E. Pfannerstill, J.T. Creed, D.T. Heitkemper, J.A. Caruso and S.E. Pratsinis, Evaluation of three low-volume interfaces for

organic solvent introduction to the inductively coupled plasma—applications to flow injection, *J. Anal. At. Spectrom.*, 4 (1989) 341.
154 D. Pollmann, C. Pilger, R. Hergenröder, F. Leis, P. Tschöpel and J.A.C. Broekaert, Noise power spectra of inductively coupled plasma mass spectrometry using a cooled spray chamber, *Spectrochim. Acta*, 49B (1994) 683.
155 S. Maestre, J. Mora, J.-L. Todoli and A. Canals, Evaluation of several commercially available spray chambers for use in inductively coupled plasma atomic emission spectrometry, *J. Anal. At. Spectrom.*, 14 (1999) 61.
156 R.I. McCrindle and C.J. Rademeyer, Electron density and hydrogen distribution in an ethanol-loaded inductively coupled plasma, *J. Anal. At. Spectrom.*, 9 (1994) 1087.
157 B.S. Ross and G.M. Hieftje, Selection of solvent load and first-stage pressure to reduce interference effects in inductively coupled plasma–mass Spectrometry, *J. Am. Soc. Mass Spectrom.*, 3 (1992) 128.
158 R. Tsukahara and M. Kubota, Studies with desolvation in inductively coupled plasma mass spectrometry, *Spectrochim. Acta*, 45B (1990) 581.
159 J.A. Koropchak, L. Allen, J.M. Davis, Aerosol interfacing effects on discrete sample introduction coupled with spectrometric detection, *Appl. Spectrosc.*, 46 (1992) 682.
160 L. Darbha, M. V. Suryanarayana, S. Natarajan, Studies on the improvement of sensitivity by heating the spray chamber in ICP-OES, *Spectrochim. Acta*, 47B (1992) 581.
161 R.H. Clifford, H. Tan, H. Liu and A. Montaser, F. Zarrin and P.B. Keady, Particle size measurements in the submicron range by the differential electromobility technique: comparison of aerosols from thermospray, ultrasonic, pneumatic and frit-type nebulizers, *Spectrochim. Acta*, 48B (1993) 1221.
162 T.J. Brotherton, W.L. Shen and J.A. Caruso, Use of the Hildebrand grid nebuliser for analysis of high matrix solutions containing easily ionisable elements with inductively coupled plasma mass spectrometry, *J. Anal. At. Spectrom.*, 4 (1989) 39.
163 E. Fatemian, J. Allibone and P.J. Walker, Use of gold as a routine and long term preservative for mercury in potable water, as determined by ICP-MS, *Analyst*, 124 (1999) 1233.
164 D.E. Nixon, M.F. Burritt and T.P. Moyer, The determination of mercury in whole blood and urine by inductively coupled plasma mass spectrometry, *Spectrochim. Acta*, 54B (1999) 1141.
165 J. Allibone, E. Fatemian and P.J. Walker, Determination of mercury in potable water by ICP-MS using gold as a stabilising agent, *J. Anal. At. Spectrom.*, 14 (1999) 235.
166 H. Vanhoe, F. Van Allemeersch, J. Versieck and R. Dams, Effect of solvent type on the determination of total iodine in milk powder and human serum

by inductively coupled plasma mass spectrometry, *Analyst*, 118 (1993) 1015.
167 H.P. Longerich, Automated two-speed peristaltic pump controller for reduced sample wash-out time and analysis of small sample volumes, *J. Anal. At. Spectrom.*, 8 (1993) 371.
168 A. Montaser, H. Tan, I. Ishii, S.-H. Nam and M. Cai, Argon Inductively Coupled Plasma Mass Spectrometry with Thermospray, Ultrasonic, and Pneumatic Nebulization, *Anal. Chem.*, 63 (1991) 2660.
169 S.P. Osborne, Ultrasonic nebulization-inductively coupled plasma spectroscopy, Optimization for various matrices, *Spectroscopy*, 7 (1992) 37.
170 H. Wildner, Application of inductively coupled plasma sector field mass spectrometry for the fast and sensitive determination and isotope ratio measurement of non-metals in high-purity process chemicals, *J. Anal. At. Spectrom.*, 13 (1998) 573.
171 R.I. Botto and J.J. Zhu, Use of an ultrasonic nebulizer with membrane desolvation for analysis of volatile solvents by inductively coupled plasma atomic emission spectrometry, *J. Anal. At. Spectrom.*, 9 (1994) 905.
172 M.G. Minnich and R.S. Houk, Comparison of cryogenic and membrane desolvation for attenuation of oxide, hydride and hydroxide ions and ions containing chlorine in inductively coupled plasma mass spectrometry, *J. Anal. At. Spectrom.*, 13 (1998) 167.
173 T. Prohaska, S. Hann, C. Latkoczy and G. Stingeder, Determination of rare earth elements U and Th in environmental samples by inductively coupled plasma double focusing sectorfield mass spectrometry (ICP-SMS), *J. Anal. At. Spectrom.*, 14 (1999) 1.
174 L.C. Alves, D.R. Wiederin and R.S. Houk, Reduction of polyatomic ion interferences in inductively coupled plasma mass spectrometry by cryogenic desolvation, *Anal. Chem.*, 64 (1992) 1164.
175 L.C. Alves, M.G. Minnich, D.R. Wiederin and R.S. Houk, Removal of organic solvents by cryogenic desolvation in inductively coupled plasma mass spectrometry, *J. Anal. At. Spectrom.*, 9 (1994) 399.
176 G.P. Thomas and L.C. Goldstone, Advances in environmental inorganic analysis, *Am. Lab.*, 26(17) (1994) 32I.
177 A.P. Weber, R. Keil, L. Tobler and U. Baltensperger, Sensitivities of inductively coupled plasma optical emission spectrometry for dry and wet aerosols, *Anal. Chem.*, 64 (1992) 672.
178 Y.Q. Tang and C. Trassy, Inductively coupled plasma: the role of water in axial excitation temperatures, *Spectrochim. Acta*, 41B (1986) 143.
179 B.R. LaFreniere, D.R. Wiederin, V.A. Fassel and R.S. Houk, Molecular hydrogen emission in the vacuum ultraviolet from an inductively coupled plasma, *Spectrochim. Acta*, 51B (1996) 3.
180 G.R. Peters and D. Beauchemin, Effect of pre-evaporating the solvent on the analytical performance of inductively coupled plasma mass

spectrometry, *Spectrochim. Acta*, 48B (1993) 1481.
181 P.E. Walters and C.A. Barnardt, The role of desolvation and hydrogen addition on the excitation features of the inductively coupled plasma, *Spectrochim. Acta*, 43B (1988) 325.
182 M.A. Tarr, G. Zhu and R.F. Browner, Microflow ultrasonic nebulizer for inductively coupled plasma atomic emission spectrometry, *Anal. Chem.*, 65 (1993) 1689.
183 N. Jakubowski, I. Feldmann and D. Stuewer, Analytical improvement of pneumatic nebulization in ICP-MS by desolvation, *Spectrochim. Acta*, 47B (1992) 107.
184 S.J. Hill, J. Hartley and L. Ebdon, Determination of trace metals in volatile organic solvents using inductively coupled plasma atomic emission spectrometry and inductively coupled plasma mass spectrometry, *J. Anal. At. Spectrom.*, 7 (1992) 23.
185 A.G.T. Gustavsson, Liquid Sample Introduction into Plasmas, In: A. Montaser and D.W. Golightly (Eds), *Inductively Coupled Plasmas in Analytical Atomic Spectrometry*, 2nd edn., VCH Publishers Inc., New York, 1992.
186 D. Beauchemin, Plasma emission spectrometry, *Anal. Chem.*, 66 (1994) 462R.
187 J. Mora, A. Canals, V. Hernandis, E.H. van Veen and M.T.C. de Loos-Vollebregt, Evaluation of a microwave desolvation system in inductively coupled plasma mass spectrometry, *J. Anal. At. Spectrom.*, 13 (1998) 175.
188 A.J. Walder, D. Koller, N.M. Reed, R.C. Hutton and P.A. Freedman, Isotope ratio measurement by inductively coupled plasma multiple collector mass spectrometry incorporating a high efficiency nebulization system, *J. Anal. At. Spectrom.*, 8 (1993) 1037.
189 A.R. Eastgate, R.C. Fry and G.H. Gower, Radiation versus conduction in heated spray chamber desolvation for inductively coupled plasmas, *J. Anal. At. Spectrom.*, 8 (1993) 305.
190 Z. Chen, H.P. Longerich and B.J. Fryer, Recycling nebulization system with a disposable spray chamber for analysis of sub-milligram samples of geological materials using inductively coupled plasma mass spectrometry, *J. Anal. At. Spectrom.*, 7 (1992) 905.
191 S.A. Beres, P.H. Brückner and E.R. Denoyer, Performance evaluation of a cyclonic spray chamber for ICP-MS, *At. Spectrosc.*, 15 (1994) 96.
192 D.R. Wiederin, F.G. Smith and R.S. Houk, Direct injection nebulization for inductively coupled plasma mass spectrometry, *Anal. Chem.*, 63 (1991) 219.
193 M.J. Powell, E.S.K. Quan, D.W. Boomer and D.R. Wiederin, Inductively coupled plasma mass spectrometry with direct injection nebulization for mercury analysis of drinking water, *Anal. Chem.*, 64 (1992) 2253.
194 J.A. McLean, H. Zhang and A. Montaser, A direct injection high-efficiency

nebulizer for inductively coupled plasma mass spectrometry, *Anal. Chem.*, 70 (1998) 1012.
195 J.S. Becker, H.-J. Dietze, J.A. McLean and A. Montaser, Ultratrace and isotope analysis of long-lived radionuclides by inductively coupled plasma quadrupole mass spectrometry using a direct injection high efficiency nebulizer, *Anal. Chem.*, 71 (1999) 3077.
196 J.A. Koropchak and T.S. Conver, Development of a high liquid flow thermospray sample introduction system for inductively coupled plasma atomic emission spectrometry, *J. Anal. At. Spectrom.*, 9 (1994) 899.
197 N. Jakubowski, I. Feldmann, D. Stuewer and H. Berndt, Hydraulic high pressure nebulization-application of a new nebulization system for inductively coupled plasma mass spectrometry, *Spectrochim. Acta*, 47B (1992) 119.
198 F. Vanhaecke, M. Van Holderbeke, L. Moens and R. Dams, Evaluation of a commercially available microconcentric nebulizer for inductively coupled plasma mass spectrometry, *J. Anal. At. Spectrom.*, 11 (1996) 543.
199 S. Augagneur, B. Medina, J. Szpunar and R. Lobinski, Determination of rare earth elements in wine by inductively coupled plasma mass spectrometry using a microconcentric nebulizer, *J. Anal. At. Spectrom.*, 11 (1996) 713.
200 J.B. French, B. Etkin and R. Jong, Monodisperse Dried Microparticulate Injector for Analytical Instrumentation, *Anal. Chem.*, 66 (1994) 685.
201 J.A. Koropchak and M. Veber, Thermospray sample introduction to atomic spectrometry, *Crit. Rev. Anal. Chem.*, 23 (1992) 113.
202 H. Vanhoe, L. Moens and R. Dams, Thermospray nebulization as sample introduction for inductively coupled plasma mass spectrometry, *J. Anal. At. Spectrom.*, 9 (1994) 815.
203 X. Zhang and J.A. Koropchak, Thermospray methods for rapid, sensitive, and nonchromatographic speciation of chromium oxidation states, *Anal. Chem.*, 71 (1999) 3046.
204 X. Zhang and J.A. Koropchak, Direct chromium speciation using thermospray: preliminary studies with inductively coupled plasma mass spectrometry, *Microchem. J.*, 62 (1999) 192.
205 J. Yang, T.S. Conver and J.A. Koropchak, Direct speciation of selenite and selenate with thermospray sample introduction methods, *Anal. Chem.*, 68 (1996) 4064.
206 L. Bordera, J.L. Todolí, J. Mora, A. Canals and V. Hernandis, A microwave-powered thermospray nebulizer for liquid sample introduction in inductively coupled plasma atomic emission spectrometry, *Anal. Chem.*, 69 (1997) 3578.
207 M.P. Field and R.M. Sherrell, Magnetic sector ICPMS with desolvating micronebulization: interference-free subpicogram determination of rare earth elements in natural samples, *Anal. Chem.*, 70 (1998) 4480.

208 J.W. Olesik and S.E. Hobbs, Monodisperse dried microparticulate injector: a new tool for studying fundamental processes in inductively coupled plasmas, *Anal. Chem.*, 66 (1994) 3371.
209 S. Yamasaki, A. Tsumura and U. Takaku, Ultratrace elements in terrestrial water as determined by high-resolution ICP-MS, *Microchem. J.*, 49 (1994) 305.
210 R.C. Hutton, A.N. Eaton and R.M. Gosland, Application of a dual-mode detection system for ICP-MS. Part I: Determination of majors, minors, and traces in geostandards, *Appl. Spectrosc.*, 44 (1990) 238.
211 E.R. Denoyer, Perkin Elmer Technical Summary No. TSMS-15.
212 Perkin Elmer Technical Summary No. TSMS-6.
213 M. Carré, S. Excoffier and J.-M. Mermet, A study of the relation between the limit of detection and the limit of quantitation in inductively coupled plasma spectrochemistry, *Spectrochim. Acta*, 52B (1997) 2043.
214 C. Latkoczy, T. Prohaska, G. Stingeder and M. Teschler-Nicola, Strontium isotope ratio measurements in prehistoric human bone samples by means of high-resolution inductively coupled plasma mass spectrometry (HR-ICP-MS), *J. Anal. At. Spectrom.*, 13 (1998) 561.
215 B.S. Duersch, J.E. Patterson and P.B. Farnsworth, The effects of a torch shield on performance of the vacuum interface of an inductively coupled plasma mass spectrometer, *J. Anal. At. Spectrom.*, 14 (1999) 615.
216 K. Hu and G. Kunselman, Inductively coupled plasma mass spectrometry with a modified offset ion lens and hyperbolic quadrupole, Paper F2, 1994 Winter Conference on Plasma Spectrochemistry, San Diego, January 1994.
217 P. De Bièvre, Isotope dilution mass spectrometry as a primary method of analysis, *Anal. Proc.*, 30 (1993) 328.
218 P. De Bièvre, Stable isotope dilution: an essential tool in metrology, *Fresenius J. Anal. Chem.*, 350 (1994) 277.
219 S.R. Koirtyohann, Precise determination of isotopic ratios for some biologically significant elements by inductively coupled plasma mass spectroscopy, *Spectrochim. Acta*, 49B (1994), 1305.
220 F. Vanhaecke, K. Moens, R. Dams and P. Taylor, Precise measurement of isotope ratios with a double-focusing magnetic sector ICP mass spectrometer, *Anal. Chem.*, 68 (1996) 567.
221 R.E. Sturgeon, Future of atomic spectrometry for environmental analysis, *J. Anal. At. Spectrom.*, 13 (1998) 351.
222 E.H. Evans and J.J. Giglio, Interferences in inductively coupled plasma mass spectrometry. A Review, *J. Anal. At. Spectrom.*, 8 (1993) 1.
223 N. Furuta, Optimization of the mass scanning rate for the determination of lead isotope ratios using an inductively coupled plasma mass spectrometer, *J. Anal. At. Spectrom.*, 6 (1991) 199.
224 A.T. Ince, J.G. Williams and A.L. Gray, Noise in inductively coupled

plasma mass spectrometry: some preliminary measurements, *J. Anal. At. Spectrom.*, 8 (1993) 899.

225 A.L. Gray, J.G. Williams, A.T. Ince and M. Liezers, Noise sources in inductively coupled plasma mass spectrometry: an investigation of their importance to the precision of isotope ratio measurements, *J. Anal. At. Spectrom.*, 9 (1994) 1179.

226 E.R. Denoyer, Optimization of transient signal measurements in ICP-MS, *At. Spectrosc.*, 15 (1994) 7.

227 B.L. Sharp, J. Batey, I.S. Begley, D. Gregson, J. Skilling, A.B. Sulaiman and G. Verbogt, Information retrieval from the inductively coupled plasma, *J. Anal. At. Spectrom.*, 14 (1999) 99.

228 I.I. Stewart and J.W. Olesik, Steady state acid effects in ICP-MS, *J. Anal. At. Spectrom.*, 13 (1998) 1313.

229 E.D. Prudnikov and R.M. Barnes, Theoretical calculation of the standard deviation in inductively coupled plasma mass spectrometry, *J. Anal. At. Spectrom.*, 14 (1999) 27.

230 E.R. Denoyer, An evaluation of spectral integration in ICP-MS, *At. Spectrosc.*, 13 (1992) 93.

231 F. Monna, J.-L. Loizeau, B. A. Thomas, C. Guéguen and P.-Y. Favarger, Pb and Sr isotope measurements by inductively coupled plasma-mass spectrometer: efficient time management for precision improvement, *Spectrochim. Acta*, 53B (1998) 1317.

232 F. Vanhaecke, L. Moens, R. Dams, L. Allen and S. Georgitis, Evaluation of the isotope ratio performance of an axial time-of-flight ICP mass spectrometer, *Anal. Chem.*, 71 (1999) 3297.

233 D.J. Douglas and L.A. Kerr, Study of solids deposition on inductively coupled plasma mass spectrometry samplers and skimmers, *J. Anal. At. Spectrom.*, 3 (1988) 749.

234 S.H. Tan and G. Horlick, Background spectral features in inductively coupled plasma/mass spectrometry, *Appl. Spectrosc.*, 40 (1986) 445.

235 L.L. Burton and G. Horlick, MS InterView: An elemental mass spectrometry spectral interference data base, *Spectrochim. Acta*, 47B (1992) E1621.

236 R.C. Hutton and A.N. Eaton, Role of aerosol water vapour loading in inductively coupled plasma mass spectrometry, *J. Anal. At. Spectrom.*, 2 (1987) 595.

237 K. Hu and R.S. Houk, Inductively coupled plasma mass spectrometry with an enlarged sampling orifice and offset ion lens. II. Polyatomic ion interferences and matrix effects, *J. Am. Soc. Mass Spectrom.*, 4 (1993) 28.

238 D. Ashley, Polyatomic interferences due to the presence of inorganic carbon in environmental samples in the determination of chromium at Mass 52 by ICP-MS, *At. Spectrosc.*, 13 (1992) 169.

239 P. Dulski, Interferences of oxide, hydroxide and chloride analyte species in

the determination of rare earth elements in geological samples by inductively coupled plasma-mass spectrometry, *Fresenius J. Anal. Chem.*, 350 (1994) 194.

240 A.R. Date, Y.Y. Cheung and M.E. Stuart, The influence of polyatomic ion interferences in analysis by inductively coupled plasma source mass spectrometry (ICP-MS), *Spectrochim. Acta*, 42B (1987) 3.

241 J.L.M. de Boer, Possibilities and limitations of spectral fitting to reduce polyatomic ion interferences in inductively coupled plasma quadrupole mass spectrometry in the mass range 51-88, *Spectrochim. Acta*, 52B (1997) 389.

242 H. Vanhoe, J. Goossens, L. Moens and R. Dams, Spectral interferences encountered in the analysis of biological materials by inductively coupled plasma mass spectrometry, *J. Anal. At. Spectrom.*, 9 (1994) 177.

243 J.S. Crain and J. Alvarado, Hydride interference on the determination of minor actinide isotopes by inductively coupled plasma mass spectrometry, *J. Anal. At. Spectrom.*, 9 (1994) 1223.

244 W. Tittes, N. Jakubowski, D. Stüwer, G. Tölg and J.A.C. Broekaert, Reduction of some selected spectral interferences in inductively coupled plasma mass spectrometry, *J. Anal. At. Spectrom.*, 9 (1994) 1015.

245 K.E. Jarvis, A.L. Gray and E. McCurdy, Avoidance of spectral interference on europium in inductively coupled plasma mass spectrometry by sensitive measurement of the doubly charged ion, *J. Anal. At. Spectrom.*, 4 (1989) 743.

246 N.M. Reed, R.O. Cairns, R.C. Hutton and Y. Takaku, Characterization of polyatomic ion interferences in inductively coupled plasma mass spectrometry using a high resolution mass spectrometer, *J. Anal. At. Spectrom.*, 9 (1994) 881.

247 M. Morita, H. Ito, M. Linscheid and K. Otsuka, Resolution of interelement spectral overlaps by high-resolution inductively coupled plasma mass spectrometry, *Anal. Chem.*, 66 (1994) 1588.

248 F. Vanhaecke, L. Moens, R. Dams, I. Papadakis and P. Taylor, Applicability of high-resolution ICP-mass spectrometry for isotope ratio measurements, *Anal. Chem.*, 69 (1997) 268.

249 E.H. Evans and L. Ebdon, Effect of organic solvents and molecular gases on polyatomic ion interferences in inductively coupled plasma mass spectrometry, *J. Anal. At. Spectrom.*, 5 (1990) 425.

250 F. Laborda, M.J. Baxter, H.M. Crews and J. Dennis, Reduction of polyatomic interferences in inductively coupled plasma mass spectrometry by selection of instrumental parameters and using an argon-nitrogen plasma; effect on multielement analyses, *J. Anal. At. Spectrom.*, 9 (1994) 727.

251 A.L. Gray, Mass spectrometry with an inductively coupled plasma as an ion source: The influence on ultratrace analysis of background and matrix

response *Spectrochim. Acta*, 41B (1986) 151.
252 J.A. Olivares and R.S. Houk, Ion sampling for inductively coupled plasma mass spectrometry, *Anal. Chem.*, 57 (1985) 2674.
253 S.D. Tanner, ICP-MS tectonics: moving the atomic and oxide ion "mountains", Paper F1, 1994 Winter Conference on Plasma Spectrochemistry, San Diego, January 1994.
254 M.A. Vaughan and G. Horlick, Effect of sampler and skimmer orifice size on analyte and analyte oxide signals in inductively coupled plasma mass spectrometry, *Spectrochim. Acta*, 45B (1990) 1289.
255 S.D. Tanner, Experimental Studies of Ion Kinetic Energies in ICP-MS, In: G. Holland and A. N. Eaton (Eds), *Applications of Plasma Source Mass Spectrometry II*, The Royal Society of Chemistry, Special Publication No. 124, Cambridge, 1993, pp. 222–234.
256 J.A. Olivares and R.S. Houk, Suppression of analyte signal by various concomitant salts in inductively coupled plasma mass spectrometry, *Anal. Chem.*, 58 (1986) 20.
257 C.J. Pickford and P.M. Brown, Comparison of ICP-MS with ICP-ES: Detection power and interference effects experienced with complex matrices, *Spectrochim. Acta*, 41B (1986) 183.
258 H.P. Longerich, B.J. Fryer and D.F. Strong, Trace analysis of natural alloys by inductively coupled plasma-mass spectrometry (ICP-MS): Application to archaeological native silver artifacts, *Spectrochim. Acta*, 42B (1987) 101.
259 D. Beauchemin, J.W. McLaren and S.S. Berman, Study of the effects of concomitant elements in inductively coupled plasma mass spectrometry, *Spectrochim. Acta*, 42B (1987) 467.
260 D.C. Grégoire, The effect of easily ionizable concomitant elements on non-spectroscopic interferences in inductively coupled plasma-mass spectrometry, *Spectrochim. Acta*, 42B (1987) 895.
261 S.H. Tan and G. Horlick, Matrix-effect observations in inductively coupled plasma mass spectrometry, *J. Anal. At. Spectrom.*, 2 (1987) 745.
262 Y.-S. Kim, H. Kawaguchi, T. Tanaka and A. Mizuike, Non-spectroscopic matrix interferences in inductively coupled plasma-mass spectrometry, *Spectrochim. Acta*, 45B (1990) 333.
263 P. Richner, Detection limits versus matrix effects: analysis of solutions with high amounts of dissolved solids by flow injection inductively coupled plasma mass spectrometry, *J. Anal. At. Spectrom.*, 8 (1993) 927.
264 C. Vandecasteele, M. Nagels, H. Vanhoe and R. Dams, Suppression of analyte signal in inductively coupled plasma/mass spectrometry and the used of an internal standard, *Anal. Chim. Acta*, 211 (1988) 91.
265 E.H. Evans and J.A. Caruso, Optimization strategies for the reduction of non-spectroscopic interferences in inductively coupled plasma mass spectrometry, *Spectrochim. Acta*, 47B (1992) 1001.

266 J. Marshall and J. Franks, Matrix interferences from methacrylic acid solutions in inductively coupled plasma mass spectrometry, *J. Anal. At. Spectrom.*, 6 (1991) 591.
267 B. Gammelgaard and O. Jons, Determination of selenium in urine by inductively coupled plasma mass spectrometry: interferences and optimization, *J. Anal. At. Spectrom.*, 14 (1999) 867.
268 D. Beauchemin and J.M. Craig, Investigations on mixed-gas plasmas produced by adding nitrogen to the plasma gas in ICP-MS, *Spectrochim. Acta*, 46B (1991) 603.
269 P. Allain, L. Jaunault, Y. Mauras, J.-M. Mermet and T. Delaporte, Signal enhancement of elements due to the presence of carbon-containing compounds in inductively coupled plasma mass spectrometry, *Anal. Chem.*, 63 (1991) 1497.
270 S.A. Wood, D. Vlassopoulos and A. Mucci, Effects of concentrated matrices on the determination of trace levels of platinum and gold in aqueous samples using solvent extraction-Zeeman effect graphite furnace atomic absorption spectrometry and inductively coupled plasma mass spectrometry, *Anal. Chim. Acta*, 229 (1990) 227.
271 G.E.M. Hall, C.J. Park and J.C. Pelchat, Determination of tungsten and molybdenum at low levels in geological materials by inductively coupled plasma mass spectrometry, *J. Anal. At. Spectrom.*, 2 (1987) 189.
272 X. Feng, S. Wu, A. Wharmby and A. Wittmeier, Microwave digestion of plant and grain standard reference materials in nitric and hydrofluoric acids for multi-elemental determination by inductively coupled plasma mass spectrometry, *J. Anal. At. Spectrom.*, 14 (1999) 939.
273 J.S. Crain, R.S. Houk and F.G. Smith, Matrix interferences in inductively coupled plasma-mass spectrometry: some effects of skimmer orifice diameter and ion lens voltages, *Spectrochim. Acta*, 43B (1988) 1355.
274 D.C. Grégoire, Influence of instrument parameters on nonspectroscopic interferences in inductively coupled plasma–mass spectrometry, *Appl. Spectrosc.*, 41 (1987) 897.
275 I. Rodushkin, T. Ruth and D. Klockare, Non-spectral interferences caused by a saline water matrix in quadrupole and high resolution inductively coupled plasma mass spectrometry, *J. Anal. At. Spectrom.*, 13 (1998) 159.
276 J.-M. Mermet, Revisitation of the matrix effects in inductively coupled plasma atomic emission spectrometry: the key role of the spray chamber, *J. Anal. At. Spectrom.*, 13 (1998) 419.
277 P. Masson, Matrix effects during trace element analysis in plant samples by inductively coupled plasma atomic emission spectrometry with axial view configuration and pneumatic nebulizer, *Spectrochim. Acta*, 54B (1999) 603.
278 G.R. Gillson, D.J. Douglas, J.E. Fulford, K.W. Halligan and S.D. Tanner, Nonspectroscopic interelement interferences in inductively coupled

plasma mass spectrometry, *Anal. Chem.*, 60 (1988) 1472.
279 S.E. Hobbs and J.W. Olesik, Laser-excited fluorescence studies of matrix-induced errors in inductively coupled plasma spectrometry: Implications for ICP-mass spectrometry, *Appl. Spectrosc.*, 45 (1991) 1395.
280 A. Fernández, M. Murillo, N. Carrión and J.-M. Mermet, Influence of operating conditions on the effects of acids in inductively coupled plasma atomic emission spectrometry, *J. Anal. At. Spectrom.*, 9 (1994) 217.
281 K.E. Nissen, J.T. Keegan and J.P. Byrne, Characterization of plastic films from trace element analysis using inductively coupled plasma-mass spectrometry, *Can. J. Anal. Sci. Spectrosc.*, 43 (1998) 122.
282 M.Y. Pérez-Jordán, J. Soldevila, A. Salvador, A. Pastor and M. de la Guardia, Inductively coupled plasma mass spectrometry analysis of wines, *J. Anal. At. Spectrom.*, 13 (1998) 33.
283 M. Todorovic, S. Vidovic and Z. Ilic, Effect of aqueous organic solvents on the determination of trace elements by flame atomic absorption spectrometry and inductively coupled plasma atomic emission spectrometry, *J. Anal. At. Spectrom.*, 8 (1993) 1113.
284 B.S. Duersch and P.B. Farnsworth, Characterization of the ion beam inside the skimmer cone of an inductively coupled plasma mass spectrometer by laser excited atomic and ionic fluorescence, *Spectrochim. Acta*, 54B (1999) 545.
285 S.D. Tanner, Space charge in ICP-MS: Calculation and implications, *Spectrochim. Acta*, 47B (1992) 809.
286 S.D. Tanner, Space charge in inductively coupled plasma mass spectrometry: calculation and implications, *Anal. Proc.*, 29 (1992) 281.
287 S.D. Tanner, L.M. Cousins and D.J. Douglas, Reduction of space charge effects using a three-aperture gas dynamic vacuum interface for inductively coupled plasma–mass spectrometry, *Appl. Spectrosc.*, 48 (1994) 1367.
288 S.D. Tanner, D.J. Douglas and J.B. French, Gas and ion dynamics of a three-aperture vacuum interface for inductively coupled plasma–mass spectrometry, *Appl. Spectrosc.*, 48 (1994) 1373.
289 T.W. Burgoyne, G.M. Hieftje and R.A. Hites, Space charge evaluation in a plasma-source mass spectrograph, *Anal. Chem.*, 69 (1997) 485.
290 G. Li, Y. Duan and G.M. Hieftje, Space-charge effects and ion distribution in plasma source mass spectrometry, *J. Mass Spectrom.*, 30 (1995) 841.
291 B.S. Duersch, Y. Chen, A. Ciocan and P.B. Farnsworth, Optical measurements of ion density in the second vacuum stage of an inductively coupled plasma mass spectrometer, *Spectrochim. Acta*, 53B (1998) 569.
292 X. Chen and R.S. Houk, Spatially resolved measurements of ion density behind the skimmer of in inductively coupled plasma mass spectrometer, *Spectrochim. Acta*, 51B (1996) 41.
293 Y. Chen and P.B. Farnsworth, Ion deposition experiments as a tool for the

study of the spatial distribution of analyte ions in the second vacuum stage of an inductively coupled plasma mass spectrometer, *Spectrochim. Acta*, 52B (1997) 231.
294 I.B. Brenner, M. Liezers, J. Godfrey, S. Nelms and J. Cantle, Analytical characteristics of a high efficiency ion transmission interface (S mode) inductively coupled plasma mass spectrometer for trace element determinations in geological and environmental materials, *Spectrochim. Acta*, 54B (1999) 991.
295 B.C. Ross and G.M. Hieftje, Alteration of the ion-optic lens configuration to eliminate mass-dependent matrix-interference effects in inductively coupled plasma–mass spectrometry, *Spectrochim. Acta*, 46B (1991) 1263.
296 L.A. Allen, J.J. Leach and R.S. Houk, Spatial location of the space charge effect in individual ion clouds using monodisperse dried microparticulate injection with a twin quadrupole inductively coupled plasma mass spectrometer, *Anal. Chem.*, 69 (1997) 2384.
297 J.A. Borowiec, A.W. Boorn, J.H. Dillard, M.S. Cresser, R.F. Browner and M.J. Matteson, Interference effects from aerosol ionic redistribution in analytical atomic spectrometry, *Anal. Chem.*, 52 (1980), 1054.
298 C. Dubuisson, E. Poussel, J.L. Todoli and J.-M. Mermet, Effect of sodium during the aerosol transport and filtering in inductively coupled plasma atomic emission spectrometry, *Spectrochim. Acta*, 53B (1998) 593.
299 J.E. Patterson, B.S. Duersch and P.B. Farnsworth, Optically determined velocity distributions of metastable argon in the second stage of an inductively coupled plasma mass spectrometer, *Spectrochim. Acta*, 54B (1999) 537.
300 S.C.K. Shum, S.K. Johnson, H.-M. Pand and R.S. Houk, Spatially resolved measurements of size and velocity distributions of aerosol droplets from a direct injection nebulizer, *Appl. Spectrosc.*, 47 (1993) 575.
301 T. Hirata, Lead isotopic analyses of NIST standard reference materials using multiple collector inductively coupled plasma mass spectrometry coupled with a modified external correction method for mass discrimination effect, *Analyst*, 121 (1996) 1407.
302 D.C. Grégoire, B.M. Acheson and R.P. Taylor, Measurement of lithium isotope ratios by inductively coupled plasma mass spectrometry: application to geological materials, *J. Anal. At. Spectrom.*, 11 (1996) 765.
303 G.P. Klinkhammer and L.H. Chan, Determination of barium in marine waters by isotope dilution inductively coupled plasma mass spectrometry, *Anal. Chim. Acta*, 232 (1990) 323.
304 H. Klinkenberg, T. Beeren, W. Van Borm, F. van der Linden and M. Raets, The use of an enriched isotope as an on-line internal standard in inductively coupled plasma mass spectrometry: a reference method for a proposed determination of tellurium in industrial waste water by means of graphite furnace atomic absorption spectrometry, *Spectrochim. Acta*,

48B (1993) 649.
305 J. Toole, K. McKay and M. Baxter, Determination of uranium in marine sediment pore waters by isotope dilution inductively coupled plasma mass spectrometry, *Anal. Chim. Acta*, 245 (1991) 83.
306 J.S. Crain and B. L. Mikesell, Detection of sub-ng/l actinides in industrial wastewater matrices by inductively coupled plasma mass spectrometry, *Appl. Spectrosc.*, 46 (1992) 1498.
307 F. Vanhaecke, C. Vandecasteele and R. Dams, Inductively coupled plasma mass spectrometry for the determination of aluminium, calcium, chlorine, iron, magnesium, manganese and sodium in fresh water, *Anal. Lett.*, 25 (1992) 919.
308 M. Leiterer and U. Münch, determination of heavy metals in groundwater samples—ICP-MS analysis and evaluation, *Fresenius J. Anal. Chem.*, 350 (1994) 204.
309 K.J. Orians and E.A. Boyle, Determination of picomolar concentrations of titanium, gallium and indium in sea water by inductively coupled plasma mass spectrometry following an 8-hydroxyquinoline chelating resin preconcentration, *Anal. Chim. Acta*, 282 (1993) 63.
310 F. Vanhaecke, J. Goossens, R. Dams and C. Vandecasteele, The determination of molybdenum in a sea water candidate reference material by inductively coupled plasma-mass spectrometry, *Talanta*, 40 (1993) 975.
311 C. Haraldsson, B. Lyvén, M. Pollak and A. Skoog, Multi-element speciation of trace metals in fresh water adapted to plasma source mass spectrometry, *Anal. Chim. Acta*, 284 (1993) 327.
312 A. Stroh, Analysis of rare earth elements in natural waters by ICP-MS, *At. Spectrosc.*, 13 (1992) 89.
313 R. Cidu, L. Fanfani, P. Shand, W. M. Edmunds, L. Van't Dack and R. Gijbels, Determination of gold at the ultratrace level in natural waters, *Anal. Chim. Acta*, 296 (1994) 295.
314 N.C. Porteous, J.N. Walsh and K.E. Jarvis, Measurement of boron isotope ratios in groundwater studies, *Analyst*, 120 (1995) 1397.
315 K.E. Jarvis, J.G. Williams, E. Alcantara and J.D. Wills, Determination of trace and ultra-trace elements in saline waters by inductively coupled plasma mass spectrometry after off-line chromatographic separation and preconcentration, *J. Anal. At. Spectrom.*, 11 (1996) 917.
316 K. Akatsuka, J.W. McLaren, J.W. Lam and Shier S. Berman, Determination of iron and ten other trace elements in the open ocean seawater reference material NASS-3 by inductively coupled plasma mass spectrometry, *J. Anal. At. Spectrom.*, 7 (1992) 889.
317 M.B. Shabani, T. Akagi, H. Shimizu and A. Masuda, Determination of trace lanthanides and yttrium in seawater by inductively coupled plasma mass spectrometry after preconcentration with solvent extraction and back-extraction, *Anal. Chem.*, 62 (1990) 2709.

318 H. Sawatari, T. Hayashi, E. Fujimori, A. Hirose and H. Haraguchi, Multielement determination of trace elements in coastal seawater by ICP-MS and ICP-AES after aluminum coprecipitation associated with magnesium, *Bull. Chem. Soc. Jpn.*, 69 (1996) 1925.
319 T. Jickells, M.M. Kane, A. Rendell, T. Davies, M. Tranter and K.E. Jarvis, Applications of inductively coupled plasma techniques and preconcentration to the analysis of atmospheric precipitation, *Anal. Proc.*, 29 (1992) 288.
320 A. Cocherie, P. Negrel, S. Roy and C. Guerrot, Direct Determination of lead isotope ratios in rainwater using inductively coupled plasma mass spectrometry, *J. Anal. At. Spectrom.*, 13 (1998) 1069.
321 P. Möller, P. Dulski and J. Luck, Determination of rare earth elements in seawater by inductively coupled plasma-mass spectrometry, *Spectrochim. Acta*, 47B (1992), 1379.
322 Y. Lu, C.L. Chakrabarti, M.H. Back, D.C. Grégoire and W.H. Schroeder, Kinetic studies of aluminum and zinc speciation in river water and snow, *Anal. Chim. Acta*, 293 (1994) 95.
323 U. Örnemark, P.D.P. Taylor and P. De Bièvre, Certification of the rubidium concentration in water materials for the international measurement evaluation programme (IMEP) using isotope dilution inductively coupled plasma mass spectrometry, *J. Anal. At. Spectrom.*, 12 (1997) 567.
324 T. Probst, P. Zeh and J.-I. Kim, Multielement determinations in ground water ultrafiltrates using inductively coupled plasma mass spectrometry and monostandard neutron activation analysis, *Fresenius J. Anal. Chem.*, 351 (1995) 745.
325 J. Wu and E.A. Boyle, Low blank preconcentration technique for the determination of lead, copper, and cadmium in small-volume seawater samples by isotope dilution ICPMS, *Anal. Chem.*, 69 (1997) 2464.
326 S. Yamasaki, T. Yoshida, A. Tsumura and G. Murase, ICP-MS: one of the most promising techniques for comprehensive studies of heavy metals in soils, *Transactions 14th International Congress of Soil Science*, II (1990) 108.
327 W. Zhu, E.W.B. De Leer, M. Kennedy, P. Kelderman and G.J.F.R. Alaerts, Study of a partial least-squares regression model for rare earth element determination by inductively coupled plasma mass spectrometry, *J. Anal. At. Spectrom.*, 12 (1997) 663.
328 J. Arunachalam, C. Mohl, P. Ostapczuk and H. Emons, Multielement characterization of soil samples with ICP-MS for environmental studies, *Fresenius J. Anal. Chem.*, 352 (1995) 577.
329 M. Viczián, A. Lásztity and R.M. Barnes, Identification of potential environmental sources of childhood lead poisoning by inductively coupled plasma mass spectrometry. Verification and case studies, *J. Anal. At. Spectrom.*, 5 (1990) 293.

330 B. Marin, M. Valladon, M. Polve and A. Monaco, Reproducibility testing of a sequential extraction scheme for the determination of trace metal speciation in a marine reference sediment by inductively coupled plasma mass spectrometry, *Anal. Chim. Acta*, 342 (1997) 91.
331 H.M. Al-Swaidan, Determination of lead and nickel in Saudi Arabian crude oils by ICP/MS using MIBK for sample pretreatment, *Anal. Lett.*, 25 (1992) 2157.
332 L. Moberg, K. Pettersson, I. Gustavsson and B. Karlberg, Determination of cadmium in fly ash and metal alloy reference materials by inductively coupled plasma mass spectrometry and chemometrics, *J. Anal. At. Spectrom.*, 14 (1999) 1055.
333 P. Krause, M. Kriews, W. Dannecker, C.-D. Garbe-Schönberg and M. Kersten, Determination of $^{206/207}$Pb isotope ratios by ICP-MS in particulate matter from the North Sea environment, *Fresenius J. Anal. Chem.*, 347 (1993) 324.
334 M. Hamester, H. Stechmann, P. Krause and W. Dannecker, Characterization of selected sources for particulate emissions by inductively coupled plasma mass spectrometry, *Anal. Proc.*, 29 (1992) 293.
335 V. Krivan, G. Schneider, H. Baumann and U. Reus, Multi-element characterization of tobacco smoke condensate, *Fresenius J. Anal. Chem.*, 348 (1994) 218.
336 J. Goossens and R. Dams, Anion exchange for the elimination of spectral interferences caused by chlorine and sulfur in inductively coupled plasma mass spectrometry, *J. Anal. At. Spectrom.*, 7 (1992) 1167.
337 C.J. Lord III, Determination of lead and lead isotope ratios in gasoline by inductively coupled plasma mass spectrometry, *J. Anal. At. Spectrom.*, 9 (1994) 599.
338 E.S. Beary and P.J. Paulsen, Selective application of chemical separations to isotope dilution inductively coupled plasma mass spectrometric analyses of standard reference materials, *Anal. Chem.*, 65 (1993) 1602.
339 E.S. Beary, P.J. Paulsen and J.D. Fassett, Sample preparation approaches for isotope dilution inductively coupled plasma mass spectrometric certification of reference materials, *J. Anal. At. Spectrom.*, 9 (1994) 1363.
340 I. Papadakis, P.D.P. Taylor and P. De Bièvre, SI-traceable values for cadmium and lead concentration in the candidate reference material, MURST-ISS A1 Antarctic sediment, by combination of ICP-MS with isotope dilution, *Anal. Chim. Acta*, 346 (1997) 17.
341 S. Caroli, O. Senofonte, S. Caimi, P. Kárpáti, Comparative study of marine sediment from Antarctica by low-pressure discharge atomic emission spectrometry and inductively coupled plasma-based spectrometry, *J. Anal. At. Spectrom.*, 11 (1996) 773.
342 J. Yoshinaga, A. Nakama and K. Takata, Determination of total tin in sediment reference materials by isotope dilution inductively coupled

plasma mass spectrometry after alkali fusion, *Analyst*, 124 (1999) 257.
343 J.P. Valles Mota, M.R. Fernández de la Campa, J.I. Garcia Alonso and A. Sanz-Medel, Determination of cadmium in biological and environmental materials by isotope dilution inductively coupled plasma mass spectrometry: effect of flow sample introduction methods, *J. Anal. At. Spectrom.*, 14 (1999) 113.
344 J. Hinrichs and B. Schnetger, A fast method for the simultaneous determination of ^{230}Th, ^{234}U and ^{235}U with isotope dilution sector field ICP-MS, *Analyst*, 124 (1999) 927.
345 Y. Muramatsu, S. Uchida, K. Tagami, S. Yoshida and T. Fujikawa, Determination of plutonium concentration and its isotopic ratio in environmental materials by ICP-MS after separation using ion-exchange and extraction chromatography, *J. Anal. At. Spectrom.*, 14 (1999) 859.
346 C.-F. Wang, J.-Y. Yang and C.-H. Ke, Multi-element analysis of airborne particulate matter by various spectrometric methods after microwave digestion, *Anal. Chim. Acta*, 320 (1996) 207.
347 C.F. Wang, W.H. Chen, M.H. Yang and P.C. Chiang, Microwave decomposition for airborne particulate matter for the determination of trace elements by inductively coupled plasma mass spectrometry, *Analyst*, 120 (1995) 1681.
348 C.F. Wang, E.E. Chang, P.C. Chiang and N.K. Aras, Analytical procedures on multi-element determinations of airborne particles for receptor model use, *Analyst*, 120 (1995) 2521.
349 H. Lachas, R. Richaud, K.E. Jarvis, A.A. Herod, D.R. Dugwell and R. Kandiyoti, Determination of 17 trace elements in coal and ash reference materials by ICP-MS applied to milligram sample sizes, *Analyst*, 124 (1999) 177.
350 C.-F. Wang, C.Y. Chang, C.J. Chin and L.C. Men, Determination of arsenic and vanadium in airborne related reference materials by inductively coupled plasma-mass spectrometry, *Anal. Chim. Acta*, 392 (1999) 299.
351 P. Schramel, I. Wendler and S. Lustig, Capability of ICP-MS (pneumatic nebulization and ETV) for Pt-analysis in different matrices at ecologically relevant concentrations, *Fresenius J. Anal. Chem.*, 353 (1995) 115.
352 K.E. Jarvis, J.G. Williams, B.C.H. Gibson, E. Temmerman and C. De Cuyper, Rapid and accurate determination of manganese in washing powders using alkali fusion and inductively coupled plasma techniques, *Analyst*, 121 (1996) 1929.
353 K. Van den Broeck, L. Helsen, C. Vandecasteele and E. Van den Bulck, Determination and characterisation of copper, chromium and arsenic in chromated copper arsenate (CCA) treated wood and its pyrolysis residues by inductively coupled plasma mass spectrometry, *Analyst*, 122 (1997) 695.

354 C.A. Williams, F.R. Abou-Shakra and N.I. Ward, Investigations into the use of inductively coupled plasma mass spectrometry for the determination of gold in plant materials, *Analyst*, 120 (1995) 341.
355 F. Petrucci, S. Caimi, G. Mura and S. Caroli, *Artemia* as a bioindicator of environmental contamination by trace elements, *Microchem. J.*, 51 (1995) 181.
356 E.S. Beary, P.J. Paulsen, L.B. Jassie and J.D. Fassett, Determination of environmental lead using continuous-flow microwave digestion isotope dilution inductively coupled plasma mass spectrometry, *Anal. Chem.*, 69 (1997) 758.
357 A. Zurhaar and L. Mullings, Characterisation of forensic glass samples using inductively coupled plasma mass spectrometry, *J. Anal. At. Spectrom.*, 5 (1990) 611.
358 U. Rohr, L. Meckel and H.M. Ortner, Ultratrace analysis of uranium and thorium in glass Part 1: ICP-MS, classical photometry and chelate-GC, *Fresenius J. Anal. Chem.*, 348 (1994) 356.
359 M.D. Norman, W.L. Griffin, N.J. Pearson, M.O. Garcia and S.Y. O'Reilly, Quantitative analysis of trace element abundances in glasses and minerals: a comparison of laser ablation inductively coupled plasma mass spectrometry, solution inductively coupled plasma mass spectrometry, proton microprobe and electron microprobe data, *J. Anal. At. Spectrom.*, 13 (1998) 477.
360 K.A. Wolnik, D.T. Heitkemper, J.B. Crowe, B.S. Barnes and T.W. Brueggemeyer, Application of inductively coupled plasma atomic emission and mass spectrometry to forensic analysis of sodium gamma hydroxy butyrate and ephedrine hydrochloride, *J. Anal. At. Spectrom.*, 10 (1995) 177.
361 R.J. Rosenberg, R. Zilliacus and P.K.G. Manninen, Determination of transition metals in the primary water of pressurized water reactors by inductively coupled plasma mass spectrometry, *J. Anal. At. Spectrom.*, 9 (1994) 713.
362 J.I.G. Alonso, D. Thoby-Schultzendorff, B. Giovanonne, J.-P. Glatz, G. Pagliosa and L. Koch, Characterization of spent nuclear fuel dissolver solutions and dissolution residues by inductively coupled plasma mass spectrometry, *J. Anal. At. Spectrom.*, 9 (1994) 1209.
363 J.I.G. Alonso, F. Sena and L. Koch, Determination of ^{99}Tc in nuclear samples by inductively coupled plasma mass spectrometry, *J. Anal. At. Spectrom.*, 9 (1994) 1217.
364 S. Vijayalakshmi, R.K. Prabhu, T.R. Mahalingam and C.K. Mathews, Determination of trace metals in uranium oxide by inductively coupled plasma mass spectrometry combined with on-line solvent extraction, *J. Anal. At. Spectrom.*, 7 (1992) 565.
365 C. Riglet, O. Provitina, J.-L. Dautheribes and D. Revy, Determination of

traces of neptunium-237 in enriched uranium solutions using inductively coupled plasma mass spectrometry, *J. Anal. At. Spectrom.*, 7 (1992) 923.
366 P. Hepiegne, D. Dall'ava, R. Clement and J. P. Degros, The separation of ^{99}Tc from low and medium-level radioactive wastes and its determination by inductively coupled plasma mass spectrometry, *Talanta*, 42 (1995) 803.
367 J.S. Crain and D.L. Gallimore, Inductively coupled plasma-mass spectrometry of synthetic elements: ^{99}Tc, *Appl. Spectrosc.*, 46 (1992) 547.
368 R.L. Sutton, Analysis of liquid-phase tungsten hexafluoride residue by inductively coupled plasma mass spectrometry with ultrasonic nebulization, *J. Anal. At. Spectrom.*, 9 (1994) 1079.
369 A. Makishima, I. Inamoto and K. Chiba, Determination of gallium in pure aluminum by isotope dilution analysis with solvent extraction using inductively coupled plasma mass spectrometry, *Appl. Spectrosc.*, 44 (1990) 91.
370 K. Chiba, I. Inamoto and M. Saeki, Application of isotope dilution analysis-inductively coupled plasma mass spectrometry to the precise determination of silver and antimony in pure copper, *J. Anal. At. Spectrom.*, 7 (1992) 115.
371 J.A.C. Broekaert, R. Brandt, F. Leis, C. Pilger, D. Pollmann, P. Tschöpel and G. Tölg, Analysis of aluminium oxide and silicon carbide ceramic materials by inductively coupled plasma mass spectrometry, *J. Anal. At. Spectrom.*, 9 (1994) 1063.
372 C. Pilger, F. Leis, P. Tschöpel, J.A.C. Broekaert and G. Tölg, Analysis of silicon carbide powders with ICP-MS subsequent to sample dissolution without and with matrix removal, *Fresenius J. Anal. Chem.*, 351 (1995) 110.
373 X. Feng and G. Horlick, Analysis of aluminium alloys using inductively coupled plasma and glow discharge mass spectrometry, *J. Anal. At. Spectrom.*, 9 (1994) 823.
374 F. Vanhaecke, H. Vanhoe, C. Vandecasteele and R. Dams, Precise determination of boron in titanium by inductively coupled plasma mass spectrometry, *Anal. Chim. Acta*, 244 (1991) 115.
375 A. Held, P. Taylor, C. Ingelbrecht, P. De Bièvre, J.A.C. Broekaert, M. Van Straaten and R. Gijbels, Determination of scandium in high-purity Ti using inductively coupled plasma mass spectrometry and glow discharge mass spectrometry as part of its certification as a reference material, *J. Anal. At. Spectrom.*, 10 (1995) 849.
376 E. Toda, A. Kioki and M. Kubota, Determination of impurities in high-purity selenium by inductively coupled plasma mass spectrometry after acetate-form anion-exchange separation, *Anal. Chim. Acta*, 333 (1996) 51.
377 V.K. Panday, J.S. Becker and H.-J. Dietze, Determination of trace impurities in tantalum by inductively coupled plasma mass spectrometry after removal of the matrix by liquid-liquid extraction, *Anal. Chim. Acta*, 329 (1996) 153.

378 B. Li, Y. Zhang and M. Yin, Determination of trace amounts of rare earth elements in high-purity cerium oxide by inductively coupled plasma mass spectrometry after separation by solvent extraction, *Analyst*, 122 (1997) 543.
379 S.-X. Zhang, S. Murachi, T. Imasaka and M. Watanabe, Determination of rare earth impurities in ultrapure europium oxide by inductively-coupled plasma mass spectrometry, *Anal. Chim. Acta*, 314 (1995) 193.
380 S. Kozuka, Y. Yokote, K. Abe, M. Hayashi and H. Matsunaga, Determination of impurities in strontium titanate ceramics by inductively coupled plasma mass spectrometry, *Fresenius J. Anal. Chem.*, 351 (1995) 801.
381 V. Valaram, K.V. Anjaiah and M.R.P. Reddy, Comparative study on the trace and rare earth element analysis of an Indian polymetallic nodule reference sample by inductively coupled plasma atomic emission spectrometry and inductively coupled plasma mass spectrometry, *Analyst*, 120 (1995) 1401.
382 C. Pin, S. Lacombe, P. Telouk and J.-L. Imbert, Isotope dilution inductively coupled plasma mass spectrometry: a straightforward method for rapid and accurate determination of uranium and thorium in silicate rocks, *Anal. Chim. Acta*, 256 (1992) 153.
383 R. Juvonen, E. Kallio and T. Lakomaa, Determination of precious metals in rocks by inductively coupled plasma mass spectrometry using nickel sulfide concentration. Comparison with other pre-treatment methods, *Analyst*, 119 (1994) 617.
384 E.H. Bailey, A.J. Kemp and K.V. Ragnarsdottir, Determination of uranium and thorium in basalts and uranium in aqueous solution by inductively coupled plasma mass spectrometry, *J. Anal. At. Spectrom.*, 8 (1993) 551.
385 A. Rivoldini and S. Fadda, Inductively coupled plasma mass spectrometric determination of low-level rare earth elements in rocks using potassium-based fluxes for sample decomposition, *J. Anal. At. Spectrom.*, 9 (1994) 519.
386 L.J. Hulbert and D.C. Grégoire, Re-Os isotope systematics of the Rankin Inlet Ni ores: an example of the application of ICP-MS to investigate Ni-Cu-PGE mineralization, and the potential use of Os isotopes in mineral exploration, *Can. Mineral.*, 31 (1993) 861.
387 S.B. Beneteau and J.M. Richardson, Rhenium-osmium isotope ratio determination by ICP-MS, *At. Spectrosc.*, 13 (1992) 118.
388 G.S. Reddi, C.R.M. Rao, T.A.S. Rao, S.V. Lakshmi, R.K. Prabbu, and T. R. Mahalingam, Nickel sulphide fire assay—ICPMS method for the determination of platinum group elements: a detailed study on the recovery and losses at different stages, *Fresenius J. Anal. Chem.*, 348 (1994) 350.
389 B.J. Perry, J.C. Van Loon and D.V. Speller, Dry-chlorination inductively coupled plasma mass spectrometric method for the determination of

platinum group elements in rocks, *J. Anal. At. Spectrom.*, 7 (1992) 883.
390 M.M. Cheatham, W.F. Sangrey and W.M. White, Sources of error in external calibration ICP-MS analysis of geological samples and an improved non-linear drift correction procedure, *Spectrochim. Acta*, 48B (1993) E487.
391 R. Alaimo and P. Censi, Quantitative determination of major, minor, and trace elements in USGS rock standards by inductively coupled plasma-mass spectrometry, *At. Spectrosc.*, 13 (1992) 113.
392 G.E.M. Hall, J.C. Pelchat and J. Loop, Determination of zirconium, niobium, hafnium and tantalum at low levels in geological materials by inductively coupled plasma mass spectrometry, *J. Anal. At. Spectrom.*, 5 (1990) 339.
393 M. Ochsenkühn-Petropoulou, K. Ochsenkühn and J. Luck, Comparison of inductively coupled plasma mass spectrometry with inductively coupled plasma atomic emission spectrometry and instrumental neutron activation analysis for the determination of rare earth elements in Greek bauxites, *Spectrochim. Acta*, 46B (1991) 51.
394 Q. Xie and R. Kerrich, Application of isotope dilution for precise measurement of Zr and Hf in low-abundance samples and international reference materials by inductively coupled plasma mass spectrometry: implications of Zr (Hf)/REE fractionations in komatiites, *Chem. Geol.*, 123 (1995) 17.
395 J. Enzweiler, P.J. Potts and K.E. Jarvis, Determination of platinum, palladium, ruthenium and iridium in geological samples by isotope dilution inductively coupled plasma mass spectrometry using a sodium peroxide fusion and tellurium coprecipitation, *Analyst*, 120 (1995) 1391.
396 A. Behrens, Determination of gadolinium in geological samples by inductively coupled plasma mass spectrometry and multivariate calibration, *Spectrochim. Acta*, 50B (1995) 1521.
397 M. Barbaro, B. Passariello, S. Quaresima, A. Casciello and A. Marabini, *Microchem. J.*, 51 (1995), 312.
398 V. Valaram, S. L. Ramesh and K.V. Anjaiah, Comparative study of the sample decomposition procedures in the determination of trace and rare earth elements in anorthosites and related rocks by ICP-MS, *Fresenius J. Anal. Chem.*, 353 (1995) 176.
399 S.M. Eggins, J.D. Woodhead, L.P.J. Kinsley, G.E. Mortimer, P. Sylvester, M.T. McCulloch, J.M. Hergt, M.R. Handler, A simple method for the precise determination of ≥40 trace elements in geological samples by ICPMS using enriched isotope internal standardization, *Chem. Geol.*, 134 (1997) 311.
400 Y.V. Yi and A. Masuda, Simultaneous determination of ruthenium, palladium, iridium, and platinum at ultratrace levels by isotope dilution inductively coupled plasma mass spectrometry in geological samples, *Anal. Chem.*, 68 (1996) 1444.

401 L. Haichen, L. Ying and Z. Zhanxia, Determination of ultra-trace rare earth elements in chondritic meteorites by inductively coupled plasma mass spectrometry, *Spectrochim. Acta*, 53B (1997) 1399.
402 K.H. Lieser and W. Fey, Isotopic fingerprint method: assessment of the origin of rare earth compounds from the isotope ratios of lead impurities, *Fresenius J. Anal. Chem.*, 351 (1995) 129.
403 M.M. Totland, I. Jarvis and K.E. Jarvis, Microwave digestion and alkali fusion procedures for the determination of the platinum-group elements and gold in geological materials by ICP-MS, *Chem. Geol.*, 124 (1995) 21.
404 I. Jarvis, M.M. Totland and K.E. Jarvis, Assessment of Dowex 1-X8-based anion-exchange procedures for the separation and determination of ruthenium, rhodium, palladium, iridium, platinum and gold in geological samples by inductively coupled plasma mass spectrometry, *Analyst*, 122 (1997) 19.
405 I. McDonald, J.W. Harris and D.J. Vaughan, Determination of noble metals in sulphide inclusions from diamonds using inductively coupled plasma–mass spectrometry, *Anal. Chim. Acta*, 333 (1996) 41.
406 Z. Chen, B.J. Fryer, H.P. Longerich and S.E. Jackson, Determination of the precious metals in milligram samples of sulfides and oxides using inductively coupled plasma mass spectrometry after ion exchange preconcentration, *J. Anal. At. Spectrom.*, 11 (1996) 805.
407 A.J. Courtney, D.J. Die and M.J. Holmes, Discriminating populations of the eastern king prawn, *Panaeus plebejus*, from different estuaries using ICP-MS trace element analysis, *At. Spectrosc.*, 15 (1994) 1.
408 P. Twiss, R.J. Watling and D. Delev, Determination of thorium and uranium in faecal material from occupationally-exposed workers using ICP-MS, *At. Spectrosc.*, 15 (1994) 36.
409 S.F. Durrant, A. Krushevska, D. Amarasiriwardena, M.D. Argentine, S. Romon-Guesnier and R.M. Barnes, Matrix separation by chelation to prepare biological materials for isotopic zinc analysis by inductively coupled plasma mass spectrometry, *J. Anal. At. Spectrom.*, 9 (1994) 199.
410 C.J. Amarasiriwardena, A. Krushevska, H. Foner, M.D. Argentine and R.M. Barnes, Sample preparation for inductively coupled plasma mass spectrometric determination of the zinc-70 to zinc-68 isotope ratio in biological samples, *J. Anal. At. Spectrom.*, 7 (1992) 915.
411 P.G. Whittaker, J.F.R. Barrett and J.G. Williams, Precise determination of iron isotope ratios in whole blood using inductively coupled plasma mass spectrometry, *J. Anal. At. Spectrom.*, 7 (1992) 109.
412 J.F.R. Barrett, P.G. Whittaker, J.G. Williams and T. Lind, Absorption of non-heme iron in normal women measured by the incorporation of two stable isotopes into erythrocytes, *Clin. Sci.*, 83 (1992) 213.
413 A. Alimonti, F. Petrucci, B. Santucci, A. Cristaudo and S. Caroli, Determination of chromium and nickel in human blood by means of

inductively coupled plasma mass spectrometry, *Anal. Chim. Acta*, 306 (1995) 35.

414 K.Y. Patterson, C. Veillon, P.B. Moser-Veillon and G.F. Wallace, Determination of zinc stable isotopes in biological materials using isotope dilution inductively coupled plasma mass spectrometry, *Anal. Chim. Acta*, 258 (1992) 317.

415 B. Casetta, M. Roncadin, G. Montanari and M. Furlanut, Determination of platinum in biological fluids by ICP-mass spectrometry, *At. Spectrosc.*, 12 (1991) 81.

416 H. Vanhoe, R. Dams and J. Versieck, Use of inductively coupled plasma mass spectrometry for the determination of ultra-trace elements in human serum, *J. Anal. At. Spectrom.*, 9 (1994) 23.

417 L. Moens, P. Verrept, R. Dams, U. Greb, G. Jung and B. Laser, New high-resolution inductively coupled plasma mass spectrometry technology applied for the determination of V, Fe, Cu, Zn and Ag in human serum, *J. Anal. At. Spectrom.*, 9 (1994) 1075.

418 M. Higashiura, H. Uchida, T. Uchida and H. Wada, Inductively coupled plasma mass spectrometric determination of gold in serum: comparison with flame and furnace atomic absorption spectrometry, *Anal. Chim. Acta*, 304 (1995) 317.

419 P.L. Trentini, M. Ascanelli, B. Zanforlini, F. Venturini, G. Bucci and F. Fagioli, Determination of aluminium by inductively coupled plasma mass spectrometry in serum of patients treated by haemodialysis, dialysis solutions and tap water, and a comparison with atomic absorption spectrometry, *J. Anal. At. Spectrom.*, 8 (1993) 905.

420 E.T. Luong, R.S. Houk and R.E. Serfass, Chromatographic isolation of molybdenum from human blood plasma and determination by inductively coupled plasma mass spectrometry with isotope dilution, *J. Anal. At. Spectrom.*, 12 (1997) 703.

421 W.T. Buckley and M. Ihnat, Determination of copper, molybdenum and selenium in biological reference materials by inductively coupled plasma mass spectrometry, *Fresenius J. Anal. Chem.*, 345 (1993) 217.

422 J.W.H. Lam, J.W. McLaren and B.A.J. Methven, Determination of chromium in biological tissues by inductively coupled plasma mass spectrometry, *J. Anal. At. Spectrom.*, 10 (1995) 551.

423 K.E. Murphy and P.J. Paulsen, The determination of lead in blood using isotope dilution inductively coupled plasma mass spectrometry, *Fresenius J. Anal. Chem.*, 352 (1995) 203.

424 D.E. Nixon and T.P. Moyer, Routine clinical determination of lead, arsenic, cadmium, and thallium in urine and whole blood by inductively coupled plasma mass spectrometry, *Spectrochim. Acta*, 51B (1996) 13.

425 A. El-Jammal and D.M. Templeton, Measurement of platinum in biomedical silicones by ICP-MS, *Anal. Proc.*, 32 (1995) 293.

426 C.E. Sieniawska, R. Mensikov and H.T. Delves, Determination of total selenium in serum, whole blood and erythrocytes by ICP-MS, *J. Anal. At. Spectrom.*, 14 (1999) 109.
427 B.J. Perry and R.E. Balazs, ICP-MS method for the determination of platinum in suspensions of cells exposed to cisplatin, *Anal. Proc.*, 31 (1994) 269.
428 Y. Ming and L. Bing, Determination of rare earth elements in human hair and wheat flour reference materials by inductively coupled plasma mass spectrometry with dry ashing and microwave digestion, *Spectrochim. Acta*, 53B (1998) 1447.
429 S. Evans and U. Krähenbühl, Improved boron determination in biological material by inductively coupled plasma mass spectrometry, *J. Anal. At. Spectrom.*, 9 (1994) 1249.
430 R. Pepelnik, A. Prange and R. Niedergesäß, Comparative study of multi-element determination using inductively coupled plasma mass spectrometry, total reflection X-ray fluorescence spectrometry and neutron activation analysis, *J. Anal. At. Spectrom.*, 9 (1994) 1071.
431 J. Goossens, T. De Smaele, L. Moens and R. Dams, Accurate determination of lead in wines by inductively coupled plasma mass spectrometry, *Fresenius J. Anal. Chem.*, 347 (1993) 119.
432 G. Rädlinger and K.G. Heumann, Iodine determination in food samples using inductively coupled plasma isotope dilution mass spectrometry, *Anal. Chem.*, 70 (1998) 2221.
433 Y. Gélinas, A. Krushevska and R.M. Barnes, Determination of total iodine in nutritional and biological samples by ICP-MS following their combustion with an oxygen stream, *Anal. Chem.*, 70 (1998) 1021.
434 K. Günther, A. von Bohlen, G. Paprott and R. Klockenkämper, Multi-element analysis of biological reference materials by total-reflection X-ray fluorescence and inductively coupled plasma mass spectrometry in the semiquant mode, *Fresenius J. Anal. Chem.*, 342 (1992) 444.
435 C. Veillon and K.Y. Patterson, selenium, chromium and zinc content of proposed SRM 1846 infant formula, *Fresenius J. Anal. Chem.*, 352 (1995) 77.
436 M. Horn and K.G. Heumann, Comparison of heavy metal analyses in hydrofluoric acid used in microelectronic industry by ICP-MS and thermal ionization isotope dilution mass spectrometry, *Fresenius J. Anal. Chem.*, 350 (1994) 286.
437 E.J. McCurdy, The preparation of plant samples and their analysis by ICP-MS, In: K.E. Jarvis, A.L. Gray, I. Jarvis and J. Williams (Eds), *Plasma Source Mass Spectrometry*, Royal Society of Chemistry, Special Publication #85, 1990, pp. 79–93.
438 J. Toole, A.S. Hursthouse, P. McDonald, K. Sampson, M.S. Baxter, R.D. Scott and K. McKay, The determination of actinides in environmental

samples by ICP-MS, In: K.E. Jarvis, A.L. Gray, I. Jarvis and J. Williams (Eds), *Plasma Source Mass Spectrometry*, Royal Society of Chemistry, Special Publication #85, 1990, pp. 155–162.
439 B.J. Perry and J.C. Van Loon, ICP-MS analysis for weakly bound gold in humus samples: an aid to gold exploration in areas of glacially transported overburden, In: G. Holland and A. N. Eaton (Eds), *Applications of Plasma Source Mass Spectrometry*, The Royal Society of Chemistry, Cambridge, 1991, pp. 38–47.
440 Ihsanullah and B.W. East, A procedure for the determination of ^{99}Tc in environmental samples by ICP-MS, In: G. Holland and A. N. Eaton (Eds), *Applications of Plasma Source Mass Spectrometry*, The Royal Society of Chemistry, Cambridge, 1991, pp. 48–58.
441 S. Yamasaki, A. Tsumura and D. Cai, Microwave dissolution and ICP-MS for the determination of trace and ultra-trace elements in plant tissue, In: G. Holland and A.N. Eaton (Eds), *Applications of Plasma Source Mass Spectrometry*, The Royal Society of Chemistry, Cambridge, 1991, pp.110-118.
442 A. Kawasaki, K. Haga, S. Arai and S. Yamasaki, Effect of digestion methods on ICP-MS measurements of Be, V, Ga, Sb, La, Bi in organic fertilizers, In: G. Holland and A.N. Eaton (Eds), *Applications of Plasma Source Mass Spectrometry*, The Royal Society of Chemistry, Cambridge, 1991, pp. 139–148.
443 C. Vandecasteele, K. Van den Broeck, V. Dutré and H. Cooreman, Environmental applications of ICP-MS using a PQe spectrometer, In: G. Holland and A.N. Eaton (Eds), *Applications of Plasma Source Mass Spectrometry II*, The Royal Society of Chemistry, Special Publication No. 124, Cambridge, 1993, pp. 48–63.
444 S. Knoblock, H. König and G. Wünsch, ICP-MS determinations in automotive catalyst exhaust, In: G. Holland and A.N. Eaton (Eds), *Applications of Plasma Source Mass Spectrometry II*, The Royal Society of Chemistry, Special Publication No. 124, Cambridge, 1993, pp. 108–114.
445 A.R. Date and K.E. Jarvis, Application of ICP-MS in the earth sciences, In: A.R. Date and A.L. Gray (Eds), *Applications of Inductively Coupled Plasma Mass Spectrometry*, Blackie and Son Ltd., Glasgow, 1989, pp. 43–70.
446 N.I. Ward, Environmental analysis using ICP-MS, In: A.R. Date and A.L. Gray (Eds), *Applications of Inductively Coupled Plasma Mass Spectrometry*, Blackie and Son Ltd., Glasgow, 1989, pp. 189–219.
447 M.E. Ketterer, J.J. Reschl and M.J. Peters, The determination of titanium, copper, and zinc in geological materials by inductively coupled plasma mass spectrometry with multivariate calibration, In: K.E. Jarvis, A.L. Gray, I. Jarvis and J. Williams (Eds), *Plasma Source Mass Spectrometry*, Royal Society of Chemistry, Special Publication #85, 1990, pp. 52–65.

448 J.M. Richardson, A.P. Dickin and R.H. McNutt, Re-Os isotope ratio determinations by ICP-MS: a review of analytical techniques and geological applications, In: K.E. Jarvis, A.L. Gray, I. Jarvis and J. Williams (Eds), *Plasma Source Mass Spectrometry*, Royal Society of Chemistry, Special Publication #85, 1990, pp. 120–144.
449 J. Godfrey and E. McCurdy, Investigation into the feasibility of ICP-MS as an alternative to fire assay measurements for gold and the platinum group elements, In: G. Holland and A.N. Eaton (Eds), *Applications of Plasma Source Mass Spectrometry II*, The Royal Society of Chemistry, Special Publication No. 124, Cambridge, 1993, pp. 64–71.
450 B.J. Perry, R.R. Barefoot, J.C. Van Loon, A.J. Naldrett and D.V. Speller, A dry chlorination/microwave digestion/ICP-MS analytical method for the determination of the platinum group elements and gold in metallic and non-metallic fractions of rocks, In: G. Holland and A.N. Eaton (Eds), *Applications of Plasma Source Mass Spectrometry II*, The Royal Society of Chemistry, Special Publication No. 124, Cambridge, 1993, pp. 91–101.
451 Y. Ming and Y. Ningwan, Determination of trace elements in 4 Chinese bio-geochemical reference materials by ICP-MS, In: G. Holland and A.N. Eaton (Eds), *Applications of Plasma Source Mass Spectrometry II*, The Royal Society of Chemistry, Special Publication No. 124, Cambridge, 1993, pp. 115–123.
452 E. Veldeman, L. Van't dack, R. Gijbels, M. Campbell, F. Vanhaecke, H. Vanhoe and C. Vandescasteele, Analysis of thermal waters by ICP-MS, In: G. Holland and A.N. Eaton (Eds), *Applications of Plasma Source Mass Spectrometry*, The Royal Society of Chemistry, Cambridge, 1991, pp. 25–33.
453 E. Johansson and T. Liljefors, Semi-quantitative estimation of some elements in standards and drinking waters by ICP-MS, In: G. Holland and A.N. Eaton (Eds), *Applications of Plasma Source Mass Spectrometry*, The Royal Society of Chemistry, Cambridge, 1991, pp. 34–37.
454 A. Tsumura and S. Yamasaki, Determination of ultra-trace levels of rare earth elements in terrestrial water by high resolution ICP-MS with an ultrasonic nebulizer, In: G. Holland and A.N. Eaton (Eds), *Applications of Plasma Source Mass Spectrometry*, The Royal Society of Chemistry, Cambridge, 1991, pp. 119–129.
455 S. Yamasaki, A. Tsumura and T. Kobayashi, Determination of the first transition elements in terrestrial water by high resolution ICP-MS with an ultrasonic nebulizer, In: G. Holland and A.N. Eaton (Eds), *Applications of Plasma Source Mass Spectrometry II*, The Royal Society of Chemistry, Special Publication No. 124, Cambridge, 1993, pp. 1–11.
456 A.V. Mistry and D.S. Lowe, A comparison of SCIEX Total Quant results with Quantitative results of various solutions by ICP-MS, In: G. Holland and A.N. Eaton (Eds), *Applications of Plasma Source Mass Spectrometry*

II, The Royal Society of Chemistry, Special Publication No. 124, Cambridge, 1993, pp. 22–28.
457 T. Probst, P. Zeh and J.I. Kim, A comparison of ICP-MS with neutron activation analysis for multielement determination in groundwater, In: G. Holland and A.N. Eaton (Eds), *Applications of Plasma Source Mass Spectrometry II*, The Royal Society of Chemistry, Special Publication No. 124, Cambridge, 1993, pp. 29–37.
458 H.E. Taylor, Water resources, In: A.R. Date and A.L. Gray (Eds), *Applications of Inductively Coupled Plasma Mass Spectrometry*, Blackie and Son Ltd., Glasgow, 1989, pp. 71–89.
459 P. Leprovost, R. Schott and A. Vian, Impurity assay in reprocessing uranyl nitrate, In: G. Holland and A.N. Eaton (Eds), *Applications of Plasma Source Mass Spectrometry II*, The Royal Society of Chemistry, Special Publication No. 124, Cambridge, 1993, pp. 213–221.
460 J.A.F. Moore, M.J. McGuire and P.A. Hart, The application of inductively coupled plasma mass spectrometry to the analysis of iron materials, In: K.E. Jarvis, A.L. Gray, I. Jarvis and J. Williams (Eds), *Plasma Source Mass Spectrometry*, Royal Society of Chemistry, Special Publication #85, 1990, pp. 163–169.
461 F. Vanhaecke, H. Vanhoe, C. Vandecasteele and R. Dams, Determination of boron in a titanium reference material by ICP-MS, In: G. Holland and A.N. Eaton (Eds), *Applications of Plasma Source Mass Spectrometry*, The Royal Society of Chemistry, Cambridge, 1991, pp. 149–161.
462 H. Ekstroem and I. Gustavsson, Application of ICP-MS to steel, other metals, and metal alloys, In: G. Holland and A.N. Eaton (Eds), *Applications of Plasma Source Mass Spectrometry II*, The Royal Society of Chemistry, Special Publication No. 124, Cambridge, 1993, pp. 150–157.
463 I.A. Rigby and S.C. Stephen, Determination of antimony and other impurities in high purity copper by ICP-MS, In: G. Holland and A.N. Eaton (Eds), *Applications of Plasma Source Mass Spectrometry II*, The Royal Society of Chemistry, Special Publication No. 124, Cambridge, 1993, pp. 158–164.
464 B. Meddings and R. Ng, ICP-MS in a metallurgical laboratory, In: A.R. Date and A.L. Gray (Eds), *Applications of Inductively Coupled Plasma Mass Spectrometry*, Blackie and Son Ltd., Glasgow, 1989, pp. 220–241.
465 A. Walsh, D. Potter, E. McCurdy and R.C. Hutton, Direct analysis of semiconductor grade reagents by ICP-MS, In: G. Holland and A.N. Eaton (Eds), *Applications of Plasma Source Mass Spectrometry*, The Royal Society of Chemistry, Cambridge, 1991, pp. 12–24.
466 B. Gercken, J. Pavel and O. Suter, Determination of trace metal impurities in electronic chemicals by ICP-MS and other methods, In: G. Holland and A.N. Eaton (Eds), *Applications of Plasma Source Mass Spectrometry II*, The Royal Society of Chemistry, Special Publication No.

124, Cambridge, 1993, pp. 12–21.
467 A.A. van Heuzen, ICP-MS in the petroleum industry, In: A.R. Date and A.L. Gray (Eds), *Applications of Inductively Coupled Plasma Mass Spectrometry*, Blackie and Son Ltd., Glasgow, 1989, pp. 169–188.
468 H. Vanhoe, C. Vandecasteele, J. Versieck and R. Dams, Evaluation of ICP-MS for the determination of trace and ultra-trace elements in human serum after simple dilution, In: K.E. Jarvis, A.L. Gray, I. Jarvis and J. Williams (Editors), *Plasma Source Mass Spectrometry*, Royal Society of Chemistry, Special Publication #85, 1990, pp. 66–78.
469 P. Tothill, K. MacKay, L.M. Matheson, M. Robbins and J.F. Smyth, The use of ICP-MS to determine the retention and distribution of platinum in animals following the administration of cisplatin, In: G. Holland and A.N. Eaton (Eds), *Applications of Plasma Source Mass Spectrometry*, The Royal Society of Chemistry, Cambridge, 1991, pp. 89–95.
470 Th.M. Lutz, P.M.V. Nirel and B. Schmidt, Whole blood analysis by ICP-MS, In: G. Holland and A.N. Eaton (Eds), *Applications of Plasma Source Mass Spectrometry*, The Royal Society of Chemistry, Cambridge, 1991, pp. 96–100.
471 D.M. Templeton and Margaret-Anne Vaughan, Multivariate analysis of ICP-mass spectra: determination of nickel and iron in body fluids, In: G. Holland and A.N. Eaton (Eds), *Applications of Plasma Source Mass Spectrometry*, The Royal Society of Chemistry, Cambridge, 1991, pp. 101–109.
472 P. Fecher, M. Leibenzeder and C. Zizek, Decomposition temperature and its influence on trace element determination by ICP-MS and ICP-AES, In: G. Holland and A.N. Eaton (Eds), *Applications of Plasma Source Mass Spectrometry II*, The Royal Society of Chemistry, Special Publication No. 124, Cambridge, 1993, pp. 83–90.
473 L.S. Dale, J.L. Stauber, O.P. Farrell, T.M. Florence and B.L. Gulson, A comparative study of ICP-MS and TIMS for measuring skin absorption of lead, In: G. Holland and A.N. Eaton (Eds), *Applications of Plasma Source Mass Spectrometry II*, The Royal Society of Chemistry, Special Publication No. 124, Cambridge, 1993, pp. 124–129.
474 J.R. Dean, H.M. Crews and L. Ebdon, Applications in food science, In: A.R. Date and A.L. Gray (Eds), *Applications of Inductively Coupled Plasma Mass Spectrometry*, Blackie and Son Ltd., Glasgow, 1989, pp. 141–168.
475 E. Angelini, F. Rosalbino, C. Atzeni, P.F. Virdis and P. Bianco, Lead isotope analysis of nuragic bronzes and copper ores by ICP-MS, In: G. Holland and A.N. Eaton (Eds), *Applications of Plasma Source Mass Spectrometry II*, The Royal Society of Chemistry, Special Publication No. 124, Cambridge, 1993, pp. 165–174.
476 M.J. Campbell, C. Vandecasteele and R. Dams, The application of isotope dilution techniques to the accurate and precise determination of lead in

reference materials by ICP-MS, In: G. Holland and A.N. Eaton (Eds), *Applications of Plasma Source Mass Spectrometry*, The Royal Society of Chemistry, Cambridge, 1991, pp. 130–138.

477 P.J. Paulsen and E.S. Beary, ICP-MS certification of toxic metals on filters, In: G. Holland and A.N. Eaton (Eds), *Applications of Plasma Source Mass Spectrometry II*, The Royal Society of Chemistry, Special Publication No. 124, Cambridge, 1993, pp. 38–47.

478 L. Moens, H. Vanhoe, F. Vanhaecke, J. Goossens, M. Campbell and R. Dams, Application of inductively coupled plasma mass spectrometry to the certification of reference materials for the Community Bureau of Reference, *J. Anal. At. Spectrom.*, 9 (1994) 187.

479 H.M. Crews, A decade of ICP-MS analysis, *Am. Lab.*, 25(4) (1993) 34Z.

480 R.M. Barnes, Advances in inductively coupled plasma mass spectrometry: human nutrition and toxicology, *Anal. Chim. Acta*, 283 (1993) 115.

481 J.K. Friel, C. Mercer, W.L. Andrews, B.R. Simmons, S.E. Jackson and H.P. Longerich, *Biol. Trace Element Res.*, 54 (1996) 135.

482 F. Vanhaecke, G. de Wannemacker, L. Moens, R. Dams, C. Latkoczy, T. Prohaska and G. Stingeder, Dependence of detector dead time on analyte mass number in inductively coupled plasma mass spectrometry, *J. Anal. At. Spectrom.*, 13 (1998) 567.

483 L. Rottmann and K. G. Heumann, Development of an on-line isotope dilution technique with HPLC/ICP-MS for the accurate determination of elemental species, *Fresenius J. Anal. Chem.*, 350 (1994) 221.

484 M. Janghorbani and B.T.G. Ting, Stable isotope tracer applications of ICP-MS, In: A.R. Date and A.L. Gray (Eds), *Applications of Inductively Coupled Plasma Mass Spectrometry*, Blackie and Son Ltd., Glasgow, 1989, pp. 115–140.

485 J.T. Rowan and R.S. Houk, Attenuation of polyatomic ion interferences in inductively coupled plasma mass spectrometry by gas-phase collisions, *Appl. Spectrosc.*, **43** (1989) 976.

486 M.E. Ketterer, J.J. Reschl and M.J. Peters, Multivariate calibration in inductively coupled plasma mass spectrometry, *Anal. Chem.*, 61 (1989) 2031.

487 M.E. Ketterer and D.A. Biddle, Multivariate calibration in inductively coupled plasma mass spectrometry. 2. Effect of changes in abundances of interfering polyatomic ions, *Anal. Chem.*, 64 (1992) 1819.

488 A. Behrens, Multivariate calibration standardization used for the reduction of standards needed for multivariate calibration in inductively coupled plasma mass spectrometry, *Spectrochim. Acta*, 52B (1997) 445.

489 E.H. van Veen, S. Bosch and M.T.C. de Loos-Vollebregt, Spectral interpretation and interference correction in inductively coupled plasma mass spectrometry, *Spectrochim. Acta*, 49B (1994) 1347.

490 F. Vanhaecke, H. Vanhoe, R. Dams and C. Vandecasteele, The use of

internal standards in ICP-MS, *Talanta*, 39 (1992) 737.
491 M.P. Dziewatkoski, L.B. Daniels and J.W. Olesik, Time-resolved inductively coupled plasma mass spectrometry measurements with individual, monodisperse drop sample introduction, *Anal. Chem.*, 68 (1996) 1101.
492 D. Ekimoff, A.M. Van Norstrand and D.A. Mowers, Semiquantitative survey capabilities of inductively coupled plasma mass spectrometry, *Appl. Spectrosc.*, 43 (1989) 1252.
493 K.Y. Patterson, C. Veillon and T.C. O'Haver, Error propagation in isotope dilution analysis as determined by Monte Carlo simulation, *Anal. Chem.*, 66 (1994) 2829.
494 H. Klinkenberg, W. Van Borm and F. Souren, A theoretical adaptation of the classical isotope dilution technique for practical routine analytical determinations by means of inductively coupled plasma mass spectrometry, *Spectrochim. Acta*, 51B (1996), 139.
495 I.L. Gacía, P. Viñas and M.H. Córdoba, Calibration in flame atomic absorption spectrometry using a single standard and a gradient technique, *J. Anal. At. Spectrom.*, 9 (1994) 553.
496 N.N. Sesi, T.K. Starn and G.M. Hieftje, Evaluation of an automated on-line method for generating calibration curves in inductively coupled plasma atomic emission spectrometry, *Fresenius J. Anal. Chem.*, 346 (1993) 545.
497 M. Hoenig and M.-M. de Kersabiec, Sample preparation steps for analysis by atomic spectroscopy methods: present status, *Spectrochim. Acta*, 51B (1996) 1297.
498 F.E. Smith and E.A. Arsenault, Microwave-assisted sample preparation in analytical chemistry, *Spectrochim. Acta*, 51B (1996) 1207.
499 L. Ebdon, M. Foulkes and K. Sutton., Slurry nebulization in plasmas, *J. Anal. At. Spectrom.*, 12 (1997) 213.
500 J.G. Williams, A.L. Gray, P. Norman and L. Ebdon, Feasibility of solid sample introduction by slurry nebulisation for inductively coupled plasma mass spectrometry, *J. Anal. At. Spectrom.*, 2 (1987) 469.
501 H.G.M. Parry and L. Ebdon, Coal analysis by analytical atomic spectrometry (ICP-AES and ICP-MS) without sample dissolution, *Anal. Proc.* 25 (1988) 69.
502 P. Goodall, M.E. Foulkes and L. Ebdon, Slurry nebulization inductively coupled plasma spectrometry—the fundamental parameters discussed, *Spectrochim. Acta*, 48B (1993) 1563.
503 P. Heitland and J.A.C. Broekaert, Addition of small amounts of helium and hydrogen to the working gases in slurry nebulization inductively coupled plasma atomic emission spectrometry for the analysis of ceramic powders, *J. Anal. At. Spectrom.*, 12 (1997) 981.
504 L. Ebdon and P. Goodall, Slurry atomization using hydrogen-modified inductively coupled plasmas, *J. Anal. At. Spectrom.*, 7 (1992) 1111.

505 R.M. Barnes and S. Nikdel, Computer simulation of inductively coupled plasma discharge for spectrochemical analysis. II. Comparison of temperature and velocity profiles, and particle decomposition for inductively coupled plasma discharges in argon and nitrogen, *Appl. Spectrosc.*, 30 (1976) 310.
506 A. Montaser and R.L. Van Hoven, Mixed-gas, molecular-gas, and helium inductively coupled plasmas for analytical atomic spectrometry: A critical review, *CRC Crit. Rev. Anal. Chem.*, 18 (1987) 45.
507 A. Montaser, K.D. Ohls and D.W. Golightly, Inductively Coupled Plasmas in Gases Other Than Argon, In: A. Montaser, and D.W. Golightly, (Eds), *Inductively Coupled Plasmas in Analytical Atomic Spectrometry*, 2nd Edition, VCH Publishers Inc., New York, 1992.
508 A. Montaser and H. Zhang, Mass Spectrometry with mixed-gas and helium ICPs, In: A. Montaser (Ed.), *Inductively Coupled Plasma Mass Spectrometry*, Wiley-VCH, New York, 1998.
509 S.F. Durrant, Alternatives to all-argon plasma in inductively coupled plasma mass spectrometry (ICP-MS): an overview, *Fresenius J. Anal. Chem.*, 347 (1993) 389.
510 B.S. Sheppard and J.A. Caruso, Plasma mass spectrometry: consider the source, *J. Anal. At. Spectrom.*, 9 (1994) 145.
511 J. Wang, E.H. Evans and J.A. Caruso, Addition of molecular gases to argon gas flows for the reduction of polyatomic-ion interferences in inductively coupled plasma mass spectrometry, *J. Anal. At. Spectrom.*, 7 (1992) 929.
512 G. Xiao and D. Beauchemin, Reduction of matrix effects and mass discrimination in ICP-MS with optimized $Ar\text{-}N_2$ plasmas, *J. Anal. At. Spectrom.*, 9 (1994) 509.
513 M. Cai, D. A. Haydar, A. Montaser and J. Mostaghimi, Computer simulation of argon–nitrogen and argon–oxygen inductively coupled plasmas, *Spectrochim. Acta*, 52B (1997) 369.
514 H. Uchida and T. Ito, Evaluation of an inductively coupled air-argon plasma as an ion source for mass spectrometry, *J. Anal. At. Spectrom.*, 12 (1997) 913.
515 S.J. Hill, M.J. Ford and L. Ebdon, Simplex optimization of nitrogen-argon plasmas in inductively coupled plasma mass spectrometry for the removal of chloride-based interferences, *J. Anal. At. Spectrom.*, 7 (1992) 719.
516 S.J. Hill, M.J. Ford and L. Ebdon, Investigation into the application of methane addition to the nebulizer gas in inductively coupled plasma mass spectrometry for the removal of polyatomic interferences, *J. Anal. At. Spectrom.*, 7 (1992) 1157.
517 L. Ebdon, M.J. Ford, R.C. Hutton, S.J. Hill, Evaluation of ethene addition to the nebulizer gas in inductively coupled plasma–mass spectrometry for the removal of matrix-, solvent-, and support-gas-derived polyatomic ion

interferences, *Appl. Spectrosc.*, 48 (1994) 507.
518 L. Ebdon, M.J. Ford, P. Goodall and S.J. Hill, Hydrogen addition to the nebulizer gas for the removal of polyatomic ion interferences in inductively coupled plasma mass spectrometry, *Microchem. J.*, 48 (1993) 246.
519 M.J. Bloxham, P.J. Worsfold and S.J. Hill, Matrix suppression in seawater analysis using inductively coupled plasma mass spectrometry with mixed gas plasmas, *Anal. Proc.*, 31 (1994) 95.
520 G.L. Moore, P.J. Humphries-Cuff and A.E. Watson, Simplex optimization of a nitrogen-cooled argon inductively coupled plasma for multi-element analysis, *Spectrochim. Acta*, 39B (1984) 915.
521 H. Klinkenberg, A. Peng, T. Beeren and K. Flach, Modifications of the roughing pump system of an Elan 500 ICP-MS for analysing complex matrices, *Spectrochim. Acta*, 47B (1992) 585.
522 B.S. Sheppard, W.-L. Shen, T. M. Davidson and J.A. Caruso, Helium–argon inductively coupled plasma for plasma source mass spectrometry, *J. Anal. At. Spectrom.*, 5 (1990) 697.
523 D.W. Koppenaal and L.F. Quinton, Development and assessment of a helium inductively coupled plasma ionisation source for inductively coupled plasma mass spectrometry, *J. Anal. At. Spectrom.*, 3 (1988) 667.
524 S.-H. Nam, W.R.L. Masamba and A. Montaser, Helium inductively coupled plasma-mass spectrometry: studies of matrix effects and the determination of arsenic and selenium in urine, *Spectrochim. Acta*, 49B (1994) 1325.
525 S.H. Nam, W.R.L. Masamba and A. Montaser, Investigation of helium inductively coupled plasma-mass spectrometry for the detection of metals and nonmetals in aqueous solutions, *Anal. Chem.*, 65 (1993) 2784.
526 L.C. Alves, L.A. Allen and R.S. Houk, Measurement of vanadium, nickel, and arsenic in seawater and urine reference materials by inductively coupled plasma mass spectrometry with cryogenic desolvation, *Anal. Chem.*, 65 (1993) 2468.
527 U. Vollkopf, ICP-MS analysis of soya oil using the ELAN 5000 organics sampling system, Perkin-Elmer Technical Summary No. TSMS-19.
528 N.N. Sesi, A. MacKenzie, K.E. Shanks, P. Yang and G.M. Hieftje, Fundamental studies of mixed-gas inductively coupled plasmas, *Spectrochim. Acta*, 49B (1994) 1259.
529 S. Greenfield and H.McD. McGeachin, Calorimetric and dimensional studies on inductively coupled plasmas, *Anal. Chim. Acta*, 100 (1978) 101.
530 E.H. Choot and G. Horlick, Spatially resolved electron density measurements in Ar, N_2-Ar and O_2-Ar ICPs using a photodiode array detection system, *Spectrochim. Acta*, 41B (1986) 935.
531 A. Montaser, V.A. Fassel and J. Zalewski, A critical comparison of Ar and Ar–N_2 inductively coupled plasma as excitation sources for atomic emission spectrometry, *Appl. Spectrosc.*, 35 (1981) 292.

532 J.W.H. Lam and G. Horlick, A comparison of argon and mixed gas plasmas for inductively coupled plasma–mass spectrometry, *Spectrochim. Acta*, 45B (1990) 1313.
533 R.S. Houk, A. Montaser and V.A. Fassel, Mass spectra and ionization temperatures in an argon–nitrogen inductively coupled plasma, *Appl. Spectrosc.*, 37 (1983) 425.
534 I. Ishii, D.W. Golightly and A. Montaser, Radial excitation temperatures in argon–nitrogen inductively coupled plasmas, *J. Anal. At. Spectrom.*, 3 (1988) 965.
535 I. Ishii and A. Montaser, Radial excitation temperatures in argon-oxygen and argon–air inductively coupled plasmas, *J. Anal. At. Spectrom.*, 5 (1990) 57.
536 E.H. Choot and G. Horlick, Vertical spatial emission profiles in Ar–N_2 mixed gas inductively coupled plasmas—I, *Spectrochim. Acta*, 41B (1986) 889.
537 M. Cai, A. Montaser and J. Mostaghimi, Computer simulation of atmospheric-pressure helium inductively coupled plasma discharges, *Spectrochim. Acta*, 48B (1993) 789.
538 N. Shibata, N. Fudagawa, and M. Kubota, Effects of hydrogen mixed with argon carrier gas in electrothermal vaporization–inductively coupled plasma–mass spectrometry, *Spectrochim. Acta*, 47B (1992) 505.
539 R. Tsukahara and M. Kubota, Some characteristics of inductively coupled plasma–mass spectrometry with sample introduction by tungsten furnace electrothermal vaporization, *Spectrochim. Acta*, 45B (1990) 779.
540 A. Goldwasser and J.M. Mermet, Contribution of the charge-transfer process to the excitation mechanisms in inductively coupled plasma atomic emission spectroscopy, *Spectrochim. Acta*, 41B (1986) 725.
541 M. Murillo and J.M. Mermet, Transfer of energy between the surrounding plasma and the central channel in inductively coupled plasma–atomic emission spectrometry, *Spectrochim. Acta*, 42B (1987) 1151.
542 D. Beauchemin and J.M. Craig, Investigations on mixed gas plasmas produced using a sheathing device in ICP-MS, In: *Plasma Source Mass Spectrometry*, Royal Society of Chemistry, Special Publication #85, (1990) 25.
543 A. Batal, J. Jaroz and J.M. Mermet, A spectrometric study of a 40 MHz inductively coupled plasma—VII. Continuum of a binary mixture in the visible region of the spectrum, *Spectrochim. Acta*, 37B (1982) 511.
544 J.W. Lam and J.W. McLaren, Use of aerosol processing and nitrogen–argon plasmas for reduction of oxide interference in inductively coupled plasma mass spectrometry, *J. Anal. At. Spectrom.*, 5 (1990) 419.
545 J.W. McLaren and J.W. Lam, Evaluation of a membrane interface sample introduction system for ICP-MS, *Spectrochim. Acta*, 45B (1990) 1091.
546 J.M. Craig and D. Beauchemin, Reduction of the effects of concomitant

elements in inductively coupled plasma mass spectrometry by adding nitrogen to the plasma gas, *J. Anal. At. Spectrom.*, 7 (1992) 937.
547 E.H. Choot and G. Horlick, Evaluation of the analytical performance of mixed gas inductively coupled plasmas, *Spectrochim. Acta*, 41B (1986) 925.
548 E.H. Evans and L. Ebdon, Simple approach to reducing polyatomic ion interferences on arsenic and selenium in inductively coupled plasma mass spectrometry, *J. Anal. At. Spectrom.*, 4 (1989) 299.
549 S. Branch, L. Ebdon, M. Ford, M. Foulkes and P. O'Neill, Determination of arsenic in samples with high chloride content by inductively coupled plasma mass spectrometry. *J. Anal. At. Spectrom.*, 6 (1991) 151.
550 M. Ford, L. Ebdon, S.J. Hill, Further investigations into the addition of nitrogen in inductively coupled plasma mass spectrometry to reduce the argon chloride interference on arsenic, *Anal. Proc.*, 29 (1992) 104.
551 C. Haraldsson, B. Lyvén, P. Öhman and J. Munthe, Determination of mercury isotope ratios in samples containing sub-nanogram amounts of mercury using inductively coupled plasma mass spectrometry, J. Anal. At. Spectrom., 9 (1994) 1229.
552 R.C. Hutton, Application of inductively coupled plasma source mass spectrometry (ICP-MS) to the determination of trace metals in organics, *J. Anal. At. Spectrom.*, 1 (1986) 259.
553 D.S. Lowe and R.G. Stahl, Determination of trace elements in organic solvents by inductively coupled plasma mass spectrometry, *Anal. Proc.*, 29 (1992) 277.
554 R. Goguel, Group separation by solvent extraction from silicate rock matrix of Nb, Mo, Ta, and W at low levels for ICP-MS, *Fresenius J. Anal. Chem.*, 344 (1992) 326.
555 G.L. Long and J.S. Bolton, Influence of propane on the atomic emission and atomic absorption signals in an inductively coupled argon plasma, *J. Anal. At. Spectrom.*, 7 (1992) 1091.
556 I. Platzner, J.V. Sala, F. Mousty, P.R. Trincherini and A.L. Polettini, Signal enhancement and reduction of interferences in inductively coupled plasma mass spectrometry with an argon–trifluoromethane mixed aerosol carrier gas, *J. Anal. At. Spectrom.*, 9 (1994) 719.
557 F.G. Smith, D.R. Wiederin and R.S. Houk, Argon-xenon plasma for alleviating polyatomic ion interferences in inductively coupled plasma mass spectrometry, *Anal. Chem.*, 63 (1991) 1458.
558 H. Louie and S.Y.-P. Soo, Use of nitrogen and hydrogen in inductively coupled plasma mass spectrometry, *J. Anal. At. Spectrom.*, 7 (1992) 557.
559 B.S. Sheppard, W.L. Shen and J.A. Caruso, Investigation of matrix-induced interferences in mixed-gas helium–argon inductively coupled plasma mass spectrometry, *J. Am. Soc. Mass Spectrom.*, 2 (1991) 355.
560 S.J. Kumar and S. Gangadharan, Determination of trace elements in

naphtha by inductively coupled plasma mass spectrometry using water-in-oil emulsions, *J. Anal. At. Spectrom.*, 14 (1999) 967.
561 Y.Q. Tang, Y.P. Du, J.C. Shao, C. Liu, W. Tao and M.H. Zhu, Air–argon inductively coupled plasma for organic solution analysis: spectral characteristics and analytical performances, *Spectrochim. Acta*, 47B (1992) 1353.
562 A.A. Verbeek and I.B. Brenner, Slurry nebulisation of geological materials into argon, argon–nitrogen, and argon–oxygen inductively coupled plasmas, *J. Anal. At. Spectrom.*, 4 (1989) 23.
563 X.R. Liu and G. Horlick, Direct analysis of materials using direct sample insertion devices and mixed gas inductively coupled plasma atomic emission spectrometry, *J. Anal. At. Spectrom.*, 9 (1994) 833.
564 A. Economou, P.R. Fielden, A.J. Packham, E.J. Woollaston and R.D. Snook, Comparison of data smoothing techniques as applied to electrothermal atomic absorption spectrometry, *Analyst*, 122 (1997) 1331.

Chapter 2

Flow injection techniques

Diane Beauchemin

2.1 INTRODUCTION

Continuous flow (CF) analysis consists in using a peristaltic pump to aspirate samples sequentially into a tube, add various reagents to the moving sample stream and carry the products to the detector [1]. Although CF analyzers are very versatile, allowing the dialysis, extraction, filtration, heating, decantation and even distillation of samples, they are nonetheless plagued by sample carryover as a result of inter-dispersion between successive samples [1]. In 1957, Skeggs developed Autoanalysers which made CF analysis practicable by segmenting the stream with air [1]. Indeed, the air bubbles break up laminar flow patterns and thereby prevent carryover from sample to sample [2]. Air-segmentation also breaks up each sample into a series of sub-samples and allows a more efficient mixing with the reagents which may be added to the stream as it moves toward the detector. The stream then flows through mixing coils which allow the reaction to reach completion before the stream reaches the detector. The air bubbles are usually removed prior to detection.

However, Autoanalysers have some limitations. In order to minimize differences between replicate samples that can result from fluctuations in the sampling time, enough sample must be used to obtain a steady-state signal [1] which, in turn, limits the sampling frequency. Furthermore, since a liquid film does form on the tube wall, some axial dispersion results from mass transfer between this film and subsequent fluid segments [2]. The size of the air bubbles must be carefully controlled, especially at high sampling rate, because the compressibility of air induces pulsations in the carrier stream [1]. Even

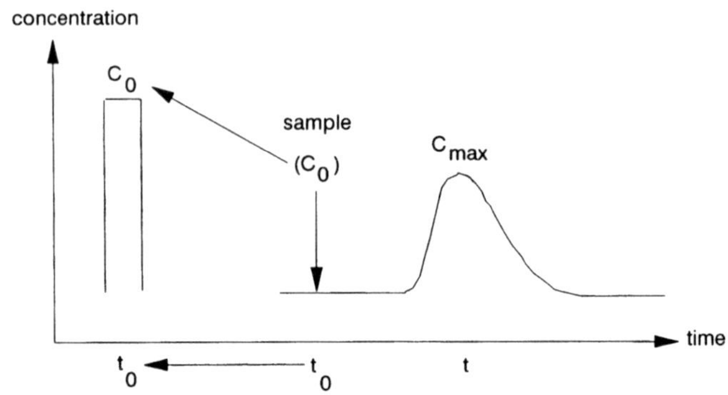

Fig. 2.1. Concentration profile resulting from the injection of a sample solution into an unsegmented carrier stream. The sample is injected at time t_0 and the resulting transient signal is observed at time t. The concentration of the undispersed sample, at time t_0, is also shown for reference. After Ref. [15].

under controlled conditions, these unavoidable pulsations induce fluctuations in the movement of the carrier stream which, in turn, prevent exact timing between sample introduction and detection [3].

Flow injection (FI) was introduced in 1975 by Ruzicka and Hansen [1] and is a variation on CF. In contrast to CF where the sample is pumped continuously, FI consists in the injection of a sample plug into an unsegmented carrier stream. Dispersion of the sample into the carrier stream occurs as the sample moves from the injection point to the detector [4]. Exact timing and synchronisation of the injection with the measurement routine allow a reproducible measurement of the dispersed sample zone [4], which is observed as a transient signal as schematically represented in Fig. 2.1. Reactions can be carried out on-line if the carrier stream contains a suitable reagent, since mixing of the injected sample with the reagent carrier stream occurs during passage to the detector. The dispersion or dilution of the sample in the carrier stream and its residence time can be adjusted and maintained reproducibly to satisfy the requirements of the chemistry taking place [5].

The success of FI lies in its ability to obtain reproducible transient signals as a result of reproducible injection volume and speed of injection, as well as reproducible dispersion of the sample into the carrier stream and adequate reproducible mixing with reagents, without significant carryover between samples [1]. The dispersion must therefore be carefully controlled. However, injecting the sample into a

TABLE 2.1

Main features of flow injection techniques

Feature	Origin
High precision	a reproducible transient signal is detected
High sampling rate	many discrete injections are possible during the time typically required to continuously nebulize one sample
Minimal sample consumption	a few injections typically require less sample than continuous nebulization
Microsample manipulation	the sample injection volume can be as small as a few microliters
Low reagent consumption	because of the reduced scale of FI systems, less reagent is required than if the same reaction is carried on off line
Ease of automation	computer-controlled FI systems are available
Rapid on-line pre-treatment chemistry	performing reactions on a micro scale speeds up the process; the low pressure applied increases the speed even further if the process normally relies on gravity; the reaction does not need to be complete prior to detection: a measurement can be made at a reproducible point of the reaction
Enhanced calibration techniques	any calibration can be carried out on-line thereby reducing time-consuming sample preparation
Minimal contamination	the whole process is carried out in a closed system
Blank controlled throughout the whole measurement process	a change in the signal from the carrier immediately indicates a change in system performance

flowing stream is more reproducible than aspirating the sample with Autoanalysers where pulsations induce fluctuations in the volume sampled. Therefore, a steady-state signal is no longer required, which means that lower sample consumption and increased sampling frequency can be achieved with FI than with Autoanalyzers [5].

Several books [3,6–9] and reviews [1,5,10–32] provide an in depth description of FI and its numerous features, some of which are highlighted in Table 2.1. In particular, FI features versatility, speed of response and miniaturization, which allows virtually any reaction or separation to be carried out on-line and on a micro-scale, with an analytical response available within typically 5–30 s of sample injection [14]. It is therefore not surprising that FI has been widely used,

TABLE 2.2

Comparison of some atomic spectrometries

	FAAS	ICP-AES	ICP-MS
Sample uptake rate	4–5 ml/min	up to 3 ml/min (free uptake)	up to 3 ml/min (free uptake)
Range of detection limit [23]	0.1 to >100 ppb	0.5–100 ppb	<1 ppt–1 ppb
Typical precision at 50 times the detection limit [23]	about 0.5%	1.5%	2-3%
Spectroscopic interferences	few	potentially numerous, depending on the matrix	some isobaric interferences; interferences from polyatomic ions from the plasma and/or matrix (see Section 1.4.2)
Matrix effects	chemical interferences from refractory matrices; ionization interferences	ionization interferences	some ionization interferences; effects of concomitant elements (see Section 1.4.3)
Effect of dissolved solids	blockage of nebulizer	blockage of nebulizer and torch injector	blockage of nebulizer, torch injector and interface (especially on the sampler)
Effect of organic matrix	enhancement of sensitivity	soot deposition on torch injector tip	soot deposition on torch injector tip and interface (especially on the sampler)

with some 10,000 refereed papers published to date (see http://www.flowinjection.com/search.html), with a significant fraction in combination to ICP-MS. Indeed, FI has been selected as an alternative to continuous nebulization with ICP-MS and other atomic spectrometric detectors [4,7,11–13,19,22–29]. In comparison to other atomic spectrometries, ICP-MS has several limitations, which were thoroughly discussed in Chapter 1, and are summarized in Table 2.2. The purpose of this chapter is to first review the principles of FI so that a firm ground is established. This will then be followed by a thorough discussion of various FI techniques which have been or could be used to circumvent the limitations and thereby enhance the analytical capabilities of ICP-MS.

2.2.1 Definition and measurement of dispersion

The extent of dispersion can be measured by relating the maximum concentration of the analyte in the sample zone at the detector C_{max} to the original undispersed analyte concentration C_0 before injection [4]. It is expressed as a dispersion coefficient, D, which is in general defined as [5]:

$$D = \frac{C_0}{C_{max}} \tag{1}$$

Within the linear dynamic range of the detector [11] which, for ICP-MS (see Tables 1.5 and 1.6), covers several orders of magnitude, a direct relationship exists between concentration and peak height, and D can therefore be easily measured as:

$$D = \frac{S_0}{S_{max}} \tag{2}$$

where S_0 is the steady-state signal observed during continuous aspiration of the sample, and S_{max} is the height of the transient peak resulting from FI of the sample under otherwise identical operating conditions [4]. Obviously, the dispersion also indicates the degree of dilution of the sample injected or ratio in which the sample was mixed with the carrier [5,11]. As a result, the FI peak broadens as the dispersion increases. It starts with an asymmetrical shape when little dispersion has occurred, and changes to a more symmetrical shape that eventually becomes Gaussian when the dispersion is extensive [5]. For a given FI system, as the length of the reactor is increased, the peak height decreases as dispersion increases, but peak area should remain constant since no analyte is created or lost during the dispersion process, unless other effects are also encountered (such as a change in the nebulization efficiency [34]).

If the carrier stream contains a reagent, then the concentration of the latter will be diluted by the sample injected, in that zone where it is mixed with the sample. Equation 1 can also apply to the reagent, with the difference that the dispersion of the reagent, D_R, would be related to the original and dispersed concentrations of the reagent. More simply, however, it is related to the dispersion of the sample by the following relationship for any single-line (i.e. without confluence or split points) FI manifold of uniform diameter [11]:

$$D_R = D/(D-1) \quad (3)$$

If a reverse-FI (rFI) manifold is used, i.e. the sample is used as the carrier and the reagent is injected into the sample carrier, as opposed to the normal FI (nFI) manifold described so far [35, and references therein], Eq. 3 would then give the dispersion of the sample (on the left-hand side) expressed in terms of the dispersion of the reagent (on the right-hand side).

Equations 1 and 2 give the total dispersion of the sample which occurs in a given FI system. As an approximation, it can also be defined as a combination of different dispersion processes:

$$D = D_i D_f D_d \quad (4)$$

where D_i is the dispersion arising from the injection system, D_f results from the FI manifold (i.e. the 'reactor' between the sample injection point and the detector) and D_d from the detector [5]. This equation is a simplification since, in the first part of the FI manifold, dispersion results from both the injection and the flow arrangement [5]. Indeed, when the content of a sample loop is delivered to the FI manifold by the carrier stream (see Section 2.2.3), the front of the sample slug will undergo less dispersion than the tail of the slug because the latter will be dispersed in both the injection loop and the manifold in contrast to the former which will be dispersed only in the manifold [36]. The contribution of D_i to D will be significant when the sample volume constitutes a large fraction of the total volume of the manifold, but will usually be negligible for small sample volumes when D_f is relatively large [36].

The last factor of this equation, D_d, arises from dispersion taking place during the nebulization process, as the dead volume of the spray chamber also contributes to the final shape of the FI peak [36,37]. More precisely, the volume of the spray chamber offers a relatively large amount of room where the droplets generated by the nebulizer can swirl around. It therefore takes a few seconds to 'fill' the chamber with a new solution and a few seconds to rinse it out [11]. Hence, the spray chamber acts as a mixing chamber whenever a switch is made from a solution to another: droplets generated from solution A are mixed with leftover droplets from preceding solution B until there is no more leftover from B, at which point the steady-state signal for A is reached.

This dispersion effect can readily be observed during continuous nebulization of solutions when, for instance, the input line is moved from the blank to a standard solution and to the blank again. Although a step change in concentration occurs during this experiment, the resulting signal is not perfectly square. Instead, an exponential increase to a steady-state signal is observed, followed by an exponential decrease when the input line is moved back to the blank solution. In the case of FI, as was mentioned in Section 2.1, a steady-state signal is not necessary and, hence, not usually reached. The dilution effect of the spray chamber will therefore affect both the rise and the fall of the transient peak, making it look smaller than it was when it reached the nebulizer. Although D_d may be negligible when the dispersion in the rest of the FI system is relatively large, it may be very significant when D_f and D_i are both close to one [5,11]. In any case, because of the contribution of other types of dispersion than mixing to the observed transient peak, Eq. 4 indicates why D is called a dispersion coefficient rather than a dilution factor [13].

In any case, each source of dispersion will induce a time variance in the flow injection peak. The time variance that is observed will actually be the sum of the individual time variances originating from the connections, mixing tees, detection system, injection volume, etc. [38].

For a single-line FI manifold of uniform diameter, the total dispersion can also be estimated by the following model:

$$D = \frac{2}{W_0} F\sqrt{T}\sqrt{t_i}(1-2^{-n}) \qquad (5)$$

where W_0 is the peak width of the undispersed sample zone (assuming a hypothetical square peak), F is the mean linear flow velocity, T is the residence time or time required for the sample to reach the detector, t_i is a constant for a given well-defined FI system (in terms of tube length, flow rate, viscosity, surface tension, temperature, etc.), and n is the ratio of the injected sample volume to the sample volume required to obtain D = 2 [5].

The value of D depends on F because of the different rate of axial and radial dispersions. In a tubular reactor, two different types of flow may result: laminar and turbulent. One or both may be present within different regions of a FI system. In a laminar flow, the liquid in the centre of the tube moves faster than that close to the wall which is almost stagnant, thereby resulting in spreading of the sample into the carrier [2]. Whereas in a turbulent flow, radial mixing is so extensive that the whole liquid front moves at more or less the same speed, and therefore the sample zone can move without considerable mixing with the carrier [2].

Although several models have been proposed to explained relatively simple systems [8,36,39], the actual value of D, as measured by Eq. 2, can in general not be predicted because of the complexity of flow systems. Any curvature in the reactor line or presence of connections will influence D as they may change the extent of radial mixing [2]. For instance, experiments with straight tubes have shown that an irreproducible dispersion pattern could be obtained unless the tube was carefully straightened [2].

Dispersion can be increased or decreased, by modifying the FI manifold to satisfy a particular application. For this reason, D is typically used to classify FI systems into four general categories: reduced ($D < 1$), limited ($D = 1–2$), medium ($D = 2–10$) and large dispersion ($D \geq 10$) systems [13]. The requirements and features of each of these systems are discussed in detail in the following sections.

2.2.2 Key experimental FI parameters

Several factors influence dispersion. The injected sample volume, the flow rate of the carrier stream, the dimensions (i.e. length and inner diameter) of the tubing connecting the sample injection valve to the nebulizer and the manifold design (for instance the presence of confluence or split points, mixing coil, mixing chamber, etc.) which influence the linear velocity and overall reactor volume, V_r, [5] determine the dispersion prior to the nebulizer.

In addition to dispersion, another important parameter is the residence time, T, of the sample in the FI system, since it determines the extent of a chemical reaction. One way of increasing T, in the case of a single-line manifold of uniform diameter, is to increase V_r (i.e. the volume of the FI system), since, in that particular case, T is defined as:

$$T = \frac{V_r}{Q} = \frac{\pi r^2 L}{Q} = \frac{L}{F} \tag{6}$$

where Q is the pumping rate, L the total length of the line and r the i.d. of tubing [5]. However, increasing T may also increase D. For instance, in FI-ICP-MS, D increased almost linearly as the tube length was increased from 65 to 350 cm [40]. Similarly, the emission signal decreased as L was increased in ICP-AES [41].

Decreasing the flow rate passing in narrow tubes (i.e. 0.2–0.5 mm i.d.) is, however, far more appropriate since it not only increases the residence time but it simultaneously reduces reagent consumption and decreases the dispersion, at least in the case of a single-line manifold of uniform diameter (see Eq. 5) [5]. This decrease of D with a decrease of Q may, however, be only true over a certain range of Q values. Indeed, a study looking at the D versus Q profiles for a wide range of Q values revealed an inflection point, i.e. a value of Q at which D was maximum [42]. Furthermore, this optimal Q value was independent of sample injection volume, length and inner diameter of the tube, and radius of the mixing coil [42]. It only changed with the diffusion coefficient of the analyte [42].

This finding may explain why the effect of Q on D for wider tubes is not as clearly defined and appears to depend on tube length. For instance, with 30 cm of 0.58-mm i.d. tubing, D decreased as Q was increased in spectrophotometry [36]. Whereas with 360 cm of 0.58-mm i.d. tubing, the D versus Q (in the 0-9 ml/min range) curve showed two maxima [36]. In FI-ICP-AES, D increased as Q decreased for 30 cm of 1.0-mm i.d. tubing and a constant injection volume, in both the normal and reverse-FI modes [43]; whereas little change occurred over the 1.0-5.0 ml/min Q range with over 40 cm of 0.8-mm i.d. tubing [44]. Results obtained in FI-ICP-MS with 100-μl injections and 65 cm of 0.8-mm i.d. tubing, showed that peak height went through a maximum at 1.5 ml/min as Q was increased from 0.5 to 2.0 ml/min, whereas peak area continuously decreased over the same Q range [40].

In any case, there is a limit to which Q can be decreased for a given nebulizer. For instance, although a pneumatic nebulizer can be starved down to an uptake rate of about a third of the free aspiration rate (see Section 1.2.3.1) without a significant degradation of sensitivity, lower uptake rates will reduce sensitivity, as a result of the significantly smaller amount of aerosol then reaching the plasma. This is illustrated in Fig. 2.2 which shows the steady-state signal obtained by continuous nebulization, followed by replicate 100-μl injections of the same 100-μg/l multielemental standard solution in a single-line FI manifold, at two different carrier flow rates under otherwise identical conditions. Although the FI peaks are significantly broader at 0.33 ml/min, the resulting D is not significantly different than at 0.67 ml/min because of the concurrent loss of sensitivity (i.e. the steady-state signal in Fig.

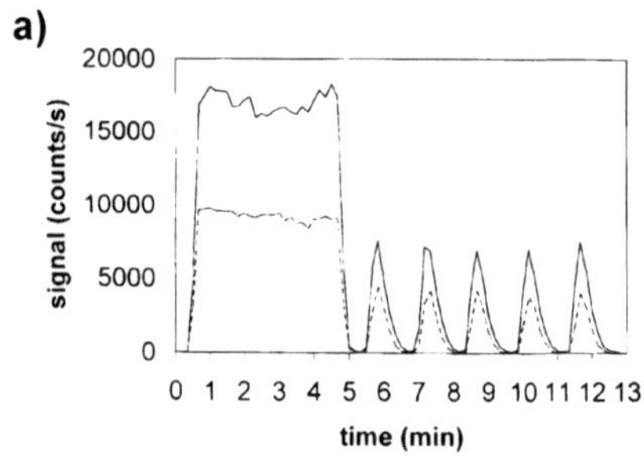

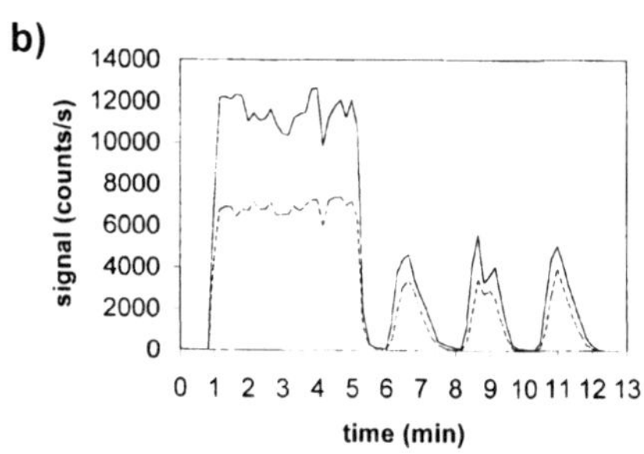

Fig. 2.2. Effect of the carrier flow rate on the signal observed for $^{60}Ni^{+}$ (solid line) and $^{75}As^{+}$ (dashed line) during continuous nebulization (from 1 to 5 min) and three to five replicate 0.1-ml injections at (a) 0.67 ml/min and (b) 0.33 ml/min carrier flow rate.

2.2b is lower than in Fig. 2.2a even though the same solution was being continuously nebulized), which results from starving the nebulizer. It should be noted that the situation is completely different with FAAS where the flow rate of the carrier must be greater than the free uptake rate of the nebulizer for optimum sensitivity and precision [45].

Equation 6 provides an estimate of T, the actual residence time depending on the length of the reactor line. Indeed, T will be under-estimated if L is very short since it will not take into account the time required for the sample zone to accelerate up to F. The residence time will also be under-estimated if L is large, but this time as a result of the decrease in Q and F which occurs with the increasing flow resistance of the reactor line [5]. Similarly, the pumping rate of a given FI set-up at a given peristaltic pump setting should be measured (this can easily be carried out by measuring the time required by the pump to aspirate a given volume of solution from a graduated cylinder, or more accurately, by weighing the solution aspirated or delivered during a measured time period) as the nominal flow rates provided by the manufacturers will not be valid if the reactor line has a high flow resistance and/or if the wear of the flexible tubing compressed by the pump is significant.

In general, a long narrow tube will result in a symmetrical FI peak whereas a shorter or wider tube will yield an asymmetrical peak shape [5]. A small tube radius, r, in the range of 0.2–0.5 mm should be selected whenever a limited volume of sample is available and low or medium dispersion is desired [5]. Indeed, since:

$$V_i = \pi r_l^{r} L_l = \pi r^2 L_s \tag{7}$$

where V_i is the sample volume injected, r_l the radius of the sample injection loop, L_l the length of the loop, and L_s the equivalent length of tube occupied by the sample in the FI manifold or reactor, the sample will occupy a greater volume of the reactor line if the reactor tube radius (r) is decreased. Furthermore, since with r = 0.2–0.5 mm, D_f essentially depends only on T which is proportional to r^2 (according to Eq. 6), a decrease in D would also result (see Eq. 5) unless the sample volume (and therefore W_0) is also decreased. This means that the volume of sample required to obtain a given D in a single-line FI manifold of reduced r will be proportionally smaller [5]. In any case, as V_i is increased, D will decrease for an otherwise identical FI system [40,46]. On the other hand, if V_i decreases, D will increase, unless Q is also changed to compensate for the change in D [43].

For many manifold designs, an exponential relationship has been observed between D and V_i [10]:

$$D = (1 - \exp(-V_i k))^{-1} \tag{8}$$

where k is a constant for a given manifold and operating conditions. In cases where L is small and Q is either moderate or large, the concentration profile of the FI peak as a function of time can often be described by the following equations, for a normal single-line flow injection set-up where an analyte-containing solution of concentration C_0 is injected into an analyte-free carrier [39,47]. The model assumes that all dispersion processes can be predicted by a single well stirred tank of volume V_r, the reactor volume:

$$C = C_0 (1 - \exp(-Qt/V_r)) \text{ for } 0 < t < t_{max} \tag{9}$$

$$C_{max} = C_0 (1 - \exp(-V_i/V_r)) \text{ for } t = t_{max} \tag{10}$$

$$C = C_0 \exp(-Q(t - t_{max})/V_r) \text{ for } t > t_{max} \tag{11}$$

where C is the concentration at time t and t_{max} is the time at which the maximum concentration is observed. From Eq. 10, it follows that k (in Eq. 8) is then given by the reciprocal of V_r. Since the ICP-MS signal is directly proportional to concentration, C and C_0 can be replaced by S and S_0 (i.e. the signal measured at time t and that obtained by continuous aspiration, respectively) in Eq. 9, and V_r can be obtained from a plot of t versus $\ln(S_0/(S_0{-}S)$ which has a slope of V_r/Q. Once V_r has been experimentally determined for a given set-up, D can therefore be calculated for any volume injected. Similarly, one can calculate C at any time, or the time required to reach a given concentration [47]. More elaborate models, such as one based on two tanks in parallel [39], or on several tanks in series [48 and references therein] can also be used to simulate the FI profile.

Nonetheless, in designing a FI system, the various parameters listed above should be optimized so that the best compromise is reached between sensitivity, sampling rate, sample and reagent consumption, reaction time and the overall time to operate the system (i.e. to start and stop it). The residence time of the sample should, in general, be kept as short as possible so that the analytical measurement is quickly available. Although the sample injection volume has an effect on D_i and therefore on D, the geometry of the sample zone is irrelevant i.e. as long as the volume injected is constant, D will not change for a given FI system [5]. For a given T, D_f is independent of the reactor tube i.d. [5].

The sampling frequency can be increased by injecting several samples at regular intervals instead of waiting for the analyte to reach the detector before making the next injection. This is possible because the sample zone broadens relatively slowly as T increases. Furthermore, as long as the sampling frequency is not higher than 1.5 W (where W is the peak width at half height), no significant carryover will result [5]. Performing multiple spaced injections is particularly important if $T > 30$ s is required for a reaction, as it would otherwise unduly decrease the sample throughput [5]. However, as T increases, so does W, which decreases the maximum sampling frequency possible (see Fig. 2.2). In general, $T \leq 30$ s for a sampling frequency of more than 100 samples per hour. The maximum sampling frequency can usually be obtained by keeping D low, V_i small and F high [5].

The sample residence time can be increased without a significant increase of D by inserting a knitted open tubular reactor, also called a three-dimensional coiled tube, between the sample injection point and the nebulizer [49]. This device was designed to allow post-column derivatization in chromatography without inducing band broadening and thereby degrading resolution [49]. As discussed in Section 2.2.1, if the configuration of a tubular reactor is such that a turbulent flow results, little dispersion occurs since radial mixing is the main process taking place. Because the three-dimensional configuration of this type of reactor induces a turbulent flow, it therefore allows mixing without increasing D.

Alternatively, T may be increased without increasing D if the carrier stream is stopped. Under stopped-flow condition and in the absence of air, when the peristaltic pump is stopped, all liquid movement also ceases and D therefore remains constant (radial molecular diffusion is assumed to be negligible), but the reaction carries on. In fact, D may change slightly, depending on the flow mode (stopped-flow or not) and manifold configuration (normal or reverse FI) which affect both the axial and radial diffusions [50]. Nonetheless, the stopped-flow mode can be useful to increase sensitivity or perform kinetic analyses [5]. The sampling frequency can be increased by inserting parallel reaction coils between the valve and the detector, so that the sample injection valve may serve different coils sequentially and thereby allow several samples to be in the FI system at any one time [5].

Although numerous configurations of FI systems can yield identical D and T, the smaller system will usually be more advantageous in terms of time and reagent consumption. When designing a FI system, preliminary experiments should therefore be carried out to establish the effect of Q, r, L and V_i on D.

2.2.3 Key components

Three basic components are required to build a FI system: a pump to provide a controlled carrier flow rate, tubing into which the carrier flows to the nebulizer and a sample injection valve which allows the injection of a discrete volume of sample into the carrier stream. As a matter of fact, if the nebulizer is used as a pump, then only some tubing and a valve are required. This was demonstrated in FAAS (and illustrated in Fig. 2.3) with a three-way valve [51]. In one position, the carrier is aspirated by the nebulizer. The valve is then temporarily switched to the other position thereby allowing the nebulizer to aspirate a sample from an autosampler. The volume injected is determined by the aspiration rate of the nebulizer and the time the valve remains in this position before being switched back to the initial position. To

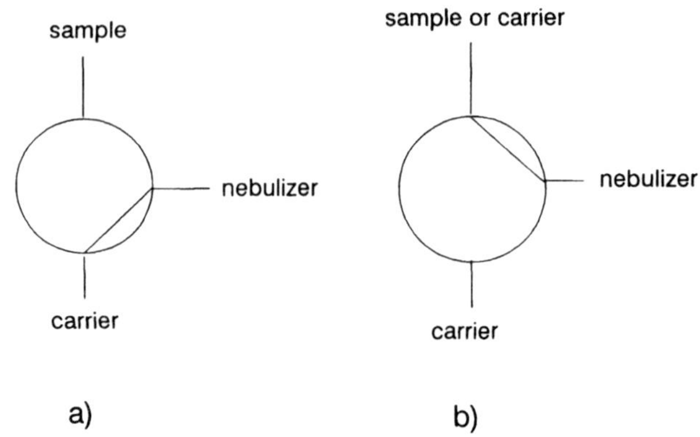

Fig. 2.3. Flow injection using a three-way valve. In position (a), the carrier is aspirated by the nebulizer. The valve is then switched temporarily to position (b) so that the sample may be aspirated and then switched back to position (a). After Ref. [51].

minimize carry-over, rinsing of the sample inlet of the valve is accomplished by inserting carrier solution between samples. Quantitation using peak area was found to be independent of viscosity differences between standards and samples (whereas that using peak height required matrix-matching or the method of standard additions) [51]. This approach should be applicable to ICP-MS. However, the precision of injections performed in this manner is not as great as that achieved with the fixed-volume rotary or slider valves described later. Furthermore, because of the pressure surge at the switching of the valve, it is unsuitable for small injection volumes.

Nonetheless, a variety of pumps can be used as long as they provide a carrier flow rate high enough so as not to starve the nebulizer to a point where sensitivity is sacrificed. These include syringe pumps, reciprocating positive displacement pumps and even gas-pressurized reservoirs [31]. The piezoelectric micropump of the MDMI (see Section 1.2.3.4) could also be used for FI if the sample injection valve is inserted between the pump and the laminar flow oven of the MDMI. Even a solenoid-driven diaphragm pump has been proposed for in situ FIA systems [52], although it remains to be seen if the pulsating flow resulting from this pump would be compatible with ICP-MS. In any case, a peristaltic pump (which is supplied with all ICP-MS instruments) is usually adequate and, thus, the most widely used [11]. Recommendations made in Section 1.2.2.1 regarding the pumping conditions should be followed even more carefully, as they may have a significant impact on the best precision which can be achieved with the FI system. In particular, the pressure of the rollers on the flexible tubing should be as small as possible to minimize pulsations (but obviously high enough to prevent the nebulizer from aspirating the carrier freely). A flexible tubing of relatively small i.d. should also be preferably selected so that the pump may be operated at a relatively high rotation rate, again to minimize pulsations. Finally, the flexible tubing should be chemically resistant to the solutions being pumped (carrier, sample, reagent, etc.). Tygon is in general adequate although Neoprene pump tubes were found to have better elasticity [53]. This tube indeed resisted deformation, even after being pressed for extended periods when the pump was devoted to sample metering, and was therefore stopped at least 90% of the time [53]. (A description of an alternative approach which can be used when no suitable tubing is available can be found in Section 1.2.3.3.)

In general, inert tubing is used to connect the pump tubing to the sample injection valve, and the latter to the nebulizer. Materials such as TEFLON, TEFZEL or PEEK are preferred since they minimize contamination of the flowing stream. The tubing can be flanged (in the case of TEFLON) or not, in which case care should be taken to select connectors with ferrules such that only inert material comes into contact with the sample. This also applies to all other components (such as tees, mixing coils, reactors, etc.) which can come in contact with the carrier and/or injected sample. Some of these components can be made in-house. For instance, coils of various diameters can be made by simply wrapping tube around a backbone tube. A detailed description has also been made of how to construct various types of knitted open tubular reactors out of TEFLON tubing without the use of special tools [49]. For μ-FIA, a micromixer that has a total internal volume of about 600 nl and dead volumes of 12.5 nl and 23 nl for the two input channels can be used for efficient laminar flow mixing at flow rates between 1 and 200 μl/min [54].

The valve is installed downstream from the pump to minimize its pulsating effect and avoid either possible contamination from the flexible tubing or retention of analyte (such as Au [55]) on the tubing. A low-pressure rotary or slider valve with an external loop is adequate [11], although more expensive valves (such as those used for HPLC) can be used, for instance, when very small injection volumes (for example, 10 μl) are required. The valve should preferably have a bypass allowing an uninterrupted flow of carrier to the nebulizer while the sample

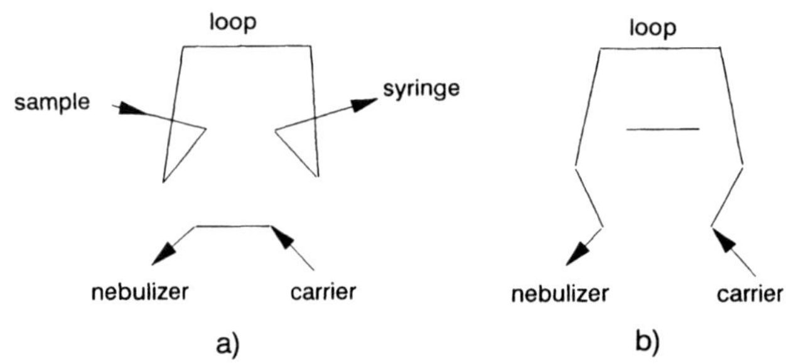

Fig. 2.4. Flow injection using a 6-port valve. In position (a), the carrier is aspirated by the nebulizer while the sample loop is filled by sucking the sample into a syringe. The valve is then switched to position (b) where the content of the loop is carried to the nebulizer.

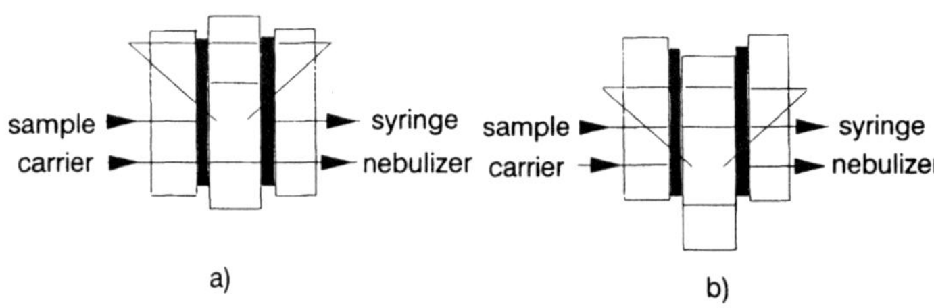

Fig. 2.5. Flow injection using a slider valve. In position (a), the carrier is aspirated by the nebulizer while the sample loop is filled by sucking the sample into a syringe. The valve is then slid to position (b) where the content of the loop is carried to the nebulizer.

injection loop is being filled. This avoids surges of the carrier stream, which improves the reproducibility of microsample injections [5].

In its least expensive version, the sample injection loop can be manually filled using a syringe. This step will require more sample solution than that required to fill the loop since the connecting lines must also be filled and/or rinsed of the previous sample. Furthermore, in order to minimize any possible contaminations from the syringe components, the sample should be sucked through the loop into the syringe, as shown in Figs. 2.4a and 2.5a. This approach also avoids the use of an additional amount of sample to simply rinse the syringe. The loop must be filled carefully to avoid air bubbles. The valve can then be manually actuated so that the carrier can push the content of the loop towards the nebulizer, as shown in Figs. 2.4b and 2.5b.

Obviously, the reproducibility of FI can be improved by automating part or all of the system. For instance, an electronically actuated valve will improve the reproducibility of the injection process. It also allows time injection, whereby the valve is switched back to the loading position before the tail of the sample slug enters the FI manifold [36]. This approach has the advantage of eliminating the tailing, asymmetrical part of the peak which is observed at low dispersion (because the tail of the sample slug undergoes more dispersion than the front, as discussed in Section 2.2.1), and the fall curve then becomes the reverse of the rise curve of the transient peak. Because less time is then required for the transient signal to return to baseline, this approach can be used to increase sample throughput [36].

In addition to the electronically actuated valve, a computer-controlled pump can be used to fill the sample injection loop so that the whole process can be automated. This approach allows reproducible injection volumes of 0.7–35 μl since the sample can then be pumped into a 100-μl loop (pre-filled with DDW carrier) for a pre-determined amount of time using a thick-wall narrow-bore (0.25-mm i.d.) Neoprene tube [53]. The sample loop is therefore only partially filled with sample, the rest of the volume being filled with carrier. To ensure the highest precision in the metering of such small volumes, the pump should only be turned on after the sample valve has been switched to the fill position [53]. After the desired volume of sample has been pumped into the loop, the valve can then be switched to the injection position, where the carrier provided by a different pump can entrain the content of the loop towards the detector [53].

Recently, a completely different approach was used to perform discrete sample introduction while avoiding some limitations of FI [56]. These include the wear of a typical sample injection valve which can develop leaks, and the relatively large amount of sample which is wasted since it is required to flush the loop and connecting tubing in addition to filling the loop itself. Controlled-dispersion flow analysis (CDFA) is a variant of FI where a probe connected to a peristaltic pump is used to aspirate a given volume of sample, then, with the pump stopped, moved to the carrier container when the pump is switched on again, thereby pulling the sample slug towards the nebulizer [56]. In this way, no sample is wasted, since the carrier rinses out the probe, in addition to dispersing the sample.

This relatively inexpensive computer-controlled system, which uses a peristaltic pump and an autosampler, was designed to maximize sample throughput and minimize memory effects [56]. It allows a selection of D by simply varying the sample volume. For instance, D

between 1 and 58.5 resulted from the injection of 800 down to 8.7 μl [56]. The reproducibility degraded significantly at lower sample volumes (i.e. smaller than 10 μl) because of the slippage of the peristaltic pump as it starts its rotation. Nonetheless, a computer-controlled linear calibration over more than three orders of magnitude could be generated from a single standard solution by simply varying D (and thereby, the dilution factor) [56]. So far, this approach was used in combination with ICP-AES only, but it is compatible with ICP-MS. Its main limitations are the dependence of the sample volume on the pumping rate, which may unfortunately vary with the wear of the flexible tubing being compressed by the pump rollers, and therefore requires monitoring, and the possibility of contamination and/or loss of analyte discussed above.

Sequential injection (SI) is another variation of FI, which can be used to decrease reagent consumption [57]. It typically requires a single-channel pump, a holding coil and a multiposition valve, as illustrated in Fig. 2.6. The pump sequentially aspirates the carrier, appropriate volumes of the sample and the reagent, and the carrier again, such that the sample and reagent are contained within the holding coil. The pump can then be stopped to allow the reaction to proceed for a given time (and thereby allow the measurement of reaction rate). Finally, the flow direction is reversed so that the composite zone is pushed towards the detector where the overlap (which is never total) between the two zones, i.e. the reaction product, can be measured. Reactions between a sample and a number of reagents can be performed in a similar fashion [57, and references therein]. As in FI, the complexity of the system can be increased by

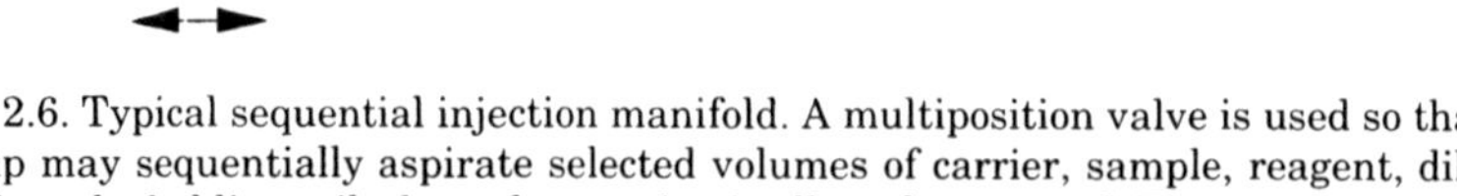

Fig. 2.6. Typical sequential injection manifold. A multiposition valve is used so that the pump may sequentially aspirate selected volumes of carrier, sample, reagent, diluent, etc. into the holding coil where the reaction is allowed to proceed. The pumping direction is then reversed and the content of the coil pushed to the nebulizer. After Ref. [57].

adding mixing chambers, membrane separators, etc., but the sampling frequency is always lower than that of the corresponding FI method as the latter uses a continuous unidirectional forward flow [57].

Another variation, which is still in its infancy and has not yet been coupled to ICP-MS, is bead injection (BI) where a precise volume of suspended beads is trapped in a flow injection manifold (such as the holding coil of a SI system) where it is perfused with sample, reagents, etc. [58]. The beads serve as carrier for reagents or reactive groups. Therefore, chemical reactions take place at the surface of the beads, and the analytes are then recovered with a suitable eluent. (With other detectors than ICP-MS, the analyte can be determined directly on the beads. However, since the beads are typically 10 to 150 µm in diameter, the use of slurry nebulization for their introduction into the ICP is not recommended.) The beads can then be discarded, rerouted or collected, depending on the application [58]. The BI approach offers several advantages. For instance, sample dilution is minimized through the use of surface-bounded reagents. Preconcentration can be performed simultaneously through accumulation of analyte on the surface of the beads [58]. Furthermore, if the beads are discarded after each injection, the approach becomes a renewable surface technique, which offers additional advantages [59]. Indeed, time, reagents and solvents are saved since there is no need to regenerate the reactive surface, which in turn eliminates problems associated with imperfect regeneration [59]. The range of applications is expanded since chemical reactions or adsorption/desorption processes no longer need to be reversible [59]. On the other hand, its implementation is not trivial. Indeed, maintaining the beads in a well-defined suspension can be challenging [58].

Although the nebulization system is not an integral component of the FI (or CDFA) system, it does contribute to the total dispersion of the sample, as discussed in Section 2.2.1, and may therefore be modified to minimize D_d. A decrease in the size of the spray chamber would decrease dispersion. For example, a 10.5-ml vertical rotary spray chamber was recently described, which drastically reduces the wash-out time compared to a standard double-pass spray chamber [60]. Similarly, a 47-ml cyclonic spray chamber offers a lower void volume than a typical 190-ml double-pass spray chamber [61]. When used at a carrier flow rate of 3.5 ml/min, the discrete frequency noise originating from the peristaltic pump, that is typically observed with a Scott-type spray chamber, even disappeared with the cyclonic spray chamber [61]. A new donut-shaped spray chamber was reported to offer higher sensitivity, lower background and improved detection limits compared to those obtained with a standard double-pass spray chamber [62]. In particular, a 2.7-fold improvement in peak height intensity was observed compared to that obtained under identical FI conditions with a Scott-type spray chamber [62].

The use of high-efficiency nebulizers (discussed in Sections 1.2.3.3 and 1.2.3.4) would also be advantageous. For instance, with 200-µl injections, a thermospray nebulizer was reported to provide a sensitivity nine times higher than that obtained with a pneumatic nebulizer [63]. On the other hand, the stability of the thermospray nebulizer is poorer, resulting in detection limits that are only improved by a factor of five over those obtained with pneumatic nebulization [63].

The DIN, which allows essentially 100% sample introduction efficiency, has a dead volume of only ca. 2 µl [64] since it nebulizes and introduces the solution directly into the heart of the plasma, without requiring a spray chamber. Under these conditions, D_d is essentially equal to unity. Alternatively, the conventional pneumatic nebulizer can be replaced by a HEN, which operates at solution uptake rates of 20–120 µl/min but, like the DIN, requires a high-pressure nebulizer gas line and a mass-flow controller capable of withstanding up to 175 psi high back-pressure. Although a peristaltic pump is adequate for 0.19-mm fused silica capillaries [38], smaller bore tubing (50-µm i.d.) will induce a high back-pressure, which will require the replacement of the peristaltic pump by a syringe pump. On the other hand, a syringe pump has the advantage of providing pulse-free delivery of carrier [65]. An even simpler approach than the DIN and HEN, is to combine a MCN with a spray chamber to perform FI on a microscale without the need for a high-pressure nebulizer gas line or high-pressure pump [66].

With a conventional nebulizer/spray chamber system, a Nafion membrane dryer can be inserted between the spray chamber and the torch to remove 97% of the total mass of water exiting the spray chamber without increasing the dispersion significantly [67]. This allowed an improvement in the detection limit for ^{56}Fe and ^{57}Fe through a substantial reduction of the ArO and ArOH background at these masses [67].

Although FI injection systems are available from some of the ICP-MS manufacturers, they can usually be put together from parts available from various manufacturers (such as those who also sell chromatographic parts). Some FI systems can also be purchased from several companies (such as Lachat, Alpkem, Fiatron, etc.).

2.2.4 Measurement conditions

A continuous monitoring of the transient signal is required since it allows instantaneous detection as well as diagnosis of any malfunction. In general, peak height is preferred, especially if the baseline is sufficiently stable [5], although peak area is often used in FI-ICP-AES [4]. Peak area has also been reported to offer better precision than peak height [40], as well as better detection limit at relatively low carrier flow rate (i.e. <5 ml/min) in FI-ICP-MS [68,69]. In fact, peak area (B) is directly related to the concentration C_0 of the sample injected by [10]:

$$B = mC_0V_i/Q \tag{12}$$

where m is the sensitivity (i.e. the slope of the calibration curve), V_i is the volume of sample injected and Q is the flow rate of the carrier.

As was seen in Section 1.4.1, single-point peak hopping (from peak centre to peak centre) has been reported to provide better S/N and therefore detection limits than measuring a full spectral profile by either peak hopping or ramp scanning, when continuous solution nebulization is used with a quadrupole-based instrument [70]. Furthermore, S/N can usually be improved by increasing the integration time [71]. Although this is easily accomplished with a steady-state signal, it is not as straightforward for a transient signal such as that generated by FI.

The situation is further complicated by the type of background on which the transient peak is superimposed. The background can indeed be of two types, random or structured, as illustrated in Fig. 2.7. In the former case, the background varies randomly around a fairly constant value, as is the case when it arises from a polyatomic species originating from either the plasma and/or the carrier. On the other hand, the background would vary in phase with the transient signal if it was caused by a polyatomic species originating from the sample matrix.

The selection of an appropriate integration time, i.e. which would maximize S/N, is very important. As illustrated in Fig. 2.8, the S/N profile for a quadrupole-based instrument goes through a maximum corresponding to an integration window of 0.9 FWHM (full width at half maximum), centred on the maximum of the transient peak, when the background is random [71]. In other words, the best S/N is obtained by integration from a time corresponding to slightly past point A to that corresponding to slightly before point B in Fig. 2.7. With a narrower

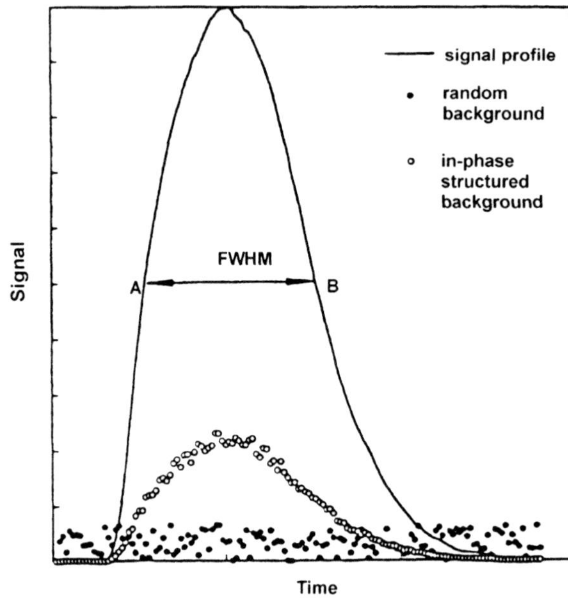

Fig. 2.7. Typical FI signal profiles showing an analyte signal (full line) as well as amplified random background and a structured background varying in phase with the analyte signal. Reproduced from Ref. [71] with permission.

window, not enough signal is integrated, whereas more noise is gained than signal for a wider window. In contrast, the S/N profile levels off as the integration window is widened, when a structured background is present. The best S/N is therefore obtained when the area under the whole peak is integrated. However, since at 0.9 FWHM more than 95% of the maximum S/N achievable is obtained, this window may be selected whatever the type of background [71]. On the other hand, although a better S/N is obtained with a partial integration, precision is better with a full integration [63].

The scanning mode of the MS, for example peak hopping or ramp scanning, is also critical when the measurement time is limited. For instance, when an ELAN 5000 was used, which has a settling time (i.e. time that the MS has to stabilize at a given mass before data acquisition begins) of 4 ms, single-point peak-hopping from spectral peak centre to spectral peak centre always provided the best S/N, no matter how large a mass range was covered, when the dwell time was

≥1.5 ms [71]. Since a better precision is achieved when more ions are counted i.e. when the dwell time is longer, as large a dwell time as possible should be selected while still allowing enough measurements across the transient peak to characterize its profile [71]. On the other hand, a full range mass scan is advantageous for semi-quantitative analysis since it allows a rapid screening of samples [23,68].

The scanning rate should be optimized even on instruments allowing simultaneous detection at several *m/z* values, such as those with multicollectors and TOF-based instruments. The latter are particularly attractive for multielement profiling of transient signals since they allow the acquisition of full mass spectra at the rate of 30,000 per second. In fact, the precisions of isotope ratios, which were measured over the *m/z* range 63–208 during 10-μl injections that generated transient signals with a duration of 6 s full width at half-maximum intensity, were similar to those obtained with a similar acquisition time during continuous nebulization [72].

Nonetheless, on sequential instruments, a 40–50 ms dwell time provided the highest integrated peak area when the settling time was 4 ms for a typical FI peak having a FWHM of 20-60 s (an even longer dwell time should be selected for instruments with longer settling time) [71]. However, the dwell time should be decreased as needed to accommodate either a large number of elements and/or a narrower FI peak, so that 10–20 readings are made across the peak. In any case, more than 50 readings are not recommended as too much overhead would result from the large number of sweeps and the short dwell time then required [71]. This would indeed degrade the duty cycle (i.e. ratio of the measurement time over the sum of measurement and overhead times) [71], and too much time would be wasted settling in comparison to the actual measurement time. For a given number of elements, the precision was reported to rapidly degrade as the dwell time was increased past a certain value [48].

All ICP-MS instruments allow the multielemental monitoring of a FI peak. However, except with TOF instruments, the maximum number of elements that can be monitored at once and the linear dynamic range will depend on the instrument. For instance, high-resolution ICP-MS can usually not scan as fast as quadrupole-based ICP-MS, depending on the mass range covered. As discussed in Section 1.3, instruments with two detectors usually require that a solution be analyzed twice if analytes at both low and high concentrations are to be determined. In contrast, the OmniRange dynamic range control of the

ELAN, which is computer-controlled, allows multielemental analysis over the whole linear dynamic range during the time of a FI peak [23]. Therefore, depending on the instrument used and the number of elements determined, a certain amount of dispersion may be required to allow enough time to measure all the analytes with adequate precision and accuracy.

In any case, the software available for processing transient signals varies from manufacturer to manufacturer. Some software includes a smoothing of the signal prior to peak height or area measurement. Depending on the amount of noise associated with a transient peak, this option may be important to eliminate some of the noise prior to further processing. Some software defines peak height as the maximum signal measured within the monitoring window. Hence, the signal obtained includes the background signal, and requires the injection of a blank, even if the latter is negligible. Obviously, any spurious signal spike observed for that blank would result in an overcompensation of the background. The same comment would apply if peak area is simply measured by integrating over the window as it would include an integration of the background as well, again requiring a blank injection to correct for the latter.

In contrast, software allowing the interpolation of the background prior to the determination of either peak height or peak area is less affected by this problem. The injection of a blank is then only required if it results in a significant signal (measured in the same way as for the sample). Therefore, depending on the software, off-line processing of the raw data on a separate computer may be advisable, and would certainly be necessary for more involved data processing. For instance, the cross-covariance function of periodic FI signals with a synchronised rectangular signal can be used to improve the detection limit by a factor of about two [73]. A reduction of noise indeed results from the fact that there is no correlation between random noise and the rectangular function [73]. Alternatively, digital filtering with discrete Fourier transform can smooth the FI peak, with the possibility of using different Fourier smoothers for asymmetrical and symmetrical peaks [74]. All these approaches efficiently reduce random noise but would have no effect on any contribution from structured background.

Although it has not yet been applied to ICP-MS, peak width may also be used for quantitation. Indeed, if the FI system can be predicted by Eqs. 9–11, then the peak width, W, at concentration C' (i.e. the

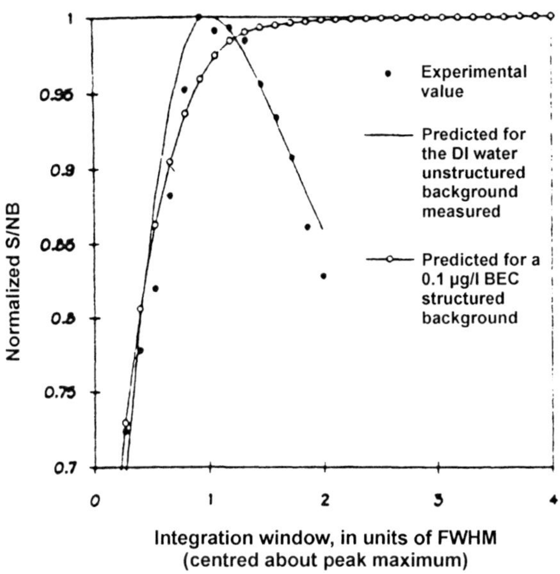

Fig. 2.8. Normalized signal-to-background noise ratio observed for a FI peak as a function of the width of the centred integration window. From Ref. [71] with permission.

concentration level chosen for the peak width measurement) is given by [10]:

$$W = (V_r/Q)\ln[(C_0/C') - 1] - (V_r/Q)\ln(D - 1) \tag{13}$$

A plot of W as a function of $\ln[(C_0/C') - 1]$ (or $\ln[(S_0/S') - 1]$, since ICP-MS has a linear response with concentration) will therefore be a straight line. This approach may be useful for concentrations outside the linear dynamic range established using either peak height or peak area.

Finally, another approach that has not yet been applied to ICP-MS, is that using mathematical models and curve-fitting to predict the steady-state signal (which would be obtained if the sample was aspirated continuously) from the leading edges of the FI peaks resulting from a series of sample injections with a range of sample volumes [75]. Although this approach requires several injections of each sample, it is reported to be less susceptible to variations in sample volumes, flow rates and/or other properties of a FI system [75].

2.3 LIMITED DISPERSION SYSTEMS ($D = 1 - 3$)

These systems simply use the carrier stream to deliver the sample to the nebulizer as quickly as possible so that little dispersion takes place, i.e. the centre of the sample zone does not undergo any significant mixing with the surrounding carrier [5].

2.3.1 Instrumental set-up

The requirements for the implementation of a limited dispersion FI system are minimal. As can be seen in Fig. 2.9, a sample injection valve can simply be inserted between the peristaltic pump and the nebulizer. Dispersion is minimized by keeping the length of the tubing between the valve and the nebulizer as short as possible. Using a tubing with a small i.d. is also important. However, as the i.d. decreases, the back-pressure will increase (it is, in fact, inversely proportional to the i.d. of the tubing [49]), making it more difficult for the peristaltic pump to provide the desired carrier flow rate.

In its least expensive version, the sample injection loop of the limited dispersion FI system can be manually filled using a syringe as described in Section 2.2.3. The system can be entirely automated as shown in Fig. 2.10, by coupling a computer-controlled autosampler to a computer-controlled sample injection valve, a channel of the peristaltic pump (which can also be computer-controlled) then being used to supply the sample selected by the autosampler to the valve. Using this approach, however, leads to sample waste. Indeed, since the carrier must be continuously supplied to the nebulizer, the pump must remain on throughout the analysis, which also means that sample is continuously pumped to waste between injections. Although the amount of sample being wasted can be reduced by pumping a washing solution

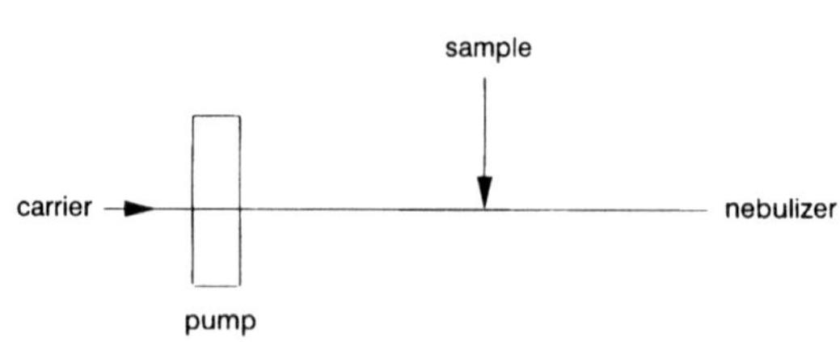

Fig. 2.9. Typical limited-dispersion FI manifold. The vertical arrow indicates sample injection using a valve.

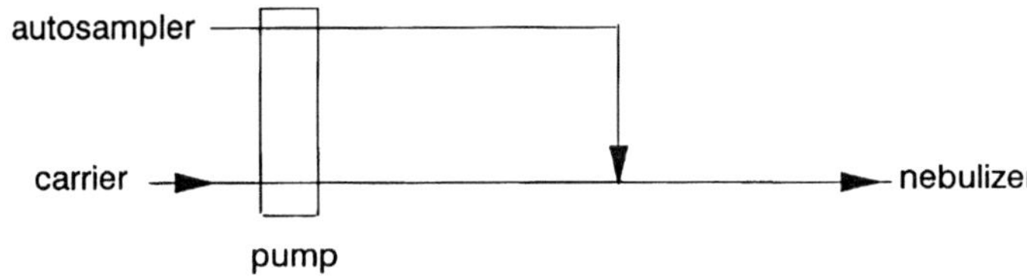

Fig. 2.10. Automated limited-dispersion FI manifold. An autosampler is used to fill the sample injection loop. Computer-controlled injection is then made into the carrier whose flow rate is selected by a computer-controlled pump.

between samples, a better approach is to use a separate pump to fill the sample loop.

This efficient alternative was implemented in commercially available FI systems, such as the computer-controlled FIAS-400 from Perkin-Elmer which is illustrated in Fig. 2.11. It uses one pump to supply the carrier and another one to fill the sample injection loop. The latter pump therefore only needs to operate during the filling step and each sample change-over while the carrier is continuously supplied to the nebulizer by a separate pump. Its flow rate is also independent of that of the carrier pump.

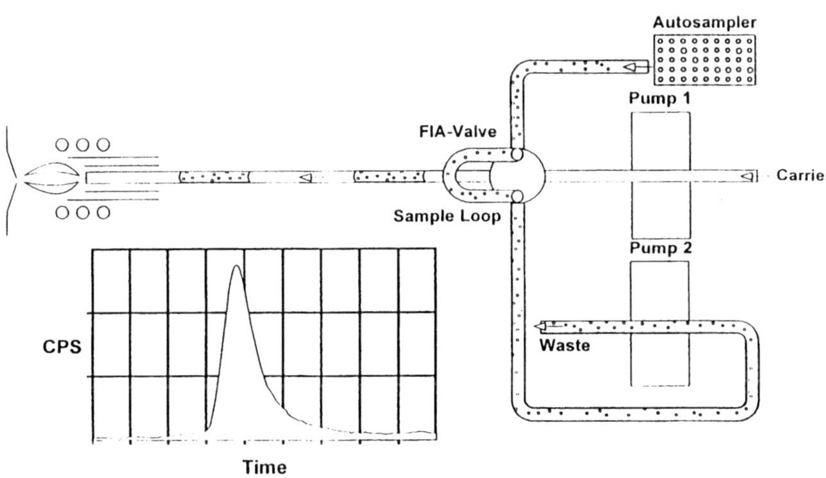

Fig. 2.11. Commercially-available FI system, the FIAS-400, where one pump fills the sample loop while the other pump continuously supplies the carrier to the nebulizer. Courtesy of Perkin-Elmer.

As mentioned in Section 2.2.3, D can be kept to a minimum by modifying the sample introduction system and using a high-efficiency nebulizer. However, if cost is a limitation, a spray chamber of smaller volume can replace the conventional double-pass spray chamber and thereby reduce D_d [76]. For instance, with 500-µl injections, D was about 1.3 with a Scott spray chamber, but was reduced to 1 with a reduced-volume spray chamber [76]. In fact, a 10.5-ml spray chamber allowed 200-µl injections with D of 1 [60]. Alternatively, a cooled mini-spray chamber or a desolvation system with a reduced volume (about 15 ml), could be used to minimize D_d and at the same time reduce solvent loading in the case of samples in organic solvents [77].

Finally, several of the manifolds described in the following sections can also result in limited dispersion. For example, the reduced dispersion systems described in Section 2.6 will result in limited dispersion if the sample volume is slightly smaller than that of the eluent, diluent or extractant, which is used to separate the analytes from the matrix. Also, a limited dispersion would result if an on-line extraction manifold (described in Section 2.6.1.3) was used to selectively extract a troublesome matrix component from an aqueous sample. The organic extract would then be discarded while the aqueous sample is pumped to the nebulizer. An example of an application that could be made into a FI manifold is that of the multielement analysis of uranium oxide samples [78].

2.3.2 ICP-MS Instrument optimization

The optimization of the plasma and MS operating conditions can be carried out as described at the end of Section 1.2.1, but using repeated injections after each modification to the conditions [79]. The sample injection loop can be increased to provide a quasi steady-state signal during which optimization can be carried out [66]. However, for aqueous carrier and samples, the optimum conditions thereby obtained are usually the same as for continuous nebulization. Therefore, the optimization process can be speeded up by performing it in the usual way, i.e. with continuous aspiration, but using a standard solution as the carrier in the FI set-up. The analyses can then be carried on by FI of the samples and standards in a DDW carrier.

Alternatively, a 2% HNO_3 carrier may be preferred since it was reported to enhance the peak height of several elements (Co, Cd, Cu, Ni, Mn, Ag and Sb) and to reduce the effect of concomitant elements compared to those observed with a DDW carrier under otherwise identical conditions [79]. In some cases, effects from concomitant elements can be significantly reduced by tuning the ion lenses for maximum analyte signal while aspirating a sample or matrix-matched standard [80]. Non-spectroscopic interferences induced by the matrix can also be minimized by using robust plasma conditions, providing a longer residence time in the plasma, i.e. reduced aerosol carrier flow rate and higher rf power [81].

As mentioned in Chapter 1, simplex optimization, which varies all the operating parameters simultaneously, can also be used to find the best operating conditions. For instance, the aerosol carrier gas flow rate, sampling position, forward power, amount of O_2 in the aerosol carrier gas flow and the FI carrier flow rate were optimized to obtain the best S/B ratio for 25-µl FI of solutions of In in diethylether [82]. An alternative is to use a factorial optimization, which also varies several parameters simultaneously, to find optimum conditions [83] and to look at the response surfaces [84]. Note that these optimization techniques can also be used to optimize the FI manifold [83,84]. The design of a FI system can also be optimized using mathematical modelling of the flow system coupled with the Powell's optimization algorithm [85].

As discussed in Chapter 1, the mixed-gas plasma resulting from an addition of O_2 greatly facilitates the introduction of organic solvent in ICP-MS by reducing soot deposition on the sampler. In addition, the greater solvent load resulting from these volatile solvents (in comparison to water) can be reduced by adding a desolvation system, such as a membrane separator [82]. This allows an increase in the sample injection volume without degrading plasma stability, which in turn improves detection limits [82]. The combination of FI with desolvation and mixed-gas Ar–O_2 plasma allows an increase in sample throughput with reduced carbon deposition and improved plasma stability. On the other hand, the use of O_2 has also been reported to visibly accelerate the corrosion of the sampler [86].

If a more extensive desolvation is accomplished, a mixed-gas plasma is then not required. For instance, the 70% efficiency of solvent removal which could be achieved with either a cooled spray chamber or a heated spray chamber/Peltier cooling desolvation system, allowed the analysis of diethyl ether [87]. The combination of such a system with FI allowed the analysis of a less volatile organic solvent, 2-ethoxyethanol [87]. In fact, the interference from CO^+ was then so reduced that the determination of Si at *m/z* 28 became possible [87].

Nonetheless, organic solvent (methanol, ethanol and propanol) can be analyzed if an on-line 1+1 dilution with water is performed and a torch with a 1.3-mm i.d. injector is used [86]. At a given rf power and carrier flow rate, the count rate maxima are expected to shift to lower aerosol carrier gas flow rate in presence of the organic solvent [86]. Depending on the rf generator, more or less optimization of the ICP conditions will be required. For instance, a free-running 40-MHz generator was reported to provide rapid, responsive tuning on-the-fly upon changes in sample matrix (such as the injection of 95% methanol in an aqueous carrier) [88]. Analyte signal enhancement is also expected, especially for elements with a high first ionization potential [86]. This enhancement may be related to the number of carbons in the alcohol but is not directly correlated [86]. The introduction of organic solvent will also affect the nebulization efficiency. For instance, the transient signals were narrower in an alcoholic medium than in aqueous solution [86]. On the other hand, a significant memory effect of the enhancement effect from alcohols is expected and should therefore be taken into account, especially since it was observed to be independent of the volume injected, the nature of the alcohol, or even the number of successive alcohol injections [86]. Since it is probable that organic vapour continues to reach the plasma from the spray chamber and its drain, following an injection, reducing the dimensions of the spray chamber may help reduce this memory effect.

2.3.3 Features

2.3.3.1 Aqueous carrier

Because dispersion is limited, the sample undergoes minimal mixing with the carrier stream and sensitivity (measured in terms of peak height) is preserved compared to what it would be if no FI was used and the solutions were continuously nebulized. A FI system with limited dispersion can therefore be used for direct sample introduction, i.e. without any treatment. For instance, the height of FI peak resulting from 100-μl injections corresponded to 50% of the continuous aspiration signal in ICP-AES [12] and ICP-MS [89, 90], i.e. $D = 2$. However, the corresponding detection limit was only slightly degraded (i.e. by less than a factor of two) [89]. With 350-μl injections, sensitivity was 70% of that obtained by continuous aspiration but the analysis time per sample was shorter by 40% [91].

When a DIN with a 1.5-µl dead volume [92] was used to directly introduce 30-µl sample slugs into the plasma without any spray chamber, such that D was unity, similar or slightly improved detection limits (based on peak heights) were observed in comparison to conventional cross-flow nebulizers/spray chamber systems in ICP-AES, with (or without) FI using 200-µl injections and a carrier flow rate such that D was very close to one [93]. Better detection limits could be achieved with USND but required 500-µl injections to obtain D close to one [93]. Relative detection limits (concentration units) deteriorated as the sample volume decreased, with USND in FI-ICP-MS, whereas absolute detection limits improved [94]. On the other hand, at a given injection volume, changing the carrier flow rate between 0.5 and 1.5 ml/min had no significant effect on detection limits [94].

The very small dead volume of the DIN means that if a short, narrow tubing connects the sample injection valve to the nebulizer, a steady-state signal will result (i.e. $D = 1$). The small dead volume is also very advantageous for the determination of elements prone to memory effects. For instance, the steady-state signal resulting from 75-µl injections of Hg showed no tailing such as that which is typically observed with pneumatic nebulization [95]. Furthermore, the elimination of several dead volume areas, such as the spray chamber and drain tube of a conventional nebulization set-up, improved the short-term precision compared to pneumatic nebulization [95].

Even without a DIN, the wash-out process is more efficient than with continuous sample aspiration since the exponential decay immediately following the transient peak occurs sooner [12]. For instance, 40-45 s was sufficient for a complete washout following 100-µl injections (D = 2) whereas 110–120 s was required following continuous nebulization [40]. In addition, the signals observed by FI for standard solutions aspirated before and after samples (with a variety of matrices) will recover more rapidly from any matrix effect than those observed by continuous nebulization [96]. This recovery can be even more complete if internal standardization is also used [96,97].

In any case, the quicker wash-out also means that if a minimum dispersion is obtained with a small sampling volume, then the sample throughput can be increased compared to what can be achieved with direct continuous aspiration of samples. For instance, in ICP-AES, 4 samples could be injected per minute, as opposed to 1.5/min using continuous aspiration [12]. The same factor of 2–4 times faster than continuous solution aspiration was also reported for FI-ICP-MS [23,40].

The continuous rinsing effect of the carrier also efficiently minimizes clogging of the sampler orifice as a result of the deposition of salt, in the presence of a high level of dissolved solids, or carbon, in the presence of organic solvents (from the combustion of organic matter). For this reason, FI-ICP-MS is ideally suited for the analysis of sample types which are difficult using continuous nebulization. These include samples with high dissolved solids content [40,89,69,98–103] such as those prepared by alkali fusion [104], with high viscosity [40], with high acid concentration [40,105], with HF [106], containing organic solvents [82] or even slurries [76,107]. For example, FI [76] allowed the analysis of 1% w/v slurries which would have resulted in blockage after about 30 min of continuous nebulization [108]. In fact, under appropriate conditions, 0.5-cm^3 injections provided the same signal as for continuous nebulization of the same solutions, while reducing sample consumption and eliminating blockage problems [76]. These conditions allowed the accurate determination of V and Cu in two reference soil slurries by ICP-AES, using a conventional external calibration with standard solutions [76]. The dispersion behaviour of slurries is dependent on several parameters, including the density of the particles, and the configuration of the manifold [109]. For instance, a broader delayed peak was observed with a straight tube compared to that obtained for a solution under identical FI conditions, while a narrower peak resulted with a horizontal coil [109].

In all these cases, clogging problems are avoided not only by rinsing with the liquid carrier but, in addition, by the dispersion taking place in the FI manifold [69]. Indeed, if $D > 1$, the concentration of the trouble matrix even at the peak maximum is lower than that which would result if the same sample was continuously nebulized. Although precision will be better than that obtained by continuous nebulization of the same samples, a degradation in precision is expected compared to that obtained with standard solutions as a result of matrix effects [104].

This type of FI system can also be used to minimize contamination of the ICP-MS instrument with radioactive samples [110]. Very small volumes can indeed be used when FI is coupled to a micronebulizer [110]. Nonetheless, even if the etching effect of HF on glassware is drastically reduced with FI, HF-resistant nebulizer, spray chamber and torch injector should be used whenever samples containing HF have to be analyzed.

Using FI-ICP-MS, solutions containing up to 10.0% salt could be analyzed without interface blockage with D of about 2 [40]. Although signal suppression was observed to increase with the salt concentration [99] as a result of the non-spectroscopic interference caused by the salt matrix, the precision of the FI signal (either peak height or peak area) remained unaffected [40]. In general, the extent of solid condensation on the interface will depend on the volatility of the salt. For instance, salts with low to medium volatility (such as Li, Na and K salts) will volatilise from the plasma without extensive condensation on the interface [98]. Whereas matrix elements forming refractory oxides (such as salts of Al, Zr, U and Th) will induce more extensive deposition on the interface [98]. Nonetheless, FI allowed the analysis of 1% w/v alumina solutions during 4 hours without any downward change in the signal (unlike continuous aspiration which would result in complete blockage of the sampler in less than an hour) [99]. Any deposition may be further reduced by increasing the wash-out time between injections at, however, the expense of sample throughput [90]. In any case, non-spectroscopic interferences will be identical to those observed by continuous nebulization if $D = 1$ [80], or slightly reduced by the dilution effect of FI if $1 < D < 3$.

When analyzing samples with high salt concentration, both the carrier flow rate (i.e. the sample delivery rate) and the sample injection volume are critical parameters. Each nebulizer has indeed a narrow range of uptake rate at which stability and repeatability are best. Furthermore, the injection volume allowing a compromise between high sensitivity and minimum solid deposition should be selected [69,100]. Undesirable diffusion effects are also avoided by using a small reactor volume (i.e. a short connection between the sample injection valve and the nebulizer), a small injection volume and/or a relatively high carrier flow rate [50]. Under optimized conditions, no significant drift was observed over 2 h with 75 injections of a spiked 3% NaCl solution in contrast to that observed during continuous nebulization [100]. This improved stability translated into detection limits for 13 elements which were lower by a factor of 3–9 compared to those achieved by continuous nebulization in the same saline matrix [100].

Similarly, the short term precision and drift observed in FI-ICP-MS with 3% Ni were the same as for 1% HNO_3 solutions, and, with internal standardization, were equivalent to those obtained by continuous nebulization of 1% HNO_3 solutions [69]. However, although a strong correlation was observed between the degree of ionization and the extent of suppression, elements with lower degree of ionization exhibiting more suppression, matrix effects varied so much across the mass range that internal standardization was impractical and the method of standard additions was required for quantitative analysis [69].

With D of about 2, the problems of cone destruction and blockage encountered with 5% H_2SO_4 matrix were also completely avoided [40]. Even with D of about 1, samples with 20% HNO_3 could be analyzed by isotope dilution analysis [40]. Similarly, the potential instabilities associated with the direct aspiration of up to 40% m/V glycerol were avoided using D of about 2 [40]. However, this did not prevent the degradation of sample transport efficiency of the nebulization system which resulted from the high viscosity of these solutions [40].

Although FI will allow the injection of small volumes of organic solvent (such as 100 µl of methanol diluted 1+1 on-line with water) without visible carbon deposits, a change in the ion lens settings may be required [111]. The injection of samples containing organic solvents may require an extensive optimization in the FI mode, as the optimal conditions will likely be quite different from those obtained by continuous nebulization of aqueous solutions. This is especially true if no other means is used to cope with the likely increase in solvent load (such as a cooled spray chamber or the addition of a desolvation system) and/or to prevent carbon deposition on the interface (such as Ar–O_2 plasma), as discussed in Section 2.3.2.

For instance, the enhancement observed with continuous nebulization in the presence of carbon-containing compounds for elements with a high first IP (such as As) (see Section 1.4.3) was also observed in the FI mode [66]. As illustrated in Fig. 2.12, the peak height for Ni and Pb was suppressed whereas a two-fold increase concurrently occurred for As when the matrix was changed from DDW to 30% MeOH/70% DDW with an appropriate adjustment of the plasma operating conditions (as discussed in Section 1.4.3). Figure 2.12b demonstrates that the increase in As sensitivity is not due to an increase in the background since the corresponding blank injections were negligible. The enhancement was caused by only a relatively small amount of organic solvent since no further enhancement was observed when the methanol content was increased to 60% (in Fig. 2.12c).

Because little dilution of the sample occurs in limited dispersion FI systems, significant non-spectroscopic (as mentioned above) as well as spectroscopic interferences may result, depending on the relative concentration of the analyte and the source of the interference. For instance, Na salts will result in $^{40}Ar^{23}Na^+$ which will interfere with

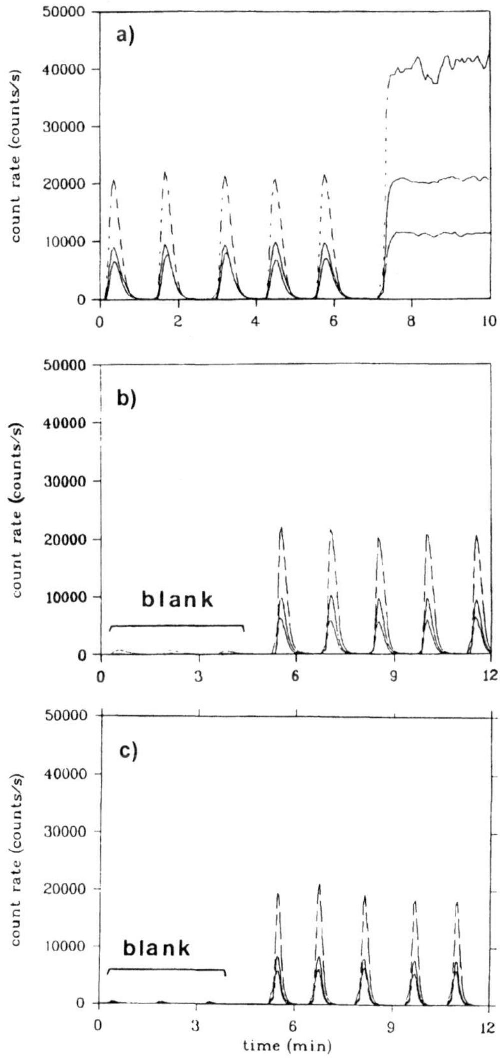

Fig. 2.12. Signal observed for (a) 5 replicate 0.1-ml injections of 100 µg/l aqueous solution of Ni, As and Pb into DDW carrier; (b) 3 blank injections and 5 replicate 0.1-ml injections of 100 µg/l Ni, As and Pb in 30% MeOH/70% DDW into DDW carrier; and (c) 3 blank injections and 5 replicate 0.1-ml injections of 100 µg/l Ni, As and Pb in 60% MeOH/40% DDW into DDW carrier. The bottom trace in both figures is that of $^{60}Ni^{+}$ but the middle trace is that of $^{75}As^{+}$ and the top trace that of $^{208}Pb^{+}$ in (a), whereas the top trace is that of $^{75}As^{+}$ and the middle trace that of $^{208}Pb^{+}$ in (b) and (c).

^{63}Cu, thereby requiring the monitoring of the less abundant ^{65}Cu instead, with an associated loss in sensitivity. The determination of monoisotopic ^{59}Co could also be seriously hampered by $^{36}Ar^{23}Na^{+}$, unless the interference from the latter is small enough that it can be corrected using matrix-matched standards.

2.3.3.2 Gaseous or gas-segmented carrier

Dispersion can be drastically reduced by surrounding the sample slug with gas bubbles, since this would obviously eliminate dilution and reduce dispersion to that occurring from mass transfer between the sample film on the tubing wall and subsequent liquid stream [2]. Houk and Thompson [112] were the first to use this approach to introduce discrete sample volumes in ICP-MS. The inlet tubing to the nebulizer was simply connected to a 50-, 100- or 200-μl pipette which was moved between a reference blank solution and sample solutions, thereby introducing an air bubble at both ends of the sample slug [112]. However, except for its small sample consumption, none of the beneficial features of this approach was then assessed. The approach was also used with FAAS [113] where improvement in sensitivity, and increased sample throughput resulted.

Surrounding both ends of each sample slug with air bubbles during the injection process is called segmented (or segmental) flow injection (SFI). Figure 2.13 shows a special valve which can be used to sandwich the sample between two air bubbles. The characteristics of SFI are intermediate between those of FI and Autoanalyzers. For instance, the number of bubbles in SFI is about 5% of that with AutoAnalyzers but

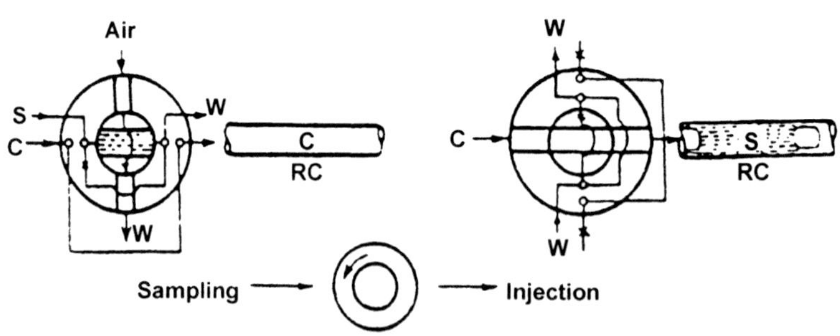

Fig. 2.13. Special valve that can be used to sandwich a sample plug between air bubbles. S: sample; C: carrier; W: waste; RC: reaction coil. Reproduced from Ref. [114] with permission.

like FI a discrete injection is made and reactions do not have to go to completion before the detection [114]. Unlike FI and Autoanalyzers, there is essentially no dispersion of the sample because of the poor wetting of PTFE by aqueous solutions. A change in the sample viscosity, which could increase the thickness of the sample film on the tubing wall, may increase *D*. In any case, the sampling frequency can be even higher than with FI, since it only depends on the sample volume injected (as well as the debubbling process if the gas is removed prior to detection) [114].

Segmented flow injection is particularly useful in a number of cases although no SFI-ICP-MS applications have yet been reported. For instance, slow reactions i.e. requiring a long residence time, can be carried out without the fear that the resulting dispersion will be too great for an accurate determination of the analyte [114]. Furthermore, temperature can be increased to speed up a reaction since the dispersion will then not be affected by any random formation of bubbles in the flow system which could result from the rise in temperature [114 and references therein]. In addition, SFI is well suited to the analysis of heterogeneous samples such as whole blood [115,116]. Indeed, if these samples were injected into an aqueous carrier, they would undergo dispersion to a different extent than homogeneous standard solutions, which would, in turn, complicate the calibration process since matrix-matched standards would then be required (unless either the method of standard additions or the isotope dilution technique are used).

Instead of sandwiching the sample between gas bubbles, FI into a gas carrier can also be carried out. This approach resulted in a 1.8-fold increase in the sensitivity of FAAS for Zn compared to that achieved by FI-FAAS with a water carrier under otherwise identical conditions [117]. The peak width at half maximum was also half of that observed in a DDW carrier which means that the sampling frequency could be doubled. Sensitivity in the air carrier was in fact slightly higher than the steady-state signal obtained by continuous nebulization [117]. It decreased by only 20% (as opposed to 70% with a water carrier) as the length of a dispersion coil was increased from 15 to 55 cm [117].

An air carrier was also used for FI-ICP-MS in an attempt to combine the benefits of FI with those of mixed-gas plasmas [118]. In the conditions used (i.e. 100-μl injections in 60 cm of 0.78-mm i.d. tubing), narrow and symmetrical peaks were observed, which allowed the sampling frequency to be easily doubled in comparison to the asymmetrical peaks resulting from FI in a water carrier [118]. Dispersion was also reduced since peak heights were higher than in a water carrier. Furthermore, the precision associated with replicate injections improved. On the other hand, the sensitivity enhancement over that obtained in a water carrier, was dependent on mass. For instance, a 5.3-fold increase in peak height occurred for Al whereas that of Pb only slightly increased (by a factor of 1.2) [118]. However, persistent and large memory effects (which could only be eliminated by thoroughly rinsing the FI system between injections) resulted because the sample introduction system was no longer being washed out by a water carrier [118].

An air-segmented water carrier stream was therefore used as an alternative. This stream simply resulted from pumping air and water through a T-piece [118]. By making the injections into the air bubbles, the above benefits of the air carrier were retained but, in addition, memory effects were substantially reduced [118]. A systematic study comparing a water carrier to an air-segmented water carrier and an air carrier showed that the latter two resulted in similar baseline peak width (which was half of that observed in the water carrier) and reproducibility of injections [34]. A significant increase in both the background and sensitivity occurred when switching from a water carrier to an air-segmented water carrier, and even further when switching to the air carrier [34]. The change in sensitivity was mass-related whether peak height or peak area was used [34,119]. This indicated that other factors were involved in addition to an elimination of dispersion which would have resulted in no change in peak area since no analyte was lost or created when changing the carrier.

In fact, as had been observed in FAAS [117], the peak height was higher (and D was smaller than 1) than with a water carrier for several elements in both the air-segmented water carrier and the air carrier [34]. This was partly explained by a concurrent increase in the nebulization efficiency: the percentage of sample solution reaching the injector of the torch indeed increased from 2.4 ± 0.2% with the water carrier to 4.2 or 4.3 ± 0.3% in the other two carriers [34]. However, it does not explain the mass-related trend that may, on the other hand, be due to a change in space-charge effects (see Section 1.4.3). Since air is less dense than water, it is indeed possible that the flux of the ion beam entering the mass spectrometer is significantly reduced when switching from a water carrier to either an air-segmented water carrier or an air carrier. In any case, the best detection limits were obtained with the air–water carrier, the memory effect mentioned earlier degrading those with the air carrier [34].

TABLE 2.3

Change in sensitivity and detection limit observed when switching from continuous nebulization to SFI, using 100-µl injections (compiled from Ref. [119])

Analyte	Ratio of sensitivity (SFI/continuous)	Ratio of detection limit (continuous/SFI)
^{27}Al	1.22	0.7
^{51}V	1.07	2.4
^{52}Cr	1.04	2.0
^{55}Mn	0.99	2.7
^{57}Fe	0.89	3.4
^{59}Co	0.99	4.5
^{60}Ni	0.96	3.4
^{63}Cu	0.91	3.1
^{66}Zn	0.71	0.5
^{69}Ga	0.92	1.0
^{75}As	0.85	1.6
^{78}Se	0.66	0.4
^{98}Mo	0.94	1.0
^{114}Cd	0.75	4.0
^{121}Sb	0.83	0.6
^{208}Pb	0.79	1.0

Under optimized conditions (in terms of aerosol carrier flow rate, forward power, air–water ratio in the carrier, carrier flow rate and injection volume), the detection limits obtained by SFI-ICP-MS using only 100-µl injections were similar or improved for a large number of elements compared to those obtained by continuous nebulization [119]. As shown in Table 2.3, these results were obtained despite the fact that the sensitivity was similar or degraded compared to that by continuous aspiration. A better precision therefore resulted from using SFI [119]. However, this improvement was not due to a mixed-gas effect since a change in the segmenting gas from air to Ar, N_2 or H_2 had no significant effect [119]. The proportion of foreign gas in the aerosol gas flow, which was about 0.1% with the SFI conditions, was probably too low.

An alternative approach to an air-segmented carrier is to only surround the sample plug with air, using an otherwise aqueous carrier [120,121]. This can be accomplished by using a two-valve two-pump manifold, with a sample injection loop [121]. The sample could also be inserted between discrete air volumes using a sandwich manifold, although this manifold has not yet been used for this purpose [122]. Alternatively, an autosampler can be modified so that the probe is connected by a T-piece to an electromagnetic vent itself connected to a tank of carrier [120]. When the vent is open, the carrier is pumped to the nebulizer while the probe is filled with air. The vent is then closed and the probe moved in and out of the sample, thereby resulting in an air–sample–air plug that is pumped to the nebulizer. This system has been called air-segmented discrete introduction [120]. The approach reduces sample consumption since no extra volume is required to fill a loop [120]. In contrast to the air-segmented carrier approach described above, this one only introduces bubbles when a sample is aspirated, which may result in better stability and washout [120]. However, changes were observed in the energy balance of the plasma when the air bubbles passed through, which resulted in a suppression of analyte signal for elements with a high ionization potential (Te, Zn, As, Cd) when sample volumes <300 μl were used [120]. This effect was not observed with the air-segmented carrier stream. Also, a second small peak is clearly observed in the tail of the transient peak [120, 121] which, although it is attributed to the sample film that is formed on the tube wall [120] and should therefore be worse with the air-segmented carrier stream, is not observed with the latter. This difference cannot be ascribed to the ICP-MS instrument since an ELAN 500 was used for both studies.

Finally, the sample plug itself could be segmented with air bubbles, by simply pumping the sample and air, using a merging manifold, into the sample loop [123]. Although this approach should yield similar results to those obtained with manifolds where the sample is sandwiched between air bubbles, it has not yet been used with ICP-MS. One potential reason is that the air bubbles within the sample would likely induce fluctuations in the analyte signal, which would require a special software for the signal processing [124]. Alternatively, a bubble gate [124] could be used to remove the bubbles before they enter the nebulizer.

2.3.4 Calibration strategies

Quantitative analysis may be performed in various ways with limited dispersion FI systems. Of course, off-line calibration strategies can be used i.e. where standard solutions (for external calibration) and/or spiked samples (for the method of standard additions or isotope dilution analysis) are prepared prior to injection. However, on-line calibration can also be carried out with great advantages. Aside from reducing the sample preparation step and minimizing contaminations, as stated in Table 2.1, the latter approach also reduces the likelihood of mistakes when a large number of solutions has to be prepared.

2.3.4.1 On-line internal standardization

Even if isotopes free of spectroscopic interferences are available for the determination of analytes, either matrix-matched standards and/or internal standardization [80,96] (as discussed in Section 1.4.3) may be required to compensate for non-spectroscopic and/or physical interferences. This is likely to be a requirement for the accurate analysis of samples of various types. As with continuous nebulization, the beneficial effect of internal standardization will be especially noticeable as the mass and/or the concentration of the matrix element increases [80]. Internal standardization may also be combined to an optimization of the ion lens settings in presence of the matrix for a further reduction of matrix effects [80,96]. On-line addition of an internal standard can easily be carried out by merging the sample with the internal standard solution through a mixing tee before filling the sample injection loop [63,91]. The same criteria which were discussed in Section 1.4 for the selection of internal standards in ICP-MS apply to FI-ICP-MS. For instance, they must be similar in mass to the analytes [125].

2.3.4.2 On-line standard additions

Whenever internal standards are not available or matrix effects are such that internal standardization across the mass range is impractical, the method of standard additions or the isotope dilution analysis technique are required for quantitative analysis. An on-line set-up may be used for this purpose. For instance, the sample was continuously merged with standard solution (supplied by an autosampler) through a tee, and mixed in 20-cm of 1-mm i.d. tubing before entering the sample injection loop [69]. In fact, both the method of standard additions and internal standardization can be carried out in this manner. Four solutions are then pumped: the sample, standard, internal standard and carrier. The sample is first merged with the standard and mixed in a coil, the outcoming mixture being itself merged to the stream of internal standard solution and mixed through another coil before entering the injection valve [105]. Alternatively, the injected sample is sequentially merged with a standard solution and the internal standard solution, resulting in a 3-fold dilution [63].

An alternative, sample-saving approach was used by Wiederin and co-workers [126], using a DIN for sample introduction into the plasma. The carrier was split into two parallel flows: one where the sample was injected from a 250- or 500-µl loop; and the other where injections of 3 standard solutions were made through a 40-or 50-µl loop. The two flows were then merged before being pumped to the DIN. Because the total volume between the valves and the ICP was about 3 µl, steady-state signals resulted as shown in Fig. 2.14: a blank one due to the carrier by-passing both loops, a long one due to the sample (diluted with the

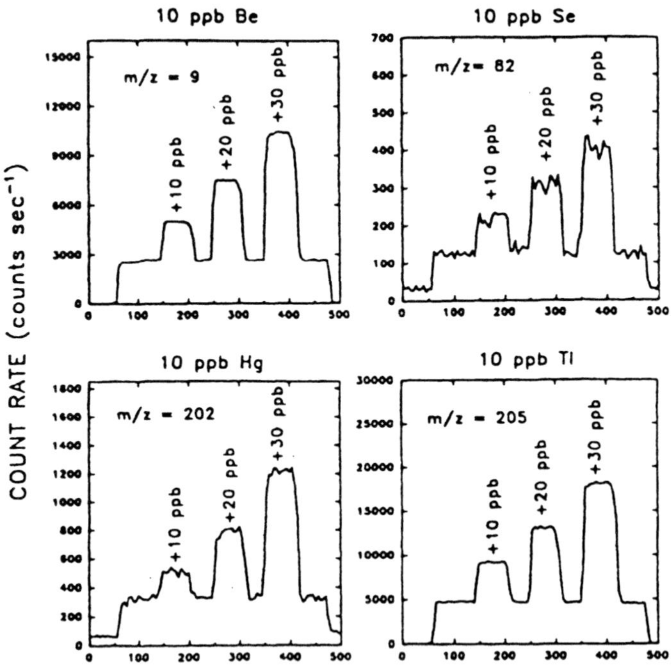

Fig. 2.14. Signals observed when a 40-µl injection of each of 10, 20 and 30 µg/l standards of Be, Se, Hg and Tl were made into a 500-µl injection of a 10 µg/l solution of the same elements in 2% HNO_3 when a DIN is used instead of the regular nebulizer/spray chamber arrangement. Reproduced from Ref. [126] with permission.

carrier) on top of which three shorter ones corresponded to each standard injection. A correction had to be made for the difference in flow rate between the two lines, since the longer tubing connecting the standard injection valve to the merging tee resulted in a slower flow rate [126]. Any significant difference in viscosity also had to be taken into account [126]. Finally, a correction was required for drift, which was simply carried out by ratioing the signal from the standard to the mean signal from the sample measured immediately before and after the standard injection [126]. The method adequately compensated the 5–10 fold suppression induced by urine, which contains about 1.3% dissolved solids [126].

All the approaches which have been discussed so far allow the regular method of standard additions to be carried out on-line. In other words, the sample concentration is found by extrapolation of the signal versus spiked concentration line, the negative concentration intercept corresponding to the unknown sample concentration. Results of highest accuracy are therefore obtained when the sample is spiked with a standard that matches the sample concentration.

A completely different approach was described by Beauchemin [127] who used a simple single-line FI manifold with a standard nebulizer/spray chamber system. Two solutions are required: a 1% HNO_3 solution and a standard solution in 1% HNO_3 which has a higher concentration than the sample. The standard and sample are sequentially injected into a carrier of 1% HNO_3. The standard is then used as the carrier and the sample injected into it. The process is then repeated in the reverse order to save rinsing time, and thereby maximize sample throughput. This cycle of two series of injections (shown in Fig. 2.15), which yields two replicate standard additions, can be repeated as required to obtain the desired number of replicates. The analyte concentration of the sample, C^0_{sam}, is then calculated using:

$$C^0_{\text{sam}} = \frac{I^{\text{p}}_{\text{sam-into-blk}} C^0_{\text{std}} (D-1)}{I^{\text{p}}_{\text{sam-into-std}} - I^{\text{p}}_{\text{sam-into-blk}} + m_1 C^0_{\text{std}}} \qquad (14)$$

where $I^{\text{p}}_{\text{sam-into-blk}}$ is the peak height of the positive peak resulting from injecting the sample into the blank carrier; $I^{\text{p}}_{\text{sam-into-std}}$ is the peak height of the negative peak resulting from injecting the sample into the standard carrier; C^0_{std} is the analyte concentration of the standard solution; D is the dispersion coefficient which is obtained from:

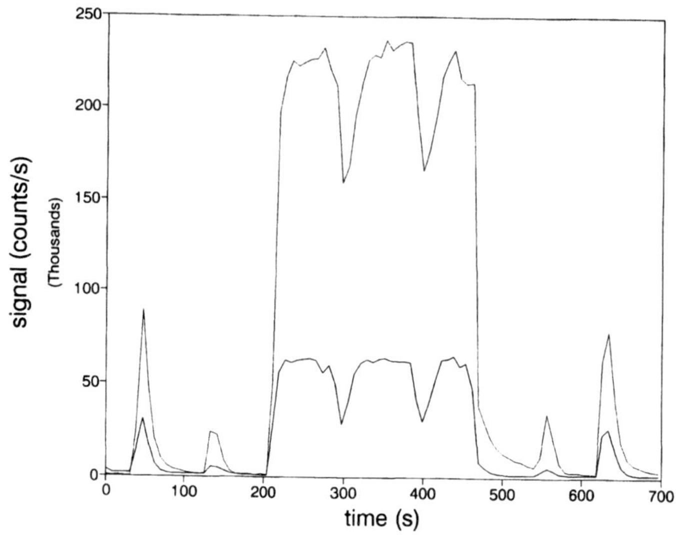

Fig. 2.15. Typical signal profile observed during two replicate on-line standard additions. The blank (1% HNO_3) is used as carrier from 0 to 200 s, into which 100 μl of 50 μg/l standard is injected followed by a 100-μl injection of the sample. Between 200 and 500 s, the 50 μg/l standard is used as carrier and two 100-μl injections of the sample are made. The carrier is then switched back to the blank and 100 μl each of the sample and 50 μg/l standard are injected sequentially. The top trace shows the $^{51}V^+$ signal whereas the bottom trace shows that of $^{62}Ni^+$.

$$D = \frac{I^0_{std} - I_{bak}}{I^p_{std}} \tag{15}$$

where I^0_{std} is the steady-state signal due to the standard, I_{bak} is the background signal (i.e. the steady-state signal due to the blank) and I^p_{std} is the peak height of the positive peak resulting from injecting the standard into the blank carrier; and m_1 is the sensitivity in presence of the standard solution which is given by:

$$m_1 = \frac{I^0_{std} - I_{bak}}{C^0_{std}} \tag{16}$$

This approach actually takes into account the change in sensitivity which occurs in the presence of the matrix. Some compensation for drift is also accomplished by the determination of D and m_1 for each replicate. The approach would not be accurate if drift was significant within the time required for one replicate, in which case an internal standard would be needed as well. It also assumes that the matrix effect is constant over the analyte concentration range 0–C^0_{std}. Indeed, for the method of standard additions to be successful, the concentration ratio of interferent to analyte must be equal to or higher than that at which the matrix effect is constant and independent of analyte concentration [47,128]. However, since ICP-MS is generally used for ultra-trace analysis, the analyte concentration is then so small compared to the matrix that a constant effect results.

In any case, the approach minimizes sample consumption, especially compared to the reversed-FI method described in Section 2.4.3.2, and preserves sensitivity which can, at the most, decrease down to half of what it would be by continuous nebulization. The only drawback of the approach is a degradation of precision compared to an external calibration. This arises from the error propagation of the various parameters which have to be determined. It is, however, a small price to pay in comparison to all the benefits.

A completely different approach, which has not yet been applied to ICP-MS, uses a single sample injection between water (or a blank) and a standard solution [129], as illustrated in Fig. 2.16. In the loop filling position, water is aspirated by the nebulizer. When the sample

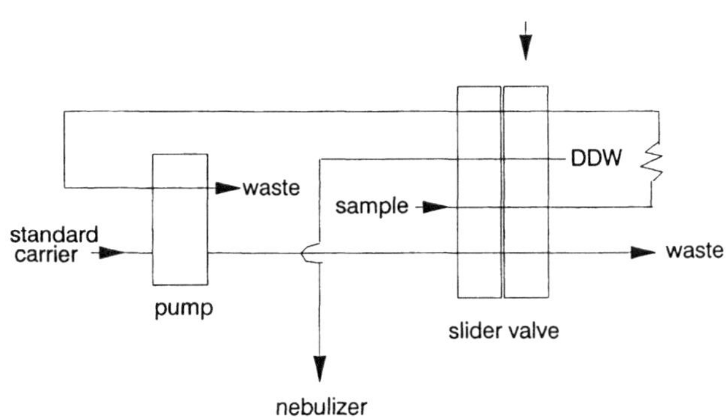

Fig. 2.16. Gradient scanning standard addition manifold. While DDW is pumped to the nebulizer, the sample is pumped through the loop. When the valve is slid down, the standard carrier pushes the sample plug against DDW, carrying both towards the nebulizer. After Ref. [129].

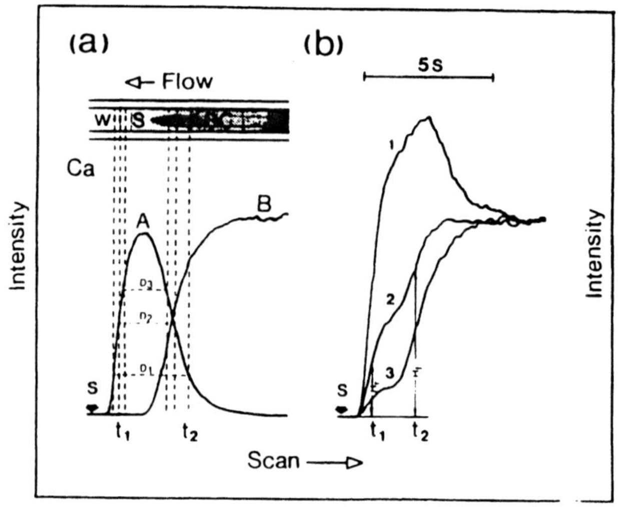

Fig. 2.17. On-line standard addition by a gradient scanning standard addition. (a) Curve A shows the signal resulting from the 100-µl injection of a Ca standard into DDW carrier while curve B shows that resulting from the injection of DDW when sandwiched between DDW and the same Ca standard now used as carrier. The dashed lines shows the times at which equal values of D are obtained on the rise and fall of the FI peak. W: water, S: sample, SC: standard carrier. (b) Profiles observed after the 100-µl injection of three different soils extracts (1–3), in order of decreasing Ca concentration. Reproduced from Ref. [129] with permission.

injection valve is switched, a standard solution pushes the sample plug into the leftover water. With a large enough sample volume, the sample is therefore dispersed into water at its front, and a standard solution at its rear, with no overlap of the water and standard zones, because dispersion is limited. If the signal is measured on both the rising (where dispersion into water takes place) and falling (where dispersion into the standard solution occurs) sides of the FI peak, at points of equal D values (see Fig. 2.17), then the sample concentration is given by:

$$C = C_{std}(D_{sm} - 1)\,[H_r/(H_f - H_r)] \qquad (17)$$

where C_{std} is the concentration of the standard solution, H_r and H_f are signals measured on the rising and falling sides of the peak, at times where the dispersion is D_{sm}, and D_{sm} is pre-determined by injecting standard solution into a water carrier and comparing the signal to that obtained by continuous nebulization of the same standard solution at the same flow rate [129]. Using this single-injection standard addition approach, the sample is injected only once per replicate standard addition (whereas two injections are required for the previous approach). However, it relies on measurements in steep regions of the FI peak which are less precise than the determination of peak height or peak area. Still, a comparison of the two approaches, especially in terms of precision, is warranted.

A manifold to carry out the generalized standard addition method (GSAM) has been described in combination with ICP-AES [130], and should also be applicable to ICP-MS. Although the proposed set-up requires the preparation of a series of standard solutions, it compensates for both matrix effects and spectroscopic interferences. It consists in a double-line manifold where 500 μl of both the sample and the standard are injected simultaneously into each line before being merged in a Y-connector and passed through a mixing coil. The resulting peak includes, at its top, a temporary steady-state signal which is long enough for a 10-s integration [130].

2.3.4.3 On-line isotope dilution analysis

Although it was described in combination with HPLC-ICP-MS [131], a simple FI manifold can be used for isotope dilution analysis. Either a standard solution (for calibration of the isotopic spike solution) or sample is injected into a carrier and merged, through a Y-junction, with a solution of isotopic spike before being fed to the nebulizer [63]. The background-corrected ratio which is measured as a function of time is then converted into a corresponding analyte mass flow profile. Integration of the resulting FI peak and normalization to the injection volume finally yields the analyte concentration [131].

An alternative approach, also based on a merging stream, has been successfully applied to the determination of Pb in a variety of materials prepared by various methods [132,133]. Basically, the injected sample is merged with the isotopic spike solution to yield either a transient (if dilution is required) or a steady-state signal.

2.3.5 Applications

Table 2.4 summarizes the results of selected applications of limited dispersion FI systems with ICP-MS. The accurate measurement of isotope ratios is straightforward [40], thereby enabling stable isotope

TABLE 2.4

Selected applications of limited dispersion FI systems with ICP-MS

Analyte	Sample	Sample preparation	Calibration strategy and/or requirements and/or features	Instrument, sample introduction, FI conditions	Ref.
V, Ni, As	river and estuarine waters, seawater and urine SRMs	sample (except river water) diluted 10-100 fold; acidified to 1% HNO_3	FI used to minimize clogging of the sampler; 5-point method of standard additions; H_2 added to aerosol carrier gas flow; excess HNO_3 required to remove Cl from NaCl as volatile HCl during cryogenic desolvation process	ELAN 250; USN with cryogenic desolvation; 500-µl injections; unspecified tube length and i.d.; unspecified carrier and flow rate	134
REEs	estuarine water SRM	no sample preparation required	external calibration with on-line internal standardization using In and Bi: analyte mass referenced to linearly interpolated amu value between In and Bi	PQ3 in S-mode; 500-µl injections; unspecified tube length and i.d.; unspecified carrier at 1.5 ml/min	101
Hg	drinking water	no sample preparation required	calibration by the method of standard additions with internal standardization using Tl; no memory effects; dilute HCl was more efficient than HNO_3 at rinsing the sample loop between samples	ELAN 5000; DIN: 75-µl injections; 25 cm of 50-µm i.d. fused silica quartz capillary tubing; acidic carrier supplied by gas displacement pump at 375 psi	95
Mo	seawater SRM	no sample preparation required	on-line standard addition (see Section 3.3.4.2 for details)	ELAN 500; 50-µl injection; 150 cm of 0.3-mm i.d. Teflon tubing; 1% v/v HNO_3 or standard carrier at 1.8 ml/min	127
Pu	seawater	co-precipitation of Pu with NdF_3; dissolution in $HClO_4/HNO_3$; evaporation to dryness; dissolution and precipitation of $Nd(OH)_3$; dissolution in HNO_3; separation of Pu on anion exchanger AG1-X4; elution with NH_4I/HCl; evaporation to dryness; dissolution in HNO_3	external calibration with aqueous Pu standards; co-precipitation with NdF_3 removed seawater matrix; precipitation of $Nd(OH)_3$ removed F^-; ion exchange removed Nd and U; final evaporation to dryness removed I^-; FI allowed a larger preconcentration factor	HP 4500; USND; 250-µl injections; unspecified tube length and i.d.; unspecified carrier at 2 ml/min	135

Analyte	Sample	Sample preparation	Calibration strategy and/or requirements and/or features	Instrument, sample introduction, FI conditions	Ref.
Au	seawater, freshwater, hydrothermal fluids	sample pH adjusted to 8 with NH_4OH; Au complexed with CN^- and preconcentrated on anion-exchange resin; further preconcentration by evaporation of eluent	FI required because final sample volume is 1.1 ml, and Au would otherwise be adsorbed on the flexible tubing of the peristaltic pump	PQ; 120-µl injection; minimum length of tubing (unspecified i.d.); 10% aqua regia carrier at unspecified flow rate	55
17 elements	rock SRM	fusion with $Li_2B_4O_7$; dissolution	calibration with a rock SRM treated in the same way as the sample	ELAN 5000; FIAS 200; 500-µl injections; acid carrier at 1-2 ml/min	103
Ir, Re and Pt	natural waters, sediments	microwave digestion of sediments with HCl/HNO_3/HF; Cl_2 bubbled through to oxidize Ir; all samples passed through AG1-X8 resin; elution with 12 M HNO_3; evaporation down	isotope dilution analysis; anion exchange separated analytes from sea salts and sources (e.g. Hf) of interferences; FI required because of small volume and high viscosity of final samples	PQ; 250 or 500-µl injection; 30 cm of 0.5-mm i.d. Teflon tubing; 0.8 ml/min unspecified carrier	136
^{226}Ra, ^{230}Th, ^{237}Np, ^{238}U, ^{241}Am	radioactive waste solutions		external calibration or ID analysis using ^{230}Th (^{232}Th as reference) with a correction for mass bias using isotope SRM	ELAN 6000; 20-µl injection; unspecified tubing length and i.d.; 2% HNO_3 at 0.085 ml/min (with micronebulizer)	110
Pb	miscellaneous SRMs (biological, food, soil, paint, sediment, etc.)	various procedures (fusion, ashing, microwave digestion, etc.) depending on the sample	on-line isotope dilution analysis (see Section 2.3.4.3 for details)	ELAN 250; FIAtron used in either merging stream mode or standard additions mode; 0.7 ml/min of sample and spike solutions	132
Ge, Pd, Pt	solution of 2.5 M H_3PO_4 and 1.5 M $NaNO_3$	no sample preparation required	sample mixed on-line with standard (for standard addition) and internal standard prior to filling the injection loop	ELAN 500; 200-µl injection; tubing for D=1.91; 1% v/v HNO_3 carrier at 1 ml/min	105
51 trace elements	solution of 15% m/m H_3PO_4, 16% m/m $NaNO_3$ and 6% m/m Na_3PO_4	no sample preparation required	sample mixed on-line with internal standard and then standard (for standard additions) prior to filling the injection loop	ELAN 500; 100-µl injection; 125 cm of unspecified i.d. tubing; unspecified carrier at 1.07 ml/min	48

continued

TABLE 2.4 (*continuation*)

Analyte	Sample	Sample preparation	Calibration strategy and/or requirements and/or features	Instrument, sample introduction, FI conditions	Ref.
13 elements	bovine liver SRM	1.0-g sample digested with 1 ml HNO_3, 1 ml H_2SO_4 and 1 ml H_2O_2 and diluted to 10.0 g with DDW	500-µl used to obtain 2-min FI peak and allow skip/scan mode of acquisition (i.e. scanning whole mass range but excluding mass regions of no analytical interest)	PQ; 500-µl injections; unspecified tube length and i.d.; unspecified carrier and flow rate	137
Pb	biological SRMs	wet digestion and dilution to 5% H_2SO_4 (for calibration) or digestion with HNO_3 and dilution to 20% HNO_3 (isotope dilution analysis)	external calibration with acid-matched standards or isotope dilution analysis (with correction for mass discrimination) provided accurate determination of Pb	PQ; high-solids nebulizer; water-cooled spray chamber; 100- or 500-µl injections; 65 cm of 0.8-mm i.d. tubing; unspecified carrier at 1.0 ml/min	40
organomercury	marine biological SRMs	sample washed with acetone and toluene, and acidified with HCl 5 M; OrgHg chloride extracted into toluene and back-extracted into saline cysteine acetate solution; ^{201}Hg isotope spike added	isotope dilution analysis compensated for 8-fold suppression induced by 4% salt matrix; no correction for mass discrimination required; results corrected for 80% recovery of extraction procedure (established by spiking the sample with MeHg prior to the extraction)	ELAN 250; 100-µl injections; unspecified tube length and i.d.; 0.1 M HNO_3 or DDW carrier at 1.1 ml/min	89
Cu, Cd, Pb	freeze-dried urine SRM	sample digested with 3 M HNO_3 in a microwave oven and diluted with DDW after cooling	isotope dilution analysis; levels of Na and Mg must be low enough to cause negligible interference on Cu (from ArNa and ArMg)	ELAN 5000; 200-µl injections; unspecified tube length and i.d.; 1% HNO_3 carrier at 1 ml/min	52
Be, Mn, Ni, Cu, Se, Cd, Pt, Hg, Pb	freeze-dried urine SRM	sample reconstituted as instructed, without further dilution	on-line standard additions by merging 40-µl standard slugs of increasing concentration with a 500-µl slug of sample; DIN modified to handle saline solution without clogging; correction for drift by ratioing to sample signal	ELAN 250; DIN; 3-µl volume between valve and the ICP; 2% HNO_3 carrier at 100 µl/min (from gas displacement pump at 1200 psi)	126

Analyte	Sample	Sample preparation	Calibration strategy and/or requirements and/or features	Instrument, sample introduction, FI conditions	Ref.
Al, V, Cr, Mn, Fe, Co, Ni, Cu, Zn, Mo, Cd, Sb, Pb	Alzheimer's amyloid plaques	plaque cores suspended in 100 µl of buffer (0.1% SDS, 150 mM NaCl and 0.2% NaN_3) and dissolved with 80 µl concentrated HNO_3 and ultrasonic shaking; dilution to 2 ml with DDW	FI required because of small sample volume and high salt content; external calibration with standards prepared in diluted salt buffer; high-resolution ICP-MS would be required to resolve potential spectroscopic interferences	ELAN 500; 50-µl injections; 150 cm of 0.3-mm i.d.; DDW at 2.0 ml/min	138
Pt	corn leaves, mouse liver and $Ca_3(PO_4)_2$ SRMs	open-vessel digestion of corn leaves and mouse liver with HNO_3, $HNO_3/HClO_4$ and aqua regia; evaporation and dilution with 0.14 M HNO_3; $Ca_3(PO_4)_2$ powder dissolved in 0.14 M HNO_3	sample mixed on-line, using two T-junctions, with carrier or standard solution and internal standard solution (of Ir)	PQ1; Meinhard or thermospray nebulizer; water-cooled spray chamber; EVA FI system; 200-µl injections; carrier at 1.33 or 1.5 ml/min	63
I	glacial acetic acid	dilution 1 + 3 with DDW and spiking with In	standard additions with internal standardization using In; peak areas decreased with increasing NH_4OH carrier concentration; different results obtained with different NH_4OH carrier concentrations	ELAN 6000; water cooled spray chamber at 5°C; 20-µl injections; 30 cm of 0.002 in i.d. tubing between valve and nebulizer; 1.7 M or 3.7 M NH_4OH at 1.2 ml/min	139
Pb, Ni, V	crude oil, residual fuel oil SRM	oil mixed with tetralin, 0.5 ml Triton X-100, 40% v/v HNO_3, In internal standard and mechanically agitated before being diluted with DDW	standard additions with internal standardization using In; 60 µl sample solution pumped to nebulizer	ELAN 250; sequential injection system (Alitea modular injection system); unspecified carrier at 60 µl/s	140
REEs (brines); 8 elements (Al)	brines (30-35% m/V NaCl); aluminium	10- to 20-fold dilution of brines; Al chip etched in HNO_3/HCl (1:1), rinsed, dried, weighed, digested in HNO_3/HCl (1:4) and diluted to 1.6% m/V Al_2O_3	external calibration with matrix-matched standards and internal standardization using In	PQ; V-groove nebulizer; spray chamber at 5°C; 500-µl injections; unspecified tube length and i.d.; unspecified carrier at 0.8 ml/min	98

continued

TABLE 2.4 (*continuation*)

Analyte	Sample	Sample preparation	Calibration strategy and/or requirements and/or features	Instrument, sample introduction, FI conditions	Ref.
Al, Cu, In, Zn (in TMG); Fe, Co, Cu, Mo, Cd, In, Ba, Pb (in ML)	trimethylgallium (TMG); methyllithium (ML)	TMG or ML stabilized in diethyl ether thereby forming TMG or ML etherate	5.86% O_2 in aerosol carrier gas flow; calibration possible with organometallic standards diluted in diethyl ether; 10% HNO_3 carrier required for ML FI to prevent blockages in the pump tubing from hydrolysis	PQ2; high solids nebulizer; membrane separator; 50-µl injections; unspecified tube length and i.d.; 2% HNO_3 carrier at 3.56 ml/min	82
12 elements	silicon wafer surfaces; dielectric silicon oxide film	HF vapour phase decomposition of SiO_2 layer; drop of HF rolled on the surface of the wafer and transferred to small sample tube	external calibration with acid-matched standards	ELAN 500; USND; unspecified FI conditions	141
Al, Ti, Cr, Mn, Fe, Cu, Nb	high-purity Ni (99.99% m/m)	Ni chips etched in 30% HNO_3, rinsed with DDW, dried, weighed, dissolved in 30% HNO_3 at 80°C and diluted with DDW to 3% m/m Ni	semiquantitative analysis for an estimate of analyte concentration; quantitative analysis by the method of standard additions performed on-line	ELAN 5000; FIAS 200; 200-µl injections; unspecified tube length and i.d.; 0.1% HNO_3 carrier at 1.5 ml/min	69
Sb, Pb, Bi	high-purity Cu	0.5-g sample dissolved in 50% v/v HNO_3 and diluted with DDW	external calibration with internal standardization using In	PQ2 Turbo +; 500-µl injection; unspecified tube length and i.d.; 1% v/v HNO_3 carrier at 0.8 ml/min	125
Mg, Mn, Nb, Ag, Sn, W	tantalum oxide	sample fused with excess potassium pyrosulphate; melt dissolved in dilute HF	external calibration with matrix-matched standards	PQ2+; specially-designed FI system; 350-µl injection; a few cm tubing of unspecified i.d.; unspecified carrier at 0.7 ml/min	91
As, Sb, Sn, Bi, Se, Te	steels and high-purity Fe SRMs	high-pressure digestion with HNO_3/HCl/HF and dilution	external calibration with internal standardization using Be (for As, Se, Te) and Rh (for Bi, Sn, Sb)	ELAN 500; FIAS 200; 200-µl injections; unspecified tube length and i.d.; 0.1% m/v HNO_3 carrier at 2.3 ml/min	97

Analyte	Sample	Sample preparation	Calibration strategy and/or requirements and/or features	Instrument, sample introduction, FI conditions	Ref.
REEs, Th, U	iron and steel SRM	microwave digestion with $HCl/HNO_3/HF/H_2SO_4$; evaporation to H_2SO_4 fumes and dilution; separation of matrix in a Hg cathode microelectrolysis cell that is connected to the sample loop of the FI system	external calibration with 0.3 M H_2SO_4 multielement solutions; amount of HF must be minimal to eliminate requirement for boric acid; H_2SO_4 concentration minimal to reduce damage to sampler and analyte signal suppression	ELAN 6000; FIAS 200; 300-μl injections; unspecified tube length and i.d.; 0.5% m/v HNO_3 carrier at 3.5 ml/min	54
B	iron and steel SRMs	microwave digestion with $HCl/HNO_3/H_2SO_4$; evaporation; double solvent extraction with $CH_3COCH_2COCH_3/CHCl_3$ at pH 1.4; evaporation; dilution	isotope dilution analysis using ^{10}B required because of matrix-induced suppression; 99.8% Fe selectively removed; FI used because of salt concentration in samples	ELAN 500; FIAS 200; 250-μl injections; unspecified carrier at 2.15 ml/min	102
Nb, V, Ti	microalloyed steels	microwave digestion with HNO_3/HF; addition of Y (internal standard) and dilution with 1% HNO_3	external calibration with Fe-free standards and internal standardization using Y; matrix effects minimized by using lower aerosol carrier gas flow rate and higher r.f. power	ELAN 500; FIAS 200; 500-μl injections; a few cm of 1.14-mm i.d. tubing; 0.1% v/v HNO_3 at 2.8 ml/min	81
Pb, La, Zn, Mo, Cr, V	soil SRM	0.25-g soil was slurried by shaking soil with 5 g zirconia beads and 10 ml DDW during 24 h; beads decanted and slurries diluted to 250 ml with 1 M HNO_3	external calibration using aqueous solutions in 1 M HNO_3; Ar-N_2 mixed -gas plasma provided compromise conditions for multielement analysis; accuracy for V & Cr dependent on soil matrix	ELAN 500; conespray nebulizer; 250-μl injections; 1 m of 0.8-mm i.d. Teflon tubing; DDW at 3 ml/min	107
Sn	sediment SRM	precombustion at 500°C for 4 h; fusion with $LiBO_2$ at 1000°C for 15 min; dissolution in 1 M HCl with heating; dilution with 1 M HCl	method of standard additions (3 point) using peak area	HP-4500; 100-μl injections; unspecified tube length and i.d.; unspecified carrier at 0.2 ml/min	104
B	silicate rock SRMs	addition of 0.1% mannitol solution and ^{10}B spike, and digestion with HF; evaporation to dryness; dissolution in 0.5 M HF; centrifugation; supernatant analyzed	isotope dilution analysis using ^{10}B; mannitol required to suppress B volatilization; glass-free HF-resistant sample introduction system	PMS 2000; 100-μl injections; unspecified tube length and i.d.; 0.5 M HF carrier at 0.65 ml/min	106

continued

TABLE 2.4 (*continuation*)

Analyte	Sample	Sample preparation	Calibration strategy and/or requirements and/or features	Instrument, sample introduction, FI conditions	Ref.
Au, Pt, Pd, Ir, Rh	ores, silicates, Fe-formation rocks, Pt ore SRM	microwave digestion with aqua regia and HF; digestion with $HF/HClO_4$; fusion with Na_2O_2; dissolution in HCl; passage through metalfix-chelamine resin; wash resin with 1.2 M HCl; microwave digestion of resin	external calibration using solutions submitted to the ion-exchange separation; method unsuitable for Os and Ru which may be lost during sample preparation; sample plug surrounded by air prior to injection	ELAN 500; FIA-200; 200-µl injections; unspecified tube length and i.d.; 0.1% HNO_3 carrier at 3.6 ml/min	121
REEs	rock SRMs	extraction of REEs from diluted rock digest into 20% diester in heptane; back-extraction; evaporation to near dryness; dilution to 1 ml with 100 µg/l Re and Ru in 5% HNO_3	external calibration with internal standardization; separation procedure removes 99.5% of Ba with complete recovery of REEs; FI allows greater preconcentration factor	PQ2+; FIA-tron; 460-µl injections; 90.6 cm PTFE tubing of 0.508 mm i.d.; unspecified carrier at 0.8 ml/min	142
17 elements	rock SRM	fusion with $Li_2B_4O_7$ (no details provided); 1.2% m/V dissolved solids in final solution	external calibration using another rock SRM, prepared as the sample; peak area was used for quantitation	ELAN 5000; FIAS-200; 500-µl injections; unspecified tube length and i.d.; unspecified carrier at 1.5 ml/min	100
Ru, Rh, Pd, Ag, Re, Ir, Pt, Au	rocks	1-g sample digested with $HCl/HClO_4$; down to fumes and digested with HCl/HNO_3; diluted with DDW	external calibration with internal standardization using In	PQ2+; 200-µl injection; unspecified tube length and i.d.; unspecified carrier at 0.8 ml/min	99
Ru, Rh, Pd, Ir, Pt, Au	rock SRM	0.5-g sample fused with 4 g Na_2O_2, leached with 40 ml DDW, heated with 40 ml HCl, cooled and diluted to 100 ml with DDW	external calibration with matrix-matched standards; internal standardization with Cs and Bi	PQ2; unspecified FI set-up and conditions	143
Cr, Co, Cr, Ce, Eu, Ta	rock SRM	sample prepared by fusion with $Li_2B_4O_7$; final sample contained 1.2% total dissolved solids	unspecified calibration procedure; FI required to counter the high concentration of dissolved solids	ELAN 5000; FIAS-200 with 500-µl injections; unspecified carrier; autosampler	23

tracers studies involving small sample volumes, and the highly accurate isotope dilution analysis technique (see Section 1.5). However, depending on the analyte and the matrix, the precision may be degraded compared to that achievable by continuous nebulization. This was observed for $^{66}Zn/^{67}Zn$, $^{66}Zn/^{68}Zn$ and $^{68}Zn/^{67}Zn$ ratios in both 1% HNO_3 matrix and blood plasma [40].

On the other hand, the accuracy and precision of $^{202}Hg/^{201}Hg$ from a 100-μg/l Hg standard in 0.1 M HNO_3, aspirated before and after saline extracts (with 4% salt) from marine biological materials whose matrix induced an 8-fold suppression of the signal, were unaffected by these samples [89]. This is consistent with the discussion in Section 1.4.3 where heavy analytes are less affected and recover more readily from effects from concomitant elements. Although isotope dilution analysis was used for the determination of organomercury (OrgHg) in these extracts [89], the ^{201}Hg stable isotope spike could not be added prior to the extraction as it was not available in the same OrgHg form as the analyte, and would have shifted equilibria, thereby modifying the speciation of the analyte [131].

2.4 MEDIUM DISPERSION SYSTEMS (D = 3–10)

These systems are used whenever moderate dilutions or reactions are carried out on-line. Obviously, an increase in dispersion corresponds to a greater dilution. Sufficient mixing of the sample and reagents is also desired to ensure that a given reaction will take place at a sufficient pace. Compromise conditions must however be chosen so that the resulting broadening will not unduly decrease the sampling frequency [5]. Nonetheless, medium dispersion systems allow a much wider range of applications than those based on limited dispersion.

2.4.1 Instrumental set-up

The same basic components as for the limited dispersion set-up are required (see Section 2.3.1). The shortest possible tubes of small and uniform i.d. are recommended in order to keep D_f as small as possible, dilution being achieved by reducing the sample injection volume [5]. When V_i is smaller than the volume required to obtain D = 2, D_f is independent of D_i [5]. Various configurations can be used to provide D between 3 and 10. The single-line manifold of Fig. 2.19 can simply be

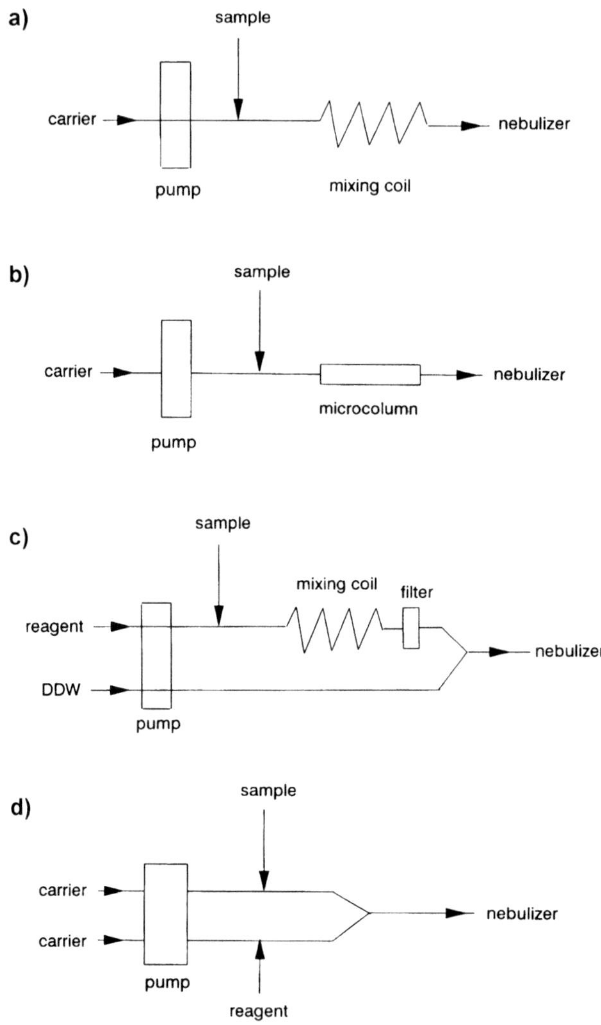

Fig. 2.18. Medium-dispersion FI manifolds. (a) A mixing coil is used to increase dispersion between the sample injection valve and the nebulizer. (b) A microcolumn is used as a mixing reactor, to increase dispersion between the sample injection valve and the nebulizer. (c) The precipitate formed between sample component and a suitable reagent is used to increase dispersion between the sample injection valve and the nebulizer. In this case, a second line (DDW) is used to also perform on-line dilution of the remaining solution. (d) Merging stream, double-line manifold where the sample is 'segmented' with the reagent or carrier at the confluence point.

modified by inserting a suitable mixing coil between the sample injection valve and the nebulizer (Fig. 2.18a).

To perform FI on a microscale, a HEN can be used. For instance, D of 7–10 was obtained using 0.5-μl sample injection in a carrier at a flow rate of 20–120 μl/min [65]. To further emphasize the microscale of this set-up, under these conditions, a 60-μl loop provided a quasi-simultaneous signal [65].

A microcolumn filled with small particles (of silica, for instance) or glass spheres [24] can also be used instead of the mixing coil as shown in Fig. 2.18b. Indeed, the flow tortuosity induced by the particles or beads will promote radial and axial mixing, thereby increasing dispersion. Similarly, the formation of a precipitate between the sample and a suitable reagent would also promote mixing. An example of a simple double-line manifold is given in Fig. 2.18c, where provision has been made for the on-line dilution of the reagent which has to be in excess to ensure quantitative precipitation [144]. If no dilution is required, then the set-up could simply involve the addition of a filter to the manifold shown in Fig. 2.18a. So far, the filters used have been either cylindrical or planar stainless steel filters of the type typically used as cleaning devices in HPLC [144].

Alternatively, a merging stream double-line manifold can be used, as illustrated in Fig. 2.18d, where a confluence point connects the line where sample injection occurs to a second line which can be used for a variety of purposes (for example, to supply a diluent, reactant, internal standard, etc.). Note that the solution provided by the second line can either be injected (at an appropriate time so that it can merge with the sample plug) or pumped continuously, and that a mixing coil may be inserted between the confluence point and the nebulizer to increase D further. Since this approach essentially involves segmentation of the sample line with the solution from the second line, mixing is increased compared to the single-line approach, although it is neither complete nor instantaneous. Nonetheless, this configuration allows sufficient mixing to promote a chemical reaction, even with a decrease in pumping rate and tube i.d. to decrease D [5].

Another frequently used approach is the zone-sampling manifold shown in Fig. 2.19. Basically, a section of the dispersed sample zone which results from the injection of the sample in the first carrier, is injected into the second carrier and transported to the nebulizer [10].

Obviously, numerous manifolds can be used to provide medium dispersion, to the point where a whole book could be devoted to them.

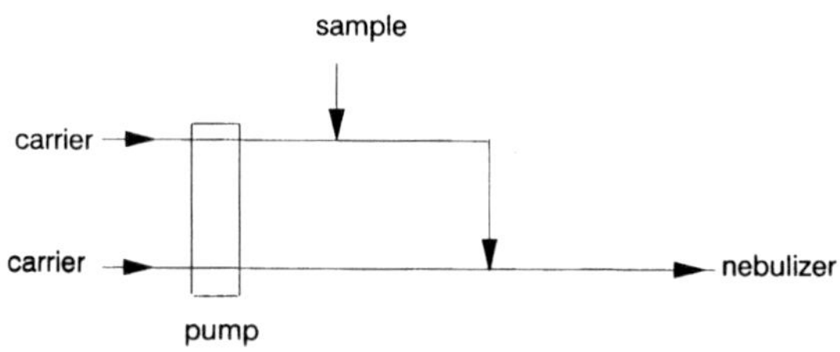

Fig. 2.19. Zone-sampling manifold. A portion of the dispersed sample zone, resulting from injection of the sample into the top carrier, is injected into the bottom carrier.

Instead, some examples are described and references are made to several others in the following sections. Whatever the set-up, optimization of the ICP-MS operating conditions is carried out as described in Section 2.3.2.

2.4.2 Features

Because of the significant mixing that occurs, medium dispersion systems are particularly useful for on-line sample pretreatment. This includes conversion reactions which transform undetectable species into detectable ones. For example, the manifold shown in Fig. 2.20 can be used for the determination of cyanide ions. When the cyanide ions traverse the column of CuS, they react to form $Cu(CN)_4^{3-}$ ions of which Cu is detected [145]. The reaction is very selective i.e. there were very few anionic interferences [145]. Although the method was so far only used with FAAS, it would be applicable to ICP-MS after an optimization of the flow rate and concentration of the carrier. In fact, the carrier flow rate would have to be lower, which would improve the conversion

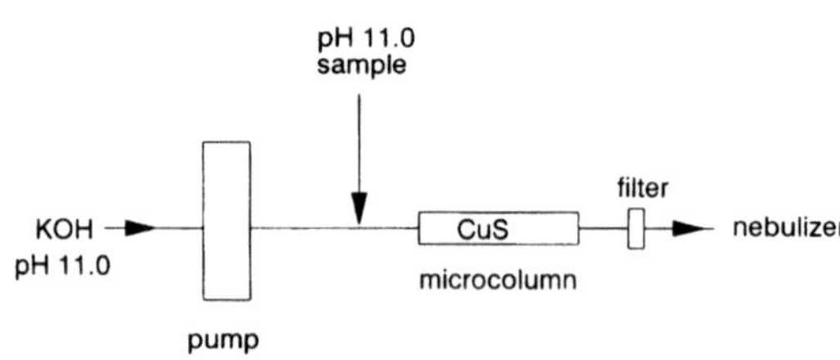

Fig. 2.20. Application of a medium-dispersion manifold to the indirect determination of CN^-. The cyanide sample is injected into basic carrier and pumped through a column of CuS chips where CN^- reacts with Cu^{2+} to form a $Cu(CN)_4^{3-}$ which is detected through the Cu signal. A filter is used to ensure that no CuS particle will reach the nebulizer. After Ref. [145].

efficiency (that was only about one-third at 6.4 ml/min [145]). Other conversion reactions could certainly be designed for the indirect detection of a variety of anions.

The precipitation set-up of Fig. 2.18c can also be used for the indirect determination of anions. Indeed, if the carrier contains a continuously monitored cation that reacts with the analyte anion injected to form a precipitate which is filtered off, then a negative peak will result from the injection of a sample as the cation concentration would be locally decreased. Under these conditions, a blank would also give rise to a negative peak, of smaller absolute magnitude than that of the analyte-containing sample, since it would only correspond to a local dilution of the carrier [144]. For instance, sulphide could be determined using a Cd^{2+} solution as carrier. The fact that the stainless steel filters may leach out some metal should not cause a problem, given the relatively high concentration of the carrier.

Of course, reverse FI could be carried out (i.e. the sample would be the carrier and the cation reagent would be injected) which would then give rise to a positive peak [144]. Using the blank as a carrier would result in a smaller positive peak. With this approach, however, any leaching of metals by the filter material would increase the background level and thereby degrade detection limits. It might therefore be essential to switch to a more inert material such as TEFLON. Whether normal or reverse FI is used, the filter must be removed and cleaned periodically, by placing it in a suitable dissolution reagent in an ultrasonic bath [144]. How often cleaning must be done (i.e. how many samples can be injected before the precipitate buildup induces too high a back-pressure and has a deleterious effect on the accuracy and/or precision of the analytical signal) will depend on the type of precipitate formed. Indeed, crystalline precipitates will require more frequent cleaning than curdy or gelatinous ones [144].

Matrix-matching, whereby a matrix element (for example Na) is added to the standard solutions so as to match the resulting Na signal to that of the sample, can be automatically carried out using zone-sampling of Na standard solutions [146]. However, this approach is unlikely to be widespread with ICP-MS, since a removal of the troublesome matrix is in general preferable to matrix-matching.

Medium dispersion systems also allow a moderate dilution of the sample which reduces non-spectroscopic interferences [65,99,147], as discussed in Section 1.5. For instance, by decreasing the sample injection volume from 100 to 25 µl, D was increased to about 4, which significantly reduced the extent of signal suppression observed in the presence of up to 10.0 % NaCl [40]. With a 10-μl loop and 10 cm of 1.0-mm i.d. tubing, which resulted in *D* of 8.7, the 95% suppression induced by 0.01 M U on 10^{-5} M Mo with continuous nebulization was improved to about 50% [147]. Obviously, since the effect of concomitant elements varies from instrument to instrument, and (as discussed in Section 1.4) depends on the mass and ionization potential of the concomitant element, a suitable dispersion must be determined for a particular application.

Tandem injection (i.e. where the time between injections is such that the sample zones overlap, thereby resulting in a quasi steady-state signal), with and without a merging stream manifold, has been used to achieve dilution factors from 7 to 900 [148]. The combination of the merging stream technique with tandem injection provided a more stable 'steady-state' signal, although zone-splitting was required to decrease the flow rate to a level compatible with the nebulizer [148].

Medium dispersion systems can also separate the analyte from the matrix. Table 2.6 gives some examples of manifolds which might be combined to ICP-MS to allow the analysis of samples with high levels of dissolved solids. As can be seen, three main approaches can be used to remove major component(s) from the matrix of the sample. These components can be retained on a microcolumn, selectively extracted into another phase which is then discarded, or precipitated and filtered off. In any case, no analyte preconcentration results since the only changes in the sample that ultimately reaches the nebulizer are a decrease in the concentration of one or more elements from the matrix and a dispersion of analyte in the carrier. As will be seen in Section 2.6.1.1, when the analytes are selectively retained, extracted or precipitated, a preconcentration then also becomes possible.

In any case, matrix separation, with or without preconcentration, can reduce the source of spectroscopic interferences, in addition to decreasing the level of dissolved solids in samples. For instance, a reduction in the concentration of Na and Mg ions has a direct impact on the level of $ArNa^+$ and $ArMg^+$ interferences on $^{63}Cu^+$ and $^{65}Cu^+$. In fact, sufficient Na must be removed if the most abundant isotope of Cu ($^{63}Cu^+$) is to be used for its determination in saline waters [149,150]; and the concentration of both Na and Mg must be substantially decreased to allow isotope dilution using $^{63}Cu^+/^{65}Cu^+$ [151]. Removing sulphate anions eliminates the polyatomic interferences of $^{32}SO_2^+$ and $^{32}S_2^+$ on $^{64}Zn^+$, the most abundant isotope of Zn [152]. The determination of platinum-group metals is facilitated by their separation from base metals, as for example, the interference of $^{40}Ar^{63}Cu^+$ on monoisotopic $^{103}Rh^+$ is eliminated [91]. On-line separations can also be carried out to remove isobaric interferences, such as that from ^{99}Ru on ^{99}Tc, or the partial overlap from ^{100}Mo on ^{99}Tc at high Mo concentrations [153]. Furthermore, they can be used to reduce and/or eliminate non-spectroscopic interferences. For example, the notorious suppressive effect of U on the signal from light elements can be significantly reduced if most of the U is selectively retained on a column while the light elements pass through to the nebulizer of the ICP-MS instrument [154]. Although this approach induces dispersion of the analytes (D was about 5 in this specific example), it also eliminates U-induced suppression (of about 80%) [154].

2.4.3 Calibration strategies

2.4.3.1 Gradient calibration

Measurements at several points on the tailing side of the transient FI peak can be used to establish a calibration curve if the dispersion at each of these points is predetermined by comparison with the transient peaks resulting from the injection of a series of standard solutions, as illustrated in Fig. 2.21 [15]. A single standard injection is thereafter

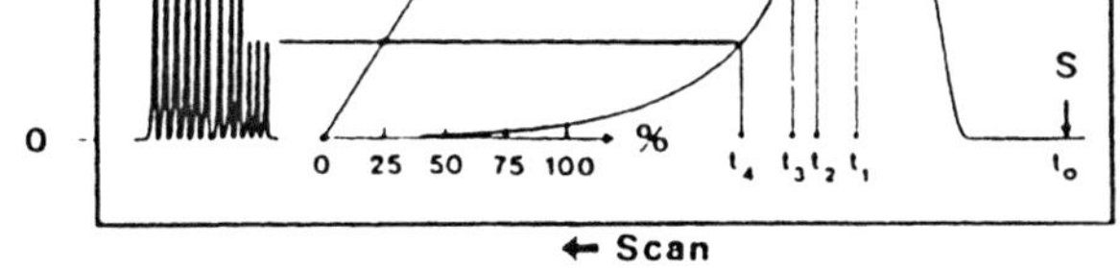

Fig. 2.21. Principle of a gradient calibration technique. Four standards are injected in triplicate on the left, with the corresponding calibration curve (absorbance versus % bromothymol blue) shown in the middle and the times t_1 to t_4 corresponding to each concentration are show on the signal profile of a single injection of the most concentrated calibration standard. Reproduced from Ref. [15] with permission.

required to establish a multipoint calibration. A number of other approaches (i.e. involving merging zones, variable tube dimensions, networks, etc.) have also been used to perform external calibrations with FAAS and should be applicable to ICP-MS [17].

2.4.3.2 On-line standard additions

A reverse-FI manifold, i.e. where the sample is the carrier and standard solutions of lower and higher concentrations than the sample are injected into it, has been used in FAAS [47,128] and ICP-AES [155] but has not yet been applied to ICP-MS. It would be advantageous if sample volume is not limited and if the sample matrix can be continuously nebulized without problem (such as clogging, corrosion, etc.). The process is illustrated in Fig. 2.22a, where negative and positive peaks are observed when the concentration of the standard injected is, respectively, smaller and greater than that of the sample. If the dispersion, i.e. mixing of the standard with the sample carrier, is such that the matrix effect is the same on the standards as in the sample, then no change in signal will result when the concentration of the standard injected matches that of the sample. Therefore, a plot of the change in signal (i.e. peak height) as a function of the concentration of the standards injected will intercept the concentration axis at the concentration of the sample (Fig. 2.22b).

The beauty of this approach resides in the fact that an *interpolation* is made to find the unknown concentration of the sample whereas an *extrapolation* would be required by the regular off-line method of standard additions. By performing the method on-line, the accuracy is therefore improved. In addition, one series of standard solutions can be used for numerous samples, which drastically decreases the sample preparation step, not to mention that there is no need to have a good idea of the sample concentration (to match the spike to the sample) prior to the analysis, as required for maximum accuracy when the method is performed off-line. Finally, because of the wide linear dynamic range of ICP-MS, only two standards (i.e. a blank and a standard of higher concentration than the sample) are really required for the analysis (although more would obviously increase the reliability of the result).

In a computer-controlled system, a zone-sampling manifold was combined to a merging streams one to perform as many as 11 standard additions within 5 min, with a single injection of standard solution [156]. Indeed, zone sampling of a trapped dispersed standard plug

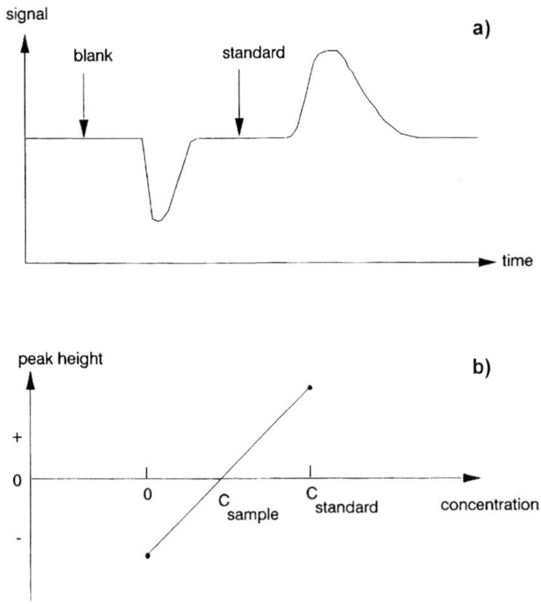

Fig. 2.22. (a) On-line standard addition using a reverse FI manifold. The sample is used as carrier and both a blank and a standard of higher analyte concentration than that of the sample are injected into it, resulting in negative and positive peaks respectively. (b) When the corresponding peak height are plotted as a function of the standard concentration, the concentration of the sample is found by interpolation at the point where the plot crosses the concentration axis.

determined the concentration of standard which was then added to the separately injected sample through the merging streams [156]. Although detection was carried out by ICP-AES, the approach is compatible with ICP-MS. The system can be used for on-line dilution of the sample if the latter is trapped instead of the standard and merging with a diluent is done [156].

2.4.3.3 On-line isotope dilution analysis

A simple single-line reverse flow injection manifold was used to perform the addition of enriched isotopes to aqueous samples on-line with ICP-MS [157]. The analysis involved one injection of a multielement solution of enriched isotopes in each of the sample carrier and a standard solution carrier [157]. This injection results in a positive peak for the enriched isotope and a negative peak for the reference isotope. The standard solution must contain one element that is absent from the enriched isotopes solution, to allow the determination of the dispersion coefficient. The same standard also allows a reverse-ID analysis, in addition to corrections for mass discrimination and any spectroscopic interference on one of the two isotopes used for the ID analysis [157]. Indeed, because the signals from the individual isotopes are monitored continuously, only one isotope free of spectroscopic interference is required for elements whose isotopic distribution does not vary in nature as a correction for the interference can be made by comparison with the signals from the standard [157].

This approach has several features that are clear advantages over both the off-line and the merging zones methods [157]. It requires only one isotope free of spectroscopic interference for elements whose isotopic distribution does not vary in nature, which leads to a broader range of applications. (Two isotopes are still needed for other elements since their 'natural' ratio must also be determined). No preliminary analysis of the sample is required prior to the actual isotope dilution analysis. The only requirement is that the concentration of the enriched isotopes be greater than that in the sample, a pre-requisite that is readily fulfilled in trace analysis with, for example, a 5-μg/l solution of enriched isotopes. This eliminates a time-consuming disadvantage of ID. Furthermore, the concentration profile resulting from the flow injection, which translates into a large range of isotopic ratios, allows the selection of the best isotopic ratio in terms of error propagation. In addition, although the sample is used as carrier, it is only aspirated long enough to appropriately surround the spike injection, which reduces sample consumption. This, in turn, leads to a higher sample throughput. Because the whole analysis is carried out in a closed system, contaminations are reduced. Finally, isotopic carryover is minimized by the fact that only discrete injections of the enriched isotopes solution are made, and the sample itself acts as a washing solution.

However, the above approach involved manual injections into two different carriers: a standard and the sample. Furthermore, because the sample was pumped continuously, sample consumption was still relatively high. An automated sandwich FI manifold was therefore developed to circumvent these two limitations [158]. The sandwich manifold (based on the nested loop approach of Rios et al. [122]) indeed reduced sample consumption since 100 μl of the enriched isotopes solution was injected in the centre of a 1.5-ml sample plug, and the resulting sample-isotopic spike-sample segment was entrained towards the nebulizer by a 1% HNO_3 carrier [158]. As with the manual method, a second injection of the enriched isotopes solution also had to be made within a standard solution containing one element that is absent from the enriched isotopes solution [158]. As mentioned above, this is necessary to allow the determination of the dispersion coefficient, a reverse ID analysis, as well as corrections for mass discrimination and/or spectroscopic interferences. Nonetheless, this automated approach significantly speeded up on-line ID-ICP-MS [158].

Although they are presently more involved in terms of data processing, both the manual and automated approaches make the physical process of isotope dilution analysis as simple as an external calibration but with the added accuracy and precision typical of ID analysis. The data processing includes blank-subtraction of the average steady-state signal observed with 1% HNO_3 from all other measurements. An average natural ratio is also computed from the steady-state signals from both the standard solution and the sample. Any spectroscopic interference is corrected point-by-point by subtracting S_i/D_{sample} from the interfered isotope, where S_i is the signal contributed by the interferent and D_{sample} is the dispersion of the sample into the enriched isotope solution [157]. A point-by-point computation of D_{sample} is accomplished using the element that is only present in the standard solution by simply ratioing the steady-state signal of the standard to each point of the FI peak. D_{sample} should not be computed when the sample is sandwiched around the isotopic spike. Indeed, any effect of concomitant element will be maximum in the steady-state region and decrease during the injection as a result of the associated dilution, thereby resulting in erroneously high dispersion values during the peak.

A point-by-point computation of the ratio R (corrected, if needed, for spectroscopic interferences as described above) is then carried out. Mass discrimination is also corrected, if needed, by multiplying R by the ratio of the isotopic ratio computed from the natural abundances to that measured for the standard solution. A point-by-point computation of C_{sample} is then done using the following equation:

$$C_{sample} = (D_{sample} - 1) C_{spike} K \frac{(A' - B' R)}{(BR - A)} \tag{18}$$

where C_{spike} is the concentration of the enriched isotopes solution; C_{sample} is the analyte concentration ($\mu g\ l^{-1}$) in the sample solution; K is the ratio of the atomic weight of the element over that of the isotopic spike; A is the natural abundance of the reference isotope; B is the natural abundance of the spike isotope; A' is the abundance of the reference isotope in the spike; B' is the abundance of the spike isotope in the spike; and R is the measured isotopic ratio (reference/spike isotopes), corrected for mass discrimination and/or spectroscopic interference. Finally, an average C_{sample} was computed in the region where a steady concentration was obtained [157,158].

2.4.4 Applications

These systems are ideally suited for on-line process monitoring [21,31]. Indeed, they are relatively simple and easy to maintain. They can be fully automated (including the calibration process). Finally, they allow chemical reactions on a short time scale, thereby providing rapid analysis with high sampling frequency [31]. In particular, reverse FI can readily be used for process control [18] since the manifold can simply be connected to a tap on the process line.

Nonetheless, few applications have so far been reported using medium dispersion manifolds in combination with ICP-MS, some of which are described in Table 2.5. However, an endless number of possibilities exists. Some examples of likely approaches are described in Table 2.6. Similarly to the selective retention of a matrix component on a column, selective precipitation of the matrix can also be carried

TABLE 2.5

Selected application of medium dispersion FI systems with ICP-MS

Analyte	Matrix	Sample preparation	Procedure/features	Instrument; sample introduction; FI conditions	Ref
Al, Be, Li, Mg	5000 mg/l U solution		pump 100-μl sample through column of TRU.Spec resin; U is retained while analytes pass through to nebulizer	ELAN 5000; FIAS 200; 1 m of 0.51 mm i.d. tubing between injection loop and column (D=5)	154
As, Pb	water, drinking water, bone ash	bone sample dissolved in HNO_3, spiked with Tl, diluted and filtered	FI required to prevent clogging from the 1% dissolved solids concentration in the bone ash digest and to partially compensate for matrix-induced analyte signal suppression; remaining 20% suppression compensated using internal standardization with Tl	PQ II; HEN in Scott spray chamber; 0.5-μl injection; 50-μm i.d. quartz capillary connected the sample loop to the HEN (D=7-10)	65

Analyte	Matrix	Sample preparation	Procedure/features	Instrument; sample introduction; FI conditions	Ref
Cr, Mn, Fe, Cu, Zn, Cd, Pb, S	plant SRMs	digestion with HNO_3, H_2O_2 and $HClO_4$; dilution 1000 fold to 0.025 M $HClO_4$	pump sample (1.4 ml/min) through column of anionic resin AG-1-X8; anions are retained while analytes pass through to nebulizer; wash column with DDW; elution of anions with 1.0 M HNO_3; removes S-containing spectroscopic interferences on Cu and Zn but not on Cr; total S obtained from elution	PQ 2; water-cooled spray chamber at 10°C; 5 cm of 0.8-mm i.d. tubing connected the column to the nebulizer	159
U	urine and serum SRMs, seawater SRM	addition of 0.1 ml concentrated HNO_3 to 10 ml of urine (no preparation required for serum samples)	100-µl injections in a carrier of 0.6% HNO_3 and 1 g/l Triton-X 100; external calibration with U standard in 0.6% HNO_3; Triton-X 100 required to minimize memory effects; internal standardization should compensate for matrix-induced suppression	ELAN 6000; FIAS 400; D=5	160
Zn, Cu, Fe	all-bran cereal; multivitamin tablet	enzymolysis of ground sample; centrifugation ; dried residue digested with HNO_3/HCl	20-µl sample solution injected; external calibration with standards in 5% HNO_3; analyte signal suppression reduced by increasing the aerosol carrier gas flow rate; salt deposition on torch injector and cones after 6 h with KCl carrier	PQ; De Galan nebulizer; 105 cm of 0.5 mm i.d. tubing connected valve to nebulizer; 0.05 M KCl (pH 2.5) at 0.91 ml/min	161
Ni, Mo	river water, saline water		on-line isotope dilution analysis; one injection of an enriched isotopes solution into each of the sample carrier and a standard solution carrier; correction for interference of ^{58}Fe on ^{58}Ni automatically included in the process	ELAN-500; 150 cm of 0.3-mm i.d. tubing; sample or standard solution at 1.0 ml/min	157
Mo	saline water		on-line isotope dilution analysis using a sandwich manifold where 100 µl of enriched isotopes solution is injected in the centre of 1.5 ml of each of the sample and of a standard solution; improved precision compared to manual approach [157]	ELAN-500; FIAS-200; 150 cm of 0.3-mm i.d. tubing connected FIAS to nebulizer; 1% HNO_3 at 1.0 ml/min	158
Pt, Pd, Ru, Rh Ir, Au		nickel sulphide collection	sample passed through column of Dowex 50W-X8 (hydrogen form) which retains the base metals but not the anionic chlorocomplexes of the analytes	PQ2+	91

TABLE 2.6

Examples of techniques that could be readily converted into medium dispersion FI systems for ICP-MS

Analytes	Matrix	Procedure	Suggested FI manifold	Ref
Ca, Cd, Cr, Cu, Fe, Mg, Mn, Mo, Ni, Pb	uranium SRM in 3 M HNO_3	pump injected sample through column of Levextrel resin of tributylphosphate; U is retained while analytes pass through	single-line manifold of Fig. 2.18b; DDW carrier	162
Ba, Bi, Cd, Co, Cu, Mn, Ni, Pb, Ti	high-purity Fe in 7 M HCl	remove Fe by extracting its chlorocomplex with isobutyl methyl ketone	double-line manifold of Fig. 2.30 with aqueous phase fed to the nebulizer; DDW carrier; IBMK reagent	163
Al, Au, Cd, Co, Cu, Fe, Mg, Mn, Ni, Pb, Sn	high-purity Ag digested in HNO_3	sample mixed on-line with HCl; AgCl precipitate filtered or accumulated in a knotted reactor	double-line manifold of Fig. 2.31; diluent only used to dissolve precipitate to waste between sample injections	164
25 trace elements	high-purity Ta in 12 M HF	extract Ta with diantipyrylmethane in dichloroethane	double-line manifold of Fig. 2.30 with aqueous phase fed to nebulizer; DDW carrier; diantipyrylmethane in dichloroethane reagent	165
25 trace elements	high-purity Ta in 4 M HF	pump injected sample through column of Dowex 1X8; Ta is retained while analytes pass through	single-line manifold of Fig. 2.18b; DDW carrier	165
Al, Ba, Be, Ca, Cd, Co, Cr, Cu, Fe, Mg, Mn, Ni, Pb, Sn, Sr, Ti, V, Zn	high-purity scandium oxide in 1 M HNO_3	pump injected sample through column of Levextrel resin of trioctylphosphine oxide; Sc is retained while analytes pass through	single-line manifold of Fig. 2.27; DDW carrier; Sc removed with 0.5 M oxalic acid (injected or carrier) and bypassed to waste	166
Al, Ca, Co, Cr, Cu, Mg, Mn, Ni, Pb, Zn	high-purity europium oxide in 0.1 M HNO_3	pump injected sample through column of Levextrel resin of di-(2-ethylhexyl) phosphoric acid; Eu is retained while analytes pass through	single-line manifold of Fig. 2.27; 0.1 M HNO_3 carrier; Eu removed with 3 M HCl (injected or carrier) and bypassed to waste	167
V, Cr, Cu, Zn, As, Se	digest of sewage, percolate water, light sandy soil, human serum	pump injected sample through Dowex-1 resin which strongly binds Cl- and S-containing anions while analytes pass through	single-line manifold of Fig. 2.18; DDW carrier	168
Al, Ag, Ba, Bi, Ca, Cd, Co, Cr, Fe, Mn, Ni, Pb, Zn	high-purity Cu salts ($CuSO_4$, $CuCl_2$, $Cu(CH_3CO_2)_2$, $CuCO_3$, $Cu(NO_3)_2$) dissolved in 4 M HCl	inject sample into 20% ascorbic acid carrier; precipitated CuCl is filtered off; analyte-containing filtrate is analyzed	single-line manifold of Fig. 2.18a with a filter added; the mixing coil may be placed in a hot water bath to speed up the precipitation	169

out. In addition to the applications detailed in Table 2.5, the on-line ID approach was also successfully applied to the analysis of radioactive waste samples, where a very small sample volume was required to minimize contamination of the mass spectrometer with radioactive material [110].

2.5 LARGE DISPERSION SYSTEMS ($D > 10$)

If either large dilution factors or reactions where the change in concentration as a function of time (for example, in kinetic studies) are to be measured, a large dispersion is required.

2.5.1 Instrumental set-up

Dilutions by more than an order of magnitude can be carried out by incorporating a mixing chamber on the carrier line as illustrated in Fig. 2.23. If the solutions, especially the carrier, are not degassed, air bubbles may be formed by the stirring in the mixing chamber. A flow restrictor may then be inserted between the chamber and the nebulizer to help prevent the generation of air bubbles. This flow restrictor can simply be a knitted open tubular reactor of smaller bore than the rest of the manifold [53]. A double-line manifold can also be used, where the sample is pumped in one line and the diluent in the other before being merged immediately preceding the mixing chamber.

Alternatively, on-line sample dilution can be carried out by decreasing the sample injection volume and increasing the tubing dimensions (i.e. its length and/or i.d.) of a single-line medium dispersion manifold (see Fig. 2.18a). For instance, D of 25.2 was obtained using a 10-μl loop and 100 cm of 1.0-mm i.d. tubing in ICP-MS [147]. With an even

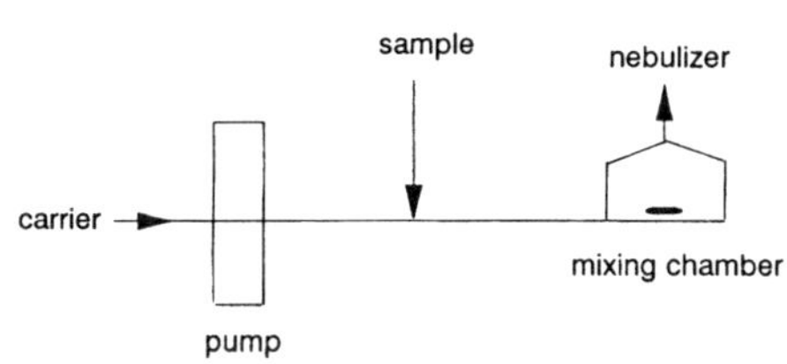

Fig. 2.23. Example of a typical large-dispersion FI manifold. The volume of the mixing chamber, where the sample plug is diluted with the carrier, determines the dispersion.

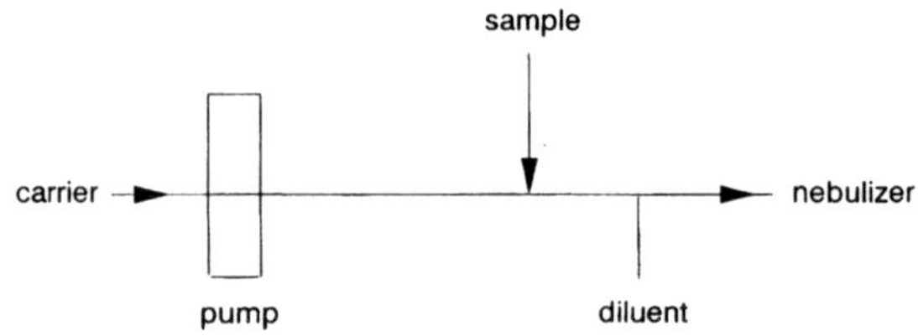

Fig. 2.24. Alternative large-dispersion FI manifold where the carrier flows at a much smaller rate than the free aspiration rate of the nebulizer which then aspirates diluent to avoid starvation.

smaller injection volume (as small as 0.7 µl) and a 160-cm long 1.3-mm i.d. mixing coil, dilutions by 2 to 3 orders of magnitude could be achieved with high sample throughput (up to 100/h) in FAAS [53]. This latter manifold could be used in ICP-MS with a decrease in carrier flow rate and a re-determination of D, since the sample uptake of the nebulizer is lower in ICP-AES and ICP-MS than in FAAS. The use of a MCN can also be advantageous for carrying out µFI. For instance, D of 10 was obtained when injecting 1.0 µl into an aqueous carrier with a flow rate of 200 µl/min that was nebulized by a MCN into a water-cooled double-pass spray chamber [66].

A variation of the above approach involves the addition of a tee between the sample injection valve and the nebulizer, as illustrated in Fig. 2.24, so that the latter can be fed at a constant flow rate [170]. Under these conditions, the carrier can then supply the sample at a small flow rate (<0.5 ml/min) since the solution aspirated through the tee then acts as a diluent [170].

As mentioned in Section 2.4.2, the combination of tandem injection with a merging stream manifold can be used to achieve dilution factors up to 900. Under these conditions, a stable 'steady-state' signal is obtained, provided the flow rate is decreased by zone-splitting to a level compatible with the nebulizer [148]. Simultaneous multiple injection is another approach, which is illustrated in Fig. 2.25, where the sample is injected, in parallel with the diluent, and air, and the resulting air–diluent–sample–diluent–air plug is then carried to the detector [171]. By choosing the volumes of the diluent and sample loop, and allowing sufficient time for mixing, large dispersion is achieved. For instance, using 500 µl for the diluent and 20 µl for the sample, dilution factors of 25 were readily achieved.

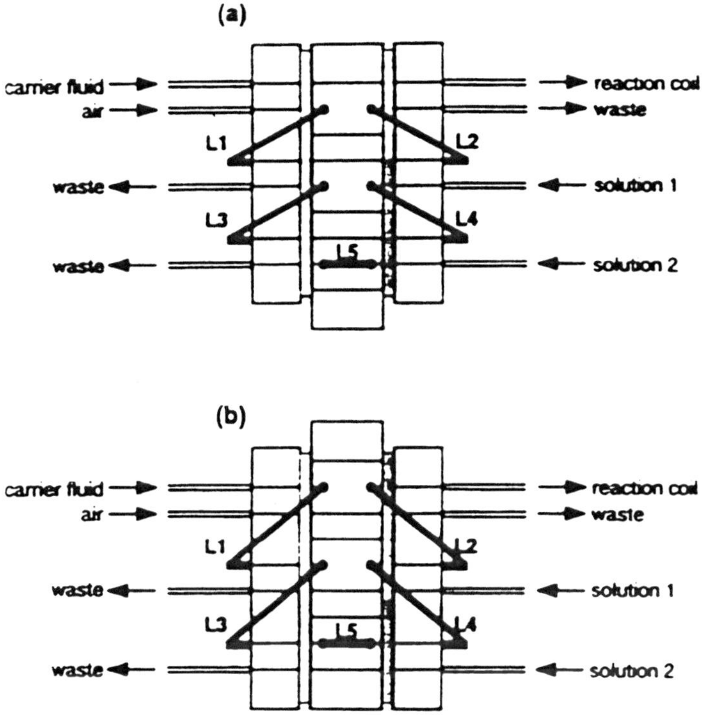

Fig. 2.25. Manifold allowing simultaneous multiple injection with a slider valve in (a) sampling position and (b) injection position. Reproduced from Ref. [171] with permission.

A recirculating loop manifold may also be used to perform successive dilutions [172]. The manifold is shown in Fig. 2.26. The two loops are first filled with sample solution. When valve 2 is closed, the sample simply recirculates inside the two loops. When valve 1 is switched to the inject position, loop 1 is then closed and its content recirculates while the content of loop 2 is pumped to the nebulizer and filled with carrier. If valve 1 is then switched back to the load position, the content of loop 1 gets diluted with the carrier in loop 2, by a dilution factor, d, equal to the total volume of the two loops divided by the volume of loop 1. With n successive actuations of valve 1, the resulting dilution factor is d^n. Hence, large dilution factors can be reproducibly obtained since the dilution factor no longer depends on the inherent dispersion coefficient of a given manifold, which may vary from day to day because of fluctuations in the flow rate provided by the peristaltic pump [172].

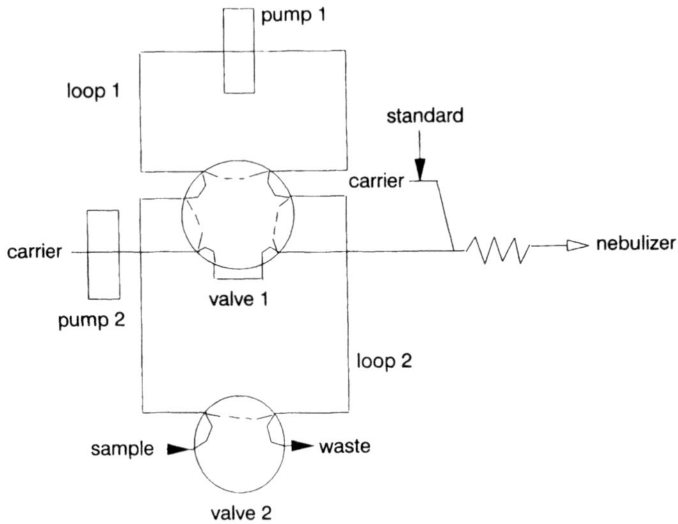

Fig. 2.26. Recirculating loop manifold. With the two valves in loading position (solid lines), the sample is pumped into loops 2 and 1. When both valves are switched (dashed lines), the content of loop 2 is pumped towards the nebulizer while that of loop 1 recirculates. If only valve 1 is switched back to the loading position, then the content of loop 1 is diluted with the content of loop 2. Provision has been made for the on-line addition of an internal standard or a standard solution. After Ref. [172].

Finally, a readout time corresponding to the tailing section of the FI peak instead of its maximum can be selected [13]. This last method of increasing dispersion has been called 'electronic dilution' since the only change made is in the readout time [13]. One or a combination of the above approaches can achieve a dilution by up to a factor of 15000, within seconds [13].

2.5.2 Features

Because the effects of concomitant elements depend on their absolute amount in solution (as discussed in Section 1.5), they can be reduced or even eliminated with a FI manifold providing enough dispersion and therefore dilution [99,147]. For instance, with D of 25.2, the 95% suppression induced by 0.01 M U on 10^{-5} M Mo with continuous nebulization was improved to only 15% [147]. Furthermore, because the effect depends on mass (getting worse as the mass of the concomitant element increases), this dispersion was sufficient to completely eliminate the effect from equimolar solutions of Li, Co, Sr, Cd and Ba on Mo [147].

Decreasing the sample injection volume can have a dramatic effect on the level of non-spectroscopic interferences in μFI-ICP-MS [66]. Indeed, the As signal enhancement in the presence of carbon (as propanol or methanol) was significantly greater with 20-μl injections (where $D < 10$) than with 0.2-μl samples (where $D > 10$) at a constant carrier flow rate of 200 μl/min, despite the fact that the organic solvent concentration was identical in both cases i.e. there was no change in nebulization efficiency [66]. An enhancement in As signal was observed as a function of the absolute amount of carbon injected up to about 1 mg carbon, after which signal suppression of As as well as In and Y also occurred, suggesting plasma overloading [66]. Keeping the injection volume small (0.2 μl) minimized the enhancement effect (although internal standardization was still required to compensate the remaining effect), which also has the advantage of maximizing sample throughput [66].

The recirculating loop manifold can be used to adjust the dilution factor so that non-spectroscopic interferences are minimized. For instance, the signal enhancement of 129% observed for Ce in wine sample could be decreased to 104% with 5 successive dilutions of $d = 3$ (i.e. a 243-fold dilution) [172]. If the analyte concentration does not allow such dilution factor, a smaller number of successive dilutions could be used together with internal standardization, which would compensate for remaining non-spectroscopic interferences. The internal standard can be added on-line at the exit of the recirculating system, provided that it is injected within the sample plug and that it is appropriately mixed with it prior to detection. The recirculating loop manifold also allows the construction of linear calibration curves through successive dilutions of a single standard solution, with and without internal standardization as just described [172].

Large dispersion systems can also be used to extend the linear dynamic range of ICP-MS towards high analyte concentrations (which would otherwise overload the detector). Furthermore, since they allow a substantial dilution on-line, little (if any) sacrifice in sample throughput is required in comparison to the direct continuous nebulization mode (which cannot be used for these high concentrations). In fact, if the time required to perform the dilutions off-line, prior to the analysis, is taken into account, then these systems allow an appreciable improvement in sample throughput.

In comparison to a mixing coil of equivalent volume, the mixing chamber provides a higher dilution factor but also results in a wider peak, with more tailing (for instance, the baseline peak width was 4 times that obtained with the mixing coil using otherwise identical FI conditions [53]). Furthermore, air bubbles could be produced in the mixing chamber by intensive stirring [53]. Thus, unless huge dilution factors are needed, this approach often results in a waste of reagents and time. It may also yield poor S/N even with peak area [5].

Nonetheless this approach can be used to perform the method of standard additions on-line, if the diluent is simply replaced by a standard solution [173]. Although this approach has not yet been combined to ICP-MS, it should be successful since it essentially consists in preparing the spiked solutions on-line. The same approach could also be used to perform isotope dilution analysis on-line, by replacing the diluent with a solution of enriched isotope.

2.5.3 Applications

Using μFI with a MCN and Scott spray chamber, wine samples were analyzed after filtration through 0.45-μm pore size membrane and dilution 1 + 1 to 0.1% v/v HNO_3 [66]. External calibration with aqueous As standards in 1% HNO_3 was carried out using In for internal standardization. With 0.2-μl injections in 200-μl/min carrier, which provided $D >$ 10, results of external calibration were in agreement with those of the method of standard additions [66], indicating that the non-spectroscopic interferences had been adequately compensated by a combination of large dispersion FI manifold and internal standardization.

A recirculating loop manifold was also used for the determination of Ce in wine, without any pretreatment [172]. Standard addition was performed on-line by injecting 156 μl of Ce standard into the sample exiting the recirculating manifold [172]. Successive dilutions were used to minimize the enhancement caused by the organic matrix [172].

2.6 REDUCED DISPERSION SYSTEMS (D < 1)

These systems allow not only a preconcentration of the analyte (resulting in $D < 1$) but they usually also involve a concurrent separation of the analyte from the matrix. Given the numerous problems that arise from the matrix (i.e. both spectroscopic and non-spectroscopic interferences) as discussed in Sections 1.4.2 and 1.4.3, it is therefore not surprising that reduced dispersion systems have received special attention in combination with ICP-MS (see Table 2.7). In addition to the books and reviews already mentioned in Section 2.1, several reviews have been specifically devoted to preconcentration and/or separation techniques [174–179,180]. In fact, many techniques that have been traditionally carried out off-line [181] can be adapted to FI with many advantages.

2.6.1 Instrumental set-up

2.6.1.1 Preconcentration on a microcolumn

There are several approaches which can both separate the analyte from the matrix and preconcentrate it. So far, the most widely used with ICP-MS (as can be judged from Table 2.7) is that involving selective retention of the analyte on a microcolumn with wash of the bulk of the matrix to waste and, subsequent elution of the analyte in a much 'cleaner' matrix. By using a smaller eluent volume than sample volume, a preconcentration is achieved concurrently with the separation, since the same absolute amount of analyte is ultimately recovered in a smaller volume. This approach allows great flexibility since the separation step can be optimized separately from the elution step. Only for the latter, is it necessary to match the flow rate of the eluent with that of the nebulizer.

A basic single-line set-up for separation of the analyte from the matrix and discard of the latter is illustrated in Fig. 2.27. It essentially involves the addition of a microcolumn and by-pass valve after the sample injection valve of the limited dispersion set-up (Fig. 2.9). If the matrix can be tolerated by ICP-MS, then the by-pass valve can be omitted. In any case, since the eluent volume is the same as the sample volume, a separation from the matrix is essentially accomplished. In order to perform a preconcentration as well, a separate valve with an injection volume smaller than that of the sample must be used for the eluent (Fig. 2.28a). In this way, a given absolute amount of analyte from the sample which is retained on the column, is recovered into a smaller volume.

As illustrated in Fig. 2.28b, time-based sample loading and elution can also be carried out instead of discrete injections using valves. In this case, the sample is pumped through the column to waste while eluent is continuously pumped to the nebulizer. For the elution step,

TABLE 2.7

Selected applications of reduced dispersion microcolumn and/or sorption FI systems with ICP-MS

Analytes	Sample	Sample preparation	On-line preconcentration strategy	Instrument	Ref.
V, Cr, Ni, Co, Cu, Mo, Pt, Hg and Bi	seawater SRM, urine SRM	seawater adjusted to pH 3.5 and diluted after addition of dithiocarbamate complexing agent; urine digested, adjusted to pH 6 and diluted after addition of standard and complexing agent	0.5-ml sample carried by 0.05 M formate buffer (pH 3.5) through non-polar XAD-4 resin where analyte complexes are selectively adsorbed; elution with 0.1 M NH_4OH i.e. salt-free matrix	ELAN 250 with continuous flow USD	285
Mn, Co, Cu, Pb and U (river water); Mn, Mo, Cd and U (seawater)	river water and seawater SRMs	sample spiked to implement either the method of standard additions or isotope dilution analysis	1-ml (for Mo and U) or 10-ml (for Mn and Cd) sample merged with 0.1 M ammonium acetate at pH 5, and passed through silica-immobilized 8-hydroxyquinoline where analytes selectively retained; elution with 1 ml 2 M HCl/0.1 M HNO_3	ELAN 250	245
Cu, Cd, Pb	riverine water, seawater SRMs	75-ml sample adjusted to pH 4.1 with ammonium acetate buffer; isotopic spike added and dilution to 100 ml with DDW	23-fold preconcentration on SO_3-CM-quinolin-8-ol-cellulose column (and separation from alkalis and alkaline earths); elution with 0.5 M HNO_3/0.5 M HCl	ELAN 5000; Tracecon system	151
Mn, Co, Ni, Cu, Cd and Pb	riverine water, seawater SRMs	75-ml sample adjusted to pH 4.7 with ammonium acetate buffer; dilution to 100 ml with DDW	same as above; external calibration for river water analysis; standard additions for seawater analysis.	ELAN 5000; Tracecon system	149
Eu, Tb, Ho, Tm and Lu	coastal seawater	100-ml sample buffered to pH 10.0 with NH_4^+/NH_3 and mixed with 1 ml of 1 mM quinolin-8-ol in ethanol	metal-oxinate complexes are retained on XAD-7 resin and separated from alkalis and alkaline earths; elution with 1 ml of 2 M HNO_3 at 1.5 ml/min; throughput of 5 samples per hour; external calibration standards also passed through column because retention efficiency was 85%	ELAN 6000	182

Analytes	Sample	Sample preparation	On-line preconcentration strategy	Instrument	Ref.
U	riverine water, seawater SRMs	2.46 g CH_3CO_2Na added to 150 ml of sample or standard and pH adjusted to 8 with 2.0 M NH_4OH	12-ml sample passed through column of activated alumina (basic form); elution of U with 250 µl 2 M HNO_3	PQ2	220
Mn, Ni, Cu, Zn, Pb	riverine water SRM, coastal seawater SRM		3-ml sample mixed on-line with 1 M NH_3/NH_4Cl buffer and passed through column of Toyopearl TSK-immobilized 8-HOQ resin; wash with buffer and then DDW; elution with 2 M HNO_3/0.1 M HCl (spiked with Sc and Bi) at 1.4 ml/min	ELAN 500; USND; high solids torch; FIAS 400	286
As(III)	porewater, river water SRM	porewater filtered through 0.45-µm pore filters, acidified to pH with HNO_3	sample mixed on-line with 0.1% m/v APDC in DDW and pumped through knotted reactor; wash with 0.02% v/v HNO_3; adsorbed complexes eluted with 1 M HNO_3	ELAN 5000; FIAS 400; 65 cm of 0.35-mm i.d. PTFE tubing; 5-ml sample	256
Cu, As, Se, Cd, In, Hg, Tl, Pb, Bi	waters SRMs, urine SRM, bovine liver and muscle SRMs	urine diluted 1+9 with same nitric concentration as that of external standards; microwave digestion of liver and muscle with HNO_3/H_2O_2	sorption of DDTP complexes on C_{18} column; elution with 0.1 ml methanol (mixed on-line 1+1 with water); cleaning column required on the sorbent line to lower blank signal; DDTP must be present in washing solution to prevent premature analyte elution; external calibration with internal standardization using In and Tl	ELAN-6000 with 1.3-mm i.d. injector tube	111
Mn, Co, Cu, Zn and Pb	estuarine water and sea water SRMs	sample pH adjusted to 8.0 with 1 M ammonia; spiked with standard solution	200-µl sample passed through Dionex Metpac CC-1 column with 0.01 M ammonium acetate buffer at pH 5.3; elution with 2 M HNO_3	PQ2+ Turbo with high solids nebulizer	150
Zn	riverine water, estuarine water, open ocean and nearshore sea waters SRMs	sample pH adjusted to 4 using 1 M ammonium acetate buffer	10.0-ml sample passed through SO_3-8-HOQ carboxymethylcellulose at 10 ml/min; wash with 0.1 M ammonium acetate buffer (pH 4); elution with 0.5 M HNO_3 at 3.0 ml/min; isotope dilution analysis using ^{67}Zn (^{66}Zn as reference); separation required to improve precision of Zn isotope ratio	ELAN 5000; USND	287

continued

TABLE 2.7 (*continuation*)

Analytes	Sample	Sample preparation	On-line preconcentration strategy	Instrument	Ref.
Cd, Co, Cu, Mn, Ni, Pb, U and Zn	seawater SRMs, water SRM	0.5 ml of 5 M ammonium acetate buffer added to 90 ml of seawater; pH adjusted to 5.5; dilution to 100 ml with DDW	1 ml of buffer passed through column of Chelite C or Muromac A-1 resin; 5-30 ml sample passed through column; wash with DDW, buffer and DDW; elution with 1 M HNO_3 (spiked with Ga, In and Tb internal standards); external calibration	PQ2 + STE; V-groove nebulizer; spray chamber cooled to 278 K	288
V, Co, Cd, Sb, Pb	riverine water SRM		sample mixed on-line with buffer was pumped through column of Si-8-HOQ; wash with DDW; elution with 50 µl of 2 M HNO_3 with DDW carrier at 1.0 ml/min; external calibration provided equivalent results to standard additions	ELAN 500; FIAS 200; 1 m of 0.3-mm i.d. tubing between column and nebulizer	289
Au	natural waters	pH adjusted to 6-8 using HNO_3 and/or ammonia	8-40 ml sample loaded at 4 ml/min through microcolumn of sulfhydryl cotton; elution with 250 µl of 0.01 M KCN at 1 ml/min with 0.01 M carrier; external calibration; both ionic and colloidal Au are retained by column; 10-1000 mg/l Na stabilized Au in solution	PQ2+	290
BrO_3^-	drinking water		0.5-ml sample passed through acidic activated alumina column with 0.01 M HNO_3 carrier; elution with 0.5 ml 2 M ammonia	HP-4500	244
'fast reactive' Al	tap water, spring water, river water	field sampling by pumping sample, merged with 0.01 M NaOAc-HOAc pH 5.0 buffer, and with 8-HOQ, through column of Amberlite XAD-2; residual fluid drawn off	330 µl of 1 M HCl used to elute Al from minicolumn with DDW carrier at 1 ml/min; external calibration using columns loaded with Al standard solutions; loaded minicolumns can be stored for at least 2 days without loss of Al	PQ2+	188
REEs	porewater, river water SRM, basalt SRMs	0.1 g rock fused with Na_2O_2 and diluted with 0.03 M HNO_3	10-ml sample mixed on-line with pH 9.0 ammonia buffer; precipitates collected in knotted reactor (no filter); dissolution with 1 M HNO_3; calibration by standard additions	ELAN 5000; FIAS 400	213

Analytes	Sample	Sample preparation	On-line preconcentration strategy	Instrument	Ref.
Mn, Co, Cu, Ce, Cd	artificial seawater		2-ml sample mixed on-line with 2.0 M ammonium acetate buffer (pH 5.5); wash with 1.0 M buffer and then DDW; elution with 1.0 M HNO_3; external calibration should be possible	PQ2+; De Galan nebulizer; water-cooled spray chamber (10°C); PrepLab unit	250
U (method should also work for other lanthanides and actinides)	groundwater		5-ml sample, immediately followed by about 100 µl air, loaded on column of TRU™ resin; elution with 0.1 M ammonium hydrogen oxalate; 30-fold preconcentration; on-line dilution with 5% HNO_3 to keep total dissolved solids %	PQ STE II; spray chamber at −5°C; autosampler	187
Al	tap water, dialysis solutions	dilution of dialysis solutions 1:36 with DDW and adjustment of pH to 7.0 with citrate buffer	0.5-10 ml of sample loaded at 1 ml/min onto a column of Chromotrope 2B immobilized on AG 1-X8 resin; elution with 180 µl of 1 M HCl; external calibration	PQ Eclipse	291
^{99}Tc, ^{230}Th and ^{234}U	soils	sample fused with Na_2O_2 and spiked with Re (for Tc) or digested with $HNO_3/HClO_4/HF$ and spiked with ^{229}Th and ^{233}U (for Th and U)	sample passed through column of TEVA-Spec resin (for Tc) or TRU-Spec resin (for Th and U); column rinsed with 0.5 or 4 M HNO_3; elution with 8 M HNO_3 (Tc) or 0.1 M ammonium oxalate (Th and U)	ELAN 5000; FIAS-200 or FIAS-400MS; AS-90 autosampler	153
Cd, Bi	high-purity Zn	digestion with 7 M HNO_3; evaporation to near dryness; dissolution in 0.5 M HNO_3	five 350-µl sample injections on column AG1-X8 (anion exchanger) with 0.1 M KI carrier at 1 ml/min; wash with 0.1 M HNO_3; elution with 2 M HNO_3 at 0.4 ml/min; external calibration with Zn-free standards	PMS 2000	292
Nb, Ta, W, Zr, Hf	high-purity Fe	microwave digestion of 0.25 g sample with HNO_3/HCl/HF; evaporation to near dryness; dissolution in 4.5 ml 2 M HF and 0.5 ml of 30% m/m H_2O_2	200 µl 2 M HF containing 50 mg/ml sample was loaded onto Dowex 1X8-100 anion exchange resin at 2 ml/min; wash with 4 M HNO_3 at 2 ml/min and 2 M HF at 0.9 ml/min; elution with 200 µl of 3:1 mixture of 0.7 M HNO_3/0.5 M HCl plus 0.05 M H_2O_2 at 2 ml/min; preconcentration achieved by increasing sample volume	ELAN 500; FIAS 400; short 1.14-mm i.d. tubing between column and nebulizer	293

continued

TABLE 2.7 (*continuation*)

Analytes	Sample	Sample preparation	On-line preconcentration strategy	Instrument	Ref.
As, Cr and V	biological SRMs	0.5-g sample digested with HNO_3/ H_2O_2 and then $K_2O_8S_2$; diluted with Tris buffer (pH 9) and NaOH; UV photolysis	200-µl sample injected on column of activated alumina (acidic form) with water carrier; elution with 200-µl 1 M HNO_3	PQ2	237
V, Mn, Cu, Zn, Cd and Pb	biological SRMs	0.3-g sample digested with HNO_3/ H_2O_2; diluted with ammonium acetate buffer at pH 5.3	175-µl sample passed through column of iminodiacetate resin with water carrier; elution with 175-µl 3 M HNO_3	PQ2+ Turbo	152

Tris = Tris(hydroxymethyl)methylamine.
DDTP = dithiophosphoric acid O,O-diethyl ester.

the column is simply switched to the eluent line. This approach has the advantages of providing total flexibility in the sampling volume, which is not restricted to available injection loop volumes, and allowing higher sample throughput since the pumping rate for sample loading can be higher than for elution. On the other hand, the precision relies heavily on the stability of the flow rate provided by the pump which tends to be somewhat sensitive to pressure changes induced by swelling and shrinking of the column material (in the case of some resins), as well as partial line blockage from any released column particles [176]. Furthermore, since the sample goes through the flexible pump tube,

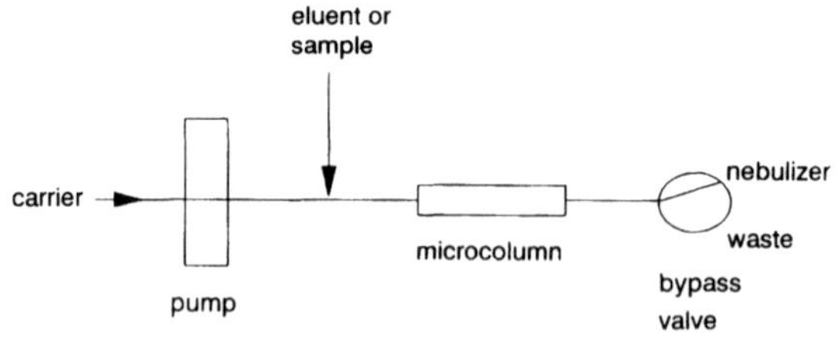

Fig. 2.27. Simple separation manifold. The analyte(s) from the injected sample is(are) selectively retained on the microcolumn while the bulk of the matrix goes to waste. Eluent is then injected to recover the analyte(s), which is then carried to the nebulizer.

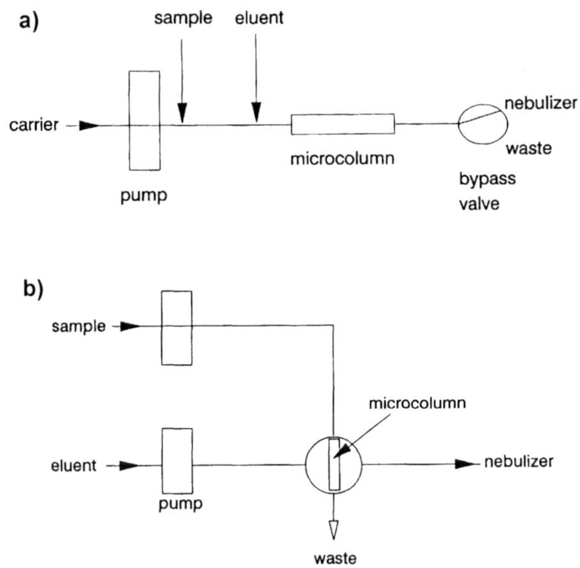

Fig. 2.28. Reduced-dispersion FI manifolds. (a) The analyte(s) from the injected sample is(are) selectively retained on the microcolumn while the bulk of the matrix goes to waste. A smaller volume of eluent is then injected to recover the analyte(s), which is then carried to the nebulizer. (b) Time-based sample injection is used to pump a discrete volume of sample through the column while the eluent is pumped to the nebulizer. When the valve is switched, the eluent recovers the selectively-retained analytes and carries them to the nebulizer.

memory effects, cross-contamination or loss of analytes through sorption on the tube walls may result [177]. Nonetheless, if the flow rate is periodically monitored and appropriate flexible pump tube is available, then good precision can be obtained.

The body of the microcolumn, which holds the chelating resin or other material used for the separation, can either be purchased from the various manufacturers of FI equipment or home-made. A simple way is to use a short length (a few cm) of tubing of wider bore than for the rest of the set-up (for example, 3-mm i.d. for the column versus 1.5-mm i.d. for the rest) and filters or plugs of glass or quartz wool at each end to retain the column material inside. The latter can be deposited inside the tube on one plug of glass wool using either a syringe or a micro-funnel made from a truncated disposable pipette. The column material can also be suspended into a slurry and pumped inside the tube, the glass wool plugs being inserted afterwards. A microbore glass column (50-mm long, 3-mm i.d.) fitted with porous 25-mm glass frits has also been used [182]. Alternatively, more elaborate and perhaps more robust home-made microcolumns have also been described in detail [176, 145, 183-185]. In general, for a given bed volume, a longer and narrower column will be more efficient at retaining analytes [13] because of the increased contact thereby provided.

In any case, care must be taken to select materials that are not contaminated with the analyte. For example, during the construction of a column of silica-immobilized 8-hydroxyquinoline, filters made of polyurethane, glass or Teflon all leached out significant amounts of Fe, even after repeated acid cleanings [186]. Only Teflon mesh or quartz wool were found suitable for the determination of Fe [186]. Additional filters may also be used to ensure that no tiny particle from the column can reach the nebulizer and clog it.

Finally, the particle size of the column material should be as small as possible to provide the maximum retention capacity, but not so small that it induces too much back-pressure [182]. For instance, in the author's experience, silica gel with particle size smaller than 40 μm simply overloads the peristaltic pump. A 40–70 μm range provides a good compromise in terms of induced back-pressure and column capacity, as long as the pumping rate is below 3 ml/min. For sample loading rate of 8–9 ml/min, a particle size of 150–200 μm is recommended [177]. Note that silica gel is a hard material which, unlike resins, does not swell. Obviously, any swelling must also be taken into account when building a column. Finally, the dimensions of the column must be optimized. Indeed, if too small a volume of column material is used, overloading of the sites may occur [182,187].

Several microcolumn FI manifolds are well suited for field sampling [188, and references therein]. Indeed, a battery-operated peristaltic pump can be used to pump the sample, with or without on-line mixing with reagents, buffer, etc., for a given time through the microcolumn, which can then be taken back to the laboratory for on-line elution to the ICP-MS instrument [188].

An alternative to a microcolumn filled with particles is to use a hollow fibre membrane onto which the chelating agent is immobilized [189]. The separation/preconcentration can then be carried on by passing the sample through the membrane [189] or on the surface of the membrane [190]. The latter approach decreases the risk of plugging and reduces the back-pressure [190].

2.6.1.2 Vapour generation

As its name implies, vapour generation consists in transforming the analyte into a volatile species. Performing vapour generation on-line with ICP-MS typically requires the addition of a gas–liquid separator to a double-line FI manifold. The chemical vapour can also be generated electrochemically, in which case an electrochemical cell will be required [191]. These can be made in-house using different layers of Perspex [191]. The design and dimensions of the gas–liquid separator should be selected carefully as it will contribute to the overall dispersion and therefore affect peak height sensitivity [192]. Although the same separators can be used for continuous and FI analyses, a gas–liquid separator with a smaller internal volume (also called headspace) is preferable for FI [193] since it minimizes dispersion in the gaseous phase without leading to incomplete gas–liquid separation [194]. An example of a fully automated set-up is shown in Fig. 2.29. The injected sample is mixed with a suitable reagent, and allowed to react before passage through a gas–liquid separator from which the generated vapours are entrained by a flow of argon into the plasma. Heat may be applied (through a water bath or a microwave oven) to the mixing coil to speed up the conversion reaction. Also, no nebulizer and

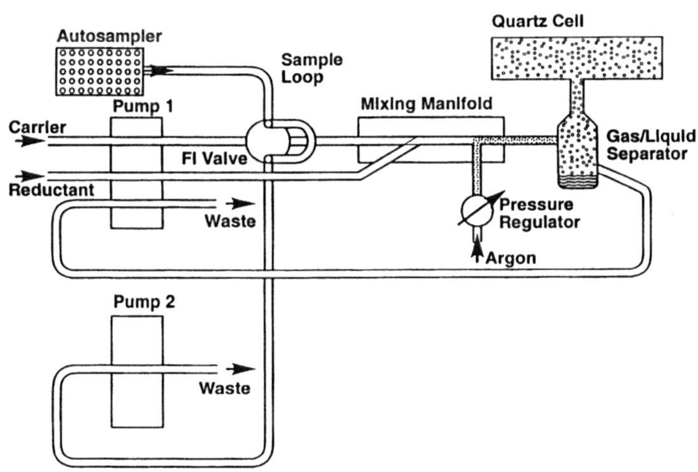

Fig. 2.29. Reduced-dispersion FI manifold involving vapour generation. The injected sample is mixed with an appropriate reagent, and the vapour emerging of a gas-liquid separator is directed towards the ICP torch (instead of the quartz cell in this figure). Courtesy of Perkin-Elmer.

spray chamber are needed with this approach since the gaseous hydride can flow directly into the injector of the plasma torch.

However, some type of water-removing device (ice-bath, cold ethanol water trap, filter, dryer, etc.) may also be needed between the gas–liquid separator and the ICP. There indeed tends to be a carry-over of liquid when a gas–liquid separator with a small internal volume is used in combination with certain borohydride concentration and carrier gas flow rate [195]. Although this secondary nebulization is minimal with gas–liquid separators using a membrane, the latter generally lack long-term stability, especially when real samples are involved, requiring too frequent changes of the membrane for practical routine analysis [196]. Those based on a tubular silicone rubber membrane, for example, can only be used with extremely low $NaBH_4$ concentrations because the membrane would otherwise be unable to quantitatively separate arsine from the flowing hydrogen gas–liquid stream [197]. The use of a drift standard is also required for quantitative analysis with the latter [197]. Although it has not yet been used with ICP-MS, perhaps a pervaporation device would provide a better performance since its membrane never touches the sample solution [198].

An alternative is to generate the vapour within the spray chamber, which then acts as gas–liquid separator [199, and reference therein]. A spray chamber has even been designed with a small inlet through which chemical vaporization mixtures can be introduced [62]. In fact, even less modification was required for a recently reported approach to minimize interferences from transition elements during the continuous generation of Se hydride [196]. Indeed, the only change involved the insertion of a capillary tube into the sample introduction tube of a Meinhard nebulizer, a conventional Scott-type double-pass spray chamber acting as gas–liquid separator [196]. The sample is then pumped into the capillary whereas the borohydride reagent is pumped into the normal sample inlet of the nebulizer [196]. Because the two were mixed immediately prior to nebulization, and gas–liquid separation took place upon nebulization, an important reduction in interferences and an increase in the level of tolerance for transition elements resulted [196]. An obvious advantage of this approach is that the switch between chemical vapour generation and solution nebulization is extremely rapid.

A compromise between removing the nebulizer/spray chamber and using the latter for the generation and separation of chemical vapour is to connect the gas–liquid separator to a tee between the torch and the spray chamber [200]. This approach would allow the simultaneous nebulization of solutions and, hence, allow as well the determination of elements that do not form volatile species. Furthermore, this approach would allow the introduction of chemical modifiers to help reduce memory effects. For example, an Au solution could be nebulized, which would help reduce the long-term drift observed during the determination of Hg, when it was introduced as a vapour directly into the ICP torch [200].

Whatever the vapour generation manifold, the use of glass in the whole system should, in general, be minimized. Losses by adsorption onto the walls of glass reaction vessels have indeed been reported [201]. These can be reduced by silanising the glass with a 10% DMCS solution in toluene [202]. Even quartz has been reported to result in high Hg blanks, especially when the quartz was new [203]. In any case, the tubing connecting the gas–liquid separator to the torch should also be made of an inert material and be as short as possible [204]. For instance, the loss of Hg vapour on Tygon was reported [205]. On the other hand, another study, which used Tygon to connect a glass collection chamber to the torch, did not observe significant Hg memory effects [206]. In the latter approach, however, the cold vapour was continuously accumulated in a 125-ml glass chamber before being swiftly injected into the plasma using 0.7 l/min Ar [206].

2.6.1.3 Liquid–liquid extraction

An approach which has yet to be used in FI-ICP-MS is that involving liquid–liquid extraction (also called solvent extraction). An example of a simple manifold is shown in Fig. 2.30. Three essential components are the segmenter, which merges the streams of the two phases, the reaction coil which allows the extraction to take place, and the phase separator which continuously splits the two phases in separate streams [175]. It is basically the same set-up as for FI-vapour generation, except that the gas–liquid separator is replaced by a liquid–liquid phase separator (and, of course, the nebulizer and spray chamber are needed).

Care must be taken to use a peristaltic pump tubing (such as silicone rubber) which is resistant to the organic phase. Similarly, any other component which comes in contact with the organic phase should be made of inert material (such as TEFLON, PEEK, etc.). If no adequate tubing is available, the displacement technique may be used (i.e. water is pumped into a closed container filled with the organic phase

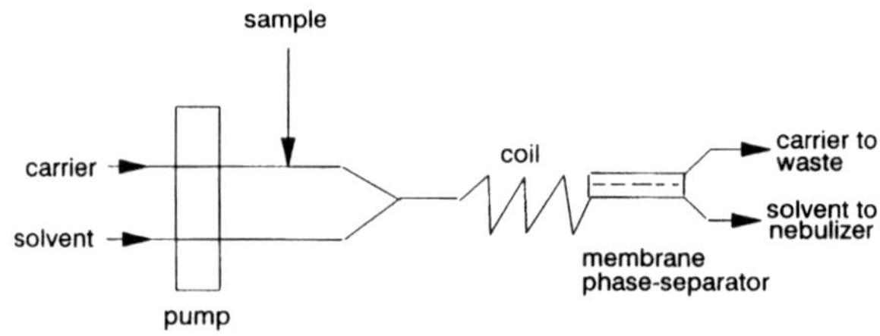

Fig. 2.30. Reduced-dispersion FI manifold involving liquid–liquid extraction. The injected sample is segmented with the solvent containing an appropriate extractant and thoroughly mixed prior to phase separation where the solvent, now containing analyte complexes, is carried to the nebulizer.

which is then pushed towards the FI manifold) [175]. Alternatively, a constant inert gas pressure may be applied to the container to force the organic phase towards the manifold [175].

Several types of segmenters are available, with various geometries [207]. The simple tee or Y-shape type are popular, although the best shape will depend on the physical properties of the two phases being segmented. In general, a mixing coil is preferably used to mix the two liquid phases. With a knitted open tubular reactor, the mixing is so vigorous that very small droplets of one phase are formed in the other, which then make the phase separation more difficult [49]. In any case, the formation of a film of organic liquid can be expected on the tube wall especially if the liquid has a high viscosity and low interfacial tension [16]. This has an additional impact on sample dispersion since it favours analyte transfer between segments and thereby increases dispersion [16]. Therefore, the thicker this film, the higher the dispersion. In general, organic solvents with high viscosity and low interfacial tension will form thick films on hydrophobic materials, such as TEFLON [208]. The film thickness will also increase with an increase in linear velocity [208].

Various phase separators are available [207,209]. Popular ones are based on porous membranes, some of which can be constructed in-house [210]. Even a microporous PTFE tube which is used for gas–liquid separation, can completely separate two liquid phases if it possesses a high enough porosity and is long enough [211].

A much simpler approach than that described above, involves a simple sequential injection system and does not require any phase separator [212]. Indeed, the organic phase, preceded by an air segment, is first loaded into the extraction coil and immediately followed by the aqueous sample which is then propelled through the organic phase by pumping air towards the extraction coil. By changing the direction of the pump, the aqueous phase is pushed back through the organic phase and can be discarded if the analytes were extracted into the organic phase. A back-extraction can then readily be performed by repeating the process with an aqueous phase at the appropriate pH [212]. This passage of the aqueous phase through the organic phase is possible because the organic phase moves more slowly as a result of hydrophobic interactions with the walls of the Teflon extraction coil [212]. Although this simple approach has not been implemented with ICP-MS yet, it is very promising.

2.6.1.4 Precipitation / dissolution

Another approach is that involving the selective precipitation of the analyte, followed by its dissolution. As when using a microcolumn, preconcentration is achieved if the volume of solution used to dissolve the precipitate is smaller than that of the sample injected. An example of a manifold is shown in Fig. 2.31. With only pump 1 on, the injected reagent (which may be buffered at the appropriate pH) is mixed on-line with the injected sample and pumped through a knotted reactor where analyte precipitates accumulate while the rest of the soluble matrix continues to waste (meanwhile, DDW is fed to the nebulizer by pump 2). With a sufficiently long reactor, no filter is required [213, and references therein]. The knotted reactor indeed not only promotes radial mixing of sample and reagent, thereby facilitating the pre-

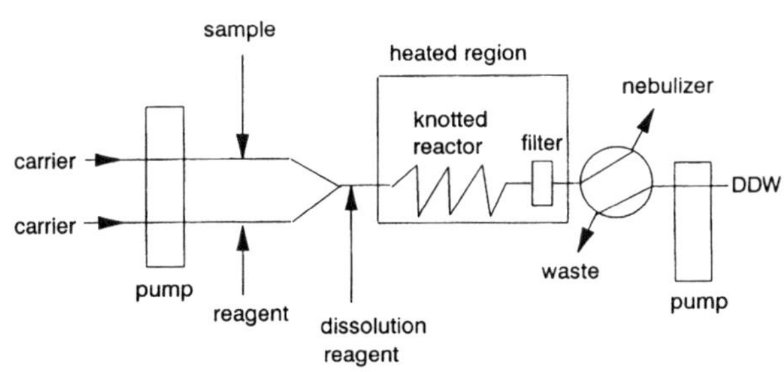

Fig. 2.31. Reduced-dispersion FI manifold involving selective precipitation. The injected sample is mixed with an appropriate injected precipitating agent and carried through a knotted reactor where selective precipitation of the analyte occurs. While the analyte precipitate is retained in the reactor and/or on the filter, the matrix is directed to waste. A suitable reagent is then injected to dissolve the analytes in a smaller volume (than the injected sample volume) which is then carried to the nebulizer.

cipitation process, but it also acts as collector because the changes in flow direction induced by the knots direct the particles towards the tubing wall [213]. A knotted reactor is therefore greatly advantageous as a filterless collector since it is easily constructed and has an unlimited lifetime [213]. The sample volume is determined by the size of the tubing and the pumping rate. After a suitable rinse time (necessary to remove loosely retained matrix components), a dissolution reagent is injected to recover the analytes and direct them towards the nebulizer. (Pump 2 is then stopped.) Depending on the trace metal(s) being determined, a metal-free filter, if one is used, may be required to minimize the blanks.

2.6.1.5 Anodic stripping voltammetry (ASV)

Anodic stripping voltammetry (ASV) is an approach which has been implemented as an on-line flow system for ICP-AES [214–217] and ICP-MS [216–218]. Figure 2.32 shows a typical set-up. Basically, while a clean solution (for instance, 0.1 M HNO_3) is fed to the nebulizer through a bypass valve, an electrolyte solution adjusted to a suitable pH flows through the ASV cell to waste. With the working electrode set at an appropriate potential, the sample is injected into the electrolyte carrier and delivered to the cell. Under these conditions, the analyte is selectively deposited on the electrode while concomitant elements are washed to waste. Following cell washout, the working electrode is ramped to an appropriate stripping potential to rapidly release the analyte into the electrolyte carrier which entrains it towards the nebulizer. Hence, both a separation from the matrix and analyte preconcentration are possible using this approach [218].

2.6.2 Optimization

Whatever the approach used, an optimization of all the FI parameters is required. In all cases, care must be taken to ensure that the blanks (i.e. results obtained by injecting high-purity water instead of the sample) are as low and reproducible as possible. Because of the very low analyte levels involved, the blanks can indeed be expected to be the limiting factor in the preconcentration. In any case, the determination of some elements may require other techniques to allow their accurate trace analysis. For instance, the determination of Fe by quadrupole-based ICP-MS is plagued by spectroscopic interferences from ArO and ArOH. Although the matrix separation that is concurrent with the

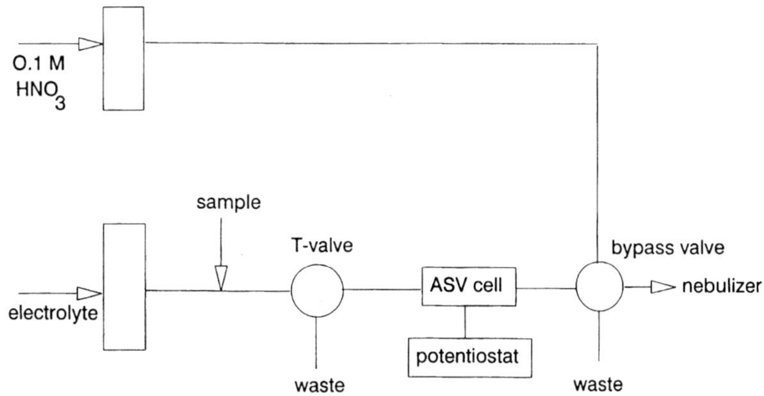

Fig. 2.32. Reduced-dispersion FI manifold involving anodic stripping voltammetry (ASV). The injected sample is entrained by an electrolyte solution, at a reduced flow rate, into the ASV flow cell where selective deposition of the analytes occurs. While the analytes are deposited onto the electrode, the matrix is directed to waste and 0.1 M HNO_3 is continuously pumped (at a conventional flow rate) to the nebulizer. The T-valve is then switched so that the analytes may be stripped into quiescent electrolyte (to prevent peak broadening that would result from stripping them into a flowing stream). Then, the speed of the electrolyte pump is increased to match that of the 0.1 M HNO_3 pump, both the T-valve and bypass valve are switched, and the released analytes are carried to the nebulizer.

preconcentration step would reduce or eliminate the contributions of CaO and CaOH, an argon–nitrogen mixed-gas plasma in combination with partial aerosol desolvation was required for the analysis of seawater by a microcolumn approach [219].

2.6.2.1 Preconcentration on a microcolumn

For a given microcolumn, the flow rate and composition of the carrier, the flow rate and concentration of the eluent, as well as the pH and composition of the sample will all influence the deposition and elution processes [220]. The back-pressure produced in the system will determine the highest carrier flow rate that can be allowed. This back-pressure may increase with time (especially with resins) if a unidirectional flow is constantly passed through a column. A widely used approach to minimize back-pressure with columns of resin is to load the sample one way and perform the elution in a countercurrent way [183,187]. In any case, a compromise will have to be made since the dynamic exchange capacity of the column increases as the flow rate is decreased [221]. A longer residence time of the analyte in the column indeed favours its retention on the column. Depending on the sample type, the column washing time prior to the elution and the reconditioning period following elution will affect both the total analysis time and the recoveries [222]. If too short a washing time and/or reconditioning time is/are used, sample-to-sample carry over may indeed result [222]. Similarly, the elution flow rate will require optimization since too fast a flow rate may result in incomplete recovery while too slow a flow rate my result in too much dispersion [190].

When, to adjust the sample pH, a buffer is mixed on-line, the cleanliness of the buffer (that may be ensured by installing a second column on the buffer line) must be monitored. As well, its complexing capacity must also be taken into consideration. Indeed, the buffer will likely act as an auxiliary complexing reagent [223], so care should be taken in selecting a reagent that does not form more stable complexes with the analytes than the separation column. On the other hand, a reagent that solubilizes the analytes at pH values where they would otherwise form precipitates is also desirable [223].

As far as the optimization of the ICP-MS instrument is concerned, standard solutions in the same matrix as the eluent can be continuously fed to the nebulizer via the bypass valve so that the optimum and/or compromise aerosol carrier flow rate, plasma position, etc., for the determination of all the analytes can be found. In cases where the eluent cannot be fed continuously to the plasma, then the optimization must be carried out in the FI mode, i.e. performing the whole preconcentration/elution process with constant FI conditions as many times as required to optimize each ICP-MS parameter. Alternatively, to save some time, a limited dispersion FI manifold can be connected to the bypass valve, and used to repeatedly inject a multielemental standard in the eluent matrix.

2.6.2.2 Vapour generation

The nebulizer and spray chamber are, in general, removed so that the gas–liquid separator can be connected directly to the torch injector with as short a transfer line as possible to minimize dispersion. Therefore, the optimization of the ICP-MS parameters must be carried out in the FI mode, by making repeated injections of a standard solution, with the FI manifold operated under constant conditions. Depending on the ICP-MS instrument, a preliminary optimization of the three-dimensional plasma position with respect to the interface may be performed while continuously nebulizing an appropriate standard solution, with the nebulizer and spray chamber in place. The plasma is then extinguished to remove the nebulizer and spray chamber and install the FI manifold/interface to the ICP while being careful not to move the torch. A fine-tuning of the plasma position can then be carried out in the FI mode.

Of course, all the FI operating conditions must also be optimized the first time around. This includes the concentration, type and flow rate of the reducing agent (some gas–liquid separators are indeed designed for reduction by sodium borohydride or tin(II) chloride but not either of them [194]), the carrier flow rate, the sweeping Ar gas flow rate (which entrains the hydride into the plasma), the presence or not of a moisture filter, etc. Depending on the FI conditions and the gas–liquid separator used, the sweeping Ar may indeed entrain a substantial amount of moisture which may condense in the transfer line. Sporadic dislodgment of droplets can then induce signal fluctuations and even extinguish the plasma. A mathematical model may be used for the optimization of the FI parameters (sample loop volume, carrier flow rate, headspace of gas–liquid separator, purge gas flow rate, etc.) [193].

Vapour generation techniques are the subject of a number of interferences where elements from the matrix may suppress the chemical vaporization process. So, the efficiency of vapour generation depends on both the composition of the matrix and experimental variables [193]. For example, in a comparison of hydride generation with ethylation for the determination of Hg^{2+} and $MeHg^{+}$ by derivatization followed by cryotrapping, gas chromatography and detection by QFAAS, hydride generation was less affected by the chloride level of samples and was therefore recommended for the analysis of saline waters (such as sea water and estuarine water) and biological tissues; whereas ethylation was a better choice for the analysis of metal-rich samples (such as sediments, waste effluents and pore waters). Similarly, hydride generation was more sensitive than ethylation for the determination of organotin compounds, but the latter was more reproducible and not affected by inorganic interferents because the mixed tetraalkyltin species produced by ethylation are more stable thermodynamically than the corresponding hydrides [224].

Chemical vaporization interferences can be removed on-line prior to the vapour generation process. For example, an anion-exchange resin was used to selectively retain the chloro complexes of Fe while the analyte continued to the vapour generation system [225]. Similarly, the interfering gold matrix was removed by reductive precipitation prior to the determination of As [226]. In the case of Pb, which is notoriously difficult to transform into plumbane, a selective separation and preconcentration of Pb dithiocarbamate complex on Chromosorb 102 may be carried out followed by hydride generation on the alcoholic column eluent since plumbane can be generated from the Pb complex [227]. A similar approach has been applied to other hydride-forming elements (As, Hg, Sb, Se, and Sn) in continuous hydride generation ICP-AES [228]. In fact, several of the separation systems described previously (for example, those in Tables 2.5, 2.6, 2.7, 2.8 and 2.9) could be successfully combined to a vapour generation manifold to remove potential sources of interferences. Separation involving medium dispersion systems would simply remove the interferent while the vapour generation manifold would perform the preconcentration. On the other hand, separation achieved concurrently with a preconcentration in reduced dispersion systems would further improve the detection limits, if they are not blank limited.

In any case, an optimization of the vapour generation reaction(s) is required if different species are to be determined. Indeed, some species, such as Se(IV), readily form hydrides whereas others, such as Se(VI) are completely inert, and organoselenium compounds react to various extents [229]. This means that for the total determination of Se, organoselenium compounds must first be decomposed, and the resulting Se(VI) must then be reduced to Se(IV) using HCl at 90–100°C. Care must be taken not to boil the sample for too long as it may reduce Se to its elemental form and/or result in losses of volatile Se compounds [230]. Automation of, at least, the reduction process on-line, through the use of FIA for example, can be used to both shorten the analysis time and minimize the risk of losses [230].

Pre-derivatization is an approach that is often used to enhance the performance of hydride generation methods. For example, the addition of the amino acid L-cysteine is becoming popular because of its many advantages. It equalises the As sensitivity for various As forms (for As(III), As(V), MMAA and DMAA) [231,232]. Furthermore, it reduces interferences from transition metals and other hydride-forming elements. Finally, the concentration of acid required for hydride generation is reduced, which, in turn, reduces the acid contribution to the blank and thereby minimizes detection limits. However, the reaction of L-cysteine with some species is slow at room temperature and requires a heating stage which adds to the complexity of on-line interfacing.

However, focused microwave heating systems are well suited for this purpose and have been successfully applied to the determination of inorganic and organometallic As [192], and Se [233].

Alternatively, thioglycollic acid, which is significantly faster than L-cysteine at room temperature, may be preferred, although its odour and toxicity are notable drawbacks [234]. Another alternative is to use a UV photoreactor to photochemically transform an inert species (such as Se(VI)) into one which can be converted into a volatile species (such as Se(IV)). This latter approach is very attractive since it does not require the addition of chemical reagents, thereby minimizing contaminations, and is relatively fast [235].

In the case of hydride-forming elements with a high ionization potential (such as As, Se and Te), on-line mixing of the sample with a methanol solution not only improved sensitivity but also reduced discrepancies between standard solutions and seawater [236]. Indeed, dissolved CO_2 in seawater, which is entrained out of the sample along with the hydrides, would otherwise induce an enhancement [236]. A continuous flow of 10% v/v methanol alleviated this problem by providing a fairly constant flow of carbon (and therefore enhancement) into the ICP, thereby allowing the use of external calibration using aqueous standards for the analysis of seawaters [236].

2.6.2.3 Liquid–liquid extraction

With on-line liquid–liquid extraction, the preconcentration efficiency depends on the ratio of the sample injection volume to that of the acceptor phase, which is proportional to the ratio of the carrier flow rate to that of the acceptor phase [16]. This ratio must therefore be selected according to the range of applications foreseen. It should be noted that most of the FI liquid–liquid extraction approaches have been used with FAAS, at flow rates much higher than that at which efficient extraction is achieved [19]. The uptake rate of ICP nebulizers is lower and therefore more compatible with the FI extraction manifold [19].

In general, the analyte will be extracted from an aqueous phase to an organic phase. The ICP-MS operating conditions must therefore be selected to handle the organic phase (see Sections 2.3.2 and 2.3.3.1). Alternatively, the analyte can be back-extracted into an aqueous phase by using a second solvent extraction manifold in series with the first one. On the other hand, this alternative would very likely decrease the analyte recovery unless 100% extraction efficiency can be achieved. In any case, the ICP-MS operating conditions can be optimized either in the FI mode, with repeated injections of a standard solution, or while continuously nebulizing standard solutions in the matrix of the final phase of the extraction, with the FI manifold disconnected from the nebulizer.

2.6.2.4 Precipitation / dissolution

Referring to Fig. 2.31, the operating conditions of the ICP-MS instrument can simply be optimized by using a standard solution of the same matrix as the dissolution solution for the carrier of pump 2 (with pump 1 off). If the instrument cannot tolerate the continuous nebulization of such a solution, then the ICP-MS optimization must be carried out in the FI mode, while performing repeated injections of the standard solution into the carrier of pump 2.

As far as the FI operation is concerned, the goal is to find conditions allowing complete precipitation almost instantaneously, and then complete dissolution. In the former case, the length of the mixing coil has been found to be critical and dependent on the sample volume [144]. In the latter case, the concentration of the dissolution solution is obviously important, as well as the pumping rate. The whole process may be improved by heating the knotted reactor since, in batch mode, heating speeds up both the precipitation and the dissolution of the precipitate.

2.6.2.5 Anodic stripping voltammetry (ASV)

Optimization of ICP-MS operating conditions can be carried out while nebulizing standard solutions (with the same matrix as the electrolyte carrier) through the bypass valve (see Fig. 2.32). The ASV conditions must also be optimized, in addition to the usual FI parameters i.e. the sample injection volume (which determines the preconcentration factor) as well as the flow rate, pH and identity of the electrolyte carrier. Important ASV parameters, in addition to the deposition and stripping potentials, include the rate at which the potential is ramped from the deposition to the stripping potential, as well as the material of the working electrode. Analyte recovery was indeed matrix dependent with a reticulated vitreous carbon electrode [217]. Furthermore, with thin-layer flow-by cells, optimum deposition efficiency is obtained at flow rates of 100 µl/min. These ASV cells are therefore best coupled to a low-flow MCN/spray chamber system if some dispersion is desired, or to a DIN if dispersion must be minimized [218].

2.6.3 Features

For samples with very high levels of dissolved solids and/or very low analyte concentrations, FI provides a means of implementing on-line separation of the analytes from the matrix and preconcentrating them. As discussed in Section 2.4.2, removing the bulk of the matrix drastically reduces potential sources of spectroscopic interferences. For example, a separation from chloride using either a microcolumn [152] or hydride generation [23], allows the determination of *V* [152,237] and As [23,152,237] without $^{35}ClO^+$ and $^{40}Ar^{35}Cl^+$ interferences, respectively. Whatever the approach used, a substantial reduction in sample and reagent consumption results from implementing it on-line with ICP-MS. This, in turn, decreases the sample-processing time and therefore increases the sample throughput. Furthermore, because everything is performed in a closed system, a significant decrease in contamination sources is also achieved.

Finally, because all the preconcentrating methods have a selectivity for certain ions, speciation is actually being carried out. For example, activated alumina in the acidic form has a high affinity for oxyanion species (such as arsenate, chromate, selenate, tungstate, vanadate, etc.) but not for cations (such as Cr(III)). This feature was applied to the speciation of Cr [238]. A sample containing both forms of Cr was injected into a 0.01 M HNO_3 carrier which maintains the alumina acidic. As the sample passed through the column, Cr(VI) was retained but not Cr(III). (A similar behaviour was reported with a column of ZnO [239].) An injection of 1 M ammonia solution was then made to elute Cr(VI). Hence, two peaks were obtained for each sample: a first one due to Cr(III), followed by one due to Cr(VI). The former is broader than the latter because Cr(III) was dispersed by the column whereas Cr(VI) was preconcentrated. Of course, depending on the injection volume, the dispersion of Cr(III) will either be medium or limited (for instance, a 1-ml injection gave rise to a steady-state signal equivalent to that obtained without FI) [238]. In any case, this approach was employed for the certification of Cr(III) and Cr(VI) in a lyophilised solution [240]. A number of FI approaches have been used to conduct speciation studies [238,241–243], but very few have yet been coupled to ICP-MS. These will be discussed in detail in the forthcoming book on ICP-MS speciation methods.

Microcolumns can also be used for sampling analytes in the field by pumping water through them. These analytes are then eluted at a later time on-line with ICP-MS. Microcolumns can therefore be used for the storage of species. For example, bromate species were stable for at least eight weeks on a microcolumn of activated alumina (in the acidic form) which was placed in a polyethylene bag in a light-tight container and stored at 4°C until analysis [244].

2.6.3.1 Preconcentration on a microcolumn

A separation from the matrix is accomplished by selectively retaining the analyte on the column; and a preconcentration results if the volume of eluent used to release the analyte from the column is smaller than that of sample originally passed through the column. In fact, since most of the analyte is eluted in the front of the eluent plug, a small preconcentration is still accomplished when the volume of eluent matches that of the sample [245].

Two different approaches can be used for analyte preconcentration using microcolumns. The most widely used involves either a chelating/ion-exchange resin or a complexing agent which has been immobilized on a solid support (such as silica, a polymer or a hollow fibre membrane). This includes activated alumina which can function as either an anion or cation exchanger, depending on the pH of the solution [4]. Indeed, under basic conditions, it has a high affinity for a wide range of cations; whereas under acidic conditions, it displays a high affinity for oxyanions [4].

In general, controlled-pore glass (CPG) particles are a popular support for immobilization of complexing agents because they do not swell and have a large surface area. On the other hand, polymeric supports lead to lower blanks [246]. Nonetheless, selective reactions have been designed to bind a variety of chelating agents (dithiocarbamate, quinolin-8-ol-5-sulfonic acid, dithizone, cysteine) to CPG [247]. The synthesis of silica-immobilized 8-hydroxyquinoline (or quinolin-8-ol or 8-quinolinol) (Si-8-HOQ) has also been described [248, and references therein]. However, the phenylazo-8-HOQ bond that is used to retain 8-HOQ on the CPG confers limited stability to acid and oxidizing media [249]. In contrast, Friedel-Crafts alkylation of 8-HOQ with a chloromethylated poly(styrene)-divinylbenzene yielded a chelation ion exchanger with excellent chemical stability [249].

One product, where iminodiacetate chelating reagent is immobilized on CPG, is commercially available [250]. It features a greater capacity than Si-8-HOQ, which should allow preconcentration factors over 100 with a mini-column [250]. The immobilization of zirconium(IV) oxide on silica gel yields a material which behaves similarly to activated alumina: under acidic conditions, anion exchange predominates; whereas cation exchange takes place at high pH [251]. Even bacteria [247,252,253], enzymes [254] and biomass [255, and references therein] can be immobilized on CPG. Although these materials do not have the high capacity of chelating agents, they are significantly cheaper, hence their use for trace metal preconcentration. The interest in immobilized enzymes is also growing because of their high selectivity [254].

An alternative is sorbent extraction [179]. It consists in mixing the sample on-line with a complexing agent to form a neutral complex with the analyte which is retained on either a reversed-phase column, such as C-18 bonded silica or on the inner walls of either a knotted reactor [256] or a loop [257]. Elution is then carried out, usually with an organic solvent such as methanol [23]. The knotted reactor or loop approaches have the advantages of providing more flexibility in terms of flow rate (as there is no back-pressure induced by a column) [257] and having an unlimited lifetime.

With a judicious selection of complexing agent (for instance diethyldithiocarbamate (DDTC) or dithiophosphoric acid O,O-diethyl ester (DDTP), which selectively react with many transitions and heavy metals), a separation from the troublesome matrix (in this example, alkalis and alkaline earths) can be accomplished [23,111]. If the complexing agent, such as DDTP, can complex the analytes in a nitric acid medium, then the need for pH adjustment using buffers is avoided, thereby eliminating a potential source of contamination [111]. Recently, the adsorption capabilities of C_{60} fullerenes for metal chelates (with ammonium pyrrolidinedithiocarbamate as ligand) were compared to those of C-18 bonded silica and activated carbon [258]. The results clearly indicated that the fullerenes had significantly higher adsorption power than the other two sorbent materials [258].

Although sorbent extraction allows total flexibility in terms of separation since the complexing agent can be changed at will while keeping the same column (unlike the former approach where only the elements which are retained by the column can be determined), some constraints are imposed by the fact that an organic solvent is usually required for elution, although some approaches [256] used an acid successfully. First, the ICP-MS instrument must be able to cope with the injection of eluent. As mentioned in Section 1.2.2.1, a cavity-tuned, free-running rf generator can easily handle a solvent gradient as extreme as going from water to 95% MeOH/5% H_2O, to water again within 30 sec [23]. Otherwise, an addition of O_2 to the plasma may be required [82].

However, even if the eluent can be handled, a blank problem may result, depending on the element to be determined. Organic solvents are indeed not available in as high a purity (in terms of metal content) as acids. These constraints explain perhaps why this approach, which has been somewhat popular with less sensitive techniques (such as FAAS), has not often been selected with ICP-MS.

Whether the complexing agent is immobilized on the column or mixed with the sample prior to sorption, the speciation of the analytes may affect the results of total analyses if the complexing agent prefers one form over another. For example, DDTP and pyrrolidine dithiocarbamate [256] only complex As(III), so low results may be obtained if part of the As is in As(V) form and/or if the sample pretreatment (such as a digestion) oxidizes As(III) into As(V) [111]. In these cases, an accurate total concentration can only be obtained if all the species are converted into the one that can be determined. On the other hand, such reagents would allow the speciation analysis of selected elements.

2.6.3.2 Vapour generation

Selectively transforming the analyte into a chemical vapour which is then separated from the liquid phase and swept directly into the plasma, is another powerful way of separating the analyte from the matrix and preconcentrating it simultaneously. It can therefore be used to remove spectroscopic interferences arising from the sample matrix, such as that of ArCl on As when the sample contains chloride [259,260]. In comparison to the off-line approach, FI offers a distinctive kinetic advantage [10]. Indeed, because the reaction time is controlled and shorter than for the off-line method, interfering reactions do not have as much time to occur and are therefore discriminated against in favour of the formation of analyte-containing vapour [10]. Furthermore, the smaller sample volume required in FI also means that a smaller absolute amount of interfering species is present which decreases the likelihood of their accumulation in the system and hence any negative impact on hydride (or other chemical vapour) formation [261].

In theory, vapour generation should provide 100% sample introduction efficiency, if the analyte is quantitatively transformed in a vapour. Furthermore, since a vapour, rather than an aerosol, is introduced in the plasma, no plasma energy is then required for the desolvation and vaporization steps, leaving more energy for atomization and ionization. In the case of Hg, whose vapour is atomic, even more energy would be available for ionization since that would be the only purpose of the ICP. Although it can be combined with a separation technique for the determination of selected species, a topic that will be discussed in the forthcoming book on ICP-MS speciation techniques, it by itself offers some selectivity. For instance, Se(VI) is not readily transformed into a hydride whereas Se(IV) is [195]. The potential of tetrahydroborate(III) for the speciation of As, Se, S and Sn was reviewed [262].

The most popular vapour generation technique is hydride generation, which is widely used for the determination of As, Bi, Ge, In, Sb, Se, Sn, Pb, Te and Tl. Hydrides can be generated from solution using a variety of reactions [263-269], the most common being a BH_4^--acid reaction [261], which is extremely fast and yields H_3BO_3 and H_2 upon mixing of the borohydride reagent with acid [193]. Hydrides can also be produced electrochemically from solution [270–272], or thermochemically [273] in the gas phase. Recent reports have even demonstrated that they could be generated from the solid phase [195,227]. The sample and borohydride reagent were indeed mixed on-line and pumped through a strongly basic anion-exchange resin (Amberlite IRA-410) where As, Se(IV) and tetrahydroborate(III) (borohydride) were co-immobilized [195]. Hydrogen arsenide and selenide were then produced by the passage of 500 µl of acid (4 M HCl) through the column [195].

The efficiency of vapour generation can be improved using surfactant-based organised assemblies (such as micelles or vesicles), through the stabilization of unstable species, increase of reaction kinetics and change in the selectivity of the reaction, which can, in turn, reduce chemical interferences [274, and references therein].

Alternatives to hydride generation include cold vapour generation, where Hg^0 is produced [204–206], halogen generation (for the determination of Cl, Br and I) [275,276], generation of AsF_3 [277], OsO_4 vapour generation [278], carbonyl vapour generation (for the determination of Ni, for example) [279], methyl borate ester generation (for the determination of B) [280], aqueous ethylation [281], and the use of Grignard reagents in adequate solvent. Some unknown volatile species of Cd and Cu have also been reported from a reaction with $NaBH_4$ [227,282].

2.6.3.3 Liquid–liquid extraction

The sensitivity may be expected to be lower in the FI mode than the batch extraction mode. Indeed, a shorter reaction time is generally used and, in any case, the sample is dispersed in the carrier [144]. However, the FI mode is more selective i.e. tolerant to interferents than the batch mode as a result of a kinetic discrimination similar to that described above, in Section 2.6.3.2. Furthermore, when the organic phase forms a sufficiently thick film on the wall of the extraction coil, the differential flow velocities which then develop between the organic and aqueous phases within the segmented flow can be used to achieve a separation in time of unextracted and extracted species [209]. In other words, if both the aqueous and organic phases can be tolerated by the ICP-MS instrument, then no phase separator is required. The two phases exiting the extraction coil can simply be homogenized by the addition of a suitable solvent (for example, methanol can be used to homogenize hexanol and water) [209]. Under these conditions, each sample injection will first result in a relatively sharp peak due to unextracted species, which will be followed by a broader peak due to extracted and dispersed species [209].

2.6.3.4 Precipitation / dissolution

The batch approach has essentially been limited to co-precipitation of the analyte(s) with a collector or carrier precipitate. Indeed, a huge sample volume is required if selective precipitation of the analyte is carried out in order to obtain an amount of precipitate which can be handled. In contrast, the FI approach readily allows the direct precipitation of the analyte, no matter how small the amount of the resulting precipitate [144]. Not only are a separation from the sample matrix and a preconcentration of the analyte achieved, but the final analyte solution does not contain an excess of collector which could be a source of spectroscopic and/or non-spectroscopic interferences. Of course, if no selective reagent is available, a co-precipitation may still be carried out in the FI mode.

With a careful control of the pH, analytes may be selectively separated from alkalis and alkaline earths, thereby substantially decreasing the total dissolved solids content [213]. The analytes may also simultaneously be separated from sources of spectroscopic interferences, such as those of BaO and BaOH on rare earth elements [213]. This separation and preconcentration can be carried out relatively quickly, requiring 5 min for a 10-ml sample [213].

Because of these features, a reduced-dispersion FI manifold has even been designed for use with a discontinuous technique i.e. GFAAS [283]. For example, Co is precipitated with ammonium pyrrolidine dithiocarbamate, and collected on a filtering cup (which also serves as dissolving vessel and sampling cup for the autosampler of the GFAAS instrument). The filtrate is dried and dissolved in methylisobutylketone which is then injected into the graphite furnace [283]. The whole process is carried out automatically by the FI manifold, the timing of the sample processing being matched to the time required by the temperature program of the graphite furnace. This manifold could therefore be coupled to ETV-ICP-MS (see Chapter 3), which would further enhance this technique.

2.6.3.5 Anodic stripping voltammetry

One important feature of ASV is its capability to preconcentrate *in situ* reducible inorganic, organometallic and even organic species from the solution into the working electrode [284]. It is applicable to a wide range of concentration [284]. Both the total metal concentration can be determined and electrospeciation (i.e. fraction of the metal concentration determined under a given set of ASV and solution conditions which are selected to match the toxic fraction of the metal) can be carried out [284]. Different valency states of a given element can also be readily distinguished because these are usually not deposited at the same potential [284]. Furthermore, samples with a high ionic content (for instance, seawater) can be analyzed directly as the matrix actually acts as a suitable electrolyte [284].

Total or partial matrix separation can be achieved, depending on the identity and concentration of concomitant elements [218]. For example, the interference of ArC^+ on $^{52}Cr^+$ was eliminated for up to 10000 µg/ml carbon whereas only up to 1000 µg/l Cl could be tolerated without inducing ClO^+ interferences on $^{53}Cr^+$ and especially $^{51}V^+$ [217].

However, multielemental analysis using ASV is limited to 6–7 elements, 8–9 at best [284]. Of all the preconcentration approaches discussed so far, the ASV one is likely the most limited in terms of sample throughput as a result of the relatively long cell washout time (4 min) which is required to remove concomitant elements in a regular cell, or the low deposition flow rate of about 100 µl/min in the case of thin-layer flow-by cells. However, it is possible that upon further optimization, a comparable sample throughput may be achieved. For instance, microelectrodes have been developed with diameters of 5–30 µm, which allow volumes as small as 5 µl [284].

2.6.4 Applications

Table 2.7 summarizes various microcolumns and/or sorption approaches which have been used with ICP-MS. Most of them involve selective retention on a microcolumn. This approach is particularly popular for the analysis of natural waters. A review has even been published on the use of FI for the determination of cationic species in seawater [30]. Another one reviews various solid phase techniques that have been used for the preconcentration of trace metals from natural water samples [294]. It includes both off-line and on-line approaches.

Most of the FI microcolumn preconcentration approaches which were developed for ICP-AES are readily compatible with ICP-MS. For example, some of them are summarized in Table 2.8. In some cases which are indicated under the comments, not all the elements can be determined by ICP-MS because of spectroscopic interferences induced by the eluent used with ICP-AES. In any case, many off-line column separation approaches can readily be converted into FI systems as demonstrated by selected examples in Table 2.9. A comparison with Table 1.10 clearly shows a substantial reduction in sample and eluent volumes, which in turn leads to a drastically decreased sample processing time and therefore a higher sample throughput. In cases where the optimal elution flow rate is higher than the uptake rate of the nebulizer, a T-junction can be inserted at the output of the column to direct part of the effluent to the nebulizer and the excess to a trap (which could later be analyzed, after mixing) [310].

Examples of FI vapour generation systems that have been developed for ICP-MS are given in Table 2.10. All the continuous vapour generation approaches that have been used with ICP-MS may

TABLE 2.8

Examples of microcolumn-based reduced-dispersion FI systems developed for ICP-AES and compatible with ICP-MS

Analytes	Matrix	Column; wash; eluent	Comments	Ref
Cd, Cu, Fe, Mn, Ni, Zn	river water, sea-water acidified with HNO_3	Amberlite XAD-2; wash with DDW; elute with 2 M HCl/0.1 M HNO_3	sorbent extraction with 8-hydroxyquinoline at pH 9.5 (mixed with the sample on-line)	295
Cd, Cu, Fe, Mn, Ni, Zn	river water, sea-water acidified with HNO_3	Amberlite XAD-2 loaded with 1-(2-thiazolylazo)-2-naphthol; wash with DDW; elute with 2 M HCl/ 0.1 M HNO_3	sample adjusted to pH 8.4 by mixing with buffer on-line prior to passage through resin	184

Analytes	Matrix	Column; wash; eluent	Comments	Ref
Mo	seawater	activated alumina (acidic form); elution with 2 M NH_4OH	bypass to waste during sample loading; column maintained acidic with 0.01 M HNO_3	296
Cu, Ni, Cd, Co, Mn, V, Pb	seawater	IDAEC membrane; elution with 2 M HNO_3	sample adjusted to pH 5.5 by mixing with 2 M ammonium acetate buffer on-line prior to passage through membrane; bypass to waste during sample loading	190
Au	refinery effluent adjusted to pH 2.5 with HNO_3 and/or NH_4OH	Amberlyst A-26 anion-exchange resin; wash with 0.01 M HNO_3; elution with 2 M NH_4OH	single-line manifold (Fig. 2.27) with 0.01 M HNO_3 carrier; presence of NaCl in sample favoured the formation of $AuCl_4^-$ which was more readily retained; should be applicable to natural waters	205
Au	refinery effluent adjusted to pH 4-7 with HNO_3 and/or NH_4OH	sulfhydryl cotton; wash with DDW; elution with 0.05 M KCN	single-line manifold (Fig. 2.27) with DDW carrier; presence of NaCl in sample improved the reproducibility of the retention and elution processes; should be applicable to natural waters	205
Cd, Co, Cu, Pb	seawater, urine; diluted 1:1; acidified to 0.01 M HNO_3	iminodiacetic acid/ ethylcellulose; wash with DDW; elution with 2 M HNO_3	DDW carrier; sample mixed with ammonium acetate buffer at pH 5.5 on-line, prior to passage through resin	183
Pt, Pd, Rh	HCl digests of biological materials spiked with $SnCl_2$ and CMDTC	XAD-4 resin; wash with 1 M HCl; elution with 0.5 M NH_4OH/1.2% m/v CMDTC	sorbent extraction of analyte-CMDTC complexes	185
Al, Cd, Co, Cu, Fe, Ni, Mn, Pb, Zn	HNO_3 digests of biological materials adjusted to pH 6.5 with 1 M NaOAc and NH_4OH	8-hydroxyquinoline-cellulose; elution with 1 M HCl/1 M HNO_3	determination of Cr and V by ICP-MS precluded because of HCl in eluent; perhaps 2 M HNO_3 could be used instead	297
Cu, Fe, Ni, Mn, Zn	HNO_3 digests of biological materials adjusted to pH 4.0 with 1 M NaOAc and NH_4OH	ethylenediaminetriacetic acid-cellulose; wash with 0.1 M NaOAc at pH 4; elution with 2 M HCl/2 M HNO_3	determination of Cr and V by ICP-MS precluded because of HCl in eluent; perhaps 4 M HNO_3 could be used instead	298
Cu, Mn, Fe, Cd, Al, Ni	human serum digest	8-hydroxyquinoline-5-sulfonic acid immobilized on active carbon-silica gel; wash with 1.0 M NaOAc/ HOAc at pH 5.6; elution with 2 M HCl	single-line manifold (Fig. 2.27) with buffer as carrier; desolvation of aerosol	299

CMDTC = bis(carboxymethyl)dithiocarbamate.
IDAEC = iminodiacetic acid-ethylcellulose

TABLE 2.9

Examples of column separation techniques that could be readily converted into reduced dispersion FI systems for ICP-MS

Analytes	Matrix	Column; wash; eluent	Suggestions	Ref
Zr, Hf	high-purity scandium oxide in 6 M HCl	Levextrel resin of 1-phenyl-3-methyl-4-benzoylpyrazol-5-one; wash with 6 M HCl; elute with 0.5 M oxalic acid	6 M HCl carrier or, to save the interface, DDW carrier and inject 6 M HCl through sample or eluent loop to remove Sc, Al, Ca	300
La, Ce, Nd, Sm, Eu, Gd, Dy, Ho, Yb, Lu	digested rocks; final solvent is 1.75 M HCl	cation-exchange Dowex 50W-X8 resin; wash with 1.75 M HCl; REEs eluted with 8 M HNO_3	1.75 M HCl carrier removes the matrix; alternately, DDW carrier and inject 1.75 M HCl through sample or eluent loop	301
Re	digested rocks, mineral samples; final solvent is 1 M HCl	anion exchange resin; elution with 8 M HCl	1 M HCl carrier; Ni, Cu and Fe are unretained and could be determined prior to elution if desired	302
REEs, Y, Sc, Hf	$LiBO_2$ fusion of geological SRMs; dissolved in 1.25 M HCl	cation-exchange AG50W-X8 resin; wash with 1.75 M HCl; elute with 6 HNO_3/5 mM oxalic acid	same comment as above	303
Fe, Cu, Hg, In, Bi, S, Pb, Ti, Zr, Ga	digested geochemical and alloy steel SRMs (at pH 2.5)	pyrocatechol violet attached to macroporous poly(vinylchloride) resin; elute with 3 M HCl/1 M HF	DDW carrier; if Cl-free eluent can be used, then the method should also be applicable to V	304
Ti	acidified seawater spiked with ^{49}Ti, equilibrated, adjusted to pH 4.0 with NaOH and NH_4OAc	8-HOQ resin; wash with 0.02 M KHP and water; elution with 2.4 M HNO_3	DDW carrier; 20-ml sample injected; KHP injected (either through sample or eluent injection loop, depending on the extent of washing required); 50-µl eluent injection	305
REEs	acidified seawater filtered and spiked with Lu	Chelex-100 (NH_4^+ form); elution with 3.5 M HNO_3	DDW carrier; 20-ml sample injected; 50-µl eluent injection	306
REEs	acidified filtered seawater	8-HOQ immobilized on PAN hollow fibre membrane; wash with DDW; elution with 5 M HCl/0.1 M HNO_3	DDW carrier; 15-ml sample injected; 50-µl eluent injection	189
Mn, Co, Cu, Zn, Cd	seawater and estuarine water SRMs	Muromac A-1; elution with 0.5 M HNO_3	1.5 M ammonium acetate buffer carrier; 100-µl sample injected; 50-µl eluent injection (even smaller scale could be used with microconcentric nebulizer)	222

Analytes	Matrix	Column; wash; eluent	Suggestions	Ref
Cd, Cu, Mn, Ni, Pb, Zn	seawater	Chelamine; wash with ammonium acetate buffer; elution with 2 M HNO_3	sample pumped through column followed by buffer solution and then eluent; column washed with DDW and reconditioned with buffer prior to next injection	307
Cu, U, Ni, Pb, Co, Cd, Yb, La, Mn	seawater, waste effluents adjusted to pH 5.4	styrene-divinylbenzene with 4-(2-thiazolylazo)resorcinol functional groups; wash with 0.01 M acetate buffer (pH 5.4) and then DDW; elution with 2 M HNO_3	the pH of the natural sample could also be adjusted on-line by mixing with buffer as described in Ref. [289] prior to passage through the column	308
Zn	digested urine and ashed milk solutions buffered with CH_3COONH_4 at pH 5.5	carboxymethylated poly(ethylene-imine)-poly(methylenepoly-phenylene)isocyanate chelating resin; elution with 2 M HNO_3	the sample could also be mixed with the buffer on-line as described in Ref. [289] prior to passage through the column	309

HOQ = hydroxyquinoline.
PAN = polyacrylonitrile

be transformed into FI manifolds by using a valve to inject the sample instead of pumping it continuously. Depending on the element, impurities in the reagents, the borohydride reagent in particular [192], may limit the detection limits attainable. The use of an electrochemical cell is advantageous in that respect since no unstable alkaline borohydride reagent is required [191]. On the other hand, the sensitivity achievable with the latter is limited by the rate of hydrogen formation in the cathodic compartment [191]. Different interferences are also observed with the latter, such as the deposition of metals on the Pt cathode [191].

Batch extraction procedures can also be converted to on-line FI approaches as illustrated in Table 2.11 with selected suggestions. Whenever the final phase is organic, desolvation is advised unless other precautions are taken (such as the use of mixed-gas plasmas) to prevent plasma extinction and/or soot deposition. Otherwise, a back-extraction into aqueous phase is required, which can readily be carried out by adding another set of segmenter, extraction coil and phase separator to the basic set-up of Fig. 2.30. However, a single extraction with the basic set-up can be performed on organic samples, as suggested in Table 2.11.

TABLE 2.10

Selected applications of reduced-dispersion vapour generation FI systems with ICP-MS

Analytes	Sample	Sample preparation	On-line preconcentration strategy	Instrument	Ref.
As, Sb	riverine water, sea-water SRMs	As and Sb pre-reduced by adding KI and ascorbic acid solution to sample and then concentrated HCl; diluted 15 min later	hydride generation using 3% HNO_3 carrier, 0.5-ml injection, 0.2% w/v $NaBH_4$ reductant and a purge gas flow rate of 1.05 l/min	ELAN 5000; FIAS-200; AS-90 auto-sampler	311
As	natural waters	reducing solution (10% m/v each thiourea, KI and ascorbic acid) added to samples to reduce As to As(III)	1-ml sample passed through electrochemical cell (2 M H_2SO_4 electrolyte being continuously recycled) or mixed on-line with 1% m/v $NaBH_4$ in 0.1 M NaOH before passing through gas diffusion chamber; external calibration	PQ2; 10 cm of 4-mm i.d. PTFE tubing inserted inside the spray chamber	191
As, Se, Sb, Hg, Bi	water and sea water SRMs	no sample preparation required	1-2 g/l $NaBH_4$ in 0.4% NaOH used to transform the analytes from 500-µl sample into volatile hydrides; for Hg, preconcentration of the Hg vapor was also carried out by amalgamation on Au/Pt net	ELAN 5000; FIAS 200; AS-90 autosampler	312
Cd	riverine, estuarine and sea-waters SRMs, sediment SRMs, urine SRMs	waters diluted 1:2 with 0.1 M HCl; microwave digestion of solid samples with HNO_3 (and HF for sediments) and dilution with 0.1 M HCl (after evaporation of HF)	100-µl sample injected into 0.1 M HCl/2% m/v thiourea/2 mg/l Co carrier and mixed on-line with 4% m/v $NaBH_4$ in 0.1 M NaOH; isotope dilution analysis required because of many interferences on vapour generation	ELAN 5000; gas-liquid separator with 120 glass beads (greater surface area)	272
As	Au	dissolution in 1:3 HNO_3/HCl; dilution with 1% v/v HCl; precipitation of Au by addition of 10% m/v l-ascorbic acid, 10% m/v KI and HCl; filtration	500-µl sample injected in 10% HCl carrier and mixed on-line with $NaBH_4$ in 0.05% m/m NaOH; Au quantitatively removed without co-precipitation of As; external calibration	ELAN 5000; FIAS-200; 1 m of 1.24-mm i.d. Teflon tubing between separator and ICP torch	226
As, Sb, Se	waters	acidification with HCl to 5 M HCl; pre-reduction of Se by heating in water-bath at 80-85°C for 90 min	200-µl sample mixed on-line with 0.3 m/v thiourea (to reduce As and Sb) and, after 180 s reaction, with 1.2% $NaBH_4$; simultaneous determination of Se, As and Sb; method of standard additions	ELAN 5000; FIAS-400	313

TABLE 2.11

Selected extraction techniques that could be readily converted into reduced dispersion FI systems for ICP-MS

Analytes	Matrix	Reagent and final phase	Comments	Ref
P	brines treated with $HClO_4$ to convert P into orthophosphate	extraction of P as phospho-antimonylmolybdate into IBMK	indirect determination of P by monitoring the Mo signal; extraction may suffer from interferences by As, V and Si	314
Ni	biological SRMs digested and adjusted to pH 3.6 using glycine-HCl buffer, 1 M NaOH and 1 M $NaClO_4$	extraction with 1,5-bis(di-2-pyridylmethylene)thiocarbonohydrazide (DPTH) in butan-1-ol	best recoveries off-line with an aqueous/organic phase ratio of 8; numerous foreign species could be tolerated	315
Cd	biological SRMs digested and adjusted to pH 5.6 with CH_3COOH-acetate buffer	extraction with 1,5-bis(di-2-pyridylmethyl-ene)thiocarbonohydrazide into IBMK	volume ratio of water to IBMK of 30 gave best results off-line	316
Ni	biological SRMs, human urine	extraction at pH 8.8 with 1,5-bis[phenyl(2-pyridyl)methylene]thiocarbonohydrazide (BPTH) into IBMK	BPTH synthesis procedure outlined; best off-line results with volume ratio of aqueous to organic phase ≤ 37	317
Zn	digests of biological materials adjusted to pH 2.5-3 with HCl or ammonia	sample segmented with 0.05% w/v DDDC in CCl_4; after phase separation, organic phase segmented with 1.2 M HNO_3; aqueous phase analyzed	DDW wash of organic phase prior to back-extraction may not be necessary on-line; if final sample is too concentrated, it can easily be merged with DDW to dilute it	318
Pb, Ni	crude oils	sample mixed with IBMK and extracted into 40% v/v HNO_3	output from phase separator merged with DDW to decrease HNO_3 concentration to <5%	319
Pb	gasoline	sample mixed with 3% m/v iodine in toluene and extracted into 10% v/m HNO_3	output from phase separator may be merged with internal standard solution if desired	320
Ga	high-purity Al dissolved in 5 N HNO_3, dried and dissolved in 7 N HCl	sample segmented 1:1 with isopropyl ether; after phase separation, organic phase segmented 2:1 with DDW; aqueous phase analyzed	proposed flow rate ratios based on off-line procedure	321

[a]Isobutyl methyl ketone.

TABLE 2.12

Selected precipitation techniques that could be readily converted into reduced dispersion FI systems for ICP-MS

Analytes	Matrix	Suggested procedure	Comments	Ref
Zr, Nb, Hf, Ta	$LiBO_2$ melt of geological materials dissolved in HCl/HF	sample mixed with 6% cupferron solution; precipitate dissolved with 16 M HNO_3/30% H_2O_2	place reaction coil in cold water to speed precipitation; several injections of dissolution solution (or another one) may be required; merge resulting solution with DDW if required to decrease analyte and acid concentration	322
As, Co, Cu, Fe, Mn, Ni, Ti, V, Zn, Zr	digested high-purity Cr and chromium disilicide; Cr(III) oxidized to dichromate(VI) with $HCl/HClO_4$	coprecipitate trace elements with $La(OH)_3$ at pH 11.0; dissolve precipitate for analysis	Cr(VI) does not precipitate with $La(OH)_3$; better Al recovery at pH 10.0, but some Ni is lost	323
Mo	seawater SRM and water	coprecipitation with cobalt-dithiocarbamate at pH 4; dissolve precipitate in diluted HNO_3 (1:2)	selective precipitation of dithiocarbamate salts of Mo and other trace elements could probably be carried out	324

A few examples of precipitation techniques which could be converted to on-line FI methods are given in Table 2.12. As already mentioned in Section 2.6.3-4, co-precipitations can be implemented but selective precipitations provide more advantages. A number of organic precipitants are available for the selective precipitation of various elements [181].

Finally, FI-ASV-ICP-MS has been applied to the monoelemental determination of Cr(VI) and V(V) in water [217], V(V) [217], Cu and Cd [216] in 10-fold diluted urine and U(VI) in seawater [325]. In general, the method of standard additions was required to compensate the variations in recoveries which likely arose from the adsorption of organic components on the carbon electrode [217], unless a Hg-coated carbon electrode was used [216]. A total determination of Cr required prior oxidation of Cr(III) to Cr (VI), which could not be achieved in urine using ammonium persulfate [217].

2.7 OTHER POSSIBILITIES

So far, all the set-ups described in the previous sections required samples in solution, i.e. solid samples first had to be dissolved. However, on-line dissolution can also be performed. For instance, on-line electrolytic dissolution of Al alloys was carried out by placing the sample in an electrolytic flow cell and using 1.0 M HNO_3 as carrier/electrolyte in ICP-AES [326]. This approach, however, can hardly be automated since the flow cell must be opened for each new sample which must be properly positioned inside [326].

A more attractive alternative is the on-line microwave digestion of slurries, which only requires grinding and suspension of the sample particles in an appropriate dispersing agent. Although it has not yet been carried out on-line with ICP-AES or ICP-MS, this approach has great potential since it would truly minimize sample manipulation, hence further decreasing potential sources of contamination. So far, the sample slurry was pumped, along with concentrated acid, into a glass reactor installed in a microwave oven [327]. The flow was then stopped for an appropriate length of time to allow total digestion of the sample. After cooling and venting, the digest was collected into a calibrated flask and then diluted [327]. Continuous monitoring of the system pressure allowed appropriate changes to the microwave energy and prevented over-pressurization of the manifold [327].

The direct coupling of this microwave digestion manifold with a dilution one (i.e. to either a medium or large dispersion FI manifold, depending on the dilution factor required) should be readily possible. Instead of diluting the digest, the microwave digestion manifold could also be directly combined with a separation manifold such as hydride generation, as was recently carried out on-line with FAAS [328]. Thus, although this approach is still in its infancy, it will likely become widespread, especially when commercially available systems are introduced.

Finally, there is a clear trend towards miniaturization of instrumentation and techniques [180]. This includes the development of integrated μFI systems on a single monolithic device [329]. It is therefore possible that, some day, very compact and even portable FI-ICP-MS instruments will become available. Nevertheless, whether a 'lab-on-a-valve' concept is pursued [330] or more conventional FI set-ups, the possibilities are limitless since not only are the individual approaches described in this chapter powerful at addressing specific problems, but they can also be combined for more selectivity, better detection limits, etc. Furthermore, these simple systems can readily provide speciation information, a feature that will be discussed in detail in the forthcoming book on ICP-MS speciation methods.

ACKNOWLEDGEMENT

D.B. is grateful to the Natural Sciences Engineering Research Council for financial assitance.

REFERENCES

1 J. Ruzicka and E.H. Hansen, Flow injection analyses Part I. A new concept of fast continuous flow analysis, *Anal. Chim. Acta*, 78 (1975) 145.
2 G. Horvai and E. Pungor, Theoretical backgrounds of flow analysis, *CRC Crit. Rev. Anal. Chem.*, 17 (1987) 231.
3 J. Ruzicka and E.H. Hansen, *Flow Injection Analysis*, John Wiley & Sons Inc., New York, 1981.
4 C.W. McLeod, Flow injection techniques in inductively coupled plasma spectrometry, *J. Anal. At. Spectrom.*, 2 (1987) 549.
5 J. Ruzicka and E.H. Hansen, Flow injection analyses Part X. Theory, Techniques and Trends, *Anal. Chim. Acta*, 99 (1978) 37.
6 Z. Fang, *Flow Injection Separation and Preconcentration*, VCH Publishers, Inc., New York, 1993.
7 J.L. Burguera (Ed), *Flow Injection Atomic Spectroscopy*, Marcel Dekker, Inc., New York, 1989.
8 W.E. van der Linden, In J.L. Burguera (Ed), *Flow Injection Atomic Spectroscopy*, Marcel Dekker, Inc., New York, 1989.
9 M. Trojanowicz, *Flow Injection Analysis Instrumentation and Applications*, World Scientific Publishing Co. Inc., New Jersey, 2000.
10 J.F. Tyson, Flow injection atomic spectrometry, *Spectrochim. Acta. Rev.*, 14 (1991) 169.
11 J.F. Tyson, Flow injection analysis techniques for atomic-absorption spectrometry: a review, *Analyst*, 110 (1985) 419.
12 R.F. Browner and A.W. Boorn, Sample introduction techniques for atomic spectroscopy, *Anal. Chem.*, 56 (1984) 875A.
13 G.D. Christian and J. Ruzicka, Flow injection analysis: a novel tool for plasma spectroscopy, *Spectrochim. Acta*, 42B (1987) 157.
14 J. Ruzicka, Flow injection analysis—A survey of its potential as solution handling and data gathering technique in chemical research and industry, *Fresenius Z. Anal. Chem.*, 329 (1988) 653.
15 E.H. Hansen, Exploitation of gradient techniques in flow injection

analysis, *Fresenius Z. Anal. Chem.*, 329 (1988) 656.
16 B. Karlberg, Flow injection extraction in theory and practice, *Fresenius Z. Anal. Chem.*, 329 (1988) 660.
17 J.F. Tyson, Flow injection calibration techniques, *Fresenius Z. Anal. Chem.*, 329 (1988) 663.
18 W. Frenzel, Potential of modified reverse flow injection analysis for continuous monitoring and process control, *Fresenius Z. Anal. Chem.*, 329 (1988) 668.
19 J.F. Tyson, Atomic spectrometry detectors for flow-injection analysis, *Anal. Chim. Acta*, 234 (1990) 3.
20 J. Ruzicka and G.D. Christian, Flow Injection Analysis and Chromatography: Twins or siblings?, *Analyst*, 115 (1990) 475.
21 M. Valcárcel and M.D. Luque de Castro, Flow injection analysis: A useful alternative for solving analytical problems, *Fresenius J. Anal. Chem.*, 337 (1990), 2877.
22 D. Beauchemin, Inductively coupled plasma mass spectrometry in hyphenation: a multielemental analysis technique with almost unlimited potential, *Trends Anal. Chem.*, 10 (1991) 71.
23 E.R. Denoyer and A. Stroh, Expanding ICP-MS capabilities using flow injection, *Am. Lab.* 24(3) (1992) 74.
24 J. Tyson, Flow-injection techniques for analytical atomic spectrometry, *Spectroscopy*, 7 (1992) 14.
25 M.F. Giné, B.F. Reis, F.J. Krug, A.O. Jacintho, H. Bergamin Filho and E.A.G. Zagatto, Flow injection plasma spectrometry, *ICP Inf. Newslett.*, 20 (1994) 413.
26 J.L. Burguera and M. Burguera, Liquid sample introduction devices in flow injection atomic spectroscopy, *J. Anal. At. Spectrom.*, 10 (1995) 473.
27 M.D. Luque de Castro and M.T. Tena, Hyphenated flow injection systems and high discrimination instruments, *Talanta*, 42 (1995) 151.
28 J.F. Tyson, Flow injection techniques in atomic spectroscopy, *Anal. Proc.*, 29 (1992) 436.
29 Z. Fang, S. Xu and G. Tao, Developments and trends in flow injection atomic absorption spectrometry, *J. Anal. At. Spectrom.*, 11 (1996) 1.
30 J. Atienza, M. A. Herrero, A. Maquieira and R. Puchades, Flow injection analysis of seawater. Part II. Cationic species, *CRC Crit. Rev. Anal. Chem.*, 23 (1992) 1.
31 K.N. Andrew, N.J. Blundell, D. Price and P.J. Worsfold, Flow injection techniques for water monitoring, *Anal. Chem.*, 66 (1994) 917A.
32 E.H. Hansen, Flow-injection analysis: leaving its teen-years and maturing. A personal reminiscence of its conception and early development, *Anal. Chim. Acta*, 308 (1995) 3.
33 W. Slavin, A Comparison of Atomic Spectroscopic Analytical Techniques, *Spectroscopy*, 6 (1991) 16.

34 J.M. Craig and D. Beauchemin, Characterization of inductively coupled plasma mass spectrometry with segmented-flow injection, *Analyst*, 119 (1994) 1677.
35 A.G. Fogg and A.E. Davies, Reverse flow injection or reagent injection? Normal flow injection, reverse flow injection and single-boundary measurements, *Analyst*, 117 (1992) 1055.
36 D.C. Stone and J.F. Tyson, Models for dispersion in flow injection analysis, Part 1. Basic requirements and study of factors affecting dispersion, *Analyst*, 112 (1987) 515.
37 H.C. Smit and P.J.H. Scheeren, Characterization of responses to injection of microliter samples and to peak-shaped input signals for inductively-coupled plasma/atomic emission spectrometry, *Anal. Chim. Acta*, **215** (1988) 143.
38 D.M. Spence and S.R. Crouch, Factors affecting zone variance in a capillary flow injection system, *Anal. Chem.*, 69 (1997) 165.
39 D.C. Stone and J.F. Tyson, Models for dispersion in flow injection analysis, Part 2. Two tanks in parallel model, *Analyst*, 114 (1989) 1453.
40 J.R. Dean, L. Ebdon, H.M. Crews and R.C. Massey, Characteristics of flow injection inductively coupled plasma mass spectrometry for trace metal analysis, *J. Anal. At. Spectrom.*, 3 (1988) 349.
41 J.L. Manzoori, A. Miyazaki and H. Tao, Rapid differential flow injection of phosphorus compounds in wastewater by sequential spectrophotometry and inductively coupled plasma atomic emission spectrometry using a vacuum ultraviolet emission line, *Analyst*, 115 (1990) 1055.
42 Y. Li and H. Ma, Two trends of sample dispersion variation with carrier flow rate in a single flow-injection manifold, *Talanta*, 42 (1995) 2033.
43 Y. Israel and R.M. Barnes, Flow injection dispersion characteristics with inductively coupled plasma atomic emission spectrometry. Part 1. Methodology, *Analyst*, 119 (1994) 1011.
44 A.O. Jacintho, E.A.G. Zagatto, H. Bergamin F., F.J. Krug, B.F. Reis, R.E. Bruns, B.R. Kowalski, Flow injection systems with inductively-coupled plasma atomic emission spectrometry, Part 1. Fundamental considerations, *Anal. Chim. Acta*, 130 (1981) 243.
45 M.W. Brown and J. Ruzicka, Parameters affecting sensitivity and precision in the combination of flow injection analysis with flame atomic-absorption spectrophotometry, *Analyst*, 109 (1984) 1091.
46 J.F. Tyson, C.E. Adeeyinwo, J.M.H. Appleton, S.R. Bysouth, A.B. Idris and L.L. Sarkissian, Flow injection techniques of method development for flame atomic-absorption spectrometry, *Analyst*, 110 (1985) 487.
47 J.F. Tyson and A.B. Idris, Flow injection sample introduction for atomic-absorption spectrometry: applications of a simplified model for dispersion, *Analyst*, 106 (1981) 1125.
48 H. Klinkenberg, T. Beeren and W. Van Borm, Multielement analysis using

flow injection inductively coupled plasma mass spectrometry: analytical aspects of multielement determinations in highly concentrated solutions of phosphoric acid, sodium phosphate and sodium nitrate, *Spectrochim. Acta*, 49B (1994) 171.
49 C.M. Selavka, K.-S. Jiao and I.S. Krull, Construction and comparison of open tubular reactors for postcolumn reaction detection in liquid chromatography, *Anal. Chem.*, 59 (1987) 2221.
50 Y. Israel and R.M. Barnes, Flow injection dispersion characteristics with inductively coupled plasma atomic emission spectrometry. 2. Influence of flow modes and manifold configuration, *Anal. Chem.*, 66 (1994) 3937.
51 K.W. Simonsen, B. Nielsen, A. Jensen and J.R. Andersen, Direct microcomputer controlled determination of zinc in human serum by flow injection atomic absorption spectrometry, *J. Anal. At. Spectrom.*, 1 (1986) 453.
52 D.A. Weeks and K.S. Johnson, Solenoid pumps for flow injection analysis, *Anal. Chem.*, 68 (1996) 2717.
53 Z. Fang, B. Welz and M. Sperling, Expansion of dynamic range of flame atomic absorption spectrometry by an efficient flow injection dilution system based on dispersion of microliter-volume samples, *Anal. Chem.*, 65 (1993) 1682.
54 F.G. Bessoth, A.J. deMello and A. Manz, Microstructure for efficient continuous flow mixing, *Anal. Commun.*, 36 (1999) 213.
55 K.K. Falkner and J.M. Edmond, Determination of gold at femtomolar levels in natural waters by flow injection inductively coupled plasma quadrupole mass spectrometry, *Anal. Chem.*, 62 (1990) 1477.
56 M. Wu, R. Ensman and G.M. Hieftje, Simple, inexpensive system for controlled-dispersion flow analysis in inductively coupled plasma spectrometry, *Appl. Spectrosc.*, 48 (1994) 448.
57 A. Ivaska and J. Ruzicka, From flow injection to sequential injection: comparison of methodologies and selection of liquid drives, *Analyst*, 118 (1993) 885.
58 J. Ruzicka and L. Scampavia, From flow injection to bead injection, *Anal. Chem.*, 71 (1999) 257A.
59 R. Ruzicka, Flow injection-renewable surface techniques, *Anal. Chim. Acta*, 308 (1995) 14.
60 M. Wu, Y. Madrid, J.A. Auxier and G.M. Hieftje, New spray chamber for use in flow-injection plasma emission spectrometry, *Anal. Chim. Acta*, 286 (1994) 155.
61 T.D. Hettipathirana and D.E. Davey, Analytical performance in flow injection-simultaneous multielement-inductively coupled plasma-optical emission spectrometry employing a cyclonic spray chamber, *Appl. Spectrosc.*, 50 (1996) 1015.
62 J. Liu, B. Huang and X. Zeng, Donut-shaped spray chamber for inductively coupled plasma spectrometry, *Spectrochim. Acta*, 53B (1998) 1469.

63 M. Parent, H. Vanhoe, L. Moens and R. Dams, Evaluation of a flow injection system combined with an inductively coupled plasma mass spectrometer with thermospray nebulization for the determination of trace levels of platinum, *Anal. Chim. Acta*, 320 (1996) 1.
64 R.S. Houk, S.C.K. Shum, D.R. Wiederin, Frontiers in elemental analysis by mass spectrometry, *Anal. Chim. Acta*, 250 (1991) 61.
65 S.A. Pergantis, E.M. Heithmar and T.A. Hinners, Microscale flow injection and microbore high-performance liquid chromatography coupled with inductively coupled plasma mass spectrometry via a high-efficiency nebulizer, *Anal. Chem.*, 67 (1995) 4530.
66 S. Wankarn and S.A. Pergantis, Determination of arsenic in organic solvents and wines using microscale flow injection inductively coupled plasma mass spectrometry, *J. Anal. At. Spectrom.*, 14 (1999) 657.
67 N. Fitzgerald, J.F. Tyson and D.A. Leighty, Reduction of water loading effects in inductively coupled plasma mass spectrometry by a Nafion membrane dryer device, *J. Anal. At. Spectrom.*, 13 (1998) 13.
68 E.R. Denoyer, Q. Lu, A. Stroh and P. Jordan, Rapid microsampling with flow injection ICP–mass spectrometry, *FIAS Application Report*, Perkin Elmer SCIEX, Norwalk, 1993.
69 P. Richner, Detection limits versus matrix effects: Analysis of solutions with high amounts of dissolved solids by flow injection inductively coupled plasma mass spectrometry, *J. Anal. At. Spectrom.*, 8 (1993) 927.
70 E.R. Denoyer, An evaluation of spectral integration in ICP-MS, *At. Spectrosc.*, 13 (1992) 93.
71 E.R. Denoyer, Optimization of transient signal measurements in ICP-MS, *At. Spectrosc.*, 15 (1994) 7.
72 F. Vanhaecke, L. Moens, R. Dams, L. Allen and S. Georgitis, Evaluation of the isotope ratio performance of an axial time-of-flight ICP mass spectrometer, *Anal. Chem.*, 71 (1999) 3297.
73 K. Doerffel, J. Dost and H. Müller, Evaluation of noisy FIAS using the cross-covariance function, *Fresenius J. Anal. Chem.*, 348 (1994) 255.
74 O. Lee, A.P. Wade and G.A. Dumont, Generalized Fourier smoothing of flow injection analysis data, *Anal. Chem.*, 66 (1994) 4507.
75 J.A. Roussin, J. Li and H.L. Pardue, Evaluation of a predictive steady-state flow-injection method adapted to an open flow tube with a tracer, *Anal. Chim. Acta*, 303 (1995) 199.
76 A.J. Ambrose, L. Ebdon, M.E. Foulkes and P. Jones, Direct atomic spectrometric analysis by slurry atomisation. Part 8. Flow injection inductively coupled plasma atomic emission spectrometry, *J. Anal. At. Spectrom.*, 4 (1989) 219.
77 T.J. Brotherton, P.E. Pfannerstill, J.T. Creed, D.T. Heitkemper, J.A. Caruso and S.E. Pratsinis, Evaluation of three low-volume interfaces for organic solvent introduction to the inductively coupled plasma— applica-

tions to flow injection, *J. Anal. At. Spectrom.*, 4 (1989) 341.
78 S. Vijayalakshmi, R.K. Prabhu, T.R. Mahalingam and C.K. Mathews, Determination of trace metals in uranium oxide by inductively coupled plasma mass spectrometry combined with on-line solvent extraction, *J. Anal. At. Spectrom.*, 7 (1992) 565.
79 H.M. Al-Swaidan, N. Lacy and G.D. Christian, Flow Injection analysis–inductively coupled plasma/mass spectrometry: study of interelement effects, *Anal. Lett.*, 22 (1989) 2653.
80 J. Wang, W.-L. Shen, B.S. Sheppard, E.H. Evans, J.A. Caruso and F.L. Fricke, Effect of ion-lens tuning and flow injection on non-spectroscopic matrix interferences in inductively coupled plasma mass spectrometry, *J. Anal. At. Spectrom.*, 5 (1990) 445.
81 A.G. Coedo and M.T. Dorado, Use of flow injection inductively coupled plasma–mass spectrometry for the determination of niobium, vanadium, and titanium in microalloyed steels, *Appl. Spectrosc.*, 49 (1995) 115.
82 S.J. Hill, J. Hartley and L. Ebdon, Determination of impurities in organometallic compounds dissolved in diethyl ether by flow injection inductively coupled plasma mass spectrometry, *J. Anal. At. Spectrom.*, 7 (1992) 895.
83 A. Matousek de Abel de la Cruz, J.L. Burguera, M. Burguera and C. Rivas, Factorial design for the response exploration of a flow injection system, *Talanta*, 42 (1995) 701.
84 A.P. Wade, P.M. Shiundu and P.D. Wentzell, Automated exploration and exploitation of flow-injection response surfaces, *Anal. Chim. Acta*, 237 (1990) 361.
85 A. Araújo, J.L.F.C. Lima, J. Gracia, M. Poch, J. Alonso, J. Bartrolí and M. del Valle, An integrated design strategy for flow-injection analysis based on the coupling of mathematical modelling and optimization algorithms, *Anal. Chim. Acta*, 310 (1995) 289.
86 V.L. Dressler, D. Pozebon and A.J. Curtius, Introduction of alcohols in inductively coupled plasma mass spectrometry by a flow injection system, *Anal. Chim. Acta*, 379 (1999) 175.
87 S.J. Hill, J. Hartley and L. Ebdon, Determination of trace metals in volatile organic solvents using inductively coupled plasma atomic emission spectrometry and inductively coupled plasma mass spectrometry, *J. Anal. At. Spectrom.*, 7 (1992) 23.
88 P. Bruckner, U. Vollkopf and E. Denoyer, The Elan 40-MHz, free-running RF generator: state-of-the-art engineering for ICP-MS, Technical summary TSMS-20.
89 D. Beauchemin, K.W.M. Siu and S.S. Berman, Determination of organomercury in biological reference materials by inductively coupled plasma mass spectrometry using flow injection analysis, *Anal. Chem.*, 60 (1988) 2587.
90 P. van de Weijer, P.J.M.G. Vullings, W.L.M. Baeten and W.J.M. de Laat,

Determination of uranium and thorium in aluminium with flow injection and laser ablation inductively coupled plasma mass spectrometry, *J. Anal. At. Spectrom.*, 6 (1991) 609.
91 S.T.G. Anderson, M.J.C. Taylor, S.J.S. Williams, Application of flow injection analysis to ICP-MS, In: G. Holland and A.N. Eaton (Eds), *Applications of Plasma Source Mass Spectrometry* II, Special Publication No. 124 of the Royal Society of Chemistry, Cambridge, 1993, pp. 72–82.
92 K.E. Lawrence, G.W. Rice and V.A. Fassel, Direct liquid sample introduction for flow injection analysis and liquid chromatography with inductively coupled argon plasma spectrometric detection, *Anal. Chem.*, 56 (1984) 289.
93 K.E. LaFreniere, G.W. Rice and V.A. Fassel, Flow injection analysis with inductively coupled plasma-atomic emission spectroscopy: critical comparison of conventional pneumatic, ultrasonic and direct injection nebulization, *Spectrochim. Acta*, 40B (1985) 1495.
94 J.J. Thompson and R.S. Houk, Inductively coupled plasma mass spectrometric detection for multielement flow injection analysis and elemental speciation by reversed-phase liquid chromatography, *Anal. Chem.*, 58 (1986) 2541.
95 M.J. Powell, E.S.K. Quan, D.W. Boomer and D.R. Wiederin, Inductively coupled plasma mass spectrometry with direct injection nebulization for mercury analysis of drinking water, *Anal. Chem.*, 64 (1992) 2253.
96 J. Wang, E.H. Evans and J.A. Caruso, Minimization of non-spectroscopic matrix interferences for the determination of trace elements in fusion samples by flow injection inductively coupled plasma mass spectrometry, *J. Anal. At. Spectrom.*, 6 (1991) 605.
97 A.G. Coedo and M.T. Dorado, Application of flow injection inductively coupled plasma mass spectrometry to the simultaneous determination of arsenic, antimony, tin, bismuth, selenium and tellurium in steels, *J. Anal. At. Spectrom.*, 9 (1994) 1111.
98 R.C. Hutton and A.N. Eaton, Analysis of solutions containing high levels of dissolved solids by inductively coupled plasma mass spectrometry, *J. Anal. At. Spectrom.*, 3 (1988) 547.
99 A.N. Eaton, R.C. Hutton and J.G. Holland, Application of flow injection sample introduction to inductively coupled plasma-mass spectrometry for geochemical analysis, *Chem. Geol.*, 95 (1992) 63.
100 A. Stroh, U. Voellkopf and E.R. Denoyer, Analysis of samples containing large amounts of dissolved solids using microsampling flow injection inductively coupled plasma mass spectrometry, *J. Anal. At. Spectrom.*, 7 (1992) 1201.
101 I.B. Brenner, M. Liezers, J. Godfrey, S. Nelms and J. Cantle, Analytical characteristics of a high efficiency ion transmission interface (S mode) inductively coupled plasma mass spectrometer for trace element deter-

minations in geological and environmental materials, *Spectrochim. Acta*, 54B (1999) 991.
102 A.G. Coedo, T. Dorado, B.J. Fernandez and F.J. Alguacil, Isotope dilution analysis for flow injection ICPMS determination of microgram per gram levels of boron in iron and steel after matrix removal, *Anal. Chem.*, 68 (1996) 991.
103 A. Stroh and U. Völlkopf, Analysis of difficult samples by flow injection inductively coupled plasma mass spectrometry, *Anal. Proc.*, 29 (1992) 274.
104 J. Yoshinaga, A. Nakama and K. Takata, Determination of total tin in sediment reference materials by isotope dilution inductively coupled plasma mass spectrometry after alkali fusion, *Analyst*, 124 (1999) 257.
105 Z. Peng, H. Klinkenberg, T. Beeren and W. Van Borm, Determination of germanium, palladium and platinum in highly concentrated solutions of phosphoric acid and ammonium nitrate by flow injection inductively coupled plasma mass spectrometry, *Spectrochim. Acta*, 46B (1991) 1051.
106 A. Makishima, E. Nakamura and T. Nakano, Determination of boron in silicate samples by direct aspiration of sample HF solutions into ICPMS, *Anal. Chem.*, 69 (1997) 3754.
107 A.T. Persaud, D. Beauchemin, H.E. Jamieson and R.J.C. McLean, Partial leaching as an aid to slurry nebulization for the analysis of soils by ICP-MS with flow injection and mixed-gas plasmas, *Can. J. Chem.*, 77 (1999) 409.
108 J.G. Williams, A.L. Gray, P. Norman and L. Ebdon, Feasibility of solid sample introduction by slurry nebulisation for inductively coupled plasma mass spectrometry, *J. Anal. At. Spectrom.*, 2 (1987) 469.
109 M. Hulsman, M. Bos, W.E. van der Linden, Dispersion behaviour of solid particles in flow-injection analysis, *Anal. Chim. Acta*, 346 (1997) 351.
110 J.S. Becker, R.S. Soman, K.L. Sutton, J.A. Caruso and H.-J. Dietze, Determination of long-lived radionuclides by inductively coupled plasma quadrupole mass spectrometry using different nebulizers, *J. Anal. At. Spectrom.*, 14 (1999) 933.
111 V.L. Dressler, D. Pozebon and A.J. Curtius, Determination of heavy metals by inductively coupled plasma mass spectrometry after on-line separation and preconcentration, *Spectrochim. Acta*, 53B (1998) 1527.
112 R.S. Houk and J.J. Thompson, Trace Metal Isotopic Analysis of Microliter Solution Volumes by Inductively Coupled Plasma Mass Spectrometry, *Biomed. Mass Spectrom.*, 10 (1983) 107.
113 B.F. Reis, M.A.Z. Arruda, E.A.G. Ziggatto and J.R. Ferreira, An improved monosegmented continuous-flow system for sample introduction in flame atomic spectrometry, *Anal. Chim. Acta*, 206 (1988) 253.
114 L.-C. Tian, X.-P. Sun, Y.-Y. Xu and Z.-L. Zhi, Segmental flow-injection analysis: device and applications, *Anal. Chim. Acta*, 238 (1990) 183.
115 T. Buch-Rasmussen, Determination of D-glucose in undiluted whole blood using chemically modified electrode and segmented sample injection in a

flow system, *Anal. Chim. Acta*, 237 (1990) 405.
116 T. Buch-Rasmussen, Flow system for direct determination of enzyme substrate in undiluted whole blood, *Anal. Chem.*, 62 (1990) 933.
117 A.S. Attiyat and G.D. Christian, Discrete microsample injection into a gaseous carrier, *Talanta*, 31 (1984) 463.
118 D. Beauchemin, Preliminary characterization of inductively coupled plasma mass spectrometry with flow injection into a gaseous (air) carrier, *Analyst*, 118 (1993) 815.
119 J.M. Craig and D. Beauchemin, Univariate optimization of segmented-flow injection for inductively coupled plasma mass spectrometry, *J. Anal. At. Spectrom*, 9 (1994) 1341.
120 V. Stefanova, V. Kmetov and L. Futekov, Air segmented discrete introduction in inductively coupled plasma mass spectrometry, *J. Anal. At. Spectrom.*, 12 (1997) 1271.
121 A.G. Coedo, M.T. Dorado, I. Padilla and F. Alguacil, Preconcentration and matrix separation of precious metals in geological and related materials using metalfix-chelamine resin prior to inductively coupled plasma mass spectrometry, *Anal. Chim. Acta*, 340 (1997) 31.
122 A. Rios, M.D. Luque de Castro and M. Valcarcel, Sandwich standardization in flow-injection analysis, *Talanta*, 36 (1989) 612.
123 Y. Hsieh and S.R. Crouch, Air-segmented flow injection: a hybrid technique for automated, low dispersion determinations, *Anal. Chim. Acta*, 303 (1995) 231.
124 Y. Hsieh and S.R. Crouch, Flow reversal and flow recycle air-segmented flow injection for simultaneous determination of a binary mixture, *Anal. Chim. Acta*, 304 (1995) 333.
125 I.A. Rigby and S.C. Stephen, Determination of antimony and other impurities in high purity copper by ICP-MS, In G. Holland and A.N. Eaton (Eds.), *Applications of Plasma Source Mass Spectrometry II*, The Royal Society of Chemistry, Special Publication No. 124, Cambridge, 1993, pp. 158–164.
126 D.R. Wiederin, R.E. Smyczek and R.S. Houk, On-line standard additions with direct injection nebulization for inductively coupled plasma mass spectrometry, *Anal. Chem.*, 63 (1991) 1626.
127 D. Beauchemin, On-line standard addition method with ICPMS using flow injection, *Anal. Chem.*, 67 (1995) 1553.
128 J.F. Tyson and A.B. Idris, Determination of chromium in steel by flame atomic-absorption spectrometry using a flow injection standard additions method, *Analyst*, 109 (1984) 23.
129 Z. Fang, J.M. Harris, J. Ruzicka and E.H. Hansen, Simultaneous flame photometric determination of lithium, sodium, potassium, and calcium by flow injection analysis with gradient scanning standard addition, *Anal. Chem.*, 57 (1985) 1457.

130 E.A.G. Zagatto, A.O. Jacintho, F.J. Krug, B.F. Reis, R.E. Bruns and M.C.U. Araújo, Flow injection systems with inductively-coupled argon plasma atomic emission spectrometry, Part 2. The generalized standard addition method, *Anal. Chim. Acta*, 145 (1983) 169.
131 L. Rottmann and K.G. Heumann, Development of an on-line isotope dilution technique with HPLC/ICP-MS for the accurate determination of elemental species, *Fresenius J. Anal. Chem.*, 350 (1994) 221.
132 A. Lásztity, M. Viczián, X. Wang and R.M. Barnes, Sample analysis by on-line isotope dilution inductively coupled plasma mass spectrometry, *J. Anal. At. Spectrom.*, 4 (1989) 761.
133 M. Viczián, A. Lásztity and R.M. Barnes, Identification of potential environmental sources of childhood lead poisoning by inductively coupled plasma mass spectrometry. Verification and case studies, *J. Anal. At. Spectrom.*, 5 (1990) 293.
134 L.C. Alves, L.A. Allen and R.S. Houk, Measurement of vanadium, nickel, and arsenic in seawater and urine reference materials by inductively coupled plasma mass spectrometry with cryogenic desolvation, *Anal. Chem.*, 65 (1993) 2468.
135 A.E. Eroglu, C.W. McLeod. L.S. Leonard and D. McCubbin, Determination of plutonium in seawater using co-precipitation and inductively coupled plasma mass spectrometry with ultrasonic nebulization, *Spectrochim. Acta*, 53B (1998) 1221.
136 D.C. Colodner, E.A. Boyle and J.M. Edmond, Determination of rhenium and platinum in natural waters and sediments, and iridium in sediments by flow injection isotope dilution inductively coupled plasma mass spectrometry, *Anal. Chem.*, 65 (1993) 1419.
137 Application of flow injection techniques with the VG PlasmaQuad 1. Determination of ng/g levels of uranium in NBS 1577A bovine liver, Technical Information PQ 704 (1987).
138 D. Beauchemin and R. Kisilevsky, A method based on ICP-MS for the analysis of Alzheimer's amyloid plaques, *Anal. Chem.*, 70 (1998) 1026.
139 K.L. Ackley, J.A. Day, K.L. Sutton and J.A. Caruso, Investigation of problems associated with the determination of iodine in glacial acetic acid samples using flow injection analysis-inductively coupled plasma mass spectrometry, *Anal. Commun.*, 36 (1999) 295.
140 H.M. Al-Swaidan, The determination of lead, nickel and vanadium in Saudi Arabian crude oil by sequential injection analysis/inductively-coupled plasma mass spectrometry, *Talanta*, 43 (1996) 1313.
141 J. Fucsko, S.S. Tan and M.K. Balazs, Measurement of trace metallic contaminants on silicon wafer surfaces in native and dielectric silicon oxides by vapor phase decomposition flow injection inductively coupled plasma-mass spectrometry, *J. Electrochem. Soc.*, 140 (1993) 1105.
142 J.K. Aggarwal, M.B. Shabani, M.R. Palmer and K.V. Ragnarsdottir,

Determination of the rare earth elements in aqueous samples at sub-ppt levels by inductively coupled plasma mass spectrometry and flow injection ICPMS, *Anal. Chem.*, 68 (1996) 4418.
143 J. Godfrey and E. McCurdy, Investigation into the feasibility of ICP-MS as an alternative to fire assay measurements for gold and the platinum group elements, in G. Holland and A.N. Eaton (Eds.), *Applications of Plasma Source Mass Spectrometry II*, The Royal Society of Chemistry, Special Publication No. 124, Cambridge, 1993, pp. 64–71.
144 M. Valcárcel and M. Gallego in J.L. Burguera (Ed), *Flow Injection Atomic Spectroscopy*, Marcel Dekker, Inc., New York, 1989.
145 A.T. Haj-Hussein, G.D. Christian and J. Ruzicka, Determination of cyanide by atomic absorption using a flow injection conversion method, *Anal. Chem.*, 58 (1986) 38.
146 F. Giné, H. Bergamin F., B.F. Reis and R.L. Tuon, Intelligent flow-injection-inductively coupled plasma system for matrix matching, *Anal. Chim. Acta*, 234 (1990) 207.
147 G.H. Vickers, B.S. Ross and G.M. Hieftje, Reduction of mass-dependent interferences in inductively coupled plasma–mass spectrometry by using flow-injection analysis, *Appl. Spectrosc.*, 43 (1989) 1330.
148 Y. Israel, A. Lásztity and R.M. Barnes, On-line dilution, steady-state concentrations for inductively coupled plasma atomic emission and mass spectrometry achieved by tandem injection and merging-stream flow injection, *Analyst*, 114 (1989) 1259.
149 K.S. Huang and S.J. Jiang, Determination of trace levels of metal ions in water samples by inductively coupled plasma mass spectrometry after on-line preconcentration on SO3-oxine CM-cellulose, *Fresenius' J. Anal. Chem.*, 347 (1993) 238.
150 M.J. Bloxham, S.J. Hill and P.J. Worsfold, Determination of trace metals in sea-water and the on-line removal of matrix interferences by flow injection with inductively coupled plasma mass spectrometric detection, *J. Anal. At. Spectrom.*, 9 (1994) 935.
151 P.-L. Lu, K.-S. Huang and S.J. Jiang, Determination of traces of copper, cadmium and lead in biological and environmental samples by flow-injection isotope dilution inductively coupled plasma mass spectrometry, *Anal. Chim. Acta*, 284 (1993) 181.
152 L. Ebdon, A.S. Fisher, P.J. Worsfold, H. Crews and M. Baxter, On-line removal of interferences in the analysis of biological materials by flow injection inductively coupled plasma mass spectrometry, *J. Anal. At. Spectrom.*, 8 (1993) 691.
153 M. Hollenbach, J. Grohs, S. Mamich, M. Kroft and E.R. Denoyer, Determination of technetium-99, thorium-230 and uranium-234 in soils by inductively coupled plasma mass spectrometry using flow injection preconcentration, *J. Anal. At. Spectrom.*, 9 (1994) 927.

154 P. Becotte-Haigh, J.F. Tyson and E. Denoyer, Flow injection manifold for matrix removal in inductively coupled plasma mass spectrometry by solid phase extraction: determination of Al, Be, Li and Mg in a uranium matrix, *J. Anal. At. Spectrom.*, 13 (1998) 1327.
155 Y. Israel and R.M. Barnes, Standard Addition Method in flow injection analysis with inductively coupled plasma atomic emission spectrometry, *Anal. Chem.*, 56 (1984) 1188.
156 F. Reis, M.F. Giné, F.J. Krug and H. Bergamin F., Multipurpose Flow Injection System Part 1. Programmable Dilutions and Standard Additions for Plant Digests Analysis by Inductively Coupled Plasma Atomic Emission Spectrometry, *J. Anal. At. Spectrom.*, 7 (1992) 865.
157 D. Beauchemin and A.A. Specht, On-line isotope dilution analysis with ICPMS using reverse flow injection, *Anal. Chem.*, 69 (1997) 3183.
158 A.A. Specht and D. Beauchemin, Automated on-line isotope dilution analysis with ICP-MS using sandwich flow injection, *Anal. Chem.*, 70 (1998) 1036.
159 A.A. Menegário and M.F. Giné, On-line removal of anions for plant analysis by inductively coupled plasma mass spectrometry, *J. Anal. At. Spectrom.*, 12 (1997) 671.
160 A. Lorber, Z. Karpas and L. Halicz, Flow injection method for determination of uranium in urine and serum by inductively coupled plasma mass spectrometry, *Anal. Chim. Acta*, 334 (1996) 295.
161 M.S. Dundar and S.J. Haswell, Use of a model gut system to study the effects of dietary fibre and multivitamins on the speciation of copper, zinc and iron, *Analyst*, 120 (1995) 2085.
162 X.J. Yang and J.S. Guan, Application of inductively coupled plasma atomic emission spectrometry for the certificate analysis of uranium certified reference material: determination of trace impurity elements in U_3O_8 by extraction chromatography, *Anal. Chim. Acta*, 274 (1993) 237.
163 J.-S. Chen, H. Berndt and G. Tölg, A contribution to the trace analysis of high-purity iron by atomic spectrometry (flame-AAS, furnace-AAS, ICP-OES), *Fresenius J. Anal. Chem.*, 344 (1992) 526.
164 Y.-C. Sun, J. Mierzwa, C.-F. Lin, T. I. Yeh and M.-H. Yang, Selective precipitation separation and inductively coupled plasma mass spectrometric determination of trace metal impurities in high purity silver, *Analyst*, 122 (1997) 437.
165 V. Krivan, R. Hausbeck, P. Willhartitz, R. Krismer and H.M. Ortner, Analysis of tantalum by ICP-AES involving trace-matrix separation, *Fresenius J. Anal. Chem.*, 341 (1991) 550.
166 X.-J. Yang and J.-S. Guan, End-on viewed inductively coupled plasma for the determination of trace impurities in high-purity scandium oxide by extraction chromatography, *Anal. Chim. Acta*, 279 (1993) 261.
167 X.-J. Yang, Extractive chromatographic separation and inductively

coupled plasma atomic emission spectrometric determination of trace impurities in high purity europium oxide, *Talanta*, 41 (1994) 1807.
168 J. Goossens and R. Dams, Anion exchange for the elimination of spectral interferences caused by chlorine and sulfur in inductively coupled plasma mass spectrometry, *J. Anal. At. Spectrom.*, 7 (1992) 1167.
169 S. Arpadjan, E. Vassileva and, S. Momchilova, Preconcentration methods for determination of trace amounts of impurities in high-purity copper salts by atomic absorption spectrometry and inductively coupled plasma atomic emission spectrometry, *Analyst*, 117 (1992) 1933.
170 I. López Garcia, P. Viñas and M. Hernández Córdoba, Flow injection dilution system for the analysis of highly concentrated samples using flame atomic absorption spectrometry, *J. Anal. At. Spectrom.*, 9 (1994) 1167.
171 V.O. Brito and I.M. Raimundo Jr, Simultaneous multiple injection in monosegmented flow analysis, *Anal. Chim. Acta*, 371 (1998) 317.
172 J.F. Tyson, H. Ge and E.R. Denoyer, On-line dilution for ICP-MS with a flow injection recirculating loop manifold, *J. Anal. At. Spectrom.*, 12 (1997) 1163.
173 E.C. Silva, M.C.U. Araújo, R.S. Honorato, J.L.F.C. Lima, E.A.G. Zagatto and S.M. Boscolo Brienza, Standard additions in flow injection analysis based on merging zones and gradient exploitation: application to copper determination in spirits, *Anal. Chim. Acta*, 319 (1996) 153.
174 M.J. Bloxham, S.J. Hill and P.J. Worsfold, Analysis of trace metals in sea-water by inductively coupled plasma mass spectrometry and related techniques, *Anal. Proc.*, 30 (1993) 159.
175 M. Valcarcel and M. D. Luque de Castro, Continuous separation techniques in flow injection analysis A review, *J. Chromatogr.*, 393 (1987) 3.
176 Z. Fang, S. Xu and S. Zhang, Fundamental and practical considerations in the design of on-line column preconcentration for flow-injection atomic spectrometric systems, *Anal. Chim. Acta*, 200 (1987) 35.
177 Z. Fang, Flow-injection on-line column preconcentration in atomic spectrometry, *Spectrochim. Acta Rev.*, 14 (1991) 235.
178 Z. Horváth, A. Lásztity and R.M. Barnes, Preconcentration and separation techniques for inductively coupled plasma atomic and mass spectrochemical analyses, *Spectrochim. Acta Rev.*, 14 (1991) 45.
179 J. Ruzicka and A. Arndal, Sorbent extraction in flow injection analysis and its application to enhancement of atomic spectrometry, *Anal. Chim. Acta*, 216 (1989) 243.
180 M.D. Luque de Castro and L. Gámiz-Gracia, Miniaturisation: a well-defined trend in separation and preconcentration techniques, *Anal. Chim. Acta*, 351 (1997) 23–40.
181 Yu.A. Zolotova and N.M. Kuz'min, In G. Svehla (Ed.), *Wilson and Wilson's Comprehensive Analytical Chemistry*, Vol. XXV, Elsevier, Amsterdam, 1990.

182 O. Vicente, A. Padró, L. Martinez, R. Olsina and E. Marchevsky, Determination of some rare earth elements in seawater by inductively coupled plasma mass spectrometry using flow injection preconcentration, *Spectrochim. Acta*, 53B (1998) 1281.
183 S. Caroli, A. Alimonti, F. Petrucci and Zs. Horváth, On-line preconcentration and determination of trace elements by flow injection-inductively coupled plasma atomic emission spectrometry, *Anal. Chim. Acta*, 248 (1991) 241.
184 V. Porta, C. Sarzanini, O. Abollino, E. Mentasti and E. Carlini, Preconcentration and inductively coupled plasma atomic emission spectrometric determination of metal ions with on-line chelating ion exchange, *J. Anal. At. Spectrom.*, 7 (1992) 19.
185 M.L. Lee, G. Tölg, E. Beinrohr and P. Tschöpel, Preconcentration of palladium, platinum and rhodium by on-line sorbent extraction for graphite furnace atomic absorption spectrometry and inductively coupled plasma atomic emission spectrometry, *Anal. Chim. Acta*, 272 (1993) 193.
186 H. Obata, H. Karatani and E Nakayama, Automated determination of iron in seawater by chelating resin concentration and chemiluminescence detection, *Anal. Chem.* 65 (1993) 1524.
187 J.A. Aldstadt, J.M. Kuo, L.L. Smith and M.D. Erickson, Determination of uranium by flow injection inductively coupled plasma mass spectrometry, *Anal. Chim. Acta*, 319 (1996) 135.
188 B. Fairman, A. Sanz-Medel and P. Jones, Field sampling technique for the 'fast reactive' aluminium fraction in waters using a flow injection minicolumn system with inductively coupled plasma atomic emission spectrometric and inductively coupled plasma mass spectrometric detection, *J. Anal. At. Spectrom.*, 10 (1995) 281.
189 B. Wen, X.-Q. Shan and S.-G. Xu, Preconcentration of ultratrace rare earth elements in seawater with 8-hydroxyquinoline immobilized polyacrylonitrile hollow fiber membrane for determination by inductively coupled plasma mass spectrometry, *Analyst*, 124 (1999) 621.
190 X. Wang, Z. Zuang, C. Yang and F. Zhyu, The application of flow injection iminodiacetic acid-ethycellulose membrane preconcentration and separation technique to atomic spectrometry, *Spectrochim. Acta*, 53B (1998) 1437.
191 L.F.R. Machado, A.O. Jacintho, A.A. Menegário, E.A.G. Zagatto and M.F. Giné, Electrochemical and chemical processes for hydride generation in flow injection ICP-MS: determination of arsenic in natural waters, *J. Anal. At. Spectrom.*, 13 (1998) 1343.
192 D. Beauchemin, Hydride generation interface for the determination of inorganic As and organoarsenic by ICP-MS using open-focused microwave digestion to enhance the pre-reduction process, *J. Anal. At. Spectrom.*, 13 (1998) 1.

193 J.T. van Elteren, Z. Šlejkovec and H.A. Das, A mathematical model for optimisation of flow injection-hydride generation atomic fluorescence spectrometry, *Spectrochim. Acta*, 54B (1999) 311.
194 C.P. Hanna, P.E. Haigh, J.F. Tyson and S. McIntosh, Examination of separation efficiencies of mercury vapour for different gas–liquid separators in flow injection cold vapour atomic absorption spectrometry with amalgam preconcentration, *J. Anal. At. Spectrom.*, 8 (1993) 585.
195 P.E. Carrero and J.F. Tyson, Determination of selenium by atomic absorption spectrometry with simultaneous retention of selenium(IV) and tetrahydroborate(III) on an anion-exchange resin followed by flow injection hydride generation from the solid phase, *Analyst*, 122 (1997) 915.
196 G.-H. Tao and R.E. Sturgeon, Sample nebulization for minimization of transition metal interferences with selenium hydride generation ICP-AES, *Spectrochim. Acta*, 54B (1999) 481.
197 J.T. Creed, M.L. Magnuson, C.A. Brockhoff, I. Chamberlain and M. Sivaganesan, Arsenic determination in saline waters utilizing a tubular membrane as a gas–liquid separator for hydride generation inductively coupled plasma mass spectrometry, *J. Anal. At. Spectrom.*, 11 (1996) 505.
198 M.D. Luque de Castro and I. Papaefstathiou, Pervaporation: a useful tool for speciation analysis, *Spectrochim. Acta*, 53B (1998) 311.
199 L.S. Zhang and S.M. Combs, Determination of selenium and arsenic in plant and animal tissues by hydride generation inductively coupled plasma mass spectrometry, *J. Anal. At. Spectrom.*, 11 (1996) 1049.
200 R. Knight, S.J. Haswell, S.W. Lindow and J. Batty, Determination of mercury in hair by coupled CVAA-ICP-MS, *J. Anal. At. Spectrom.*, 14 (1999) 127.
201 D.C. Reamer, C. Veillon and P.T. Tokousbalides, Radiotracer techniques for evaluation of selenium hydride generation systems, *Anal. Chem.*, 53 (1981) 245.
202 J.C. Ng, D. Johnson, P. Imray, B. Chiswell and M.R. Moore, Speciation of arsenic metabolites in the urine of occupational workers and experimental rats using an optimised hydride cold-trapping method, *Analyst*, 123 (1998) 929.
203 C. Haraldsson, S. Westerlund and P. Öhman, Determination of mercury in natural samples at the sub-nanogram level using inductively coupled plasma/mass spectrometry after reduction to elemental mercury, *Anal. Chim. Acta*, 221 (1989) 77.
204 R. Brown, D.J. Gray and D. Tye, Hydride generation ICP-MS (HG-ICP-MS) for the ultra lowlevel determination of mercury in biota, *Water Air Soil Pollut.*, 80 (1995) 1237.
205 M.M. Gomez and C.W. McLeod, Trace enrichment and determination of gold by flow injection inductively coupled plasma spectrometry, Part 1. Inductively coupled plasma atomic emission spectrometry, *J. Anal. At. Spectrom.*, 8 (1993) 461.

206 D. Karunasagar, J. Arunachalam and S. Gangadharan, Development of a 'collect and punch' cold vapour inductively coupled plasma mass spectrometric method for the direct determination of mercury at nanograms per litre levels, *J. Anal. At. Spectrom.*, 13 (1998) 679.
207 J.F. van Staden, In: J.L. Burguera (Ed.), *Flow Injection Atomic Spectroscopy*, Marcel Dekker, Inc., New York, 1989.
208 C.A. Lucy and K. K.-C. Yeung, Solvent extraction-flow injection without phase separation through the use of the differential flow velocities within the segmented flow, *Anal. Chem.* 66 (1994) 2220.
209 V. Kubán, *Crit. Rev. Anal. Chem.*, 22 (1991) 477.
210 J.L. Manzoori and A. Miyazaki, Indirect inductively coupled plasma atomic emission determination of fluoride in water samples by flow injection solvent extraction, *Anal. Chem.*, 62 (1990) 2457.
211 M. Yamamoto, Y. Obata, Y. Nitta, F. Nakata and T. Kumamaru, Microporous polytetrafluoroethylene tube separator for the determination of beryllium by flow injection—or suction-flow—solvent extraction followed by inductively coupled plasma atomic emission spectrometry, *J. Anal. At. Spectrom.*, 3 (1988) 441.
212 K.L. Peterson, B.K. Logan, D.G. Christian and J. Ruzicka, Sequential-injection extraction for sample preparation, *Anal. Chim. Acta*, 337 (1997) 99.
213 X.-P. Yan, R. Kerrich and M.J. Hendry, Flow injection on-line group preconcentration and separation of (ultra)trace rare earth elements in environmental and geological samples by precipitation using a knotted reactor as a filterless collector for inductively coupled plasma mass spectrometric determination, *J. Anal. At. Spectrom.*, 14 (1999) 215.
214 J.R. Pretty and J.A. Caruso, Signal enhancement of lead and thallium in inductively coupled plasma atomic emission spectrometry using on-line anodic stripping voltammetry, *J. Anal. At. Spectrom.*, 8 (1993) 545.
215 J.R. Pretty, E.H. Evans, E.A. Blubaugh, W.L. Shen, J.A. Caruso and T.M. Davidson, Minimisation of sample matrix effects and signal enhancement for trace analytes using anodic stripping voltammetry with detection by inductively coupled plasma atomic emission spectrometry and inductively coupled plasma mass spectrometry, *J. Anal. At. Spectrom.*, 5 (1990) 437.
216 J.R. Pretty, E.A. Blubaugh, E.H. Evans, J.A. Caruso and T.M. Davidson, Determination of copper and cadmium using an on-line anodic stripping voltammetry flow cell with detection by inductively coupled plasma mass spectrometry, *J. Anal. At. Spectrom.*, 7 (1992) 1131.
217 J.R. Pretty, E.A. Blubaugh, J.A. Caruso and T.M. Davidson, Determination of chromium(VI) and vanadium(V) using an on-line anodic stripping voltammetry flow cell with detection by inductively coupled plasma mass spectrometry, *Anal. Chem.*, 66 (1994) 1540.
218 J.R. Pretty, D.C. Duckworth and G.J. Van Berkel, Anodic stripping

voltammetry coupled on-line with inductively coupled plasma mass spectrometry: optimization of a thin-layer flow cell system for analyte signal enhancement, *Anal. Chem.*, 69 (1997) 3544.
219 K. Akatsuka, J.W. McLaren, J.W. Lam and S.S. Berman, Determination of iron and ten other trace elements in the open ocean seawater reference material NASS-3 by inductively coupled plasma mass spectrometry, *J. Anal. At. Spectrom.*, 7 (1992) 889.
220 S. Dadfarnia and C. McLeod, On-line trace enrichment and determination of uranium in waters by flow injection inductively coupled plasma mass spectrometry, *Appl. Spectrosc.*, 48 (1994) 1331.
221 C.-R. Lan and M.-H. Yang, Synthesis, properties and applications of silica-immobilized 8-quinolinol Part 2. On-line column preconcentration of copper, nickel and cadmium from sea water and determination by inductively-coupled plasma atomic emission spectrometry, *Anal. Chim. Acta*, 287 (1994) 101.
222 S.D. Lofthouse, G.M. Greenway and S.C. Stephen, Miniaturisation of a matrix separation/preconcentration procedure for inductively coupled plasma mass spectrometry, *Anal. Comm.*, 35 (1998) 177.
223 B. Pasullean, C.M. Davidson and D. Littlejohn, On-line preconcentration of chromium(III) and speciation of chromium in waters by flame atomic absorption spectrometry, *J. Anal. At. Spectrom.*, 10 (1995) 241.
224 F.M. Martin and O.F.X. Donard, Aqueous ethylation of organotin compounds in simple solution for speciation analysis by cryofocussing and detection by atomic absorption spectrometry—comparison with hydride generation, *Fresenius J. Anal. Chem.*, 348 (1994) 1768.
225 L.D. Martinez, E. Saidman, E. Marchevsky and R. Olsina, Determination of selenium in geochemical samples by flow injection hydride generation inductively coupled plasma atomic emission spectrometry following on-line removal of iron interference by anion exchange, *J. Anal. At. Spectrom.*, 12 (1997) 487.
226 P. Becotte-Haih, J.F. Tyson, E. Denoyer and M.W. Hinds, Determination of arsenic in gold by flow injection inductively coupled plasma mass spectrometry with matrix removal by reductive precipitation, *Spectrochim. Acta*, 51B (1996) 1823.
227 J.F. Tyson, High-performance, flow-based, sample pre-treatment and introduction procedures for analytical atomic spectrometry, *J. Anal. At. Spectrom.*, 14 (1999) 169.
228 L. Vuchkova and S. Arpadjan, Behaviour of the dithiocarbamate complexes of arsenic, antimony, bismuth, mercury, lead, tin and selenium in methanol with a hydride generator, *Talanta*, 43 (1996) 479.
229 A. Chatterjee and K.J. Irgolic, Behaviour of selenium compounds in FI-HG-AAS, *Anal. Commun.*, 35 (1998) 337.
230 K. Pyrzynska, Speciation Analysis of Some Organic Selenium Com-

pounds, *Analyst*, 121 (1996) 77R.
231 X.C. Le, W.R. Cullen and K.J. Reimer, Effect of cysteine on the speciation of arsenic by using hydride generation atomic absorption spectrometry, *Anal. Chim. Acta*, 285 (1994) 277.
232 H. Chen, I.D. Brindle and X.C. Le, Prereduction of arsenic(V) to arsenic(III), enhancement of the signal, and reduction of interferences by L-cysteine in the determination of arsenic by hydride generation, *Anal. Chem.*, 64 (1992) 667.
233 M.M. Gómez, T. Gasparic, M.A. Palacios and C.Camara, Determination of five selenium compounds in urine by liquid chromatography with focused microwave assisted digestion and hydride generation-atomic absorption spectrometric detection, *Anal. Chim. Acta*, 374 (1998) 241.
234 A.G. Howard and C. Salou, Arsenic speciation by cryogenic trap hydride generation atomic absorption spectroscopy: performance enhancement by pre-derivatization, *J. Anal. At. Spectrom.*, 13 (1998) 683.
235 R. Rubio, A. Padró and G. Rauret, Photoreduction-hydride generation: a new on-line system for the determination of selenate and selenite, *Anal. Chim. Acta*, 353 (1997) 91.
236 S.J. Santosa, H. Mokudai and S. Tanaka, Automated continuous-flow hydride generation with inductively coupled plasma mass spectrometric detection for the determination of trace amounts of selenium(IV), and total antimony, arsenic and germanium in sea-water, *J. Anal. At. Spectrom.*, 12 (1997) 409.
237 L. Ebdon, A.S. Fisher and P.J. Worsfold, Determination of arsenic, chromium, selenium and vanadium in biological samples by inductively coupled plasma mass spectrometry using on-line elimination of interference and preconcentration by flow injection, *J. Anal. At. Spectrom.*, 9 (1994) 611.
238 A.G. Cox, I.G. Cook and C.W. McLeod, Rapid sequential determination of chromium(III)–chromium(VI) by flow injection analysis–inductively coupled plasma atomic–emission spectrometry, *Analyst*, 110 (1985) 331.
239 A.C. Sahayam, J. Arunachalam and S. Gangadharan, Determination of Cr(VI) in potable waters after selective separation of Cr(III) using ZnO, *Can. J. Anal. Sci. Spectrosc.*, 43 (1998) 4.
240 K. Vercoutere, R. Cornelis, L. Mees and Ph. Quevauviller, Certification of the contents of the chromium(III) and chromium(VI) species and total chromium in a lyophilised solution (CRM 544), *Analyst*, 123 (1998) 965.
241 L. Campanella, K. Pyrzynska and M. Trojanowicz, Chemical speciation by flow-injection analysis. A review, *Talanta*, 43 (1996) 825.
242 M.D. Luque de Castro, Speciation studies by flow-injection analysis, *Talanta*, 33 (1986) 45.
243 M.D. Luque de Castro, D.W. Bryce and A. Izquierdo, Speciation with unsegmented continuous systems, *Talanta*, 42 (1995) 1215.

244 A.R. Elwaer, C.W. McLeod and K.C. Thompson, Stability of bromate species immobilised on microcolumns of activated alumina, *Analyst*, 123 (1998) 981.
245 D. Beauchemin and S.S. Berman, Determination of trace metals in reference water standards by inductively coupled plasma mass spectrometry with on-line preconcentration, *Anal. Chem.*, 61 (1989) 1857.
246 S.N. Willie, R.E. Sturgeon and S.S. Berman, Comparison of i-quinolinol-bonded polymer supports for the preconcentration of trace metals from seawater, *Anal. Chim. Acta*, 149 (1983) 59.
247 G.M. Greenway and A. Townshend, On-line immobilized chelating agents in flow injection systems, *Anal. Proc.*, 30 (1993) 438.
248 C.-R. Lan and M.-H. Yang, Synthesis, properties and applications of silica-immobilized 8-quinolinol Part 1. Characterization of silica-immobilized 8-quinolinol synthesized via a Mannich reaction, *Anal. Chim. Acta*, 287 (1994) 101.
249 A. Seubert, G. Petzold and J.W. McLaren, Synthesis and application of an inert type of 8-hydroxyquinoline-gased chelating ion exchanger for seawater analysis using on-line inductively coupled plasma mass spectrometry detection, *J. Anal. At. Spectrom.*, 10 (1995) 371.
250 G.M. Greenway, S.M. Nelms and D. Koller, Application of a novel iminodiacetate chelating material to automated matrix separation for inductively coupled plasma mass spectrometry, *Anal. Commun.*, 33 (1996) 57.
251 C.R.M. Peixoto, Y. Gushikem and N. Baccan, Selective spectrophotometric determination of trace amounts of chromium(VI) using a flow injection system with a microcolumn of zirconium(IV) oxide modified silica gel, *Analyst*, 117 (1992) 1029.
252 A. Maquieira, H.A.M. Elmahadi and R. Puchades, Immobilized cyanobacteria for on-line trace metal enrichment by flow injection atomic absorption spectrometry, *Anal. Chem.*, 66 (1994) 3638.
253 A. Maquieira, H.A.M. Elmahadi and R. Puchades, Use of *Saccharomyces cerevisiae* in flow injection atomic absorption spectrometry for trace metal preconcentration, *Anal. Chem.*, 66 (1994) 1462.
254 M.B. de la Calle-Guntiñas, Y. Madrid ad C. Cámara, Fructose-6-phosphate kinase immobilized on controlled-pore glass as a substrate for selective separation of antimony(III), *J. Anal. At. Spectrom.*, 8 (1993) 745.
255 L.R. Drake and G.D. Rayson, Plant-derived materials for metal ion-selective binding and preconcentration, *Anal. Chem.*, 68 (1996) 22A.
256 X.-P. Yan, R. Kerrich and M.J. Hendry, Determination of (ultra)trace amounts of arsenic(III) and arsenic(V) in water by inductively coupled plasma mass spectrometry coupled with flow injection on-line sorption preconcentration and separation in a knotted reactor, *Anal. Chem.*, 70 (1998) 4736.
257 A. Gáspár and J. Posta, On-line sorption preconcentration of chro-

mium(VI) and its determination by flame atomic absorption spectrometry, *Anal. Chim. Acta*, 354 (1997) 151.
258 M. Gallego, Y. Petit de Peña and M. Valcárcel, Fullerenes as sorbent materials for metal preconcentration, *Anal. Chem.*, 66 (1994) 4074.
259 W.C. Story, J.A. Caruso, D.T. Heitkemper and L. Perkins, Elimination of the chloride interference on the determination of arsenic using hydride generation inductively coupled plasma mass spectrometry, *J. Chromatogr. Sci.*, 30 (1992) 427.
260 S.T.G. Anderson, R.V.D. Robért and H.N. Farrer, Determination of total and leachable arsenic and selenium in soils by continuous hydride generation inductively coupled plasma mass spectrometry, *J. Anal. At. Spectrom.*, 9 (1994) 1107.
261 Z. Fang, In: J.L. Burguera (Ed), *Flow Injection Atomic Spectroscopy*, Marcel Dekker, Inc., New York, 1989.
262 A.G. Howard, (Boro)hydride techniques in trace element speciation, *J. Anal. At. Spectrom.*, 12 (1997) 267.
263 T. Nakahara, Hydride generation techniques and their applications in inductively coupled plasma–atomic emission spectrometry, *Spectrochim. Acta Rev.*, 14 (1991) 95.
264 R.G. Godden and D.R. Thomerson, Generation of covalent hydrides in atomic absorption spectroscopy, *Analyst*, 105 (1980) 1137.
265 C.H. Branch and D. Hutchison, Simultaneous determination of arsenic and selenium in geochemical samples by hydride evolution and atomic-absorption spectrometry: success and failure, *Analyst*, 110 (1985) 163.
266 J. Agterdenbos and D. Bax, Mechanisms in hydride generation AAS, *Fresenius' Z. Anal. Chem.*, 323 (1986) 783.
267 S. Tesfalidet and K. Irgum, Polymer-bound tetrahydroborate for arsine generation in a flow injection system, *Anal. Chem.*, 61 (1989) 2079.
268 H. Narasaki, Y. Kato and H. Kimura, *Anal. Sci.*, 8 (1992) 893.
269 S. Tesfalidet and K. Irgum, Heterogeneous hydride generation in a packed membrane cell, *Fresenius' Z. Anal. Chem.*, 341 (1991) 532.
270 A. Brockmann, C. Nonn and A. Golloch, New concept for hydride generation technique: electrochemical hydride generation, *J. Anal. At. Spectrom.*, 8 (1993) 397.
271 Y. Lin, X. Wang, D. Yuan, P. Yang, B. Huang and Z. Zhuang, Flow injection-electrochemical hydride generation technique for atomic absorption spectrometry, *J. Anal. At. Spectrom.*, 7 (1992) 287.
272 T.-J. Hwang and S.-J. Jiang, Determination of cadmium by flow injection inductively coupled plasma mass spectrometry with vapour generation sample introduction, *J. Anal. At. Spectrom.*, 12 (1997) 579.
273 J.S. Blais, G.M. Momplaisir and W.D. Marshall, Determination of arsenobetaine, arsenocholine, and tetramethylarsonium cations by liquid chromatography-thermochemical hydride generation-atomic absorption

spectrometry, *Anal. Chem.*, 62 (1990) 1161.
274 M.F. Fernández de la Campa, E. Sgovia García, M.C. Valdés-Hevia y Temprano, B. Aizpún Fernández, J.M. Marchante Gayón and A. Sanz-Medel, Effects of organised media on the generation of volatile species for atomic spectrometry, *Spectrochim. Acta*, 50B (1995) 377.
275 F. Camuña, J.E. Sanchez Uria and A. Sanz Medel, Continuous flow and flow injection halogen generation for chloride, bromide and iodide determinations by microwave induce plasma atomic emission spectroscopy, *Spectrochim. Acta*, 48B (1993) 1115.
276 R.J. Cox, C.J. Pickford and M. Thompson, Determination of iodine-129 in vegetable samples by inductively coupled plasma mass spectrometry, *J. Anal. At. Spectrom.*, 7 (1992) 635.
277 A. Lopez-Molinero, M. Benito, Y. Aznar, A. Villareal and J. R. Castillo, Volatile species of arsenic(III) with fluoride for gaseous sample introduction into the inductively coupled plasma, *J. Anal. At. Spectrom.*, 13 (1998) 215.
278 D.C. Grégoire, Sample introduction techniques for the determination of osmium isotope ratios by inductively coupled plasma mass spectrometry, *Anal. Chem.*, 62 (1990) 141.
279 C.J. Park and S.A. Yim, Determination of nickel in water samples by isotope dilution inductively coupled plasma mass spectrometry with sample introduction by carbonyl vapor generation, *J. Anal. At. Spectrom.*, 14 (1999) 1061.
280 T. Wilke, H. Wildner and G. Wünsch, Ester generation for the determination of ultratrace amounts of boron in volatile high-purity process chemicals by inductively coupled plasma mass spectrometry, *J. Anal. At. Spectrom.*, 12 (1997) 1083.
281 S. Rapsomanikis, Derivatization by ethylation with sodium tetraethylborate for the speciation of metals and organometallics in environmental samples, *Analyst*, 119 (1994) 1429.
282 R.E. Sturgeon, J. Liu, V.J. Boyko and V.T. Luong, Determination of copper in environmental matrices following vapor generation, *Anal. Chem.*, 68 (1996) 1883.
283 Z. Zhuang, X. Wang, P. Yang, C. Yang and B. Huang, Flow-injection on-line Co-APDC coprecipitation preconcentration and determination of lead and copper in seawater with graphite furnace atomic absorption spectrometry, *Can. J. Appl. Spectrosc.*, 39 (1994) 101.
284 P.M. Bersier, J. Howell and C. Bruntlett, Advanced electroanalytical techniques *versus* atomic absorption spectrometry, inductively coupled plasma atomic emission spectrometry and inductively coupled plasma mass spectrometry in environmental analysis, *Analyst*, 119 (1994) 219.
285 R. Plantz, J.S. Fritz, F.G. Smith and R.S. Houk, Separation of trace metal complexes for analysis of samples of high salt content by inductively

coupled plasma mass spectrometry, *Anal. Chem.*, 61 (1989) 149.
286 K.W. Warnken, G.A. Gill, L.-S. Wen, L.L. Griffin and P.H. Santschi, Trace metal analysis of natural waters by ICP-MS with on-line preconcentration and ultrasonic nebulization, *J. Anal. At. Spectrom.*, 14 (1999) 247.
287 T.-J. Hwang and S.-J. Jiang, Determination of trace amounts of zinc in water samples by flow injection isotope dilution inductively coupled plasma mass spectrometry, *Analyst*, 122 (1997) 233.
288 D.B. Taylor, H.M. Kingston, D.J. Nogay, D. Koller and R. Hutton, On-line solid-phase chelation for the determination of eight metals in environmental waters by inductively coupled plasma mass spectrometry, *J. Anal. At. Spectrom.*, 11 (1996) 187.
289 D. Beauchemin and A.A. Specht, Analysis of river water by ICP-MS with on-line preconcentration using flow injection, *Can. J. Anal. Sci. Spectrosc.*, 43 (1998) 43.
290 M.M. Gómez Gómez and C.W. McLeod, Trace enrichment and determination of gold by flow injection inductively coupled plasma spectrometry Part II. Inductively coupled plasma mass spectrometry, *J. Anal. At. Spectrom.*, 10 (1995) 89.
291 A. Martín-Esteban, P. Fernández, C. Pérez-Conde, A. Gutiérrez, and C. Cámara, On-line preconcentration of aluminium with immobilized Chromotrope 2B for the determination by flame atomic absorption spectrometry and inductively coupled plasma mass spectrometry, *Anal. Chim. Acta*, 304 (1995) 121.
292 Y. Sayama, Y. Hayashibe and M. Fukuda, Determination of cadmium and bismuth in high-purity zinc metal by inductively coupled plasma mass spectrometry with on-line matrix separation, *Fresenius J. Anal. Chem.*, 353 (1995) 162.
293 A.G. Coedo, T.D. López and F. Alguacil, On-line ion-exchange separation and determination of niobium, tantalum, tungsten, zirconium and hafnium in high-purity iron by flow injection inductively coupled plasma mass spectrometry, *Anal. Chim. Acta*, 315 (1995) 331.
294 R.A. Nickson, S.J. Hill and P.J. Worsfold, Solid phase techniques for the preconcentration of trace metals from natural waters, *Anal. Proc.*, 32 (1995) 387.
295 V. Porta, C. Sarzanini, E. Mentasti and O. Abollino, On-line preconcentration system for inductively coupled plasma atomic emission spectrometric with quinolin-8-ol and Amberlite XAD-2 resin, *Anal. Chim. Acta*, 258 (1992) 237.
296 N. Furuta, K.R. Brushwyler and G.M. Hieftje, Flow-injection analysis utilizing a spectrally segmented photodiode-array inductively coupled plasma emission spectrometer—I. Microcolumn preconcentration for the determination of molybdenum, *Spectrochim. Acta*, 44B (1989) 349.
297 P. Schramel, L. Xu, G. Knapp and M. Michaelis, Multi-elemental analysis

in biological samples by on-line preconcentration on 8-hydroxyquinoline-cellulose microcolumn coupled to simultaneous ICP-AES, *Fresenius J. Anal. Chem*, 345 (1993) 600.
298 P. Schramel, L.-Q. Xu, G. Knapp and M. Michaelis, Application of an on-line preconcentration system in simultaneous ICP-AES, *Mikrochim. Acta*, 106 (1992) 191.
299 X. Peng, Z. Jiang and Y. Zen, On-line microcolumn preconcentration with desolvation and determination of trace elements by flow-injection inductively coupled plasma atomic emission spectrometry, *Anal. Chim. Acta*, 283 (1993) 887.
300 X.-J. Yang, J.-S. Guan and T.-J. Shen, Determination of trace zirconium and hafnium in high-purity scandium oxide by inductively coupled plasma atomic emission spectrometry and extraction chromatography, *Anal. Chim. Acta*, 277 (1993) 157.
301 P. Voldet, From neutron activation to inductively coupled plasma-atomic emission spectrometry in the determination of rare-earth elements in rocks, *Trends Anal. Chem.*, 12 (1993) 339.
302 S. B. Beneteau and J. M. Richardson, Rhenium-osmium isotope ratio determination by ICP-MS, *At. Spectrosc.*, 13 (1992) 118.
303 P.J. Watkins and J. Nolan, Determination of rare-earth elements, yttrium, scandium and hafnium using cation-exchange separation and inductively coupled plasma-atomic emission spectrometry, *Chem. Geol.*, 95 (1992) 131.
304 X.-Y. Luo, Z.-X. Su, W.-Y. Gao, G.-Y. Zhan, X.-J. Chang, Separation and preconcentration of traces of gallium, titanium and zirconium by means of a new chelating resin containing pyrocatechol violet as functional group Determination by ICP-OES, *Fresenius J. Anal. Chem*, 344 (1992) 252.
305 K.J. Orians and E.A. Boyle, Determination of picomolar concentrations of titanium, gallium and indium in sea water by inductively coupled plasma mass spectrometry following an 8-hydroxyquinoline chelating resin pre-concentration, *Anal. Chim. Acta*, 282 (1993) 63.
306 P. Möller, P. Dulski and J. Luck, Determination of rare earth elements in seawater by inductively coupled plasma–mass spectrometry, *Spectrochim. Acta*, 47B (1992), 1379.
307 M.T.S.D. Vasconcelos and M.F.C. Leal, Speciation of Cu, Pb, Cd and Hg in waters for the Oporto coast in Portugal, using pre-concentration in a Chelamine resin column, *Anal. Chim. Acta*, 353 (1997) 189.
308 C.H. Lee, J.S. Kim, M.Y. Suh and W. Lee, A chelating resin containing 4-(2-thiazolylazo)resorcinol as the functional group. Synthesis and sorption behaviour for trace metal ions, *Anal. Chim. Acta*, 339 (1997) 303.
309 S.F. Durrant, A. Krushevska, D. Amarasiriwardena, M.D. Argentine, S. Romon-Guesnier and R.M. Barnes, Matrix separation by chelation to prepare biological materials for isotopic zinc analysis by inductively

coupled plasma mass spectrometry, *J. Anal. At. Spectrom.*, 9 (1994) 199.
310 Y. Israel, A.P. Krushevska, H. Foner, L.J. Martines and R.M. Barnes, Flow configurations for interfacing automatic ion-exchange devices with sequential inductively coupled plasma atomic emission spectrometry: application to trace metal determination in pure alkali metal salts, *J. Anal. At. Spectrom.*, 8 (1993) 467.
311 A. Stroh, E.R. Denoyer, Q. Lu and U. Völlkopf, Optimization and use of flow injection vapor generation for ICP-MS, *FIAS Application Report*, Perkin Elmer SCIEX, Norwalk, 1993.
312 U. Voellkopf, A. Guensel, M. Paul and H. Wiesmann, Applications of ICP-MS with sample introduction by electrothermal vaporization and flow injection techniques, In: G. Holland and A. N. Eaton (Eds), *Applications of Plasma Source Mass Spectrometry*, The Royal Society of Chemistry, Cambridge, 1991, pp.162-177.
313 J. Bowman, B. Fairman and T. Catterick, Development of a multi-element hydride generation–inductively coupled plasma mass spectrometry procedure for the simultaneous determination of arsenic, antimony and selenium in waters, *J. Anal. At. Spectrom.*, 12 (1997) 313.
314 H. Ibrahim, Y.M. Issa and R. Kamal, *Mikrochim. Acta* 109 (1992) 201.
315 E.V. Alonso, A.G. de Torres and J.M.C. Pavón, Determination of nickel in biological materials after microwave dissolution using inductively coupled plasma atomic emission spectrometry with prior extraction into butan-1-ol, *Analyst*, 117 (1992) 1157.
316 J.M.L. Almendro, C.B. Ojeda, A.G. de Torres and J.M.C. Pavón, Determination of cadmium in biological samples by inductively coupled plasma atomic emission spectrometry after extraction with 1,5-bis(di-2-pyridyl-methylene)thiocarbonohydrazide, *Analyst*, 117 (1992) 1749.
317 E.V. Alonso, A.G. de Torres and J.M.C. Pavón, Determination of nickel in biological samples by inductively coupled plasma atomic emission spectrometry after extraction with 1,5-bis[phenyl-(2-pyridyl)methylene]thiocarbonohydrazide, *J. Anal. At. Spectrom.*, 8 (1993) 843.
318 K.Y. Patterson, C. Veillon, P.B. Moser-Veillon and G.F. Wallace, Determination of zinc stable isotopes in biological materials using isotope dilution inductively coupled plasma mass spectrometry, *Anal. Chim. Acta*, 258 (1992) 317.
319 H.M. Al-Swaidan, Determination of lead and nickel in Saudi Arabian crude oils by ICP/MS using MIBK for sample pretreatment, *Anal. Lett.*, 25 (1992) 2157.
320 C.J. Lord III, Determination of lead and lead isotope ratios in gasoline by inductively coupled plasma mass spectrometry, *J. Anal. At. Spectrom.*, 9 (1994) 599.
321 A. Makishima, I. Inamoto and K. Chiba, Determination of gallium in pure aluminum by isotope dilution analysis with solvent extraction using

inductively coupled plasma mass spectrometry, *Appl. Spectrosc.*, 44 (1990) 91.

322 G.E.M. Hall, J.C. Pelchat and J. Loop, Determination of zirconium, niobium, hafnium and tantalum at low levels in geological materials by inductively coupled plasma mass spectrometry, *J. Anal. At. Spectrom.*, 5 (1990) 339.

323 O. Kujirai, K. Yamada and R. Hasegawa, Simultaneous determination of trace impurities in high-purity chromium metal and chromium disilicide by inductively coupled plasma atomic emission spectrometry after coprecipitation, *J. Anal. At. Spectrom.* 8 (1993) 481.

324 J. Zucheng and P. Schramel, Determination of trace amounts of molybdenum in water samples by inductively coupled plasma atomic emission spectrometry after cobalt–dithiocarbamate coprecipitation, *Fresenius J. Anal. Chem.* 343 (1992) 600.

325 J.R. Pretty, D.C. Duckworth and G.J. Van Berkel, Electrochemical sample pretreatment coupled on-line with ICP-MS: analysis of uranium using an anodically conditioned glassy carbon working electrode, *Anal. Chem.*, 70 (1998) 1141.

326 D. Yuan, X. Wang, P. Yang and B. Huang, On-line electrolytic dissolution of alloys and multi-element determination by inductively coupled plasma atomic emission spectrometry, *Anal. Chim. Acta*, 251 (1991) 187.

327 T.J. Gluodenis, Jr. and J.F. Tyson, Flow injection systems for directly coupling on-line digestions with analytical atomic spectrometry, Part 2. Reactions in a microwave field, *J. Anal. At. Spectrom.*, 8 (1993) 697.

328 C. Cabrera, Y. Madrid and C. Camara, Determination of lead in wine, other beverages and fruit slurries by flow injection hydride generation atomic absorption spectrometry with on-line microwave digestion, *J. Anal. At. Spectrom.*, 9 (1994) 1423.

329 S.J. Haswell, Development and operating characteristics of micro flow injection analysis systems based on electroosmotic flow, *Analyst*, 122 (1997) 1R.

330 J. Ruzicka and E.H. Hansen, Flow injection analysis: From beaker to microfluidics, *Anal. Chem.*, 72 (2000) 212A.

Chapter 3

Electrothermal vaporization sample introduction for inductively coupled plasma–mass spectrometry

D. Conrad Grégoire

3.1 INTRODUCTION

Four decades have passed since electrothermal vaporization (ETV) was first introduced [1] as a means of producing analyte vapour for analytical atomic spectrometry. During this time, graphite furnace atomic absorption spectrometry (GFAAS) has become firmly established and has reached maturity as a technique capable of determining about 50 elements at trace levels in a wide variety of sample types. One of the attractive features of GFAAS is the very small sample size requirement where only microlitre volumes can be successfully analysed.

In 1974, Nixon et al. [2] coupled an ETV to an inductively coupled plasma–atomic emission spectrometer (ICP-AES). The promise of truly multi-element analysis using ETV did result in a significant number of publications, but the full potential of the technique was not realized and popularization of the technique has been slow. The widespread use of ETV-ICP-AES has been hindered by the difficulty conventional spectrometers have with accurately measuring transient signals, especially when background correction is required. As well, limits of detection for most elements were not competitive with those available by GFAAS. The advent of solid-state array detectors for ICP-AES spectrometers allows for the measurement (with background correction) of fast transient signals, and may result in renewed interest in this technique.

Inductively coupled plasma–mass spectrometry (ICP-MS) offered a second opportunity for ETV sample introduction to successfully couple with argon plasmas. Although quadrupole based ICP-MS is not a simultaneous technique, it is a fast sequential technique capable of scanning spectra on the ms time scale, thus allowing for the accurate measurement of transient signals. The high sensitivity of ICP-MS, which rivalled GFAAS in detection power, also offered the capability of measuring isotope ratios of individual elements. This combination of attributes set the stage for the development of a new technique which could significantly extend the reach of the analytical chemist, especially for the trace analysis of micro-samples.

Shortly after the introduction of ICP-MS, Gray and Date [3] pointed out that the origin of most of the "permanent" background peaks in ICP-MS originate from water and result in high hydrogen and oxygen populations in the plasma which are typically 10^8 times greater than that of analyte ions. These authors suggested that ETV could eliminate water from the sample stream and diminish these interferences. Detection limits of 1 to 12 pg (absolute) were quoted [3] and the measurement of Cd isotope ratios was demonstrated. In a later paper Gray [4], by way of background spectra, showed how ETV could be used to reduce or eliminate the interference of $^{56}ArO^+$ on $^{56}Fe^+$.

The first comprehensive application of ETV-ICP-MS was published by Date and Cheung [5] who reported on the use of the technique for the determination of Pb isotope ratios. These authors showed that Pb isotope ratios could be determined with a precision of about 1% in NBS 981 (Pb isotope standard reference material) and a series of Pb mineral concentrates. Good agreement was obtained for these ratios when compared with values obtained by thermal ionization mass spectrometry.

In a series of papers, Park et al. [6–11] reported on the use of a custom designed ETV device based on the use of a metal (W, Re) filament or a small flat graphite platform (1.1 × 0.4 cm) as the vaporization surface. The relevance of vaporizer design and its influence on analyte mass transport efficiency [7] was used to construct an ETV, oriented vertically, and having a 5 ml volume above the vaporization surface. This design, along with a tangentially oriented argon carrier gas flow, was reported to maximize the transport of analyte vapour to the argon plasma resulting in transport efficiencies on the order of 80%.

The analytical potential for this device was demonstrated [6,8] by the successful determination of As, Cu, Mn, Pb, Rb, V, Zn and Ag in orchard leaves (NBS SRM 1571) and oyster tissue (NBS SRM 1566) as well as measurement of Pb isotope ratios in NBS SRM 981. The determination of Mo, W [10] and Tl [11] by isotope dilution in a number of geological reference materials was also reported. A comparison of the Park-design ETV unit [12] with the direct insertion device showed that ETV was more versatile and amenable to the analysis of samples high in salt content.

ETV-ICP-MS was found useful for the determination of the platinum group elements in rocks [13] and in a series of 27 [14] geological reference materials including silicates, iron formation rocks and ores. Grégoire [15] determined Os isotope ratios by ETV-ICP-MS and found that the addition of Te as a chemical modifier increased the sensitivity of the determination by a factor of 18. Osmium isotope ratios were measured with a precision of 0.5% compared to a precision of 0.2% for solution nebulization sample introduction. Gold was determined in seawater by Falkner and Edmond [16] who reported an ETV-ICP-MS limit of detection of 1 fM in one litre of sample.

Imakita et al. [17] determined trace amounts of Bi in iron and steel using ETV-ICP-MS. Femtogram limits of detection were reported [18] for the determination of Pu and U in urine using ETV-ICP-MS. Two papers have appeared on the analysis of semiconductor materials by ETV-ICP-MS. Trace amounts of U and Th in silicon and silicon dioxide [19] and trace impurities of As, Fe, Sb, Mg, Mo, K, Na and V [20] were determined in semiconductor grade HCl.

Newman et al. [21] determined Te in biological fluids with an absolute limit of detection of 57 pg (5.7 ng ml^{-1}) Te in blood. Whittaker et al. [22] determined Fe isotope ratios in unprocessed human blood as part of a metabolic study. The analysis of petroleum products (C_5–C_{32} alkanes) for Hg was reported by Osborne [23]. A limit of detection of 3 ng g^{-1} for Hg was obtained. In a novel application of the technique, Richner and Wunderli [24] used ETV-ICP-MS to differentiate between organic and inorganic chlorine for the determination of polychlorinated biphenyls in waste oils. Organic chlorine was vaporized at a temperature of 400°C whilst the inorganic chlorine containing compounds were volatilized at 2650°C. Detection limits between 0.5 and 10 μg^{-1} PCB in oils were reported.

A series of papers on the use of a tungsten electrothermal vaporization device was published by Tsukahara and Kubota [25] and by Shibata et al. [26–28]. Extensive studies were carried out on the optimization of instrumental parameters [25] and the use of hydrogen mixed with the argon carrier gas [27]. Hydrogen [27] not only prevented the oxidation of the W filament, but also produced a rise in the excitation temperature and an increase in the electron number density at the interface, resulting in enhanced ionization of the analyte. This system [26] was applied to the determination of the rare-earth elements (REE) with absolute limits of detection ranging from 2 fg (10^{-15} g) for La to 12 fg for Gd. Shibata et al. [28] published an extensive study of REE oxide formation in ETV-ICP-MS. Their findings were applied to the analysis of high-purity REE oxide samples for traces of other REEs. Under optimized conditions, Tb and Lu impurities were determined at concentration levels of 0.01 μg^{-1} in high-purity Gd_2O_3.

Shen et al. [29] and Carey et al. [30] used a modified HGA-300 ETV system for the determination of Pb (d.l.= 10 fg), of Fe (d.l.= 0.2 pg) and As (d.l.= 1.5 pg). The system incorporated a specialized coolant argon flow designed to isolate the analyte vapour from the vaporizer walls, thus promoting efficient transport to the argon plasma.

Grégoire et al. [31] compared and contrasted GFAAS and ETV-ICP-MS. These authors pointed out that these two techniques are complementary in nature and that data obtained from each can be combined, by repeating the same experiment on both systems, to provide fundamental information on vaporization and atomization processes and mechanisms of matrix modification. This work was followed up by two papers by Byrne et al. [32,33] in which the mechanism of chloride interferences in GFAAS was elucidated by combining information obtained from GFAAS and ETV-ICP-MS. Also studied was the mechanism of chemical modification of ascorbic acid on the elimination of the interference of magnesium chloride on Mn in GFAAS [33] as well as the pyrolysis of magnesium chloride itself [34].

The virtues of ETV-ICP-MS have been reported in a number of papers by several authors [31,35–42] and the advantages of the technique will be discussed in detail later in the chapter. A number of reports have appeared on fundamental aspects of ETV-ICP-MS including the role of chemical modifiers on analyte sensitivity [43–45], transport loss [46–50], particle size distribution of analyte aerosol produced in an ETV [51], the background spectral features of ETV mass spectra [52] and the formation of oxides in ETV-ICP-MS [28]. Studies have also been published on the mechanism of vaporization of B [53], W [54], Ra and other alkaline earth elements [55], Cr, [56], Y [57], the rare earth elements (REE) [57,58], U [59], Al [60], the platinum group elements [61], P [62] and S [63] in ETV devices as well as the effect of acids on analyte signals [64].

Applications of the technique have appeared on the determination of isotope ratios in rocks and minerals [65] and biological tissue [66], the determination of trace elements in Arctic snow [67], the determination of REE in single zircons [68], the direct analysis of solids and solids introduced as slurries [69]. A more complete listing of published applications is given at the end of this chapter.

Further information on ETV-ICP-MS can be obtained from a book chapter written by Williams [70]. An excellent literature review covering publications to 1992 on both ETV-ICP-AES and ETV-ICP-MS was published by Carey and Caruso [71] and is recommended as companion reading.

3.2 INSTRUMENTATION

Research on the development of an 'ideal' electrothermal vaporization device for use in plasma spectrochemistry is unfolding in much the same manner as did research on the development of the best atomization device for GFAAS during the early 70s. A number of configurations have been investigated including filament or platform [9,13,22] based systems as well as tube-type [20,52,67,72,73] devices and ETV systems based on the insertion of a metal wire loop into a tube-type [29] ETV and an electrically heated wire loop [74] placed directly into the argon plasma. Vaporization substrates have included Re [8,73], W [25,73], Ta [15,73], Mo [73] and a variety of graphites ranging from pyrolytically coated crystalline graphite [14,32] to glassy carbon [15]. Low voltage high current power supplies are generally used to resistively heat the metal or graphite vaporizer.

Those using metal vaporization surfaces for ETV work report that these are less likely (than graphite) to react with analyte to form refractory compounds. However, there are reports [75] on the formation of involatile intermetallic compounds between V [76], Mo [77] and Ni [76] with hot Ta metal. Metal filaments or strips can lose their ductility (W) once heated to incandescence and are prone to contamination.

A major shortcoming of metal vaporization surfaces is the rate of evaporation of metal at high temperatures. Microgram quantities of metal entering the plasma can result in loading effects and serious matrix effects, both spectroscopic and non-spectroscopic [26,73]. Grégoire [52] has shown, however, that carbon volatilized at 2600°C from a graphite tube did not result in the suppression of analyte signals. Carbide formation can be a problem when using graphite, but the judicious selection of vaporization temperature along with the use of chemical modifiers can minimize the effects of carbide formation for most elements [53,54].

Today, most commercially available systems use graphite as the vaporization substrate and approximately 75% of the ETV-ICP-MS literature is based on the use of a tube-type system. No publication has appeared reporting on a comprehensive comparison of the analytical performance of filament or platform systems with tube-type ETV devices.

An 'ideal' ETV system for ICP-MS would incorporate design features that promote complete vaporization of analyte followed by the efficient (100%) transport of analyte vapour from the ETV to the argon plasma. Other desirable features include versatile and accurate temperature control, an inert vaporization surface capable of being rapidly heated to high temperatures (3000°C), ease of automation and low cost.

Early work in ETV-ICP-MS [3,5,36] was carried out on a device similar to the system developed by Gunn et al. [78] which used a carbon rod containing a 5 µl sample volume cavity. The graphite rod was enclosed in a glass envelope of 1 l volume. The first ETV device specifically designed for use as a sample introduction system for ICP-MS was the filament type device built by Park et al. [9], shown in Fig. 3.1. The main performance criterion behind this design was the isolation of vaporized sample from surfaces (where condensation could take place) to ensure good analyte transport efficiency. The introduction of carrier argon gas (1 l min^{-1}) at a tangent to the filament or platform support base resulted in a sheathing effect which isolated analyte vapour from the surface of the glass cover long enough to allow for condensation of analyte into solid particles. These particles could then be transported to the plasma with a reported efficiency of up to 80%. The effective volume above the vaporization surface was only 5 ml. This reduced analyte dilution by argon gas and also served to promote collisions of analyte vapour/aerosol particles during condensation resulting in the formation of easily transported particulates. A novel spray chamber assembly [79] was designed for use with this system to accommodate fast conversion from ETV to solution nebulization (SN) sample introduction without plasma shut-down.

The filament shown on the ETV in Fig. 3.1 can easily be substituted for a pyrolytic graphite coated platform. Later designs of the platform

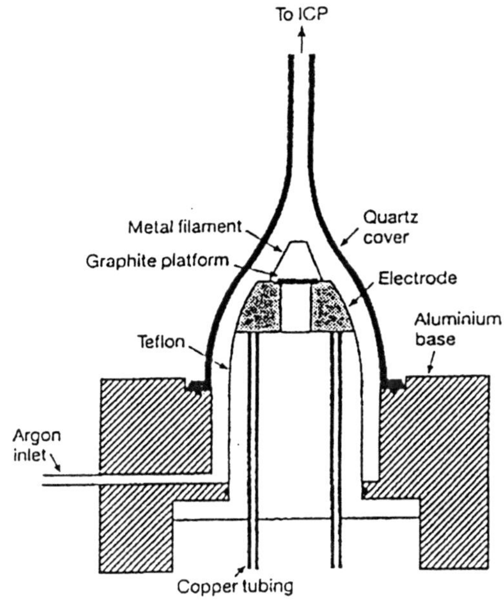

Fig. 3.1. Filament or platform type electrothermal vaporizer (from Ref. [7] with permission).

incorporated a small depression in the centre [13] to prevent spreading of the analyte solution during the drying step. When highly acidic solutions or organic solvents were used as sample solutions, the sample tended to spread towards (and at times into) the contact electrodes. This situation became more serious as the surface of the platform aged and became more porous. Control of the problem was achieved by using small volumes (5 µl) and relatively low drying temperatures (80°C).

Two manufacturers have developed ETV systems based on the Park design. Tsukahara and Kubota [25] have characterized the Seiko Instruments Micro Sampling (ETV) System which uses a W ribbon in which a small cavity has been impressed to contain up to 50 µl of sample solution. The glass chamber covering the W ribbon has a volume of 300 ml.

The second platform-type ETV device [80] is the EV2000 System and is marketed by Hewlett-Packard Co. This device uses a graphite platform contained in a glass chamber 90 ml in volume. Gas flows are carefully controlled to provide laminar flow conditions preventing analyte loss on the chamber walls.

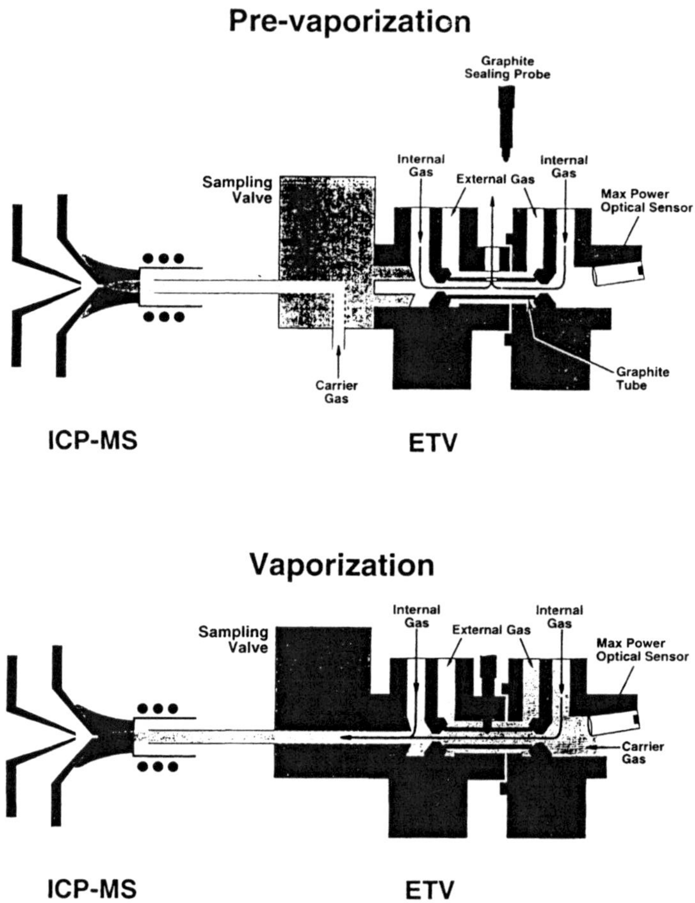

Fig. 3.2. Perkin-Elmer HGA-600MS electrothermal vaporization system.

Two tube-type ETV systems for ICP-MS are currently commercially available: the Perkin-Elmer HGA-600MS (Fig. 3.2) and the VG Elemental MicroTherm system. Both instruments use graphite tubes of approximately the same dimensions (28 × 8 mm o.d., 6 mm i.d.) which can be obtained pre-coated with pyrolytic graphite. These tubes are identical to those which are standard for GFAAS applications.

The MicroTherm System uses an Ar sheath gas which flows between the graphite tube and a quartz envelope surrounding the tube.

The sheath gas protects the graphite from air-oxidation at high temperatures. Using either manual or automated pipetting, a sample of up to 100 µl in volume can be deposited into the graphite tube through a hole cut in the quartz envelope which is aligned with the dosing hole. During the temperature pre-treatment steps (dry, pyrolysis), water and other waste vapours are removed from the dosing hole of the graphite tube by an argon carrier flow of about 1 l min^{-1} During the high-temperature or vaporization step a graphite rod is placed into the dosing hole of the graphite tube essentially sealing the tube and thus allowing analyte-containing vapour to be swept into the argon plasma.

The Perkin-Elmer Sciex HGA-600MS (Fig. 3.2) uses a modified HGA system originally developed for GFAAS. Argon sheath gases continuously flow over the exposed graphite parts to prevent oxidation. During the dry and pyrolysis steps, opposing argon flows originating from both ends of the graphite tube remove water and other vapours through the dosing hole of the graphite tube. An argon flow of about 300 ml min^{-1} is generally used during the dry and pyrolysis steps. During the high-temperature or vaporization step, a graphite probe is pneumatically activated to seal the dosing hole. Once the graphite tube is sealed, a valve located at one end (downstream) of the HGA workhead directs the carrier argon flow (800–1100 ml min^{-1}), which originates from the far end of the graphite tube, directly to the argon plasma. The HGA-600MS is also equipped with an autosampler which can deliver sample volumes of from 2 µl to 100 µl, which is the practical maximum capacity of the graphite tube.

Two laboratory-built tube-type ETV systems that incorporate features not offered by commercial suppliers are those reported by Shen et al. [29], Carey et al. [30] and Lamoureux [81] and Lamoureux et al. [82,83]. The device built in Caruso's laboratory [29,30] was a standard HGA furnace incorporating specialized gas flows to help isolate analyte vapour from any surfaces on leaving the graphite tube. A tungsten loop [29] or graphite platform [30] set within the graphite tube were used to contain sample. The second system built in Grégoire's laboratory [81–83] also used a modified HGA system. Unlike other tube-type configurations, all vapours produced are carried out through the dosing hole rather that out of one of the ends of the graphite furnace. This was done by installing a quartz ring to surround the graphite tube. The quartz ring was fitted with a ball-joint which was connected via a length of PTFE tubing directly to the argon plasma. The opening in the quartz ball joint was aligned directly above the dosing hole of the graphite tube. Argon flows could be carefully controlled and laminar flow conditions were obtained for analyte vapour leaving the graphite tube. This ETV device features both the vapour isolating conditions characteristic of filament-type systems and the advantages of a graphite tube ETV.

The interface between each of these ETV devices and the argon plasma was generally a 1 m (or less) length of plastic or PTFE tubing. The diameter of the tubing can be an important parameter, but in general, a diameter equal to the inner diameter of the graphite tube is used to minimize turbulent flow.

Comparison of the performance of the ETV systems discussed in this section is difficult. Because the ETV is only a sample introduction device, reported limits of detection contain information on both the performance of the ETV and the ICP-MS. A valid comparison of different ETV designs and types can only be done if a single ICP-MS instrument is interfaced with different ETVs and operated under identical experimental conditions. Table 3.1 contains reported limits of detection for eight elements determined using a variety of ETV and ICP-MS systems. Although the limits of detection for individual elements can vary by two orders of magnitude, virtually every system

TABLE 3.1

Comparison of absolute limits of detection in picograms

	As	Co	Cu	Fe	Mn	Ni	Pb	Zn
Tube-type ETV systems								
Ref. 66–68	–	0.14	0.42	–	0.120	0.47	0.023	0.27
Ref. 39	–	0.030	0.020	–	0.020	0.070	0.001	–
Ref. 83	–	0.075	0.169	–	0.072	–	–	0.617
Ref. 29	1.5	–	–	0.20	–	–	0.010	–
Ref. 84*	–	0.025	0.50	1.0	0.10	0.50	–	0.50
Ref. 37	0.005	–	–	–	–	–	0.002	–
Filament or platform type ETV systems								
Ref. 8	0.1	–	–	–	–	0.93	0.3	0.4
Ref. 25,26	–	0.02	–	0.3	–	–	0.008	–
Ref. 80	–	0.14	0.42	–	0.120	0.47	0.023	0.27

*Calculated using 50 μl sample volume.

listed reports the lowest limit of detection for at least one element. Taking into consideration that more than one generation of ICP-MS instrumentation was used to produced these data and widely different experimental and data collection parameters were also used, the overall impression is that all systems can, under certain conditions, perform equally well. The real test for each of the systems developed thus far will be in their use in actual applications for the analysis of complex matrices.

3.3 ANALYTICAL DOMAIN OF ETV-ICP-MS

Most analytical techniques have associated with them some applications which no other technique can handle as well or as efficiently. A legitimate question is whether or not ETV-ICP-MS has a unique domain in analytical chemistry. Electrothermal vaporization ICP-MS is a microanalytical technique capable of analysing μl volumes of sample solution. Table 3.2 lists some of the applications which are well suited to ETV-ICP-MS. Graphite furnace AAS, however, is also a microanalytical technique and is, in fact, very sensitive with limits of detection for many elements that compare well with solution nebuli-

TABLE 3.2

Applications well suited to ETV-ICP-MS

1. Analysis of small samples
— blood, body fluids
— rare materials
— pre-concentrates
2. Analysis of radioactive and toxic samples
3. Analysis of difficult matrices
— organic samples
— high-salt samples
— solids and slurries
4. Isotope ratio determinations
— Br, S, Se
— tracer studies
— isotope dilution calibration

zation ICP-MS [85]. Although both ETV-ICP-MS and GFAAS can handle applications listed in the first three categories of Table 3.2, only ETV-ICP-MS is practical for the measurement of isotope ratios. This inherent capability of mass spectrometric techniques allows for the measurement of isotope ratios important to geochemical studies [65], human and animal metabolic studies [66] and the application of isotope dilution calibration [11,13,14] for quantitative analysis.

Table 3.3 compares absolute limits of detection obtained by both ETV-ICP-MS and GFAAS. With the exception of Cr and Zn, limits of detection for all elements determined by ETV-ICP-MS are better than those reported by GFAAS. Except for the transition metals for which ETV-ICP-MS is about 0.3 (Cr) to 150 times more sensitive, ETV-ICP-MS surpasses GFAAS in detection power by four or more orders of magnitude. Another noteworthy feature is that for most elements, ETV-ICP-MS limits of detection are in the low femtogram (10^{-15} g) range and do not show the very large variations in detection limits across the periodic table as does GFAAS. The ETV-ICP-MS data contained in Table 3.3 was obtained using a Perkin-Elmer ELAN 5000. Modern ICP-MS instruments are from 10 to 100 times more sensitive than the ELAN 5000 giving even better limits of detection for ETV-ICP-MS than those reported here.

The comparison of limits of detection can be taken a step further if we examine Table 3.4 which lists sensitivities for certain elements determined by GFAAS. Sensitivities are expressed as the number of pg required to give 1% absorbance. For elements B through Y, these sensitivities range from 240 to over 26,000. Elements from S to Zr in practice cannot be determined at all by GFAAS. Only under extraordinary experimental conditions [86] have some of these elements been determined by GFAAS. Most elements successfully determined by GFAAS have sensitivities ranging from 0.3 to 30 pg. All of the elements listed in Table 3.4 can be determined with fg to pg sensitivities using ETV-ICP-MS.

Clearly, ETV-ICP-MS not only has superior sensitivity when compared to GFAAS, this technique can be used for the determination of analytes, including the rare earth elements, refractory platinum group elements, non-metals and elements such as B, Th, U, W and Zr which are difficult or impossible to handle by GFAAS. The added capability of ICP-MS for multi-element analysis and the determination of isotope ratios of individual elements promises to make ETV-ICP-MS one of the most sensitive and versatile analytical techniques available.

TABLE 3.3

Comparison of limits of detection

Element	ETV-ICP-MS[1]	GFAAS[2]
Cd	0.230 pg	0.3 pg
Ce	0.023 pg	>50 ng
Co	0.140 pg	2.0 pg
Cu	0.420 pg	1.0 pg
Cr	3.200 pg	1.0 pg
Cs	0.057 pg	5.0 pg
La	0.018 pg	≈20 ng
Mn	0.120 pg	1.0 pg
Nd	0.013 pg	≈13 ng
Ni	0.470 pg	10.0 pg
Pb	0.023 pg	5.0 pg
Sc	0.044 pg	≈1.2 ng
Sm	0.036 pg	≈3.3 ng
Th	0.038 pg	–
Tl	0.029 pg	10.0 pg
U	0.016 pg	–
V	0.360 pg	20.0 pg
Y	0.030 pg	≈4.8 ng
Yb	0.004 pg	≈38 pg
Zn	0.27 pg	0.1 pg

[1]This laboratory.
[2]Ref. [86]

When compared to solution nebulization ICP-MS, electrothermal vaporization sample introduction has many advantages. The removal of water and other solvent allows for the measurement at low levels of several important elements including Fe and S. The advent of micro-concentric nebulizers has, in some cases, made this device more

TABLE 3.4

Elements not easily determined by GFAAS (Ref. [86])

Element	Sensitivity (pg for 0.0044 A·s)
B	700
Gd	11000
Ir	250
La	26000
Nd	1800
Os	1400
P	4200
Sm	240
U*	12000
Y	13000
S	–
Ce	–
Pr	–
Tb	–
Ho	–
Lu	–
Th	–
W	–
Zr	–

Note: For most other elements, sensitivities range from 0.3 to 30 pg for 0.0044 A·s.
*Ref. [86].

practical for the analysis of small volume solutions. However, microconcentric nebulizers are prone to clogging and all have a sample solution dead volume greater than 0.

As the remaining sections of this chapter will show, ETV-ICP-MS has unique capabilities for the analysis of difficult samples (Table 3.2) including slurries and solids. By using selective vaporization of analytes, unresolvable (by quadrupole mass spectrometry) spectral interferences can be avoided. ETV-ICP-MS is a fully quantitative technique which stands ready to take advantage of any developments in the field of mass spectrometry, including instruments having increased sensitivity, resolution, the ability to remove (by gas phase collisions and/or reactions) spectral interferences and to measure the entire mass spectrum simultaneously.

3.4 PRINCIPLE OF OPERATION OF ETV-ICP-MS

Sample introduction by ETV techniques is based on the conversion of solid sample (dried solution, slurry, etc.) to vapour using heat, followed by transport of the sample using an argon stream, to the argon plasma. Although this process may appear to be trivial, complex chemical reactions and physical effects can take place making application of this technique difficult for quantitative chemical analysis. A more detailed examination of some of these factors will serve to illustrate the point.

The heating cycle during which liquid sample is converted to analyte vapour comprises several separate steps. Each of the steps is designed to perform a specific function, but at the same time, other undesirable effects may occur. Table 3.5 lists heating steps in an ETV

TABLE 3.5

Processes occurring during ETV heating steps

A.	Drying step	Evaporation of water Evaporation of acid Evaporation of organic solvent
B.	Pyrolysis step	Loss of water of hydration Loss of residual acids Loss of hydrolysis products Volatile matrix removal Pre-vaporization analyte loss Chemical modification Analyte–matrix interaction Analyte–surface interaction
C.	Vaporization step	Loss of matrix components Decomposition of analyte salt Carbon reduction of analyte salt Volatilization of analyte Analyte–matrix interaction Analyte–surface interaction Gas-phase reactions Vapour-phase chemical modification
D.	Cleaning step	Removal of residual matrix components Removal of residual analyte

program and some of the processes that are associated with each of these.

The lowest temperature step is the drying step which is used to drive off solvent including acids and volatile organics. Loss of sample solution and hence analyte is possible during the drying step if the rate of heating or the maximum temperature is too high. This generally results in explosive boiling of the sample causing a physical loss or redistribution of sample material within the graphite tube. Low sensitivity, poor reproducibility and broad or multiple analyte peaks indicate too vigorous a drying step.

The pyrolysis step can vary in temperature from 100°C to 1800°C depending upon the specific element being determined and the composition of the matrix. The pyrolysis step is used to complete solvent removal, to volatilize some of the matrix components especially carbon and volatile salts and to convert analyte to a form that can be easily volatilized during the high temperature vaporization step. The maximum pyrolysis temperature that can be used for an element is limited by the volatility of the element and/or the volatility of the compound containing analyte. Table 3.6 contains the maximum pyrolysis temperature for a number of elements, some of which have been thermally stabilized by the addition of a chemical modifier. The maximum pyrolysis temperatures given were obtained from GFAAS studies, and it is likely that many of these will also apply to ETV-ICP-MS work. Measurement of pyrolysis temperatures is done by determining the temperature beyond which a measurable loss in analytical signal is obtained during the subsequent high temperature atomization step. Since losses occurring during a GFAAS heating step will also occur during an ETV heating step (these are essentially identical except for internal argon gas flow), it follows that the temperatures given in Table 3.6 can be used at least as an upper limit for ETV-ICP-MS studies.

The wide range of maximum pyrolysis temperature imposes a limitation on ETV-ICP-MS when used for multi-element analysis. It may not be possible to determine both volatile and involatile elements during a single ETV run under optimum pyrolysis conditions for a given matrix type.

Several other processes can occur during the pyrolysis step. One that has been explored in detail is the co-volatilization of Mn in the presence of hydrolysable salts such as magnesium chloride. Magnesium chloride is common in saline samples such as seawater and brines and would be present as a major component. Using ETV-ICP-MS

TABLE 3.6

Maximum pyrolysis temperatures for elements determined by GFAAS

Element	Modifier	Pyrolysis temperature (°C)	Element	Modifier	Pyrolysis temperature (°C)
Ag	Pd+$Mg(NO_3)_2$	1000	Mn	$Mg(NO_3)_2$	1400
Al	$Mg(NO_3)_2$	1700	Mo	–	1800
As	Pd+$Mg(NO_3)_2$	1300	Na	–	900
Au	Pd+$Mg(NO_3)_2$	1000	Nd	–	1500
B	Ca	1000	Ni	–	1400
Ba	–	1200	Os	–	200
Be	$Mg(NO_3)_2)$	1500	P	Pd+$Ca(NO_3)_2$	1350
Bi	Pd+$Mg(NO_3)_2$	1200	Pb	PO_4	700
Ca	–	1100	Pd	–	900
Cd	Pd+$Mg(NO_3)_2$	850	Pt	–	1300
Co	$Mg(NO_3)_2$	1400	Rb	–	800
Cr	$Mg(NO_3)_2$	1650	Rh	–	1300
Cs	H_2SO_4	900	Ru	–	1400
Cu	–	1000	Sb	Pd+$Mg(NO_3)_2$	1100
Dy	–	1500	Se	Pd+$Mg(NO_3)_2$	1100
Er	–	1700	Si	–	1400
Eu	–	1300	Sm	–	1400
Fe	$Mg(NO_3)_2$	1400	Sn	Pd+$Mg(NO_3)_2$	1400
Ga	Pd+$Mg(NO_3)_2$	1200	Sr	–	1300
Gd	–	1600	Te	Pd+$Mg(NO_3)_2$	1200
Ge	Pd+$Mg(NO_3)_2$	1500	Ti	–	1400
Hg	Pd+$Mg(NO_3)_2$	250	Tl	Pd+$Mg(NO_3)_2$	1000
In	Pd+$Mg(NO_3)_2$	1200	Tm	–	1700
Ir	–	1300	U	–	1000
K	–	950	V	$Mg(NO_3)_2$	1100
La	–	1600	Y	–	1400
Li	–	900	Yb	–	1300
Mg	–	900	Zn	$Mg(NO_3)_2$	700

From Refs. [86,87].

techniques, Byrne et al. [32] showed that during the pyrolysis step at temperatures above 700°C, Mn is co-volatilized with HCl which is produced as a reaction product from the hydrolysis of magnesium chloride. A similar loss of Mn would not occur for aqueous reference standards volatilized alone illustrating the importance of proper method development in ETV-ICP-MS. These authors also reported that the addition of ascorbic acid can control the hydrolysis reaction and hence the pre-vaporization loss of analyte.

The high-temperature vaporization step can range from 800°C to a maximum of 2700°C. The maximum temperature is limited by the thermal stability of the vaporization surface. For graphite, temperatures above 2800°C result in the vaporization of large quantities of C which can overload the plasma and cause premature wear of the graphite tube. The optimum pyrolysis and vaporization temperatures for many elements determined by ETV-ICP-MS have been reported in individual papers studying the mechanism of vaporization of elements. Generally, the optimum vaporization temperature is the lowest temperature that provides complete vaporization of analyte or analyte-containing compound, while at the same time minimizing unwanted reactions such as carbide formation. When a complex sample matrix is present in relatively large quantities, co-volatilization occurs and the vaporization characteristics of the analyte can take on the vaporization properties of the matrix. Well defined atomization temperatures are known for GFAAS, but these are not generally transferable to ETV-ICP-MS because for GFAAS, the atomization of analyte is a primary requirement, whereas for ETV-ICP-MS only vaporization of analyte or analyte-containing molecules is necessary. Literature reports to date indicate that temperatures significantly lower than those used in GFAAS are optimum for ETV-ICP-MS work. The reasons for this will be more fully discussed in the section on vaporization mechanisms in this chapter.

During the vaporization step, analyte-containing compounds are heated to reactive temperatures resulting in the thermal decomposition of these compounds, carbon reduction and volatilization. One of the undesirable chemical reactions that can also occur is the interaction of analyte with the metal filament or graphite tube to form refractory alloys and/or compounds. For graphite-based ETV systems, the formation of carbides is perhaps most important particularly for elements such as B, U, Th and the REEs. Most of the elements listed in Table 3.4 as being not easily determinable by GFAAS are in this category because of their propensity to form involatile carbides.

TABLE 3.7

Typical ETV-ICP-MS heating program

Heating step	Temperature (°C)	Ramp time (s)	Hold time (s)	Internal gas flow (ml min^{-1})	Valve setting
Drying	100	10	30	300	vent
Pyrolysis	300	10	20	300	vent
Vaporization	2500	0	6	1000	plasma
Cooling	20	–	10	1000	plasma
Cleaning	2700	1	5	1000	plasma
Cooling	20	–	20	1000	plasma

Chemical modifiers and physical carriers are also used during the vaporization step. Improvement in sensitivity by orders of magnitude is possible, particularly for volatile elements.

The final ETV heating step is the cleaning step which is normally a maximum temperature (2700°C) step. Matrices high in organic compounds such as blood, or those containing refractory compounds such as dissolved zircons require high temperatures to get rid of residual quantities of these materials from the graphite tube. Solid material left behind in the graphite tube can affect the drying characteristics of subsequent samples injected into the graphite tube or may provide active sites for undesirable chemical reactions such as those discussed above. The cleaning step is not always required and should only be used when necessary because this step is tantamount to an additional tube firing and contributes to shortening tube lifetime.

Table 3.7 gives a typical heating program for the determination of trace metals by ETV-ICP-MS using the Perkin-Elmer HGA-600MS system. The entire cycle takes about 2 min to complete. Including the time required to load sample into the ETV, approximately one determination can be completed every 3 min or about 20 h^{-1}. A cooling step is required after the vaporization step to prevent overheating of the graphite tube during the cleaning step. A second cooling step following the cleaning step is needed to allow graphite parts to cool while still being protected by the argon sheath gas. This is done because the graphite probe is heated to incandescence following the cleanup step. Withdrawing the probe before cooling allows rapid air oxidation of the probe and eventual failure of the seal between the probe and the dosing hole. The sealing probe is left in place sealing the dosing hole during

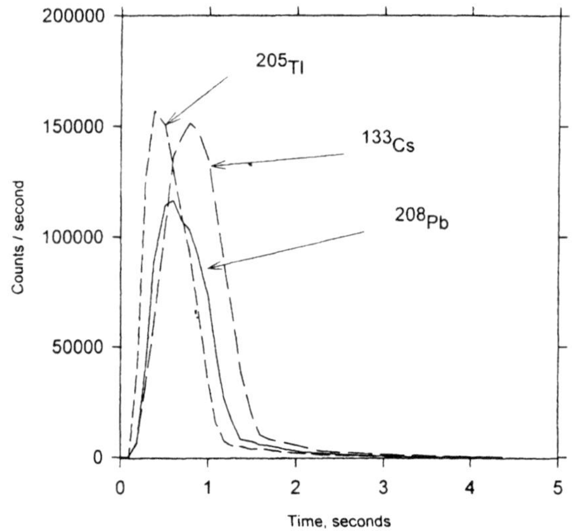

Fig. 3.3. ETV-ICP-MS signals for 20 pg Tl, 30 pg Cs and 50 pg Pb.

the cleaning step to remove any material that may have condensed onto the tip of the probe during previous heating steps.

Internal gas flows should be relatively low during the drying and pyrolysis steps to prevent expulsion of solution or analyte-containing salt crystals and to avoid the excessive build-up of back-pressure within the graphite tube which could blow out the argon plasma or significantly change the effective sampling depth at the interface. The internal gas flow can also be a mixture of gases incorporating air during the pyrolysis step to promote oxidation of organic matrices or added freon during the vaporization step which acts as a chemical modifier.

Typical ETV-ICP-MS signals for several elements are shown in Fig. 3.3. Signals were obtained using conditions given in Table 3.7 and are about 1 to 2 s in width. The relative temporal variation in the appearance of analyte signals is a reflection of the different volatility characteristics of each element being determined.

Because of the transient nature of ETV-ICP-MS signals, it is important that the analyte signal intensity be monitored frequently during the high temperature vaporization step. Too few readings results in an inaccurate measurement of the shape of the analyte signal profile leading to error whether peak-height or area counts are recorded. Most practitioners restrict the number of isotopes monitored to about 5 and use a dwell time of from 10 ms to 50 ms depending upon the number of isotopes monitored and the intensity of the analyte signal. Spectra are generally collected in peak-hopping mode with one measurement point per m/z.

3.5 BACKGROUND SPECTRAL FEATURES

When solution nebulization sample introduction is used for ICP-MS measurements, large quantities of water and water vapour are generally introduced into the argon plasma. The presence of liquid aerosol can have a profound effect on the excitation characteristics of the plasma itself [88] in addition to contributing to the formation of polyatomic ions which may interfere with the determination of certain elements [89]. For example, $^{40}Ar^{16}O^+$ interferes with the determination of the major isotope of iron, $^{56}Fe^+$, which has an abundance of 91.52% [3,22]. In the presence of Cl, $^{35}Cl^{16}O^+$ interferes with the determination of ^{51}V which is nearly monoisotopic with an abundance of 99.76%. Studies on the background spectral features of ETV-ICP-MS and polyatomic ions produced from the vaporization of common matrix components have been published [52,90].

Figure 3.4 shows the background spectra obtained using both SN and ETV sample introduction techniques. The virtual elimination of water from the sample results in an ETV background spectrum which is qualitatively simpler than the SN background spectrum. Marked reductions in the background signal at m/z = 19,32,33,34 and 56 allow for the determination of ^{19}F (100% abundance), ^{32}S (95.02%), ^{33}S (0.75%). ^{34}S (4.22%) and ^{56}Fe (91.52%) by ETV-ICP-MS. In a paper on the application of ICP-MS to the determination of isotope ratios important in geochemical studies, Grégoire [65] has shown that the use of a dry plasma allows for the determination of isotope ratios of S, Cl and Se at trace levels and aids in the determination of Br isotopes. For example, using SN, the background spectrum for the three S isotopes free of interference from other elements gave count rates (Cs^{-1}) which were over-range for $^{32}S^+$, 142,400 for $^{33}S^+$ and 45040 for $^{34}S^+$. Using ETV, count rates were reduced to 60,300 for $^{32}S^+$, 600 for $^{33}S^+$ and 4100 for $^{34}S^+$.

Interfering species eliminated or reduced using ETV-ICP-MS compared to SN include oxygen and hydroxide species as well as oxides and hydrides of Ar which combine to form a fairly large number of well documented polyatomic ions [89].

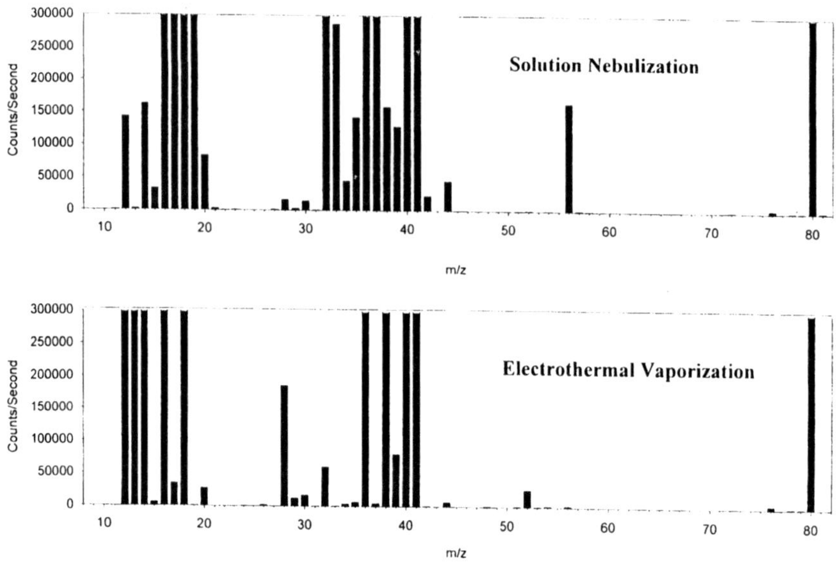

Fig. 3.4. Background spectra for solution nebulization and electrothermal vaporization sample introduction.

When using ETV, the major ionic species other than argon found in the plasma are nitrogen and oxygen entrained from the air surrounding the torch compartment. When heating carbon at temperatures up to 2700°C, sublimation occurs and carbon becomes a fourth species found in abundance in the plasma. Table 3.8 lists the four major elemental sources of ions involved in formation of polyatomic ions in the argon plasma using electrothermal vaporization sample introduction. It is fortuitous that three of these four elements are nearly monoisotopic whereas the minor carbon isotope is only 1.11%

TABLE 3.8

Isotopes originating from graphite electrothermal vaporizer, plasma and atmospheric gases

Carbon	Nitrogen	Oxygen	Argon
^{12}C (98.89)*	^{14}N (99.6)	^{16}O (99.76)	^{40}Ar (99.6)
^{13}C (1.11)	^{15}N (0.4)	^{18}O (0.20)	^{36}Ar (0.34)
		^{17}O (0.04)	^{38}Ar (0.06)

*Percent natural abundance.

abundant. However, since the ion concentration of these elements in the plasma is high, the influence of the minor isotopes cannot be ignored. The net result of the introduction of vaporized carbon into the argon plasma is evident when comparing the two spectra on Fig. 3.4. The appearance of several new intense peaks on the ETV background spectra warrant further discussion.

Table 3.9 lists the polyatomic ions formed as a result of the introduction of carbon into the plasma as well as other major species present as a result of Ar flowing through the graphite tube. The quantity of carbon vaporized from the surface is dependent on the type of graphite used as well as the physical condition of the tube surface and, at times, the composition of the sample. For a given graphite tube, the temperature of its surface is the most significant parameter influencing the absolute quantity of carbon reaching the plasma since the vapour pressure of carbon increases by 25 orders of magnitude between 1000 and 3000 K. Therefore, it is important to study the background spectrum as a function of steady-state temperature. Table 3.10 lists the possible polyatomic ions produced when the graphite tube is heated to temperatures ranging from 20°C to 2500°C. Polyatomic ions not observed or remaining at background noise levels (such as C_3^+) were omitted from this table. A number of polyatomic ions not associated with carbon but which are included for information and possible comparison with intensities obtained using solution nebulization sample introduction are given in parentheses. Minor isotopes of C and O were monitored to obtain values that were on scale.

Ion intensity levels for all species not containing carbon were relatively insensitive to vaporization temperature. It is interesting to note, however, that levels of O_2^+ did not decrease with the introduction of C, indicating that any scavenging effect of C on oxygen either took place primarily with O^+ (which decreased somewhat with increasing temperature) or that diffusion of O_2 from the surrounding atmosphere was rapid enough to replace any oxygen scavenged by carbon.

The argon dimer ($^{80}Ar_2^+$) is a species produced only in the plasma (and/or interface region of the mass spectrometer) and therefore can be used as a diagnostic tool to monitor the plasma for loading and/or interference effects. Additionally, it has been used [91] as an internal standard reference isotope to correct for non-spectroscopic or matrix interferences. Although the intensity of the argon dimer does not quantitatively reflect the severity of matrix effects for all ions across the entire mass range, it was useful in this study (see below) as a

TABLE 3.9

Count rate for ions produced during heating of graphite vaporizer at steady-state temperatures

Mass	Ion(s)	Temperature (°C)			
		20	800	1500	2500
13	$^{13}C^+$	7,730	26,600	177,700	432,200
[14]*	$^{14}N^+$	1,140,000	1,128,900	1,075,300	1,045,600
[18]	$^{18}O^+$	1,390,000	1,430,700	1,151,000	1,202,800
24	$^{12}C^{12}C^+$	75	100	200	300
26	$^{12}C^{14}N^+$	50	100	300	1,900
28	$^{14}N^{14}N^+$, $^{12}C^{16}O^+$	14,000	38,100	101,700	186,000
29	$^{14}N^{14}N^1H^+$, $^{12}C^{16}O^1H^+$ $^{13}C^{16}O^+$	1,450	10,500	7,500	12,800
30	$^{14}N^{16}O^+$, $^{13}C^{16}O^1H^+$	10,000	21,900	17,800	17,200
[31]	$^{14}N^{16}O^1H^+$	75	100	200	500
[32]	$^{16}O^{16}O^+$	44,500	40,200	42,600	60,300
[33]	$^{16}O^{16}O^1H^+$	100	200	100	600
44	$^{12}C^{16}O^{16}O^+$	2,050	4,700	5,800	7,050
48	$^{12}C^{36}Ar^+$	–	–	190	790
52	$^{12}C^{40}Ar^+$	200	500	6,400	26,200
53	$^{13}C^{40}Ar^+$	–	–	50	300
[54]	$^{40}Ar^{14}N^+$	1,500	1,500	1,500	1,500
[80]	$^{40}Ar^{40}Ar^+$	963,200	930,900	963,800	945,300

*Masses in parentheses arise from the presence of plasma and/or atmospheric gases and are included for comparison purposes.

general indicator of plasma loading, particularly for the lighter analytes studied. Table 3.9 shows that the ion intensity for the argon dimer did not change throughout the range of vaporization temperatures studied indicating no serious plasma loading effects from carbon. It is therefore unlikely that carbon caused the suppression of analyte ion signals even at the highest vaporization temperature of 2500°C.

Most carbon containing polyatomic ion intensities increase rapidly with temperature although the greatest increase occurs for polyatomic ions containing both O and C. The carbon dimer increased only slightly with temperature in accord with the fact that the volume composition of the vapour in equilibrium over solid carbon is only about 0.5% C_2 at

3000 K, and the likelihood that this species is very reactive toward entrained oxygen to form CO. This supposition was supported by an experiment in which particulate activated carbon was injected into the carrier argon flow to the argon plasma. The addition of large quantities of carbon to the gas stream resulted in the production of relatively small signal intensities for C^+ compared to massive signals for CO^+ despite the lower ionization energy of C. Higher mass carbon ion clusters involving three or more carbon atoms were not observed.

The growth of carbon ion count rate with temperature closely followed the temperature–time curve of the graphite tube as measured by optical pyrometry. Thus, the repeatability of the heating program and/or the condition of the graphite tube could be assessed by monitoring the carbon ion intensity during the vaporization step.

It is interesting to note that the growth of the C ion count rate on increasing the temperature of the graphite tube to 2500°C was orders of magnitude less than predicted by the vaporization characteristics of C. This implies that only a fraction of the C volatilized in the graphite tube is converted to ions and detected by the mass spectrometer. Several mechanisms [92] can account for this observation including the formation and deposition of soot within the graphite tube or at the contact cones, the deposition of C on the transfer line between the ETV and the plasma and the traverse of unionized carbon and carbon particles through the argon plasma. The deposition of C on the surface of the transfer line is evident by the gradual build-up of a carbon deposit over several hundred firings. The carbon deposit is visible particularly on the first 5 cm of PTFE tubing located at the exit and downstream of the graphite tube.

Table 3.10 lists the background ions observed at a graphite tube temperature of 2500°C along with analyte isotopes which are isobaric with these polyatomic ions. Table 3.11 summarizes the analytical importance of these interferences expressed as background equivalent mass of analyte. Primarily light elements were interfered by carbon-containing polyatomic ions. These elements included Mg, Si, Ti, Ca and Cr.

Magnesium: Two Mg isotopes are isobaric with carbon-containing ionic species leaving ^{25}Mg (10.11%) available for analytical purposes. However, the BEM for ^{24}Mg (78.6%) was only 0.15 pg which may make this isotope attractive for certain applications.

Silicon: All three Si isotopes are interfered with carbon-containing polyatomic ions making this element difficult to determine at trace

TABLE 3.10

Spectral interferences arising from graphite electrothermal vaporizer heated to 2500°C

Mass	Background ion(s)	Analyte (% abundance)
24	$^{12}C^{12}C^+$	Mg(78.6)
26	$^{12}C^{14}N^+$	Mg(11.3)
28	$^{14}N^{14}N^+$, $^{12}C^{16}O^+$	Si(92.2)
29	$^{14}N^{14}N^1H^+$, $^{12}C^{16}O^1H^+$ $^{13}C^{16}O^+$	Si(4.7)
30	$^{14}N^{16}O^+$, $^{13}C^{16}O^1H^+$	Si(3.1)
[31]*	$^{14}N^{16}O^1H^+$	P(100)
[32]	$^{16}O^{16}O^+$	S(95.0)
[33]	$^{16}O^{16}O^1H^+$	S(0.8)
44	$^{12}C^{16}O^{16}O^+$	Ca(2.1)
48	$^{12}C^{36}Ar^+$	Ca(0.18),Ti(73.5)
52	$^{12}C^{40}Ar^+$	Cr(83.8)
53	$^{13}C^{40}Ar^+$	Cr(9.6)
[54]	$^{40}Ar^{14}N^+$	Cr(2.4),Fe(5.9)
[80]	$^{40}Ar^{40}Ar^+$	Se(49.97),Kr(2.3)

*Masses in parentheses arise from the presence of plasma and/or atmospheric gases and are included for comparison purposes.

levels using ETV-ICP-MS. Note that a significant proportion of the BEM is due to nitrogen containing polyatomic ions which interfere with the determination of Si by all modes of sample introduction.

Titanium: ^{48}Ti (73.5%), which is isobaric with a minor Ca isotope, was interfered by a carbon argide polyatomic ion. For the determination of levels of Ti approaching an absolute mass of 0.42 pg, use of ^{47}Ti (7.75%) or ^{49}Ti (5.51%) is recommended.

Calcium: ^{48}Ca (0.18) is isobaric with a carbon argide species with a BEM of 171 pg. ^{44}Ca (2.13%) was isobaric with CO_2^+ with a BEM of 122 pg. ^{42}Ca (0.64%) and ^{43}Ca (0.13%) are free of interferences. The major calcium isotope (^{40}Ca) is isobaric with the massive ^{40}Ar peak and cannot be used for analysis.

Chromium: ^{52}Cr (83.3%) and ^{53}Cr (9.6%) are both interfered by carbon argides leaving only ^{54}Cr (2.38%) which was free of carbon argide but isobaric with ^{54}Fe (5.9%) and argon nitride. Since most real

TABLE 3.11

Background equivalent analyte mass for spectral interferences arising from graphite electrothermal vaporizer heated to 2500°C

Analyte	Background equivalent mass (pg)
^{24}Mg	0.15
^{26}Mg	6.6
^{28}Si	839
^{29}Si	1127
^{30}Si	2296
^{31}P	2.0
^{32}S	150*
^{33}S	150*
^{48}Ti	0.42
^{44}Ca	129
^{48}Ca	171
^{52}Cr	12.3
^{53}Cr	1.2
^{54}Cr	24.4
^{54}Fe	9.9

*Nanograms.

samples will contain significant quantities of Fe, the use of ^{53}Cr with a BEM of 1.2 pg is recommended for use.

Each of the elements discussed above is relatively involatile and therefore will be vaporized from the graphite surface at a temperature near the pre-programmed maximum or steady-state temperature. Consequently, the seriousness of interferences from carbon-containing polyatomic ions will be dependent on the actual appearance and peak temperatures of the analyte and thus it is important to consider the temporal characteristics of the background species. The appearance temperature is defined as the temperature at which the analyte signal can be detected above the baseline noise and the peak temperature is

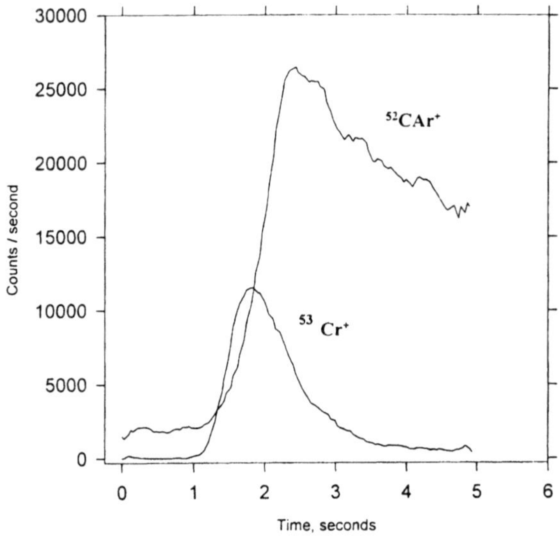

Fig. 3.5. ETV-ICP-MS signals for chromium and carbon argide ^{52}Cr: 20 pg (from Ref. [52] with permission).

the temperature at the maximum of the analyte signal. For example, Fig. 3.5 shows the temporal characteristics of Cr^+ relative to that of the carbon argide species produced during rapid heating of the HGA-600MS to a maximum temperature of 2500°C. Clearly, the Cr and the carbon argide signals are nearly temporally resolved and if a means could be found to volatilize Cr at a lower temperature (with a chemical modifier) thus shifting the Cr peak to the left (more volatile), this element could then be easily determined using the major chromium isotope (^{52}Cr). Alternatively, the use of peak-height measurements rather than integrated signals will reduce the BEM significantly for the determination of ^{52}Cr, especially in view of the very broad signal produced by the carbon argide.

For all elements interfered by carbon-containing polyatomic ions, optimum analytical conditions should include the use of the lowest possible vaporization temperature combined with peak-height measurements and chemical modification to increase analyte volatility. Integration of analyte signals will only be practical if the analyte signal can be totally resolved from that of the interfering polyatomic ion.

Since the metal filaments used for ETV studies have a relative high atomic mass (e.g. Re = 186.2; W = 183.8; Ta = 180.95), few important polyatomic species are formed that can interfere with the determination of elements of m/z lower than that of the metal itself, with the possible exception of doubly-charged ionic species. Argides of each of these three elements do not interfere with any element of the periodic table. Oxides, however, can form with oxygen entrained into the argon plasma and serious interference could result. For example, the oxides of Ta are isobaric with ^{196}Pt (35.3%), ^{196}Hg (0.15%) and ^{197}Au, which is monoisotopic. Tungsten oxides are isobaric with ^{198}Pt (7.21%) and three isotopes of Hg: ^{199}Hg (16.84); ^{200}Hg (23.13%); ^{201}Hg (13.22%). Rhenium oxides interfere with the determination of ^{201}Hg (13.22%) and ^{203}Tl (29.5%). Interferences on Hg isotopes, however, would not be severe since this element is volatile and would vaporize at temperatures well below the appearance temperatures of W or Re.

Although the effect of evaporated metal ETV filament material on analyte signals has been reported [8,25,73], no systematic study of possible filament metal oxide spectral interferences has been completed with the exception of a study by Karanassios and Horlick [93]. These authors report on the presence of metal ions and their oxide species in the background spectrum of direct insertion devices made from graphite, Mo, Ta and W. A tantalum oxide at m/z 197 was observed when using a Ta filament and significant contaminant metals such as Sn and Ta in Mo cups and Zr and Cs in W wire loops were also reported.

3.6 MECHANISM OF VAPORIZATION OF ANALYTE

When sample material deposited within a graphite tube is heated during the vaporization step, analyte can be converted to vapour via several pathways. Understanding the mechanism of analyte vaporization is important in determining optimum experimental conditions for quantitative analysis. As discussed earlier in this chapter, both GFAAS and ETV-ICP-MS use graphite tubes to produce analyte vapour. During the past 30 or more years, much information has been published in the GFAAS literature concerning the mechanisms of atomization of a number of elements. An extensive review of this literature has been published by Styris and Redfield [94] and should be consulted by anyone conducting research in this field. In an earlier paper, Sturgeon and Chakrabarti [95] proposed that there are four potential routes by which atomic vapour (M_g) is produced in the graphite furnace:

A. $MO_{s/l} + C \longrightarrow M_s + CO_g \longrightarrow M_g$

$\longrightarrow MM_g \longrightarrow 2\ M_g$

B. $MO_s \longrightarrow M_g + O_g$

C. $MO_s \longrightarrow MO_g \longrightarrow M_g + O_g$

D. $MX_s \longrightarrow MX_g \longrightarrow M_g + X_g$

In mechanism A, analyte atoms are produced by the reduction of analyte-containing salt by C followed by the volatilization of metal from the graphite surface. An alternative pathway involves the sublimation of metal dimers which further dissociate within the graphite tube into monomeric species. Mechanism B represents the thermal dissociation of analyte oxide followed by the sublimation of its decomposition products from the graphite surface. The vaporization of oxide to the gaseous state to produce analyte atoms by thermal dissociation of the vapour-phase oxide is represented in mechanism C. The last mechanism describes the volatilization of analyte in another chemical form such as a chloride which, on reaching the gas-phase, thermally dissociates. Table 3.12 lists some of the elements which are atomized via each of mechanisms A, B and C.

Although much information of use to ETV-ICP-MS mechanistic studies can be gleaned from the GFAAS literature, there are important differences between these two techniques that may change the nature of the vaporization and atomization process. This is especially true for reaction mechanisms which do not occur on the surface of the graphite

TABLE 3.12

Mechanisms Controlling Atomization in GFAAS (from Ref. [86])

A. Vaporization of the Metal:
Ag, Au, Co, Cu, Fe, Hg, Ni, Pb, Sn, Pt, Pd, Ru, Ir, Rh, Os
B. Thermal Dissociation of the Oxide:
Al, Ba, Be, Ca, Cd, Cr, Mg, Mn, Sr, Zn
C. Thermal Dissociation of the Carbide:
Hf, Mo, Nb, Ta, Ti, Th, V, Zr

tube. Clearly, surface reactions such as carbon reduction will be similar in either GFAAS or ETV-ICP-MS systems. A detailed comparison of GFAAS and ETV-ICP-MS has been published by Grégoire et al. [31].

During the vaporization (atomization) step in GFAAS, no argon gas is normally allowed to flow within the graphite tube. This is done to increase the vapour-phase temperature of volatilized analyte and analyte species and to help confine atomic vapour within the graphite tube for as long as possible. Ideal conditions for GFAAS include 100% vaporization and atomization of analyte and long residence times for analyte atoms within the analytical volume where atomic absorption takes place.

Conditions within the graphite tube during an ETV-ICP-MS experiment are quite different than for GFAAS work. The most important parameter change for ETV-ICP-MS is the use of a carrier gas flow of about 1 l min^{-1} which continually flushes out the graphite tube of vaporized sample. The requirement for complete vaporization of analyte is the same for ETV-ICP-MS as it is for GFAAS, but for ETV-ICP-MS the analyte can be in any form including atoms and molecules. Whereas the measurement (atomic absorption) step occurs within the graphite tube in GFAAS, in ETV-ICP-MS sample-containing aerosol is atomized and ionized at a remote location (argon plasma) as much as a metre away from the graphite tube.

The argon stream continuously passing through the plasma during the vaporization step is constantly cooling and removing vaporized material from the graphite surface. Reduction reactions and solid state thermal decomposition will occur and atoms released from the surface may quickly re-condense as dimers, clusters or in polyatomic form as oxides and chlorides. As is shown for reactions C and D (above), the vaporization of analyte oxides and chlorides as well as carbides can become important. Many elements have volatile chlorides and oxides and the vaporization of these within the graphite tube is likely the dominant mechanism by which analyte-containing vapour is produced. Table 3.13 lists the boiling point of elements and their chlorides, oxides and carbides. Once volatilized and carried to the argon plasma by the gas stream, oxides and carbides of all of the elements listed in Table 3.13 are easily dissociated. Table 3.14 lists the bond strengths for these oxides and carbides. The strongest bonds listed are for the carbides of W and U and experiment [54,59] has shown that vaporized oxides of these elements resulted in UO^+ and WO^+ ion levels well below those observed when nebulizing aqueous U and W solutions. For ETV-ICP-MS, it is

TABLE 3.13

Boiling points of elements and their chlorides, oxides and carbides (from Ref. [96])

Element	Boiling point (°C)	Boiling point of chloride (°C)	Boiling point of oxide (°C)	Boiling point of carbide (°C)
Ag	2212	1550	d 230	–
Al	2467	d 262	2980	d 2200
As	s 613	130	s 193	–
Au	2808	s 265	–	–
B	2550	12.5	≈1860	>3500
Ba	1640	1560	≈2000	–
Be	2970	520	≈3900	–
Bi	1560	447	1890	–
Ca	1484	>1600	2850	2300
Cd	765	140	d 950	–
Ce	3443	1727	–	–
Co	2870	1049	–	–
Cr	2672	s 1300	4000	3800
Cs	669	1290	–	–
Cu	2567	1490	–	–
Dy	2567	1500	mp 2340	–
Er	2868	1500	–	–
Eu	1527	>2000	–	–
Fe	2750	≈290	mp 1594	mp 1837
Ga	2403	535	–	–
Gd	3273	–	mp 2330	–
Ge	2830	84	s 710	–
Hf	4602	s 319	≈5400	mp 3890
Hg	357	302	d 500	–
Ho	2700	1500	–	–
In	2080	600	≈850	–
Ir	4130	–	d 1100	–
K	760	s 1500	d 350	–
La	3464	>1000	4200	–
Li	1342	1350	1200	–
Lu	3402	s 750	–	–
Mg	1107	1412	3600	–
Mn	1962	1190	–	–
Mo	5560	268	s 1155	mp 2692
Na	883	1413	s 1275	–

Element	Boiling point (°C)	Boiling point of chloride (°C)	Boiling point of oxide (°C)	Boiling point of carbide (°C)
Nb	5127	254	–	mp 3500
Nd	3074	1600	mp ≈1900	–
Ni	2370	s 973	mp 1984	–
Os	>5300	d ≈550	130	–
P	280	180	173	–
Pb	1740	950	mp 886	–
Pd	2970	d 500	–	–
Pr	3520	1700	–	–
Pt	3827	d 581	d 550	–
Rb	686	1390	d 400	–
Re	≈5627	>500	d 400	–
Rh	3727	s 800	–	–
Ru	3900	d 500	d 108	–
S	445	–	–	–
Sb	1750	283	s 1550	–
Sc	2836	s 800	–	–
Se	684	d 130	s 340	125
Si	2355	57	2230	s 2700
Sm	1794	–	–	–
Sn	2270	652	–	–
Sr	1384	1250	≈3000	–
Ta	5425	242	mp 1872	5500
Tb	3230	–	–	–
Te	1390	327	1245	–
Th	–	d 928	4400	≈5000
Ti	3287	d 475	≈2750	4820
Tl	1457	720	1865	–
Tm	1950	1440	–	–
U	3818	792	mp 2878	4370
V	3380	–	d 1750	3900
W	5660	–	s 1430	6000
Y	3338	1507	mp 2410	–
Yb	1196	1900	–	–
Zn	907	732	mp 1975	–
Zr	4377	d 350	≈5000	5100

s = sublimes; d = decomposes; mp = melting point.

TABLE 3.14

Bond strengths of diatomic molecules (from Ref. [96])

Oxides	Bond strength (eV)	Carbides	Bond strength (eV)
Al-O	5.3	B-C	4.6
B-O	8.4	Hf-C	5.6
Ba-O	5.8	Mo-C	5.0
Be-O	4.5	Nb-C	5.9
Ca-O	4.2	Ta-C	–
Cd-O	2.4	Ti-C	4.4
Cr-O	4.5	Th-C	4.7
Mg-O	3.8	U-C	4.7
Mn-O	4.2	V-C	4.4
Sr-O	4.4	W-C	–
U-O	7.9	Zr-C	5.8
W-O	7.0		
Zn-O	<2.8		

possible that observed oxide to metal ratios (<0.2%) originated from the mass spectrometer interface region rather than from undissociated oxide from the plasma.

Because carbides are relatively involatile, the volatilization of these cannot be relied upon as a means of producing useful analyte vapour, but rather the formation of these should be avoided by selecting a vaporization temperature and experimental conditions that favour the production of the more volatile oxide or halide species. For some elements such as the platinum group elements [61] which are known to be reduced by graphite to metallic form, the very high boiling point of some of these elements requires that means be taken to either prevent reduction or promote the formation of volatile compounds with the use of chemical modifiers [13,61].

When selecting ETV operating conditions for the determination of any element, reference should be made to known vaporization (atomization) pathways and to analyte physical constants such as those given in Table 3.14.

Recent studies have appeared on the ETV-ICP-MS vaporization mechanisms of more than 30 elements [53–63]. One of the main conclusions that can be drawn from these studies is that it is almost always advantageous to vaporize analyte in the combined form as an oxide or a halide while at the same time avoiding the production of very refractory compounds such as carbides which tend to remain incompletely vaporized from the graphite surface and can also result in a memory or carryover effect.

To illustrate how the mechanism of vaporization of an analyte can be elucidated, the vaporization characteristics of two refractory elements, B and W are summarized below.

3.6.1 Vaporization of boron in ETV-ICP-MS

The sensitivity for the determination of B by GFAAS is very poor. One explanation for the low sensitivity for B implicates formation of non-volatile refractory carbides or the pre-atomization loss of volatile B_2O_3 [86]. Following up on thermodynamic calculations published by Frech et al. [97], Byrne et al. [53] investigated the mechanism of B vaporization using both GFAAS and ETV-ICP-MS. Graphite furnace AAS was used to obtain data on atomic species produced within the graphite tube and ETV-ICP-MS could be used to detect the volatilization of both polyatomic and atomic B species.

The thermodynamic equilibrium calculations by Frech [97] predicted that a number of volatile B polyatomic species should be formed in the graphite furnace at temperatures well below the appearance temperature of B atoms (≈ 2200°C). These species include $B_2O_{3(l)}$, $HBO_{2(g)}$, $B_2O_{2(g)}$ and $BO_{(g)}$ as well as involatile $B_4C_{(s)}$. According to Frech's calculations these species should form in various temperature ranges below 2300°C. The species listed above are in the order of increasing temperature of formation, i.e. the first gaseous species listed, HBO_2 forms between 900°C and 1600°C, while solid boron carbide is not predicted to form until about 2000°C, just prior to the appearance of B atoms (GFAAS). If this is the case, then B should begin to vaporize and be lost from the graphite tube as gaseous polyatomic species at temperatures above 900°C. This being so, then the determination of B by ETV-ICP-MS should be possible and not limited in sensitivity as is GFAAS. The volatilization of B species at temperatures below that required for the formation of B_4C may provide a means of sensitive detection for this element by ETV-ICP-MS.

Figure 3.6 shows the ETV-ICP-MS vaporization curve as well as the GFAAS atomization curve for B. Above this figure is shown the various

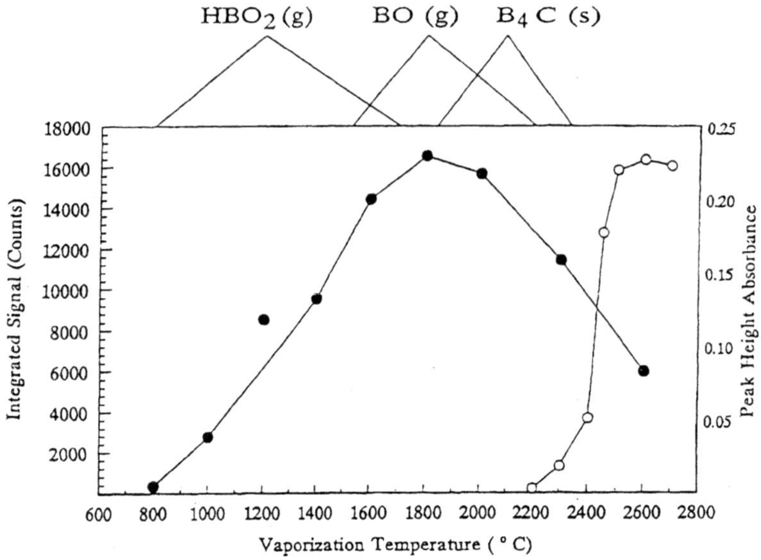

Fig. 3.6. ETV-ICP-MS vaporization curve for 2.5 ng of boron and GFAAS atomization curve for 25 ng of boron. The temperature regions in which molecular species are predicted to form are shown above the figure (from Ref. [53] with permission).

boron species that are predicted to form with lines drawn to indicate the temperature range over which these species occur. Data on vaporization and atomization curves were obtained by increasing the vaporization/atomization temperature and recording either the integrated ion intensity (ICP-MS) or the peak-height absorbance (GFAAS).

The most striking feature of Fig. 3.6 is the marked difference in the appearance temperature for boron in the GFAAS and ETV-ICP-MS experiments. In GFAAS, where only boron atoms are detected, boron is not observed below 2200°C. In the ETV-ICP-MS experiments, the boron signal appears at a vaporization temperature of about 800°C. Since atomic boron is not detected by GFAAS below 2200°C, the boron observed by ETV-ICP-MS at vaporization temperatures between 800 and 2200°C clearly originates from boron polyatomic species. These would be vaporized and lost from the graphite furnace at some stage prior to atomization, either in the pyrolysis step or the atomization ramp normally employed in GFAAS. This pre-atomization loss, commencing at temperatures as low as 800°C, accounts for the low sensitivity for boron analysis in GFAAS.

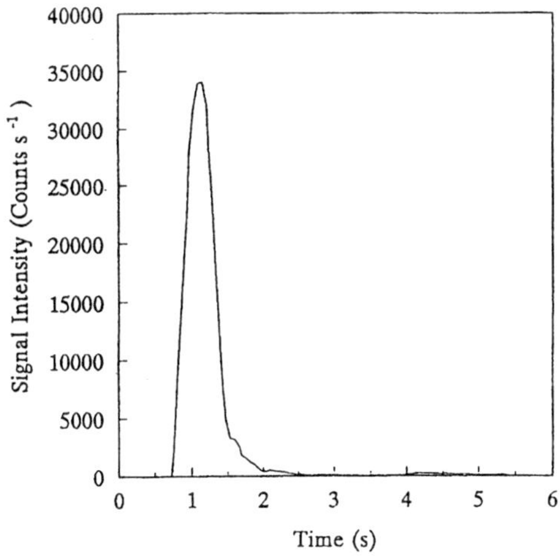

Fig. 3.7. ETV-ICP-MS signal for 2.5 ng of boron vaporized at 1600°C (from Ref. [53] with permission).

With regard to ETV-ICP-MS, these results show that the optimum vaporization temperature for the determination of boron is about 1800°C. Above this temperature the boron peak areas decrease, presumably because of the formation of polyatomic species, either BO or BC. Figure 3.7 shows an ETV-ICP-MS signal obtained for the vaporization of 2.5 ng of B at 1600°C. The signal is narrow and shows no tailing effect from the vaporization of less volatile species later in the vaporization cycle. A 10-fold increase in sensitivity for B was also reported [53] when Ni (0.5 µg) was used as a chemical modifier.

3.6.2 Vaporization of tungsten in ETV-ICP-MS

The vaporization mechanism of W in ETV-ICP-MS was reported by Byrne et al. [54]. A similar approach was used as for the B study except that GFAAS data was not obtainable. Tungsten is one of the elements (Table 3.4) not determinable by GFAAS because of the formation of the refractory carbide (Table 3.13) and the loss of volatile W polyatomic species from the graphite furnace prior to the high temperature measurement step. The latter effect would not hinder the determination of W by ETV-ICP-MS because W oxide species can be atomized and ionized on reaching the argon plasma.

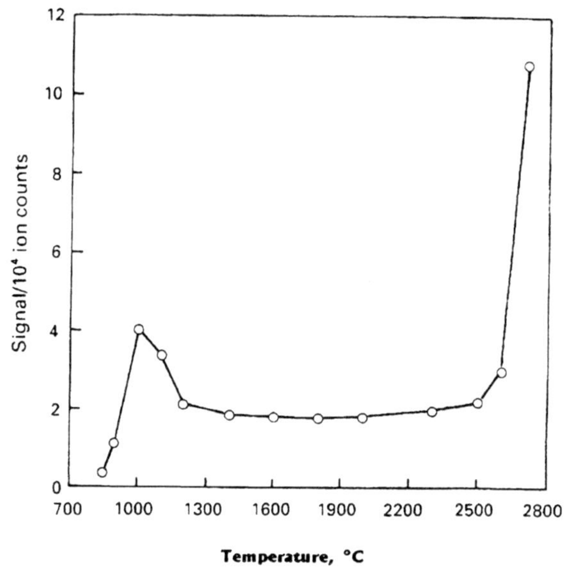

Fig. 3.8. ETV-ICP-MS vaporization curve for 250 pg of tungsten (from Ref. [54], with permission).

Figure 3.8 shows the ETV-ICP-MS vaporization curve for tungsten which, as discussed earlier, is a plot of the integrated signal intensity for $^{184}W^+$ as a function of vaporization temperature. In order to interpret the vaporization process for tungsten, this curve should be considered in conjunction with the series of ETV-ICP-MS signals shown in Fig. 3.9. The vaporization curve of Fig. 3.8 shows that tungsten begins to volatilize from the graphite surface at temperatures around 800°C. The amount of tungsten vaporized increases with increasing temperature up to 1000°C. Since the melting and boiling points of tungsten metal are extremely high, atomic tungsten could not be volatilized at these low temperatures. However tungsten oxide, WO_3, which would be expected to form when a tungstate solution is dried and heated, is much more volatile and has a vapour pressure of 9.1 torr at 980°C and is reported to sublime at around 1100°C. Thus the single peak observed at a vaporization temperature of 1000°C (Fig. 3.9a) could be attributed to the volatilization of WO_3.

As the vaporization temperature is raised above 1000°C the ETV-ICP-MS signal intensity, rather unexpectedly, begins to decrease and then levels off at around 20,000 counts when the vaporization tempera-

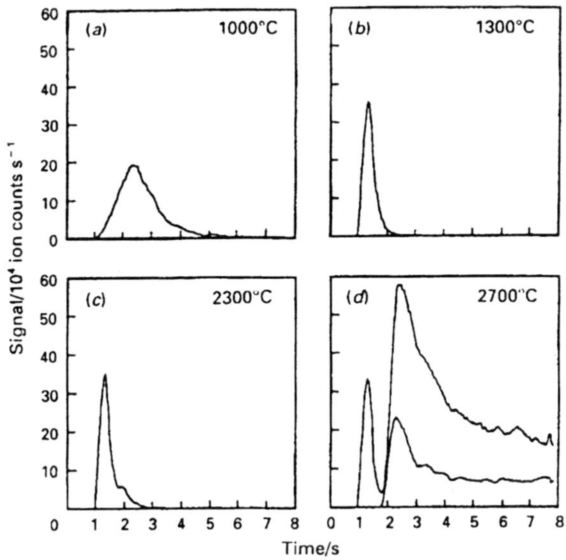

Fig. 3.9. ETV-ICP-MS signals for 250 pg of W vaporized at various temperatures: (a) 1000°C; (b) 1300°C; (c) 2300°C; (d) 2700°C (from Ref. [54] with permission).

ture reaches 1400°C (Fig. 3.8). This decrease can be explained by the onset of tungsten carbide formation which is reported to commence at about 850°C and be complete at 1410°C. As the temperature of the graphite surface rises above 1000°C any residual WO_3 which has not been vaporized will be converted to refractory WC which should not be then released until much higher vaporization temperatures. The amount of WO_3 converted to WC remains approximately constant for vaporization temperatures between 1400 and 2200°C. This is reflected by the flat portion of the vaporization curve in this temperature region and by the fact that the ETV-ICP-MS signal consists of a single peak throughout this temperature range. The peak shape at 1300°C, shown in Fig. 3.9b, is typical for this temperature range.

At a temperature around 2300°C the vaporization curve of Fig. 3.8 begins to rise as more tungsten is released from the graphite surface. This coincides with the appearance of a small shoulder on the trailing edge of the ETV-ICP-MS signal at 2300°C (Fig. 3.9c). As the temperature is further increased to 2700°C the vaporization curve rises steeply and a large second peak occurring between 2 and 8 s now appears in the ETV-ICP-MS signal (Fig. 3.9d). This second peak can be explained by the volatilization of WC at temperatures above 2500°C. Also, WC melts to form a eutectic with W_2C at 2525°C and then decomposes at 2600°C to form a W-rich liquid and solid carbon. Thus when tungsten is vaporized from a graphite substrate there are two distinct vaporization processes. At low temperatures WO_3 is volatilized to give a single peak in ETV-ICP-MS; at temperatures above about 2500°C this signal becomes double peaked as WC is released at the higher temperature. Associated with this second and much larger carbide peak is the problem of memory effect. Figure 3.9d (second smaller signal starting at 1.8 s) shows a typical blank signal from a second heating of the graphite tube at a temperature of 2700°C. For this blank firing there is no memory effect associated with the early WO_3 peak between 1 and 2 s, but the memory effect from the larger carbide peak, between 2 and 8 s, amounts to about one third of the signal for 250 pg of tungsten. Double peaks and memory effects pose practical problems for the analysis of tungsten by ETV-ICP-MS. Possible strategies to overcome these problems include: (i) prevention of formation of carbide and use of suitable chemical modifiers, and (ii) use of lower vaporization temperatures which release only WO_3 giving a single but smaller peak without memory effects. Reported limits of detection for W by ETV-ICP-MS were 0.51 pg using a vaporization temperature of 1100°C.

3.7 CHEMICAL MODIFICATION, PHYSICAL CARRIERS AND MASS TRANSPORT EFFICIENCY

3.7.1 Chemical modifiers and physical carriers

Chemical modification refers to the deliberate alteration of the thermal properties of the analyte, matrix components or the vaporization surface by the addition of a foreign substance. Within the context of ETV-ICP-MS, chemical modification also refers to the alteration of analyte vapour transport properties. Thus chemical modifiers can have both chemical effects and physical effects, each improving conditions for the determination of an analyte in a specific matrix.

It is at times beneficial to add modifiers to stabilize the analyte to allow for the elimination of more volatile matrix components, to improve the temporal separation of analyte and matrix during the volatilization step and to reduce losses prior to the high temperature step by preventing hydrolysis reactions which can expel analyte by co-volatilization with hydrolysis reaction products. On the other hand, modifiers can be added to *increase* the volatility of the analyte to promote the vaporization of analyte before matrix components and to prevent or reduce the interaction of analyte with graphite to form refractory compounds. As well, chemical modification can be useful in converting analyte to a single chemical form, eliminating double signals which can complicate quantitative analysis.

Chemical modification is generally accomplished by the addition of a foreign substance to the sample solution or the introduction of a gas other than argon to the transport (Ar) gas flow. Modifiers in solution are added just prior to or at the same time as sample solutions are added to the graphite tube and gas-phase modifiers are used during the high-temperature vaporization step. Chemical modifiers used in GFAAS have been reviewed in a paper by Tsalev et al. [87]. The quantities used range from 25 to 200 μg depending on the application (Table 3.6). Modifiers which improve the mass transport efficiency of analyte from the graphite tube to the argon plasma are called physical carriers and will be discussed later in this part of the chapter.

A significant number of papers have been published which report on the use of chemical modifiers. The use of Freon to increase the volatility of W vaporized on a carbon surface was first reported by Kirkbright and Snook [98] for ETV-ICP-AES studies and was later applied [10] to the determination of W in geological materials. Ediger [99] reported that the use of Freon could dramatically improve the sensitivity, especially for the determination of refractory elements such as U, Th and the REE. For the determination of the platinum group metals, Grégoire [13] demonstrated that the addition of Ni chemical modifier could improve the sensitivity by up to a factor of 9.4. Table 3.15 shows the effect of adding a variety of chemical modifiers on the determination of Os by ETV-ICP-MS. Osmium is reduced to metal on the graphite surface during heating and remains on the surface as a refractory metal. The boiling point of Os is greater than 5300°C and it therefore has a relatively low vapour pressure at the maximum vaporizer temperature of 2700°C. The presence of Ni, Se or Te likely results in the formation of relatively more volatile alloys or compounds [13, 61]. An improvement of 17-fold in Os sensitivity was realized when 2 μg of Te was used as chemical modifier. Byrne et al. [32] reported that the use of ascorbic acid as a chemical modifier reduced analyte loss during the pyrolysis heating step by slowing down the hydrolysis of magnesium

TABLE 3.15

Effect of matrix modifier on Os (0.25 ng) ion count rate (from Ref. [15] with permission)

Matrix modifier (2 μg)	$^{192}Os^+$ Ion count rate (cs^{-1})
–	36,450
Sodium chloride	36,800
Thiourea	50,100
8-Hydroxyquinoline	63,400
Nickel	176,900
Selenium	183,700
Tellurium	653,000

chloride thus preventing loss of analyte (Mn) by co-volatilization with released HCl. This example illustrates the use of a chemical modifier to stabilize matrix components.

The influence of the condition of the graphite substrate on analyte signal was reported by Majidi and Miller-Ihli [100]. These authors concluded that the use of an oxygenated pyrolytic graphite surface and chemical modifiers gave the best precision for ETV-ICP-MS signals. Ediger and Beres [46] and Grégoire et al. [47] reported on the use of NaCl and Pd/Mg chemical modifiers. In a comprehensive paper by Hughes et al. [43], the use of seawater stripped of trace metals was investigated as a universal mixed chemical modifier. Other papers report on the use of mannitol as a modifier for B [101], halogen-containing compounds for Zn, Cd and Pb [44] and a series of ten modifiers ranging from HCl to freon for the determination of B, La and U [102]. Eleven modifiers were studied by Grünke et al. [45] and applied to the determination of Mn, Cu, Zn, Cd and Pb. Fonseca and Miller-Ihli [103] reported on the influence of sample matrix components on analyte signals and on the selection of calibration strategy, particularly as applied to the analysis of slurries.

The use of a physical carrier is essential if aqueous external standards are to be used for quantitative analysis of multi-component sample solutions [104]. For a physical carrier to work, it must be present in the gas-phase at the same time (and place) as the analyte vapour. Any single substance used as a physical carrier will have a limited temperature range over which it will volatilize, and therefore it is unlikely that a physical carrier could be found that would work equally well for a volatile element such as Cd and relatively involatile elements such as the REE. Therefore, a mixed physical carrier of known composition and of high purity is required.

A mixed physical carrier that could meet all of the criteria mentioned above was proposed by Grégoire and Sturgeon [52] and Hughes et al. [43]. These authors suggested that seawater be used as a mixed physical carrier. A standard reference seawater (NASS-3) used in trace analysis studies and available from the National Research Council of Canada was selected. NASS-3 reference seawater is readily available and is well characterized for its trace metal content. Before use, trace metals were stripped from the NASS-3 by chelation on a column of immobilized ligand and diluted 500-fold with ultra-pure water. The resulting solution provides ≈0.07 $\mu g\ \mu l^{-1}$ dissolved solids of which 0.055 $\mu g\ \mu l^{-1}$ is NaCl, 0.011 $\mu g\ \mu l^{-1}$ is $MgCl_2$, 0.0022 $\mu g\ \mu l^{-1}$ is $CaCl_2$ and 0.0015 $\mu g\ \mu l^{-1}$ is KCl.

Figure 3.10 shows ETV-ICP-MS signals for a number of analytes (parts a and b) and for the major components of NASS-3 (part c). A comparison of these signals shows that individual components of NASS-3 volatilize at different times and together span the time during which all of the analytes in parts a and b (Fig. 3.10) are being volatilized. Significant (>×10) sensitivity enhancements using NASS-3 physical carrier were reported [67] for several analytes studied.

In separate study, Hughes et al. [43] reported on the mechanism by which NASS-3 improves the mass transport of analyte from the graphite furnace to the argon plasma. The determination of 10 elements using NASS-3 as a physical carrier as well as studies using separate salts (NaCl and $MgCl_2$) in different acids (HCl and HNO_3) were reported. Table 3.16 summarizes results obtained for differing masses of physical carrier using HCl and HNO_3 as diluent. The results show that for both volatile and relatively involatile elements, the greatest enhancements (increase in mass transport efficiency) in integrated counts were obtained using $MgCl_2$. For elements of intermediate volatility, NaCl gave the best results. Elements of high volatility (e.g. Cd) benefited most from the use of HCl and it was concluded that HCl itself is a powerful physical carrier. This acid may promote the formation of simple chlorides of analytes which are volatilized from the graphite surface in elemental form. Single atomic particles or small clusters of atoms are reported to have a low mass transport efficiency [105] over distances greater than a few centimetres.

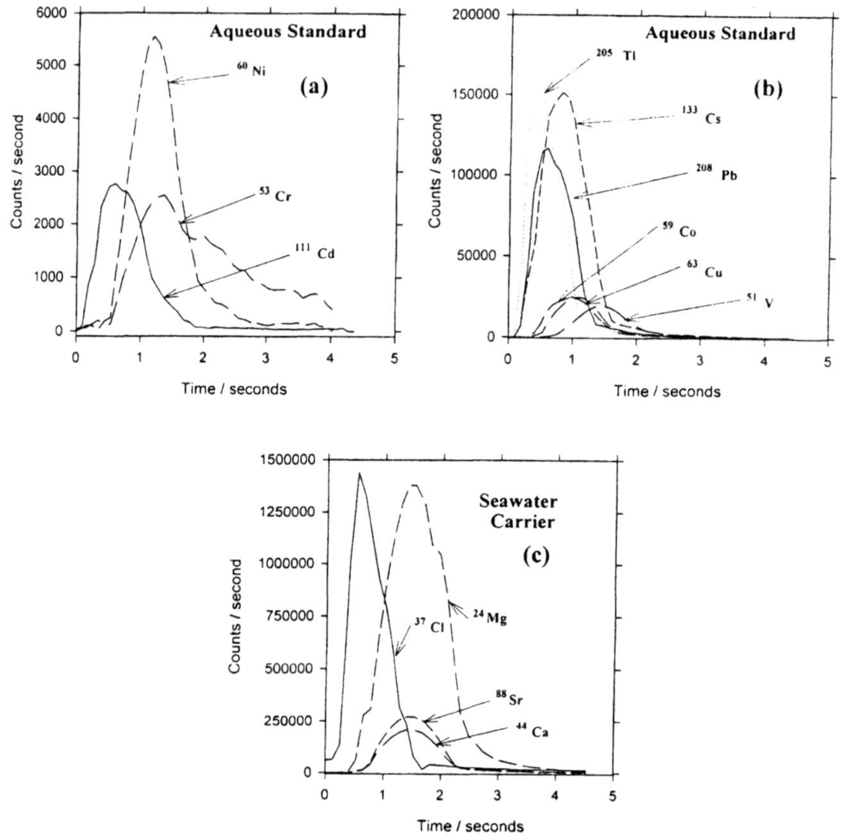

Fig. 3.10. Comparison of ETV-ICP-MS signals derived from analyte (a and b) and NASS-3 seawater components (c) (from Ref. [67] with permission).

The mechanisms proposed by Hughes et al. [43] define three zones of action to explain the effect of NASS-3 on analyte signals. Figure 3.11 schematically illustrates these three zones and also shows the elements whose appearance temperatures are within these zones. The appearance temperature is defined as the temperature (of the graphite surface) at which the analyte signal becomes measurable above the background noise.

In the first zone, which is the lowest temperature zone, hydrolysis of magnesium chloride occurs. A reaction product of hydrolysis is HCl

TABLE 3.16

Effect of acids, physical carrier and mass of physical carrier on integrated analyte (20 pg) signal (from Ref. [43])

Analyte	Acid (1%)	NASS-3 (μg)			NaCl (μg)			$MgCl_2$ (μg)		
		0.07	0.35	1.05	0.07	0.35	1.05	0.07	0.35	1.05
^{114}Cd	HNO_3	2.9	3.2	2.5	2.2	1.9	1.8	1.8	4.5	6.1
	HCl	1.2	4.2	8.5	1.0	0.4	0.2	3.5	8.5	11.5
^{205}Tl	HNO_3	1.8	1.6	1.5	2.0	1.7	1.6	0.8	2.1	2.4
	HCl	0.8	0.7	0.8	0.9	1.0	1.0	0.9	1.8	2.6
^{208}Pb	HNO_3	1.8	1.5	1.3	1.8	1.7	1.6	1.6	1.9	2.2
	HCl	1.0	0.9	0.9	1.0	0.8	0.7	1.7	1.2	1.1
^{133}Cs	HNO_3	1.3	1.2	1.2	1.1	1.2	1.2	1.0	1.1	1.3
	HCl	2.6	2.2	2.1	2.5	2.2	2.2	1.6	2.5	2.8
^{85}Rb	HNO_3	1.3	1.1	0.9	1.2	1.1	1.0	1.0	1.0	1.2
	HCl	1.3	0.9	0.8	1.2	0.9	0.9	1.1	1.6	1.8
^{209}Bi	HNO_3	1.8	1.1	1.0	2.1	1.4	1.3	0.9	1.1	1.6
	HCl	1.2	1.1	0.9	1.3	1.0	1.0	1.0	0.9	1.1
^{107}Ag	HNO_3	1.5	0.7	0.5	1.4	0.6	0.5	1.0	1.1	1.5
	HCl	1.3	0.7	0.5	1.2	0.7	0.5	1.0	1.2	1.8
^{115}In	HNO_3	1.7	1.3	0.9	1.7	1.5	1.2	1.3	1.2	0.9
	HCl	0.8	0.9	1.0	0.9	0.8	1.0	1.0	1.0	1.0
^{69}Ga	HNO_3	1.2	0.8	0.7	1.3	1.0	0.8	0.8	0.7	0.7
	HCl	1.2	1.0	0.9	1.3	1.0	0.8	0.9	0.7	0.7
^{59}Co	HNO_3	1.4	1.2	1.0	1.3	1.2	1.1	1.2	1.1	1.1
	HCl	1.4	1.2	1.0	1.1	0.9	0.8	1.2	1.1	1.1

which acts as an effective carrier for volatile elements. The second zone, which is of intermediate temperature, spans the time during which most of the NaCl is being volatilized and serves as a physical carrier. In the third and highest temperature zone, residual MgO remaining from the hydrolysis reaction which occurred in zone 1 is volatilizing. Several phenomena could be responsible for signal enhancement in zone 3. First, volatilized MgO could simply act as physical carrier and interact with analyte only in the gas phase. Secondly, MgO could occlude analyte within its matrix and co-volatilize with the analyte and finally, MgO can serve as an effective barrier

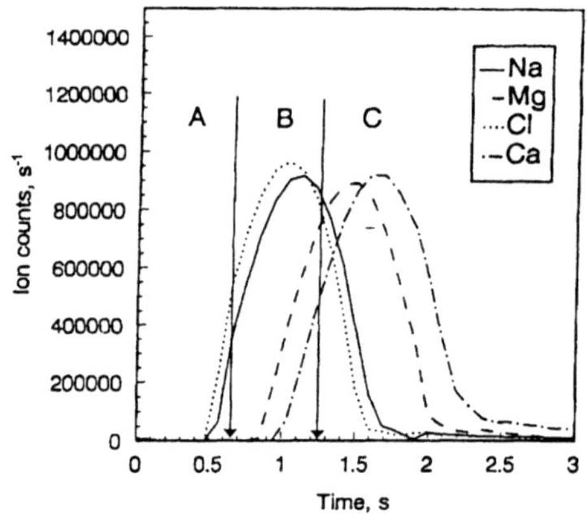

Fig. 3.11. ETV-ICP-MS signals for major ions derived from NASS-3 seawater carrier. Zone A: Carrier HCl; volatile elements: Cd, Ag, Tl, Cs, Rb, Pb, Zn. Zone B: Carrier NaCl; medium volatility elements: Bi, In, Ga. Zone C: Carrier Mg/MgO, Ca/CaO; low volatility elements: Co, Ni, Dy, Eu,Tm, Yb.

preventing interaction of analyte with the graphite surface inhibiting the formation of refractory compounds such as carbides.

In a related study, Grégoire et al. [64] reported on the vaporization properties of HCl and HNO_3 vaporized under various experimental conditions. Drying step temperatures of 140°C (50 s) and pyrolysis step temperatures of 400°C (10 s) were effective in volatilizing most of the chloride from a 10 µl 1% (v/v) HCl solution, however, a small amount (40 ng) of acid was retained on the graphite even after pyrolysis at 400°C. Under the same experimental conditions, HNO_3 was completely volatilized from the graphite tube. The effect of a range of concentrations of HCl, HNO_3, H_2SO_4 and H_3PO_4 on analyte signals for Co, Cu, Ag, Cs, Pb, Bi and U were studied. Analyte signals were enhanced by as much as a factor of two in the presence of 1% (v/v) HNO_3 and H_2SO_4. Phosphoric acid suppressed analyte signals for Ag and Bi and the use of HCl resulted in relatively small changes in analyte sensitivity. The use of a pyrolysis step in the heating program reduced effects associated with acid matrices, but at the expense of signal intensity. NASS-3 reference seawater carrier attenuated the matrix effects associated with H_2SO_4 and H_3PO_4 and essentially eliminated them for HNO_3 and HCl. Table 3.17 shows the effect of these acids at two different concentrations and Table 3.18 illustrates the power of NASS-3 to control these differences. As a consequence of the above studies, one can

TABLE 3.17

Effect of acid concentration on integrated analyte signal pulses (50 pg) (from Ref. [64], with permission)

	Integrated signal (counts)						
	Co	Cu	Ag	Cs	Pb	Bi	U
0.05% HNO_3	16275	7144	12507	163790	122529	174329	136369
RSD % (n=5)	2.5	6.4	3.8	4.5	7.6	8.5	7.7
1.0% HNO_3	19659	11417	22294	196151	196731	263292	225123
RSD % (n=5)	5.6	8.1	9.9	3.6	7.7	5.6	5.4
0.05% HCl	19833	10530	20060	187167	250100	212667	106140
RSD % (n=5)	1.1	3.6	1.00	0.9	0.70	6.1	6.3
1.0% HCl	19616	10155	19863	187006	248335	199736	99589
RSD % (n=5)	4.8	4.7	3.1	3.6	6.3	4.3	12.0
0.05% H_2SO_4	25211	15720	13497	122575	129097	136053	116543
RSD % (n=5)	6.5	6.5	6.3	5.1	5.7	4.2	5.3
1.0% H_2SO_4	34887	21455	18065	268796	239324	368462	217413
RSD % (n=5)	5.1	5.6	5.0	8.8	7.0	7.4	8.4
0.05% H_3PO_4	45040	–	23254	188548	–	206831	335407
RSD % (n=5)	9.7	–	9.9	12.6	–	5.6	10.1
1.0% H_3PO_4	27139	–	13225	371608	–	192756	239306
RSD % (n=5)	10.6	–	2.5	9.4	–	4.0	6.4

conclude that the use of HCl rather than HNO_3 is recommended for ETV-ICP-MS work and in particular for the determination of volatile elements. This is in marked contrast to SN-ICP-MS where HNO_3 is the preferred acid. The formation of analyte chlorides in the argon plasma however, remains a concern for both ETV and SN work when samples high in chloride concentration are being analysed. In a related study, the retention of water and hydrogen from graphite tubes and its effect on analyte signals for Al, V, Cu and Pb was reported by Grünke et al. [106].

The very high sensitivity of ETV-ICP-MS and its use for multi-element analysis places some requirements on the chemical modifiers

TABLE 3.18

Effect of acid concentration on integrated analyte signal pulses (50 pg) using NASS-3 as a physical carrier (from Ref. [64] with permission)

	Integrated signal (counts)						
	Co	Cu	Ag	Cs	Pb	Bi	U
0.05% HNO_3	30382	10148	29399	125492	220425	373365	53410
RSD % (n=5)	9.4	8.1	3.3	8.1	5.4	2.4	6.3
1.0% HNO_3	28464	11165	27839	115206	225277	366476	48990
RSD % (n=5)	7.9	3.3	4.1	3.3	7.3	2.8	5.4
0.05% HCl	23085	9297	24848	149903	256772	223700	30720
RSD % (n=5)	2.3	10.4	5.5	7.9	4.7	4.2	8.3
1.0% HCl	21648	9409	19402	144490	256704	181660	48030
RSD % (n=5)	2.9	5.6	1.3	5.4	6.9	5.1	8.4
0.05% H_2SO_4	40437	29286	37877	322094	284412	300002	58469
RSD % (n=5)	3.42	4.1	7.8	3.5	6.0	3.1	5.3
1.0% H_2SO_4	42471	30117	27204	394135	381171	480804	114642
RSD % (n=5)	5.34	4.3	7.0	3.7	1.6	6.7	6.8
0.05% H_3PO_4	46581	–	31027	181483	–	209078	469163
RSD % (n=5)	9.1	–	4.6	5.9	–	17.8	9.5
1.0% H_3PO_4	27581	–	13626	419118	–	389964	188118
RSD % (n=5)	5.6	–	2.0	4.4	–	5.0	13.6

used. For example, modifiers must be of the highest purity, free of elements that could interfere with target analytes. The modifier should be of benefit to the largest number of analytes avoiding the use of several different modifiers for multi-element schemes.

An important consideration when using modifiers such as those mentioned above, is the contribution that microgram quantities of these substances can have to the background spectrum. For example, if sufficient quantities of an element are vaporized into the plasma, it is known that polyatomic ions can be formed such as argides, oxides, nitrides and for ETV, carbides. Grégoire and Sturgeon [52] and also Björn et al. [90] addressed this problem and reported on the polyatomic

TABLE 3.19

Isotopes from graphite and sodium chloride, magnesium nitrate, nickel nitrate and palladium nitrate chemical modifiers (from Ref. [52] with permission)

Carbon	Sodium	Magnesium	Chlorine
^{12}C (98.89)*	^{23}Na (100)	^{24}Mg (78.6)	^{35}Cl (75.4)
^{13}C (1.11)		^{25}Mg (10.11)	^{35}Cl (24.6)
		^{26}Mg (11.29)	
Nickel	Palladium		
^{58}Ni (67.77)	^{106}Pd (27.1)		
^{60}Ni (26.16)	^{108}Pd (26.7)		
	^{105}Pd (22.6)		
	^{110}Pd (13.5)		
	^{104}Pd (9.3)		
	^{102}Pd (0.8)		

*Percent natural abundance.

ions formed in the plasma from the vaporization chemical modifiers. The results obtained by Grégoire and Sturgeon [52] using ascorbic acid, sodium chloride, magnesium nitrate, nickel nitrate and palladium nitrate are summarized below. These modifiers were selected for study because these are the most commonly used for chemical modification in ETV-ICP-MS studies. For each modifier, the impact of the polyatomic ions formed on the determination of individual elements is discussed and alternative analyte isotopes are suggested for use.

Table 3.19 lists the source isotopes (carbon and modifiers) from which interfering polyatomic ions can be formed in the argon plasma. Grégoire and Sturgeon [52] report that for all of the matrix modifiers studied, no carbides or nitrides were detected in the plasma and therefore the formation of these polyatomic ions will not be considered further.

3.7.1.1 Sodium chloride

Table 3.20 lists the possible oxide and argide polyatomic ions that can be formed when NaCl chemical modifier is vaporized into an argon plasma. The detection of sodium oxide was not possible since this polyatomic ion at a *m/z* of 39 daltons is overlain by the wings of the intense ^{40}Ar ion. Sodium oxide (*m/z* = 39), should it exist, would be

TABLE 3.20

Possible major interfering polyatomic ions produced when using sodium chloride chemical modifier (from Ref. [52] with permission)

	m/z of Polyatomic ion	Isobaric analyte isotopes (abundance)
Sodium oxide	39	K (93.1)
Chlorine oxides	51	V (99.8)
	53	Cr (9.6)
Sodium argide	63	Cu (69.1)
Chlorine argides	75	As (100)
	77	Se (7.5)

isobaric with the major isotope of K (93.1%), an element not generally determinable at trace levels by ICP-MS.

Copper: No sodium argide was observed at masses below 10 μg of added sodium. Sodium argide is isobaric with ^{63}Cu (69.1%), the major Cu isotope. The BEMs for ^{63}Cu resulting from the addition of 10 μg and 20 μg of Na were 1.5 pg and 3.6 pg, respectively.

Arsenic and Selenium: The use of up to 10 μg of NaCl did not result in the production of any carbides, nitrides, oxides or argides of chlorine. At elevated levels of Cl (20 μg) only the chlorine argides are observed, resulting in BEMs of 0.4 pg and 1.1 pg for ^{75}As and ^{77}Se, respectively. The use of such high levels of Cl as NaCl chemical modifier is unnecessary [46] and therefore this interference would not normally be encountered. However, it is possible that such high levels of Cl may exist in a real sample solution or slurry. Correction for these polyatomic ions may be necessary, particularly for As which is monoisotopic and also for ^{77}Se (7.5%) which is the only Se isotope that is free of isobaric interferences from isotopes of other elements.

The formation of natrides with analyte ions was also investigated. Copper natrides at *m/z* of 86 and 88 daltons were not detected even when 5 μg of NaCl were vaporized with 1000 pg of Cu. The corresponding copper chlorides were also not observed during vaporization of 8 μg of Cl as sodium chloride.

3.7.1.2 Magnesium nitrate

Table 3.21 lists the oxide and argide polyatomic ions that can be formed when magnesium nitrate chemical modifier is vaporized into an argon

TABLE 3.21

Possible major interfering polyatomic ions produced when using magnesium nitrate chemical modifier (from Ref. [52] with permission)

	m/z of polyatomic ion	Isobaric analyte isotopes (abundance)
Magnesium oxides	40	Ar (99.6), K (0.01); Ca (96.9)
	41	K (6.9)
	42	Ca (.64)
Magnesium argides	64	Ni (1.2), Zn (49.8)
	65	Cu (30.9)
	66	Zn (27.8)

TABLE 3.22

Background equivalent analyte mass for spectral interferences produced when vaporizing 5 μg of magnesium nitrate chemical modifier (from Ref. [52] with permission)

Analyte	Background equivalent mass (pg)
^{42}Ca	149
^{64}Ni	28.5
^{65}Cu	0.14
^{64}Zn	0.68
^{66}Zn	0.18

plasma. When heated, magnesium nitrate decomposes to form the oxide which is subsequently vaporized from the graphite surface during the high temperature step.

Magnesium oxides and argides were detected when 5 μg of Mg as nitrate were vaporized in the HGA-600MS. The BEMs for these polyatomic ions for the affected analytes are given in Table 3.22 and are discussed below.

Calcium: Magnesium oxides occur at three *m/z* ranging from 40 to 42 daltons. The oxides at *m/z* 40 and 41 are isobaric with $^{40}Ar^+$ and $^{40}Ar^1H^+$ and hence detection of these oxides was not possible. However, a magnesium oxide isobaric with ^{42}Ca (0.64%) was detected giving a

BEM of 149 pg Ca for this isotope. ^{44}Ca (2.13%) is isobaric with the small but significant CO_2^+ ion (see background spectrum and ascorbic acid discussions) and ^{48}Ca (0.18%) is isobaric with the major isotope of Ti. The only Ca isotope free from isobaric interference was ^{43}Ca with an abundance of 0.13%.

Nickel: ^{64}Ni (1.2%) is isobaric with the major zinc isotope and also isobaric with a magnesium argide. The three remaining Ni isotopes including ^{60}Ni (21.16%) are free of isobaric interferences from magnesium argide polyatomic ions.

Copper: ^{65}Cu (30.91%) is isobaric with a magnesium argide giving a BEM of 0.14 pg. However, the major copper isotope (^{63}Cu) was free of magnesium argide polyatomic ions.

Zinc: Both ^{64}Zn (49.8%) and ^{66}Zn (27.8%) are isobaric with magnesium argides giving BEMs of 0.68 pg and 0.18 pg, respectively. Of the remaining Zn isotopes, ^{64}Zn (48.9%) is isobaric with a minor Ni isotope and ^{70}Zn is isobaric with a Ge isotope leaving ^{67}Zn (4.1%) and ^{68}Zn (18.56%) free of interference.

3.7.1.3 Nickel nitrate

Table 3.23 lists the oxide and argide polyatomic ions that can be formed when nickel nitrate chemical modifier is vaporized into an argon plasma. When heated, nickel nitrate decomposes to form the oxide which is subsequently either thermally decomposed to elemental nickel or rapidly reduced by carbon during the high temperature step.

The formation of nickel oxide in the plasma produced isobaric interferences with two isotopes each of Ge and Se and nickel argides produced isobaric interferences with two isotopes each of Mo and Ru. Background equivalent masses for these isobaric interferences are given in Table 3.24.

Germanium: A nickel oxide polyatomic ion is isobaric with ^{74}Ge (36.53%), the major Ge isotope. This isotope is isobaric with a minor Se isotope of 0.96% abundance. ^{76}Ge (7.76%) which was interfered by a nickel oxide is also isobaric with a minor Se isotope of 9.12% abundance. Of the three remaining Ge isotopes, ^{72}Ge (27.43%) and ^{73}Ge (7.76%) were free of isobaric interferences. ^{72}Ge, having the higher isotopic abundance is recommended for use when determining this element in the presence of nickel.

Selenium: A nickel oxide is isobaric with ^{74}Se (0.96%) which is also isobaric with the major Ge isotope of 36.53% abundance. A second nickel oxide interfered with ^{76}Se (9.12%) which is isobaric with ^{76}Ge

TABLE 3.23

Possible major interfering polyatomic ions produced when using nickel nitrate chemical modifier (from Ref. [52] with permission)

	m/z of Polyatomic ion	Isobaric analyte isotopes (abundance)
Nickel oxides		
74	Ge (36.5), Se (.96)	
76	Ge (7.8), Se (9.1)	
Nickel argides		
98	Mo (23.8), Ru (2.2)	
100	Mo (9.6), Ru (12.7)	

TABLE 3.24

Background equivalent analyte mass for spectral interferences produced when vaporizing 1 μg of nickel nitrate chemical modifier (from Ref. [52] with permission)

Analyte	Background equivalent mass (pg)
^{74}Ge	0.22
^{76}Ge	0.40
^{74}Se	8.4
^{76}Se	0.34
^{98}Mo	43.1
^{100}Mo	41.9
^{98}Ru	464
^{100}Ru	31.6

(7.76%). Of the four remaining isotopes of Se, only ^{77}Se (7.5%) is free of isobaric interferences. ^{80}Se (49.97%), the major Se isotope, is isobaric with the intense argon dimer. Both Se isotopes at *m/z* of 78 and 82 daltons are isobaric with Kr isotopes. Krypton is often present as a contaminant in the Ar supply with intensities ranging to hundreds of counts/s for the more abundant isotopes. Since ^{78}Se is more abundant (23.61%) than ^{82}Se (8.84%) and since ^{78}Kr is only 0.35% abundant compared to 11.56% for ^{82}Kr, then ^{78}Se is a second isotope which can be used for the determination of this element in the presence of Ni.

Molybdenum: Nickel argides are isobaric with ^{98}Mo (23.78%) and ^{100}Mo (9.63%). Both of these isotopes are isobaric with minor Ru isotopes. Of the remaining five Mo isotopes, only ^{95}Mo (15.72%) and ^{91}Mo (9.46%) are free of isobaric interferences.

Ruthenium: ^{98}Ru (2.22%) and ^{100}Ru (12.7%) are isobaric with nickel argides. Both of these Ru isotopes are also isobaric with Mo isotopes. Of the remaining five Ru isotopes, ^{99}Ru (12.81%) and ^{101}Ru (16.98%) are free of isobaric interferences.

The nickel oxides observed from the use of nickel nitrate chemical modifier were likely formed in the plasma and not vaporized as oxides from the surface of the graphite tube. This was verified by increasing the charring temperature from 300°C to beyond 1000°C, a temperature at which carbon reduction of the oxide readily takes place. No change in the nickel oxide intensity was observed with increasing charring.

3.7.1.4 Palladium nitrate

Table 3.25 lists the polyatomic ions that can be formed when palladium nitrate chemical modifier is vaporized into an argon plasma. The relatively large number of species that can arise is due to the large number of Pd isotopes each having significant natural abundances. Therefore, use of palladium nitrate as a chemical modifier can have a potentially considerable impact on a fairly large number of analytically useful isotopes.

No oxides of palladium were detected when vaporizing 1 μg and 10 μg quantities of Pd (as nitrate) chemical modifier. Pd argide polyatomic ions, however were detected and these are isobaric with six isotopes of Nd, three isotopes of Sm and one Ce isotope. The BEMs determined for these polyatomic ion interferences are given in Table 3.26.

Neodymium: All but one of the seven Nd isotopes is interfered by a palladium argide. ^{143}Nd (12.14%) is free of argide isobaric interferences and also free of isobaric interference from isotopes of other elements. A second isotope, ^{142}Nd, has an abundance of 27.4% and a relatively low BEM of 0.27 pg. ^{142}Nd can be used for Nd determinations in the presence of Pd only if sample concentration levels are well above concentrations corresponding to 0.27 pg in 20 μl of sample solution.

TABLE 3.25

Possible major interfering polyatomic ions produced when using palladium nitrate chemical modifier (from Ref. [52] with permission)

	m/z of Polyatomic ion	Isobaric analyte isotopes (abundance)
Palladium oxides	118	Sn (24.1)
	120	Sn (33), Te (0.09)
	121	Sb (57.3)
	122	Sn (4.7), Te (2.5)
	124	Te (4.6), Sn (5.98)
	126	Te (18.7), Xe (0.09)
Palladium argides	142	Ce (11.1), Nd (27.1)
	144	Nd (23.8), Sm (3.1)
	145	Nd (8.3)
	146	Nd (17.3)
	148	Nd (5.7), Sm (11.2)
	150	Nd (5.8), Sm (7.4)

TABLE 3.26

Background equivalent analyte mass for spectral interferences produced when vaporizing 1 µg of palladium nitrate chemical modifier (from Ref. [52] with permission)

Analyte	Background equivalent mass (pg)
^{142}Nd	0.27
^{144}Nd	3.6
^{145}Nd	25.0
^{146}Nd	14.4
^{148}Nd	43.0
^{150}Nd	22.0
^{144}Sm	27.0
^{148}Sm	21.6
^{150}Sm	16.4
^{142}Ce	0.65

Samarium: Three of the seven Sm isotopes are isobaric with Pd argides. ^{147}Sm (14.97%) and ^{149}Sm (13.8%) are free of Pd argide interference and are not isobaric with isotopes from other elements. ^{152}Sm (26.26%) is the major Sm isotope and is also free of Pd argide interference, however, this isotope is isobaric with ^{152}Gd (0.2%). For samples low in Gd (relative to Sm), ^{152}Sm can be used for analysis providing appropriate Gd corrections are made.

Cerium: ^{142}Ce (11.1%) is isobaric with a Pd argide. ^{140}Ce (88.48%) is free of isobaric interference and as the major Ce isotope is the analyte isotope of choice.

It is clear from the recent ETV-ICP-MS literature that the judicious selection and use of chemical modifiers is key to the successful application of this technique. Most of the elements which suffer interference from modifier polyatomic ions have suitable isotopes that are free from any isobaric interferences or have at least one isotope with a BEM close to the limit of detection. However, the loss of available free isotopes may affect the determination of isotope ratios or the use of isotope dilution as a calibration strategy.

The above discussion shows that the use of microgram quantities of a variety of chemical modifiers results in the formation of only oxide and argide polyatomic ions. The concentration levels of these interfering species correspond in most cases to relative concentrations of analyte in the parts-per-trillion range assuming the use of a 50 µl sample aliquot. The use of in-situ pre-concentration techniques such as repetitive sampling (and drying) prior to vaporization or the deposition of gaseous hydrides in the graphite furnace may reduce even further the influence of these polyatomic ions.

It is well known that the selection of plasma parameters and to some extent instrumental operating conditions of the quadrupole mass spectrometer can influence the degree of oxide and argide formation. Therefore, it is possible that instrumental operating conditions other than those used to obtain the above results [52] could reduce (or increase) the BEM values for isobaric interferences reported for each chemical modifier.

3.7.2 Analyte transport efficiency

Electrothermal vaporization as a means of sample introduction for ICP mass spectrometry has been successfully used for over fifteen years. It was clear, even in early studies, that the mass transport efficiency was

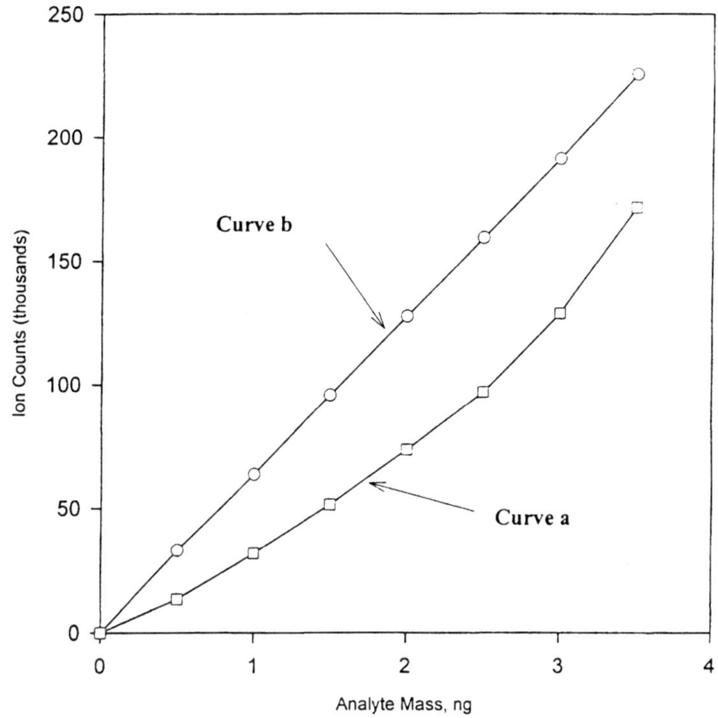

Fig. 3.12. Effect of physical carrier on ETV-ICP-MS calibration curves. Curve a: analyte alone. Curve b: analyte with carrier.

not quantitative and that a significant fraction of vaporized analyte remained within the ETV system or transfer tubing and did not reach the plasma. Electrothermal vaporization ICP-MS signals increased in intensity when nanograms of analyte were co-vaporized with microgram quantities of an added salt, such as NaCl. A second effect, observed and reported almost simultaneously in the literature for both filament type [47] and tube-type [46] ETV systems, was the nonlinearity of calibration curves when analyte was vaporized alone. Upon the addition of a salt, linear calibration curves were obtained along with a general enhancement of signal ranging from 20% to about a factor of ten, depending upon analyte and experimental conditions. This effect is illustrated in Fig. 3.12. The similarity of results obtained from the two different vaporizer systems indicated that incomplete analyte transport was not a problem specific to an ETV design, but rather an effect present (to various degrees) in all ETV systems. Ediger and Beres [46] interpreted the non-linearity of calibration curves and incomplete mass transfer using a model derived by Kántor [105] to describe similar effects in ETV ICP atomic emission studies. Kántor [105] first used the theoretical prediction (Eq. 1) to describe the transport of analyte vapour from the ETV surface to the argon plasma as being dependent on the formation of stable nuclei of size exceeding a critical diameter whose value is given by the equations:

$$d_p = 4\sigma \frac{V_m}{kT \ln S} \tag{1}$$

where, d_p is the critical diameter, σ is the surface tension of the liquid droplet, V_m, is the molecular volume of the vapour species, k is the Boltzmann constant, T is the temperature in Kelvin and S is the saturation ratio. The saturation ratio is a measure of the magnitude of analyte supersaturation and is given by:

$$S = \frac{p_{vap}}{p_{(T)}} \tag{2}$$

where p_{vap}, is the partial vapour pressure (atm) and $p_e(T)$ is the equilibrium vapour pressure at the temperature of nucleation.

For ETV sample introduction, as the vaporization surface increases in temperature and sample vaporization takes place, the partial vapour pressure of the analyte in the argon carrier gas stream increases. The vapour pressure of the analyte can be approximated by the following relationship:

$$p_{vap} = N_s T_g R / V_g t_v \tag{3}$$

where N_s is the number of moles of vapour from the evaporated sample, T_g is the initial temperature of the carrier gas (in Kelvin), R is the gas constant (cm^3 atm K^{-1} mol^{-1}), V_g is the flow rate (cm^3 s^{-1}) of the carrier gas at the initial temperature T_g, and t_v is the mean evaporation time (s) of the major fraction of the sample.

The above equations show that, as the partial vapour pressure of the analyte increases, the saturation ratio increases and the critical size shifts to smaller particles. This results in an increase in the concentration of clusters and stable nuclei. This increase in stable nuclei is much steeper than a linear increase in S; therefore, as the mass of the analyte in the vapour is increased, the formation of stable particles increases non-linearly, which accounts for the shape of the analytical calibration curve observed when no carrier is present. In other words, at very low analyte mass, the concentration of atoms or clusters in the vapour state is insufficient to promote self-nucleation or homogeneous condensation. In this case, particles are lost to cold containment or conduit surfaces before they can be transported to the argon plasma or other measurement device. At higher analyte vapour concentrations, the efficiency of formation of transportable (by the argon stream) particles increases at a faster rate than the vapour state concentration of analyte, resulting in non-linear analytical calibration curves.

The presence of a foreign substance, such as a matrix component, also serves to provide nucleation and condensation sites (heterogeneous nucleation) for analyte atoms or molecules and thus foreign substances improve the transport efficiency of analyte by providing a 'carrier' for the analyte. The presence of sufficient concentrations of analyte or matrix component vapour provides for the production of adequate numbers of transportable particles capable of carrying all or most of the analyte to the argon plasma. Kántor [105] concluded that the efficiency of formation of dry aerosols is one of the major contributing factors determining the mass transport efficiency of analyte from the vaporization surface to the argon plasma.

A number of physical carrier systems have been suggested for use in ETV-ICP-MS, including NaCl [46] palladium [49] and diluted seawater [67] stripped of trace metals. A detailed study of the use of mixed carriers, including seawater and its components, was completed by Hughes et al. [43], wherein it was shown that the addition of carrier not only served to linearize calibration curves, but also normalized analyte transport from the ETV to the plasma. Once a very small quantity of carrier was added (i.e., micrograms), analyte signal was enhanced by up to a factor of 5. Further additions of carrier did not result in corresponding enhancements and analyte signal remained constant until a point was reached when additional carrier resulted in ion signal suppression in the mass spectrometer system. Once the initial carrier was added, calibration curves remained linear even when analyte signal was suppressed at higher carrier additions. These results indicate that vaporized analyte is likely occluded (co-precipitated) within condensed particles of the carrier material.

Analyte transport efficiency in ETV-ICP-MS was first reported by Park et al. [9] using a platform-type ETV device. Solutions containing

1000 μg ml^{-1} of Cu, Pb, Cd, Fe and Ni were vaporized from the graphite ETV surface and collected in a 60 ml bottle containing 1.5 g of loosely packed cotton wool. Fifteen replicate valorizations of 2 μl of analyte solution were completed for each analytical run. The average analytical recovery of analyte found in the cotton wool was about 84%, which corresponds to the amount of analyte that would reach the plasma under the same experimental conditions. Using a tungsten furnace ETV device, Nonose et al. [44] concluded that analyte signal enhancement observed when analyte was vaporized in the presence of halide-containing modifiers was due to increased ionization of analyte within the plasma and was not a result of any physical or carrier effects.

In more recent studies, Sparks et al. [50,51] examined both the particle size distribution of sample transported from a tube-type ETV device to the argon plasma [51] and the retention of sample in the transport tubing [50]. These authors [51] used laser scattering techniques to count particles and concluded that the majority of particles produced were in the 0.1 μm to 0.2 μm range and that the number of particles produced increased by about 25% when NaCl was used as a carrier. The use of a laser system necessitated the application of instrumental conditions, in particular, gas flows, which are not typical of ETV-ICP-MS analytical conditions. Additionally, the majority of particles measured were emitted from the graphite tube at a time when all analyte would be completely vaporized and had been swept out to the argon plasma. Using Ag as a test element, Sparks et al. [50] reported that about 10% of the analyte was retained on the walls of the transfer tubing. For these experiments, from 10 to 60 μl of 10 μg ml^{-1} Ag was vaporized from the ETV surface, which is a quantity of metal orders of magnitude greater than would ordinarily be encountered in ETV-ICP-MS experiments. Langer et al. [107] reported that vaporization of high masses (20 μl of 100 μg ml^{-1}) of Ag from a graphite ETV system resulted in the agglomeration of micro-particles, suggesting homonucleation as the primary particle formation mechanism in ETV-ICP-MS. Particles were captured from the Ar carrier gas using a thermophoretic collection technique. Fonseca et al. [108] used oxygen ashing and Pd modifier to improve analyte transport efficiency, possibly by increasing the number of carbon particles produced during the vaporization step, which could act as a physical carrier for analyte.

Kántor [109] recently reported on the efficiency of graphite furnace ETV sample introduction into argon plasmas. By dissolving deposited sample material from the transport tube, a transport efficiency between 67% and 76% for the medium volatility elements (Cu, Mg, Mn) and between 32% and 38% for the volatile elements Cd and Zn was found. Using a modified commercially available ETV system (with electrostatic collection of the aerosol vaporized), Buchkamp and Hermann [110] found that the transport efficiency for Pb increased with increasing mass of vaporized Pb up to about 6 ng. For a sample containing greater than 6 ng of Pb, the transport efficiency was about 20%. Transport losses were measured by Barth et al. [111] for analytes vaporized as slurries using a tungsten coil ETV device. These authors used radio-tracers to determine the fate of vaporized elements and showed that transport losses ranged from 7% for La to 54% for Cu. For the 12 elements studied, transport losses decreased with increasing boiling point of the element or the oxide. A more recent study by Schäffer and Krivan [112] measured transport efficiencies for eight elements in an ETV-ICP-AES device using a mixture of radiotracers. The authors reported transport efficiencies ranging from 26% for Cr to 40% for K. However, as for many studies discussed above, analyte (radiotracer) masses used were orders of magnitude greater than normally determined by ETV-ICP-MS and were high enough to act as physical carriers.

Grégoire and Sturgeon [48] reported results for the quantitative determination of absolute transport efficiency in electrothermal vaporization inductively coupled plasma mass spectrometry (ETV-ICP-MS) for the Perkin-Elmer HGA-600MS electrothermal vaporizer. The absolute transport efficiencies for Mo, In, Tl and Bi were determined using experimental conditions typical of those applied to real analysis by ETV-ICP-MS. Experiments using an on-line filter trapping apparatus indicated that particles produced by the ETV device were smaller than 0.1 μm in diameter. The nature and condition of the ETV graphite surface, the length of the transfer tube, and the effect that diluted seawater and palladium matrix modifiers have on analyte transport efficiency were investigated. Transport efficiency was comparable for all elements studied and was enhanced with previously used, rather than new, graphite tubes and when seawater and palladium carriers were present. When analyte was vaporized without carrier from a new graphite tube, the transport efficiency to the plasma was approximately 10%. About 70% of the total amount of analyte vaporized was deposited within the ETV switching valve, 19% onto the transfer tubing and 1% onto the components comprising the torch assembly. These conditions however, represent the 'worst case scenario', with analyte transport to the plasma increasing to about 20% or more with the addition of carrier.

Analyte transport losses in ETV-ICP-MS are substantial, amounting to approximately 75% of the analyte vaporized from the graphite tube. These losses, however, do not limit the applicability of the technique to quantitative analysis. The published literature currently contains in excess of 70 papers detailing fully quantitative applications of ETV-ICP-MS to the analysis of real samples. Careful selection of experimental conditions, including the use of carriers and chemical modifiers, normalizes analyte transport for both samples and standards, making accurate calibration possible.

3.8 NON-SPECTROSCOPIC INTERFERENCES

Non-spectroscopic or so-called matrix interferences are well known in ICP-MS [113]. Since electrothermal vaporization is only an alternative source of analyte vapour/aerosol for the ICP-MS, clearly the same factors affecting matrix interferences in ICP-MS will also affect ETV-ICP-MS. In fact, the situation may be worse in ETV-ICP-MS because of the higher sample transport efficiency characteristic of the technique. For example, a comparison of the amount of salt reaching the plasma for SN and ETV from a 1000 μg ml^{-1} illustrates this point. If for SN, a sample uptake rate of 2 ml min^{-1}, a mass transfer efficiency of 2% and a sampling time of 2 s is assumed, then 667 ng of NaCl will reach the argon plasma. If for ETV a 10 μl sample is used and a mass transfer efficiency of 25% and a vaporization time of 2 s is assumed, then a total of 2500 ng of NaCl reaches the plasma. On the surface it would appear that the ETV loads the plasma with salt by a factor of 3.8 greater than does SN, however in reality the situation is somewhat worse than this. Electrothermal vaporization produces a transient signal and the maximum rate of vaporization occurs at the centre of this signal and for a period of time much shorter than the total signal time. Because of this, the amount of NaCl reaching the plasma during the evolution of the central part of the vaporization signal is perhaps 20 times greater than during the equivalent period for a steady-state sample introduction technique such as SN. Therefore, matrix effects can be expected to be serious and to change in severity during the lifetime of the analyte signal.

Grégoire et al. [31] showed that when Os is determined in the presence of NaCl (Table 3.27), loss of analyte signal intensity was more

TABLE 3.27

Effect of NaCl on ^{192}Os ion count rate (from Ref. [31] with permission); concentration of ^{192}Os= 100 ng g^{-1}

NaCl concentration ($\mu g\ g^{-1}$)	Solution nebulization	Electrothermal vaporization
–	1.00	1.00
1000	1.00	0.96
2000	0.95	0.82
3000	0.91	0.53
5000	0.83	0.37
7000	0.73	0.34

severe for ETV than for SN sample introduction, when both systems were operated under optimum conditions. For example, when Os was determined in the presence of 5000 µg ml^{-1} NaCl, the SN signal was suppressed by only 17%, whereas for the ETV, Os in the same solution was suppressed by 63%.

Park et al. [8] reported on the effect of added Na, Cr, Ca and Se on ETV-ICP-MS signals for As, Cd and Cu. *These authors found that 1 µg of concomitant vaporized with ng masses of analyte resulted in significant signal suppression.*

The argon dimer is a polyatomic ion produced within the argon plasma or the interface region of the ICP-MS and is known to behave in a similar manner to analyte ions with respect to non-spectroscopic interferences [114]. The argon dimer has found some use as an internal standard reference ion [91] in the determination of trace elements in marine sediments. Alary and Salin [115] also found that the argon dimer as well as other argon polyatomic ions could be used for plasma diagnostics in ETV-ICP-MS.

Figure 3.13 illustrates the effect of added concomitant salts on the shape of the Ar dimer ETV-ICP-MS signal. The concomitant compounds are arranged in order of increasing atomic mass to show the effect of concomitant mass on non-spectroscopic interferences. For NaCl, the volatilization of up to 20 µg did not appreciably change the shape of the argon dimer signal. When using $Mg(NO_3)_2$, masses in excess of 10 micrograms caused an enhancement in the argon dimer signal. Nickel nitrate caused a similar enhancement when more than 0.5 µg were used. At masses greater than 3 µg of $Ni(NO_3)_2$, serious

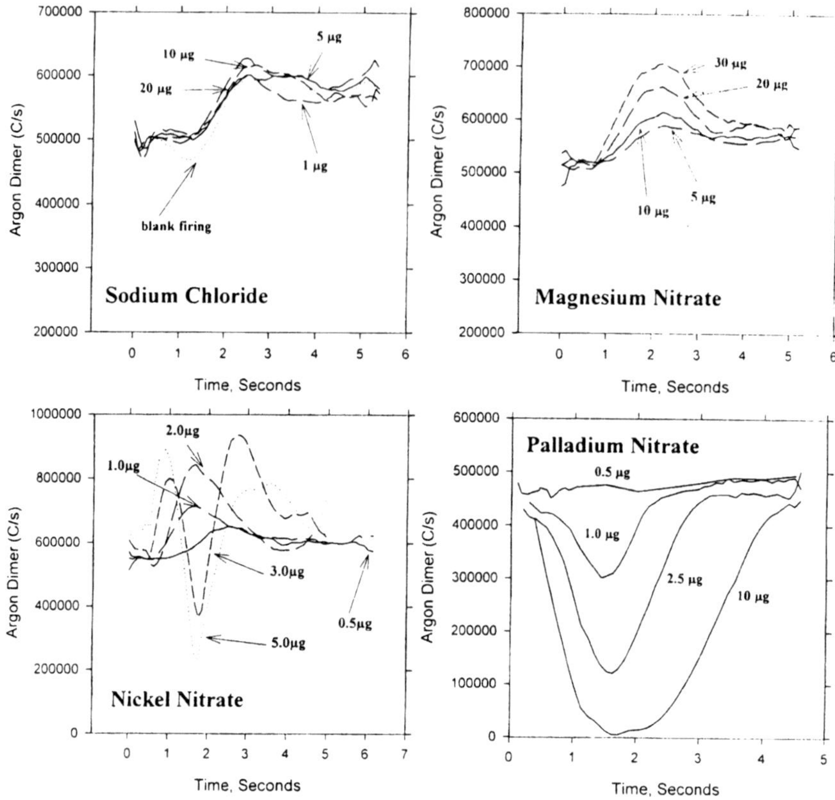

Fig. 3.13. Effect of matrix component mass on argon dimer signal (from Ref. [52] with permission).

argon dimer signal suppression was observed. The use of Pd, the concomitant with the greatest atomic mass, resulted in argon dimer signal suppression for all masses above 0.5 μg and complete suppression of the signal at a mass of 10 μg. Clearly, both the atomic mass of the analyte and the absolute quantity of concomitant present during the vaporization step have an important influence on observed matrix effects.

Although these curves can only show in a qualitative manner the effect of concomitants on analyte signals, the trends observed may be useful in predicting the quantity of concomitant that ETV-ICP-MS can tolerate. Figure 3.14 shows the analyte signal obtained for 10 pg of Sn

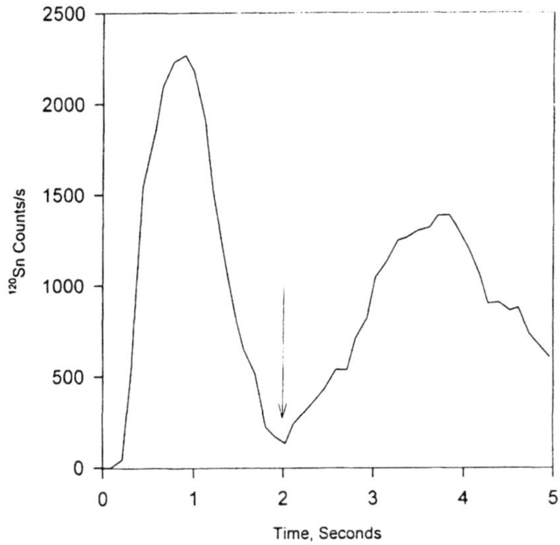

Fig. 3.14. ETV-ICP-MS signal for 10 pg of Sn vaporized with 10 µg of Pd. Arrow indicates location of signal maximum for Pd (from Ref. [52] with permission).

vaporized in the presence of 10 µg of Pd. The arrow in Fig. 3.14 indicates the location of the peak of the Pd ETV-ICP-MS signal (not shown) where the Sn intensity is completely suppressed. As the temperature of the vaporization surface increased during the high temperature ETV step, Sn was volatilized first, followed by Pd. As the vaporization step progressed, the quantity of Pd entering the gas-phase increased and caused the Sn signal to be suppressed resulting in what appears to be a double-peak. Care should be exercised when interpreting the nature of a double analyte signal especially when concomitants are present during vaporization. These results also show that in most cases no more than 0.5 µg to 1 µg of chemical modifier or physical carrier should be added to sample. This conclusion is supported by the results of Hughes et al. [43] who showed that with the addition of more than 1 µg of a mixed physical carrier, analyte signals for many elements were suppressed.

A non-spectroscopic interference of importance to ETV-ICP-MS is related to the expansion of gases within the ETV during the high temperature vaporization step. The ETV is essentially a flow-through cell with an argon flow restricted by the diameter of the injection tube within the plasma torch. The diameter at the exit of the injection tube

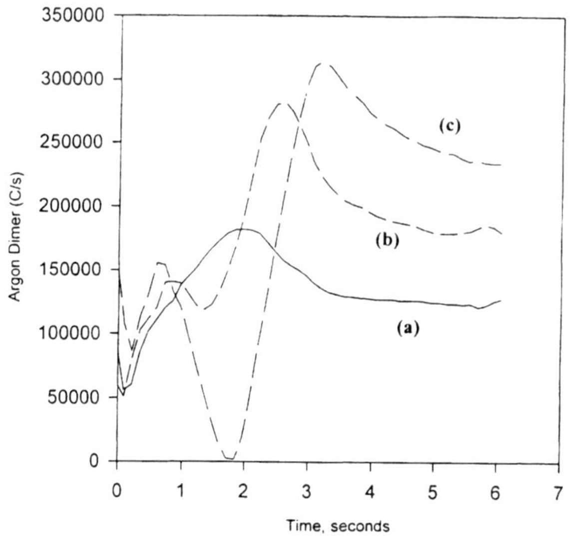

Fig. 3.15. Effect of sample matrix on argon dimer signal: (a) no sample; (b) organic sample; (c) high salt sample.

is generally about 2 mm whereas the diameter of the graphite tube and the PTFE transfer line is about 6 mm. During rapid heating of the ETV gas expansion occurs and a pressure wave is transmitted through to the argon plasma. The net effect of this is a change in the back-pressure within the PTFE transfer line which increases the velocity of gas leaving the injector tube which in turn changes the sampling depth at the plasma mass spectrometer interface. This process occurs during the first 0.5 s to 3 s into the vaporization step. After this period of time, the system recovers to the normal or pre-set argon carrier gas flow-rate.

Similar gas expansion effects can also occur during the vaporization of matrix components. Figure 3.15 shows argon dimer signal profiles [69] obtained for samples vaporized as solids. Curve a was obtained for a blank (no sample) vaporization; curve b was for a sample high in organics (lobster hepatopancreas) and curve c was for an organic sample high in salt (total diet reference material). Curve a shows a normal argon dimer profile whose intensity is changing along with the Ar gas pressure inside of the graphite tube. Curve b shows a large maximum at about 2.5 s at which time the organic matrix is volatilized and curve c shows an initial drop in sensitivity caused by suppression from NaCl vaporization followed by a net enhancement of signal, which coincides with the vaporization of the organic components of this sample. This figure shows that the nature of the sample matrix can also have a profound effect on the sensitivity of ETV-ICP-MS.

3.9 APPLICATIONS

There are currently over 80 published papers detailing applications of ETV-ICP-MS. Table 3.28 contains a list of these applications and also includes information on the analytes determined and the sample preparation method used. A cursory examination of Table 3.28 clearly shows the versatility of ETV-ICP-MS to solve many analytical problems that cannot be easily handled by other techniques. Of particular interest are the applications dealing with direct solids analysis or the analysis of solids introduced into the ETV as a slurry. Nearly 60% of ETV-ICP-MS applications involve the direct analysis of samples with little or no sample preparation required. Bauman [185] and Voellkopf [186] discussed some of the advantages and report on the direct analysis of solids by ETV-ICP-MS and the field was recently reviewed by Darke and Tyson [187]. The work of the Ghent University group [132,134,145,161,177,180,182] on the direct analysis of solids is noteworthy and is recommended reading especially for those interested in the direct solids analysis of biological materials. These researchers have written an extensive book chapter on solid sampling ETV [188] as applied to ICP-AES and MS. A recent review on the use of hydride generation techniques in ETV-ICP-MS was published by Olson et al. [189]. A new book edited by Montaser [190] on ICP mass spectrometry contains detailed information on the principles and applications of the technique including sections on electrothermal vaporization sample introduction.

The remainder of this part of the chapter is devoted to a more detailed presentation of four applications from this laboratory. For all of these studies a Perkin-Elmer HGA-600MS ETV interfaced with a Perkin-Elmer ELAN 5000a ICP mass spectrometer was used. The methods discussed were selected to show the versatility of ETV-ICP-MS in several areas of specialized application including ultra-trace analysis [67], the analysis of small samples [68], solids introduced as slurries [69] and the determination of isotope ratios [66].

TABLE 3.28

Applications of electrothermal vaporization ICP mass spectrometry

Sample type	Analyte(s)	Sample preparation	Ref.
Geological materials	Pb isotope ratios	mixed acid, fusion	5
Biological materials	As,Cu,Mn,Pb,Rb,V, Zn,Ag	mixed acid	8
Geological materials	Mo,W	fusion	10
Geological materials	Tl	mixed acids, fusion, solvent extraction	11
Geological materials	Pt,Pd,R,Ir	nickel sulphide fusion	13
Blood plasma	Fe isotope ratios	direct determination	22
Geological materials	Ru,Pd,Ir	mixed acid, ion exchange	14
Silicon dioxide	U,Th	mixed acids	19
Blood plasma, urine	Te	direct determination	21
Iridosmines	Os isotope ratios	vapour generation, fusion	15
Petroleum	Hg	direct determination	23
Iron and steel	Bi	mixed acids	17
Snow	Tl,Cs,Pb,Mn,Co,V, Cu,Ni,Cd,Cr	direct determination	67
Waste oils	Cl as polychlorinated biphenyls	dilution with xylene	24
Biological materials, coal	Ni,Cr,Cu,Pb,Mn,Co	direct determination, slurry sampling	69
Zircons	U,Th, rare earth elements	HF digestion in sealed bombs	68
Blood, liver, kidney	Cd and Zn isotope ratios	mixed acid digestion, ion exch.	66
Fresh waters	Pd,Pt	activated charcoal pre-conc.	116
Photoresist	Na,Fe	dilution with organic solvent	117
Photoresist	Na,Fe,Mn,Ni,Cu	dilution with organic solvent	118
Organic materials, sediments	extractable organic chlorine	solvent extraction	119
Blood plasma	Pb	direct determination	120
Semiconductor materials	Al,Ca,Cd,Cr,Cu,Fe, Ga,In,Mg, Mn,Ni,Pb,Si,Sn,Zn	direct determination	121
Semiconductor materials	Al,Cu,Fe,Na,Ni,Pb, Zn	direct determination	122
Basalt	Hg,Pb,Cu,Ag	direct determination	123

Sample type	Analyte(s)	Sample preparation	Ref.
Silicon wafer surface	Na,Mg,Al,Ti,Cr,Mn ,Fe,Ni,Co, Cu,Zn	HF/peroxide surface etch	124
Fly ash, total diet	As,Cd,Pb	direct determination	125
Biological materials	As,Sb,Se,Bi,Sn	hydride generation/in-situ trapping	126
Urine	As,Bi,Te	hydride generation/in-situ trapping	127
Rain and river waters	Al,V,Mn,Fe,Ni,Cu, Zn,As,Se,MoCd,Sb, Pb	direct determination	128
Trichlorosilane	P,As	CuCl treatment, evaporation	129
Sludge	Hg	direct determination	130
Biological materials	Cd	direct determination	131
Plant samples	As	direct determination	132
Plant samples	As	direct determination	133
Sea lettuce	As	direct determination	134
Silicon wafer surface	Cu,Zn,Pb	HF etching	135
Water, basalt	As,Be,Cd,Cr,Cu,Rh ,Sb	direct (water) mixed acid (basalt)	136
Pure acids	Fe,Cu,Ni,Zn	direct determination	137
Steel	S	acid digestion, solvent extraction	138
Calcite, seawater	V	acid digestion, ion exchange	139
Water	^{99}Tc,^{238}U,^{236}U,^{232}Th,^{230}Th,^{226}Ra	direct determination	140
Lubricating oils	metallo-organic Al,Mg,Fe,Y	direct determination	141
Seawater	Co,Cu,Mn,Ni,V	direct determination	142, 143
Sediments	Cu,Cd,Pb	direct determination, slurry sampling	144
Biological materials	Se,As	direct determination	145
Plants, soils	Cd	direct determination	146
Airborne particles	Cr,Fe,Mn,Cu,Zn,Sr ,Cd,Sb,Ba, Pb	direct determination	147
Lake sediment	^{90}Sr	aqua regia digestion	148
Blood plasma	Al,Ti,V	direct determination	149
Biological materials	Cu,Cd,Pb	nitric acid	150

continued

TABLE 3.26 (*continuation*)

Sample type	Analyte(s)	Sample preparation	Ref.
Fresh and saline waters	Hg,Bi	vapour generation/in-situ trapping	151
Silver alloy	Au,Bi,Cd,Pb,Sb,Sn, Zn	in-tube acid digestion	152
Waters	As,Sb,Se	direct determination	153
Fresh and saline waters	As	direct determination	154
Seawater	V,Mn,Fe,Co,Ni,Cu, Zn,Mo,Sb	complexation/solvent extraction	155
Seawater	V,Mn,Fe,Co,Ni,Cu, Zn,Mo,Sb	complexation/solvent extraction	156
Fish samples	Hg	direct determination/slurry sampling	157
Biological tissues	inorganic and total Hg	direct determination/slurry sampling	158
Quartz	Al,Ba,Co,Cr,Cu,Fe, Li,Mg,Mn, Na,Pb,Sr,U,Zn	direct determination/slurry sampling	159
Nickel alloys	Bi,Pb,Te	direct determination	160
Botanical samples, sludge	Cd,Se	direct determination	161
Synthetic samples	As,Sb	hydride generation/in-situ trapping	162
Water, urine	As,Se,Pb	direct determination	163
High-purity iron	P	acid dissolution/solvent extraction	164
Geological materials	Ra	direct determination/slurry sampling	165
Sulfamic acid	Li,Fe,Cd,Pb	direct determination	166
Urine	Hg	direct determination	167
Body fluids, tissue	lanthanide elements	acid/peroxide digestion	168
Seawater	Cu,Cd,Pb,Bi,Se(IV)	column chromatography	169
Seawater	Mn,Mo,U	direct determination	170
Urine	Cd,Pb	direct determination	171
Biological materials	As,Se,Te,Ag,Cr,Cu, V,Ni,Mn,Co,Cs,Pb, Sb,Sn,Bi	direct determination/slurry sampling	172
Soils	Cd,Hg,Pb	direct determination/slurry sampling	173

Sample type	Analyte(s)	Sample preparation	Ref.
Fish tissue	Cu,Zn,Cd,Pb	direct determination/slurry sampling	174
Aluminium oxide	Cr,Cu,Ga,Fe,Mg,M n,Na,V,Zn	direct determination/slurry sampling	175
Fish tissue	As,Se,Hg	direct determination/slurry sampling	176
Biological materials, soils	thermal decomposition products	direct determination	177
Serum	Se	direct determination	178
Sediments	Se	acid dissolution, citric acid modifier	179
Photographic materials	Ru	direct determination	180
Waters, biological materials	U, Th	acid dissolution, fusion, on-line solid phase extraction	181
Plastics	Pd	direct determination	182
Plant materials	Ge, As, Se, Cd and Pb	slurry sampling direct determination	183
Waters	Mo. U and B	direct determination	184

3.9.1 Ultratrace analysis of Arctic snow

Heavy metal concentrations of near surface Arctic snow typically range from 0.5 (Cd) to 20 (Pb, Zn, Ni, Cu) pg g^{-1}. These levels pose considerable challenges for the representative sampling of this material as well as for the analysis for the metals of interest.

Snow samples analysed in this study were collected from an ice flow north of Ellesmere Island, N.W.T., Canada. All samples were collected from the surface snow layer and taken from small 'pits' approximately 30 cm deep. A pre-cleaned plastic scoop was used to transfer snow from the sampling face into the sample containers, which were two-litre wide mouth screw capped polyethylene bottles.

All sample preparation operations were conducted in class 100 clean room facilities equipped with class 10 fume cupboards. The frozen snow samples were thawed under a heat lamp in their original bottles, acidified to pH 2 by addition of high purity HNO_3 and weighed. An aliquot of each sample was then transferred to a 30 ml pre-cleaned polyethylene screw capped bottle for subsequent analysis by ETV-ICP-

TABLE 3.29

Analysis of SLRS-2 standard reference water by ETV-ICP-MS (from Ref. [67] with permission)

Element	Concentration (ng g^{-1})	
	Determined	Certified Value
Pb	0.123 ± 0.004	0.129 ± 0.011
Cd	0.035 ± 0.005	0.028 ± 0.004
Cu	2.77 ± 0.19	2.76 ± 0.17
Ni	0.93 ± 0.03	1.03 ± 0.10
Mn	11.5 ± 0.4	10.1 ± 0.3
Co	0.071 ± 0.004	0.063 ± 0.012
Cr	0.465 ± 0.008	0.45 ± 0.07
V	0.308 ± 0.013	0.25 ± 0.06
Cs	0.016 ± 0.003	–
Tl	0.0028 ± 0.0013	–

MS. The remaining portions were reweighed, transferred to pre-cleaned PFA beakers and evaporatively concentrated (73-fold on average) in a class 10 hood to approximately 10 ml volumes. This was accomplished over a period of 6-12 h using sub-boil conditions.

Operating conditions for the ETV-ICP-MS were as given in Table 3.7. Five µl of purified NASS-3 (diluted 1:500) mixed physical carrier were added to all samples and standards. Data acquisition was done using peak-hopping mode with a dwell time of 20 ms per isotope with one reading per peak (m/z). Five isotopes (analytes) were determined during a single ETV signal measurement cycle. The number of isotopes monitored was limited to assure accurate measurement of the analyte signal profile.

Table 3.29 compares results obtained for the determination of 10 analytes in National Research Council of Canada Reference River Water, SLRS-2. Excellent agreement between certified values and determined values were obtained for all elements.

To further verify the accuracy of ETV-ICP-MS for the analysis of snow, these authors [67] used GFAAS with which to compare their results. Table 3.30 reports on the determination of nine elements in Arctic snow by both GFAAS and ETV-ICP-MS. Analysis of these materials by GFAAS could only be done using concentrates whereas

TABLE 3.30

Analysis of Arctic snow by ETV-ICP-MS (from Ref. [67])

Sample	Concentration (ng g^{-1})								
	Mn	Cd	Cu	Pb	Ni	V	Tl	Cs	
1	0.212 (0.18)	0.018 (0.015)	0.09 (0.12)	0.321 (0.270)	0.053 (0.041)	0.084	0.0070	0.0022*	0.0041*
2	0.087 (0.13)	0.024 (0.0056)	0.053 (0.069)	0.161 (0.035)	0.040 (0.030)	0.053	0.0057	–	–
3	0.139 (0.17)	0.008 (0.0083)	0.066 (0.084)	0.241 (0.20)	0.038 (0.037)	0.051	0.0007	0.0018 (0.0013)*	0.0021*
4	0.054 (0.093)	0.0014 (0.0018)	0.0022 (0.030)	0.034 (0.042)	0.005 (0.014)	0.005	0.0014	0.00024 (0.00027)*	0.0006*
5	0.0008 (0.0013)	0.0006 (0.0017)	0.014 (0.022)	0.029 (0.031)	0.013 (0.007)	0.004	0.0011	0.00028	–

Values in parentheses obtained from GFAAS analysis of concentrates (×72).
*Data from analysis of concentrates (×72).

with few exceptions, all ETV-ICP-MS determinations were completed on samples at their natural levels. Table 3.30 shows excellent agreement between ETV-ICP-MS and GFAAS results for most elements. For V, Co, Cs and Tl in some samples, GFAAS did not have adequate sensitivity even for concentrates.

Reported limits of detection for ETV-ICP-MS were in the fg (10^{-15} g) range: 29, 57, 86, 120, 140, 360, 420, 470, 870 and 3200 for Tl, Cs, Pb, Mn, Co, V, Cu, Ni, Cd and Cr, respectively.

3.9.2 Trace analysis of single zircons for REE, U and Th

The trace element composition of zircons is of interest to geologists as an aid in identifying the source of the sedimentary rock in which they were formed. Single zircons typically have masses in the μg range. The multi-element quantitative analysis of samples this small is a challenge particularly for analytes occurring in the ng g^{-1} concentration range.

Zircons were pre-cleaned by air-abrasion with pyrite to remove the outer layers of the crystals. The clearest crack and inclusion-free crystals were selected for analysis by hand-picking in ethyl alcohol, and then leached for about 30 min in warm high-purity 3 M HNO_3. Prior to weighing, the crystals were rinsed in high-purity water followed by acetone and then dried. The zircons were weighed in aluminum boats on a Mettler micro balance (± 0.2 μg) and then loaded into clean PTFE microcapsules. Samples (0.1 mg) of powdered zircon reference material BCS 388 were weighed into identical microcapsules. Approximately 0.2 ml of high purity 28.9 M HF and 0.02 ml of high purity 16 M HNO_3 were added to all of the microcapsules which were then placed in a Parr Instrument Co. (Moline, Illinois) steel-jacketed PTFE bomb. High-purity HF and HNO_3 (10:1, v/v; total volume about 5 ml) were also added to the bomb to equalise the vapour pressure between the bomb and the microcapsule. The bombs were placed in an oven and heated at 220°C for two days. After cooling, the samples were evaporated to dryness prior to adding about 0.2 ml of 3.1 M HCl to the microcapsules. Five ml of 3.1 M HCl were also added to the bomb before replacing in an oven at 210°C for about 12 hours. The clear HCl solution was then poured onto clean 50 ml polymethyl pentene beakers which had been equilibrated with double-distilled 8 M HCl and rinsed with high-purity water. The beakers were then placed on a warm hotplate and the HCl solution allowed to dry. When dry, the beakers were removed from the hotplate and allowed to cool. The samples were then taken up in 0.3 M HNO_3 prior to analysis by ETV-ICP-MS. For micro-sample analysis, the dissolved and dried down single zircon crystals and the BCS 388 sample were taken up to a volume of 500 μl. For the 100 mg sample of BCS 388, the sample was dissolved in a similar fashion to the single zircons.

Optimization of plasma and mass spectrometer experimental conditions was accomplished using solution nebulization sample introduction. Only small adjustments in the argon carrier gas flow (± 50 ml min^{-1}) were required to optimize signal output in the ETV mode. The peak hopping mode of data acquisition was used with a 1-ms dwell time for each isotope monitored. Single (one m/z per mass) sequential measurements were made for each analyte isotope on a continuous basis from the start of the vaporization cycle until the signal returned to baseline values.

Because of the significant tailing effect exhibited for several elements (La, Ce, Nd, Y, U), peak height measurements were recorded for each ETV-ICP-MS signal produced.

Table 3.31 summarises analytical results for the analysis of BCS 388 zircon reference material. Concentration values obtained for micro-sample (150 μg) analysis (first 3 rows), the analysis of a bulk sample (row 4) and reference values (row 5) for comparison purposes are

TABLE 3.31

Analysis of BCS-SRM No. 388 Zircon (from Ref. [68] with permission)

Technique	Sample mass	Concentration (μg/g)							
		Y	La	Ce	Nd	Sm	Yb	Th	U
ETV-ICP-MS (ext. calib.)	150 μg	973±30	24.9±0.8	63±1	30.5±0.5	13±3	234±10	173±13	337±17
ETV-ICP-MS (std. addition)	150 μg	937±24	24.7±0.7	61±1	29.9±0.5	13±3	226±10	165±17	322±16
FI-SN-ICP-MS (micro sample)	150 μg	954±28	26.0±0.8	65±2	30.5±0.9	13.0±0.4	231±7	165±5	319±10
SN-ICP-MS (bulk sample)	100 mg	1083±33	37±1	95±3	39±1	12.4±0.4	227±7	165±5	334±10
Reference value*	–	1071±71	30±10	80±10	40±10	10±10	300±100	158±18	288±34

*Bureau of Analyzed Samples Ltd., Newham Hall, Middlesborough, England. Revised certificate, October, 1992. Standard error on reference values for La, Ce, Nd, Sm, Yb taken as ± smallest significant figure reported (concentrations expressed as percent on certificate); Values for Y, Th and U are certified values; ETV = electrothermal vaporization; FI = flow injection; SN = solution nebulization sample introduction.

presented. BCS 388 is the only reference zircon material available for which there are reported trace-element data.

For micro-sample analysis, two calibration strategies for ETV-ICP-MS were used as well as a flow injection method (FI) for solution nebulization ICP-MS analysis of the remaining sample solution from the ETV determinations. The flow injection technique involved the nebulization of 400 μl (entire sample solution). Data acquisition commenced one second before the sample entered the spray chamber and was terminated 10 seconds after the final volume of sample solution reached the nebuliser.

The results in Table 3.31 show excellent agreement between values obtained by both external calibration and the method of standard addition indicating no matrix effects on the analyte signal intensity caused by the presence of Zr. Excellent agreement between these values and the FI-SN-ICP-MS were also obtained which provides an independent check on the ETV results for the analysis of micro-samples.

Data for the bulk sample (100 mg) determined by SN-ICP-MS agree with micro-analysis values within the stated errors for Y, Yb, Th and U but are not in agreement with La, Ce, Nd and Sm. This may be a result of inhomogeneous distribution of some of the elements in the reference zircon powder. Note that all of the analytes for which micro and bulk values agree are those of highest abundance indicating a greater degree of homogeneity for these elements.

Concentration values obtained by SN-ICP-MS for the bulk sample compare well with the reference literature values for all of the elements determined. This provides further evidence of inhomogeneity at the μg sample size scale as well as validating the results reported for the bulk sample analysed by SN-ICP-MS.

A selection of single zircons from suites previously dated using high-precision U-Pb geochronology were prepared for analysis as described above and analysed using ETV-ICP-MS. The ICP-MS was calibrated using a curve produced from aqueous standards to which was added NASS-3 (5 μl) physical carrier.

Results for the analysis of single zircons are given in Table 3.32 Sample masses ranged from 7.7 μg (Sample 4) to 75.8 μg (Sample 6). Limits of detection for the method were adequate for all of the elements studied with the exception of La on the low mass samples.

Grégoire et al. [68] reported that analyte concentrations were measured with a precision of about 6%. Limits of detection for the analysis of a 10μg zircon were 150 ng g^{-1} for Y, 90 ng g^{-1} for La, 115 ng g^{-1} for Ce, 65 ng g^{-1} for Nd, 180 ng g^{-1} for Sm, 22 ng g^{-1} for Yb, 190 ng g^{-1} Th and 80 ng g^{-1} for U. Absolute limits of detection for a 10 μl solution aliquot ranged from 4 to 36 fg.

TABLE 3.32

Analysis of single zircons by ETV-ICP-MS (from Ref. [68] with permission)

Sample	Mass (μg)	Concentration (μg g^{-1})							
		Ce	La	Sm	Nd	Th	Y	Yb	U
1	19.8	21.4	<0.05	1.7	1.1	163	371	115	284
2	12.4	30.9	<0.07	2.4	1.5	227	473	148	273
3	11.6	11.9	<0.08	0.84	0.54	149	361	150	270
4	7.7	10.3	<0.12	2.3	2.1	57.1	393	92.0	192
5	8.2	3.3	<0.11	1.4	0.81	33.4	281	71.3	68.0
6	75.8	2.5	0.07	1.6	1.7	606	379	130	890

3.9.3 Direct analysis of solids introduced as slurries

The direct analysis of solids and/or slurries by any analytical technique offers advantages over more conventional sample preparation. Among these advantages are the reduced sample preparation time, the reduced possibility of sample contamination, increased sensitivity (no dilution), decreased likelihood of analyte loss through volatilization prior to analysis and the selective analysis of micro-amounts of solids. Slurry sampling offers additional advantages for samples that occur naturally as slurries such as milk and blood and for the completion of surveys where the analysis of large numbers of samples such as contaminated soils is required. Further, slurry sampling combines the benefits of solid and liquid sampling and permits use of conventional liquid sample handling apparatus such as autosamplers. Grégoire et al. [69] reported on the direct analysis of solid (as slurries) materials including coal, lobster hepatopancreas and total diet reference materials and this application will be summarized here. The same experimental conditions as for the previous two applications were used for these experiments with the exception that the drying step time was increased to 50 s and the pyrolysis temperature changed to 400°C (for 50 s).

Slurries were prepared by taking a weighed portion of reference material (sample) and placing it in a 20 ml plastic centrifuge tube. A diluent consisting of 5% Ultrex (Baker) HNO_3 containing 0.005% Triton X-100 was added to the sample and mixed on a vortex mixer. An aliquot of sample slurry was then immediately withdrawn from the centrifuge tube using an Eppendorf pipette and placed into a clean Teflon auto-sampler cup. In preparing the slurries prior to analysis, care was taken to ensure that an appropriate ratio of sample/diluent was used by taking into account the particle size and the density of the reference material in order to minimize effects arising from poor sampling statistics.

The coal (NIST 1632a) slurry was prepared by mixing 4 mg of sample per 4 ml of diluent. The density of the coal was approximately 0.7 g/cm^3 and the particle size was apparently less than 50 µm. A 20 µl aliquot of sample slurry would require a minimum of 0.08 mg/ml of coal to give 50 particles, ensuring a representative sampling of the bulk slurry.

The NIST diet (SRM 1548) was prepared in a similar fashion except that 400 mg of sample was suspended in 4 ml of diluent. With a density of about 1 g/cm^3 and a particle size of less than 250 µm, only 20 mg/ml was required to ensure that each 20μl aliquot contained 50 particles. This criterion was met with a concentration of 100 mg/ml.

The NRC LUTS-1 lobster hepatopancreas reference material was provided as 10.3 g of homogeneous slurry. The material was diluted by adding (in a volumetric flask) diluent to a total volume of 100 ml. The suspension was sonicated for 30 min in an ultrasonic bath and, following mixing on a vortex mixer, a 1 ml aliquot of the diluted LUTS slurry was removed using an Eppendorf pipette and placed in a Teflon autosampler cup. Although no data are available characterizing the actual particle size of the LUTS material, we believe it to be less than 50 μm and certainly less than 100 μm. Looking at the guidelines for minimum mass, only 1.3 mg/ml was needed for 100 μm particles of a density of 1 g/cc. to provide the minimum 50 particles per 20 μl analytical sample aliquot.

TABLE 3.33

Analysis of Reference Materials by Ultrasonic Slurry Sampling ETV-ICP-MS (from Ref. [69] with permission)

	Concentration (μg/g)				
	Cu	Cr	Ni	Pb	Mn
N.I.S.T. Coal SRM 1632a					
Ext. Cal.	16.8 ± 0.1	22.7 ± 2.8	20.0 ± 3.5	11.0 ± 0.5	30.8 ± 2.2
Certified	16.5 ± 1.0	34.4 ± 1.5	19.4 ± 1.0	12.4 ± 0.6	28 ± 2
N.R.C. LUTS-1 lobster hepatopancreas					
Ext. Cal.	0.17 ± 0.02	0.45 ± 0.07	0.59 ± 0.03	0.013 ± 0.001	
M. of Addn.	0.048 ± 0.006	0.081 ± 0.013	0.159 ± 0.009	0.007 ± 0.001	
Certified	0.051 ± 0.006	0.079 ± 0.012	0.200 ± 0.034	0.010 ± 0.002	
N.I.S.T. Total Diet SRM 1548					
Ext. Cal.	2.1 ± 0.2	0.28 ± 0.02	0.75 ± 0.02	0.025 ± 0.005	
M. of Addn.	2.9 ± 0.2	0.11 ± 0.02	0.29 ± 0.02	0.045 ± 0.006	
Certified	2.6 ± 0.3	0.094 ± .014[1]	(0.41)[2]	(0.05)	
			0.30 ± 0.01[3]		

[1]Acid digestion, GFAAS determination.
[2]Certificate information values.
[3]Acid digestion, solution nebulization ICP-MS determination.

Table 3.33 summarizes the results obtained for the analysis of ultrasonic slurries by ETV-ICP-MS. For coal, excellent agreement between found and certified values was obtained with the exception of Cr. The authors [69] report that even the application of the method of standard addition did not give the correct result indicating that Cr remained either locked in the coal matrix or was not totally volatilized from the graphite tube. For both the lobster hepatopancreas and the total diet samples, excellent agreement was obtained between found and certified values. Because of the large quantities of matrix components (organic or salt) present during the high temperature vaporization step, suppression of analyte ion signals occurred. For these two reference materials, the use of the method of standard addition was required (except for Pb in lobster hepatopancreas) to obtain accurate results. These results were in excellent agreement with certified values. Lead in the lobster reference material could be determined using external calibration because Pb volatilized before the sample matrix components avoiding plasma/interface suppression effects.

Using reported ETV-ICP-MS limits of detection (section 3.3) for the analytes studied, the following limits of detection, in ng/g, were calculated for a slurry sample size of 2 mg: Co, 0.070; Cu, 0.21; Cr, 3.2; Mn, 0.060; Ni, 0.24 and Pb, 0.0212.

3.9.4 Determination of Cd isotope ratios in sheep's blood and organ tissue

Cadmium is a non-essential metal which because of its toxic properties has been widely studied in small animals and humans. However, the few studies reported for ruminants are based on the intake of Cd at concentrations much higher than those encountered by grazing animals. The main impediment to the completion of more realistic studies involving whole body infusion of enriched stable isotopes has been the lack of adequate analytical techniques capable of measuring Cd isotope ratios in small samples at concentration levels below 0.2 ng/g. This problem was addressed by Grégoire and Lee [66] using ETV-ICP-MS.

Plasma or red cells (7 g) were prepared by wet digestion with a minimum quantity (10 ml) of concentrated HNO_3. Digestions were performed in a Milestone MLS 1200 microwave oven system. After bringing the digest to near dryness on a hotplate, the sample was made up to 10 ml with 1% HNO_3. Liver and kidney samples were prepared in an analogous manner using 0.4 g of freeze-dried sample.

Prior to the determination of Cd isotope ratios by ETV-ICP-MS, both analytes were separated from matrix components using off-line column chromatography. The acidity of the prepared blood or organ tissue digest was adjusted to a pH of 6.0 with ammonium hydroxide before being passed through a 1 cm^3 column containing silica-immobilized 8-hydroxyquinoline. The resin was eluted with 10 ml of 1 M HNO_3–0.1 M HCl and evaporated to dryness in the microwave oven. The residue was dissolved in 0.2 ml of 1% HNO_3. For the determination of Cd isotope ratios, a pre-concentration factor of 35 was achieved increasing the concentration of Cd in solution from approximately 0.2 ng/g in plasma to 7 ng/g in the analytical sample.

The fast-transient nature of the ETV-ICP-MS signal requires a relatively short duty cycle to ensure good precision of the isotope ratio measurement. Using a dwell time of 20 ms, approximately 40 intensity readings were obtained for each isotope during the high temperature vaporization step. For ETV-ICP-MS determinations, 10 μl of NASS-3 (diluted 1 in 500) physical carrier was added to both standard and sample solutions.

Instrumental mass discrimination for both ETV and solution nebulization work was reported to be generally less than the precision of the isotope ratio measurement itself. Correction for mass discrimination was accomplished using as reference the mean of ten separate ETV-ICP-MS isotope ratio measurements for a pure aqueous standard of Cd. Comparison of this value with the accepted value for the natural abundance ratio was used to calculate the appropriate correction factor.

Table 3.34 summarizes analytical results for the determination of Cd isotope ratios in a standard reference solution. Isotope ratios were determined on 200 pg of Cd in the presence of NASS-3 physical carrier. Isotope pairs were selected to include both the reference and spike isotope used in the actual sheep study. A third isotope for each element was also selected such that a comparison could be made of the precision of measurement for isotope ratios ranging from unity to some higher value.

For Cd, peak-height measurements gave poorer precision than ratios calculated using area counts. The isotope ratio for $^{111}Cd/^{110}Cd$ (1.035) was measured with a precision which was about 1% better than the precision of the $^{111}Cd/^{106}Cd$ (10.54) isotope ratio.

Tables 3.35 summarizes results for the determination of $^{111}Cd/^{106}Cd$ isotope ratios in a number of representative plasma, red cell and organ

TABLE 3.34

Precision of Cd isotope ratio measurement by ETV-ICP-MS (from Ref. [66] with permission)

Run	$^{111}Cd/^{106}Cd$		$^{111}Cd/^{110}Cd$	
	Integrated	Peak height	Integrated	Peak height
1	10.33	10.37	1.013	1.007
2.	10.56	10.06	1.038	1.027
3.	10.75	10.97	1.034	1.049
4.	10.45	10.64	1.033	1.012
5.	10.90	10.69	1.019	1.007
6.	10.27	10.94	1.011	1.044
7.	10.07	9.84	1.053	1.051
8.	10.58	10.38	1.057	1.065
9.	10.94	11.03	1.060	1.051
10.	10.55	10.53	1.030	1.033
Mean	10.54	10.54	1.035	1.035
s.d.	0.27	0.39	0.017	0.021
RSD	2.61%	3.71%	1.67%	2.03%

TABLE 3.35

Cadmium isotope ratios in sheep's blood and organ tissue (from Ref. [66] with permission

Sample	$^{111}Cd/^{106}Cd$
Plasma (2 h)	1.531 ± 0.028
Plasma (4 h)	1.263 ± 0.045
Plasma (8 h)	0.987 ± 0.014
Plasma (17 h)	0.758 ± 0.045
Plasma (24 h)	0.691 ± 0.003
Plasma (50 h)	0.624 ± 0.018
Plasma (56 h)	0.621 ± 0.030
Plasma (120 h)	0.615 ± 0.017
Red cells	1.639 ± 0.033
Kidney	0.313 ± 0.003
Liver	0.070 ± 0.001

tissue samples. These results demonstrated that the $^{111}Cd/^{106}Cd$ isotope ratio in blood and organ tissues varied from 0.070 to 1.531 during the course of the infusion study. For plasma samples, the Cd isotope ratio decreased by 60% during the study. A measurement precision of 2% for the $^{111}Cd/^{106}Cd$ isotope ratio was more than adequate to accurately track the change in this ratio with time. A limit of detection for Cd in blood of 0.35 pg g^{-1} was reported.

3.10 CONCLUDING REMARKS

Today, ETV-ICP-MS is a well developed and proven fully quantitative technique. The contents of this chapter clearly demonstrate the potential and the promise of this technique for solving some of the most challenging analytical problems.

Since electrothermal vaporizers are attachments to ICP mass spectrometers, the capabilities of ETV-ICP-MS will grow along with new developments in ICP-MS instrumentation such as the development of high resolution systems based on magnetic sector mass spectrometers with multi-collector detectors and systems based on time-of-flight mass spectrometers [191,192]. These will provide the opportunity for ETV-ICP-MS to be used for the simultaneous multi-element determination of large numbers of elements in very small samples and for the high-precision measurement of isotope ratios.

Because of its versatility and its potential, I am confident that ETV-ICP-MS will take its place among other spectroscopic techniques [193] and be widely used by analysts everywhere.

REFERENCES

1 B.V. L'vov, Investigation of atomic absorption spectra by complete vaporization of the sample in a graphite cuvette, *Spectrochim. Acta*, 39B (1984) 159–166. Translated from Russian in Inzh. Fiz, Zh., No. 2, (1959) 44.

2 D.E. Nixon, V.A. Fassel and R.N. Kniseley, Inductively coupled plasma–optical emission analytical spectroscopy, *Anal. Chem.*, 46 (1974) 210–213.

3 A.L. Gray and A.R. Date, Inductively coupled plasma source mass spectrometry using continuum flow ion extraction, *Analyst*, 108 (1983) 1033–1050.

4 A.L. Gray, Mass spectrometry with an inductively coupled plasma as an

ion source: the influence on ultratrace analysis of background and matrix response, *Spectrochim. Acta*, 41B (1986) 151–167.

5 A.R. Date and Y.Y. Cheung, Studies in the determination of lead isotope ratios by inductively coupled plasma mass spectrometry, *Analyst*, 112 (1987) 1531–1540.

6 C.J. Park, Feasibility study of an electrothermal vaporizer inductively coupled plasma mass spectrometry system, Ph.D. Thesis, University of Toronto, pp. 144, 1985.

7 C.J. Park and G.E.M. Hall, Electrothermal vaporization as a means of sample introduction into an inductively coupled plasma mass spectrometer: a preliminary report of a new analytical technique, Geological Survey of Canada, Paper 86-1B, 767–773, 1986.

8 C.J. Park, J.C. Van Loon, P. Arrowsmith and J.B. French, Sample analysis using plasma source mass spectrometry with electrothermal sample introduction, *Anal. Chem.*, 59 (1987) 2191–2196.

9 C.J. Park, J.C. Van Loon, P. Arrowsmith and J.B. French, Design and optimization of an electrothermal vaporizer for use in plasma source mass spectrometry, *Can. J. Spectrosc.*, 32 (1987) 29–36.

10 C.J. Park and G.E.M. Hall, Analysis of geological materials by inductively coupled plasma mass spectroscopy with sample introduction by electrothermal vaporization; Part 1: Determination of Mo and W, J. *Anal. At. Spectrom.*, 2 (1987) 473–480.

11 C.J. Park and G.E.M. Hall, Analysis of geological materials by inductively coupled plasma mass spectrometry with sample introduction by electrothermal vaporization; Part 2: Determination of Tl, *J. Anal. At. Spectrom.*, 3 (1998) 355–361.

12 G.E.M. Hall, J.-C. Pelchat, D.W. Boomer and M. Powell, Relative merits of two methods of sample introduction in inductively coupled plasma mass spectrometry: electrothermal vaporization and direct sample insertion, *J. Anal. At. Spectrom.*, 3 (1988) 791–797.

13 D.C. Grégoire, Determination of platinum, palladium, ruthenium and iridium in geological materials by inductively coupled plasma mass spectrometry with sample introduction by electrothermal vaporization, *J. Anal. At. Spectrom.*, 3 (1988) 309–314.

14 J.G. Sen Gupta and D.C. Grégoire, Determination of ruthenium, palladium and iridium in 27 international reference silicate and iron-formation rocks, ores and related materials by isotope dilution inductively coupled plasma mass spectrometry, *Geostandards Newsletter*, 13 (1989) 197–204.

15 D.C. Grégoire, Sample introduction techniques for the determination of osmium isotope ratios by inductively coupled plasma mass spectrometry, *Anal. Chem.*, 62 (1990) 141–146.

16 K.K. Falkner and J.M. Edmond, Determination of gold at femtomolar

levels in natural waters by flow injection inductively coupled plasma quadrupole mass spectrometry, *Anal. Chem.*, 62 (1990) 1477–1481.
17 T. Imakita, H. Horii, T. Kawamura and K. Narita, Determination of trace amount of bismuth in iron and steel by inductively coupled plasma mass spectrometry with sample introduction by electrothermal vaporization, *Tetsu to Hagane*, 77 (1991) 212–218.
18 R.J.B. Hall, M.R. James, T. Wayman and P. Hulmston, The feasibility of the use of electrothermal vaporization inductively coupled plasma mass spectrometry for the determination of femtogram levels of plutonium and uranium, *Spec. Pub. R. Soc. Chem.*, 85 (1990) 145–154.
19 H. Matsunaga, N. Hirate and K. Nishikida, Determination of trace amounts of uranium and thorium in silicon and silicon dioxide by electrothermal vaporization ICP-MS, *Bunseki Kagaku*, 38 (1989) T21–T25.
20 S.A. Beres, The analysis of semiconductor grade hydrochloric acid by electrothermal vaporization ICP mass spectrometry, Perkin-Elmer Publication EAR-1, July, 1993.
21 R.A. Newman, S. Osborn and Z.H. Siddik, Determination of tellurium in biological fluids by means of electrothermal vaporization inductively coupled plasma mass spectrometry (ETV-ICP-MS), *Clin. Chim. Acta.*, 179 (1989) 191–196.
22 P.G. Whittaker, T. Lind, J.G. Williams and A.L. Gray, Inductively coupled plasma mass spectrometric determination of the absorption of iron in normal women, *Analyst*, 114 (1989) 675–678.
23 S.P. Osborne, Quantitation of mercury in petroleum by ETV-ICP-MS, *Appl. Spectrosc.*, 44 (1990) 1044–1046.
24 P. Richner and S. Wunderli, Differentiation between organic and inorganic chlorine by electrothermal vaporization inductively coupled plasma mass spectrometry: application to the determination of polychlorinated biphenyls in waste oils, *J. Anal. At. Spectrom.*, 8 (1993) 45–49.
25 R. Tsukahara and M. Kubota, Some characteristics of inductively coupled plasma mass spectrometry with sample introduction by tungsten furnace electrothermal vaporization, *Spectrochim. Acta.*, 45B (1990) 779–787.
26 N. Shibata, N. Fudagawa and M. Kubota, Electrothermal vaporization using a tungsten furnace for the determination of rare-earth elements by inductively coupled plasma mass spectrometry, *Anal. Chem.*, 63 (1991) 636–640.
27 N. Shibata, N. Fudagawa and M. Kubota, Effects of hydrogen mixed with argon carrier gas on electrothermal vaporization inductively coupled plasma mass spectrometry, *Spectrochim. Acta.*, 47B (1992) 505–516.
28 N. Shibata, N. Fudagawa and M. Kubota, Oxide formation in electrothermal vaporization inductively coupled plasma mass spectrometry, *Spectrochim. Acta.*, 48B (1993) 1127–1137.

29 W.-L. Shen, J.A. Caruso, F.L. Fricke and R.D. Satzger, Electrothermal vaporization interface for sample introduction in inductively coupled plasma mass spectrometry, *J. Anal. At. Spectrom.*, 5 (1990) 451–455.
30 J.M. Carey, E.H. Evans, J.A. Caruso and W.-L. Shen, Evaluation of a modified commercial graphite furnace for reduction of isobaric interferences in argon inductively coupled plasma mass spectrometry, *Spectrochim. Acta.*, 46B (1991) 1711–1721.
31 D.C. Grégoire, M. Lamoureux, C.L. Chakrabarti, S. Al-Maawali and J.P. Byrne, Electrothermal vaporization for inductively coupled plasma mass spectrometry and atomic absorption spectrometry: symbiotic analytical techniques, *J. Anal. At. Spectrom.*, 7 (1992) 79–
32 J.P. Byrne, C.L. Chakrabarti, D.C. Grégoire, M. Lamoureux and T. Ly, Mechanism of chloride interferences in atomic absorption spectrometry using a graphite furnace atomizer investigated by electrothermal vaporization inductively coupled plasma mass spectrometry; Part 1: Effect of magnesium chloride matrix and ascorbic acid chemical modifier on manganese, *J. Anal. At. Spectrom.*, 7 (1992) 371–381.
33 J.P. Byrne, M.M. Lamoureux, C.L. Chakrabarti, T. Ly and D.C. Grégoire, Mechanism of chloride interferences in atomic absorption spectrometry using a graphite furnace atomizer investigated by electrothermal vaporization inductively coupled plasma mass spectrometry; Part 2: Effect of sodium chloride matrix and ascorbic acid chemical modifier on manganese, *J. Anal. At. Spectrom.*, 8 (1993) 599–609.
34 S.I. Ima, Y. Nishiyama, T. Tanaka and T. Hayashi, Investigation of pyrolysed ascorbic acid in an electrothermal graphite furnace by inductively coupled plasma mass spectrometry and raman spectrometry, *J. Anal. At. Spectrom.*, 10 (1995) 439–442.
35 I. Ida, K. Isobe, Y. Ishibashi and N. Gunju, Application of electrothermal vaporization for inductively coupled plasma atomic emission spectrometry and inductively coupled plasma mass spectrometry, *Tetsu to Hagane*, 77 (1991) 204–211.
36 S.A. Darke, C.J. Pickford and J.F. Tyson, Study of electrothermal vaporization sample introduction for plasma spectrometry, *Anal. Proc.*, 26 (1989) 379–381.
37 P. Hulmston and R.C. Hutton, Analytical capabilities of electrothermal vaporization inductively coupled plasma mass spectrometry, *Spectroscopy*, 6 (1991) 35–38.
38 S. Beres, R. Thomas, E. Denoyer and P. Bruckner, The benefits of electrothermal vaporization for minimizing interferences in ICP-MS, *Spectroscopy*, 9 (1994) 20–26.
39 U. Voellkopf, A. Guensel, M. Paul and H. Wiesmann, Applications of ICP-MS with sample introduction by electrothermal vaporization and flow injection techniques, In: G. Holland and A.N. Eaton (Eds), *Applications of*

Plasma Source Mass Spectrometry, Royal Society of Chemistry, London, 1991, pp. 162–177.
40 J. Marshall and J. Franks, Multielement analysis and reduction of spectral interferences using electrothermal vaporization inductively coupled plasma mass spectrometry, *At. Spectrom.*, 11 (1990) 177–186.
41 E. Hoffmann, C. Lüdke and H. Scholze, Electrothermal vaporization for simultaneous multi-element determination, *J. Anal. At. Spectrom.*, 9 (1994) 1237–1241.
42 D.J. Gray, S. Wang and R. Brown, Stability and sensitivity enhancement using ETV-ICP-MS, *Appl. Spectrosc.*, 48 (1994) 1316–1320.
43 D.M. Hughes, C.L. Chakrabarti, D.M. Goltz, D.C. Grégoire, R.E. Sturgeon and J.P. Byrne, Seawater as a multi-component physical carrier for ETV-ICP-MS, *Spectrochim. Acta*, 50B (1995) 425–440.
44 N. Nonose, N. Matsuda, N. Fudagawa and E. Kubota, Signal enhancement effect of halogen matrix in electrothermal vaporization inductively coupled plasma mass spectrometry, *Spectrochim. Acta*, 51B (1996) 1551–1565.
45 K. Grünke, H.-J. Stärk, R. Wennrich and H.M. Ortner, An investigation of different modifiers in electrothermal vaporization inductively coupled plasma mass spectrometry (ETV-ICP-MS), *Fresenius J. Anal. Chem.*, 359 (1997) 465–468.
46 R.D. Ediger and S.A. Beres, The role of chemical modifiers in analyte transport loss interferences with electrothermal vaporization ICP mass spectrometry, *Spectrochim. Acta*, 47B (1992) 907–922.
47 D.C. Grégoire, S. Al-Maawali and C.L. Chakrabarti, Use of Mg/Pd chemical modifiers for the determination of volatile elements by electrothermal vaporization ICP-MS: Effect on mass transport efficiency, *Spectrochim. Acta*, 47B (1992) 1123–1132.
48 D.C. Grégoire and R.E. Sturgeon, Analyte transport efficiency with electrothermal vaporization inductively coupled plasma mass spectrometry, *Spectrochim. Acta*, 54B (1999) 773–786.
49 R.W. Fonseca and N.J. Miller-Ihli, Analyte transport studies of aqueous solutions and slurry samples using electrothermal vaporization ICP-MS, *Appl. Spectrosc.*, 10 (1995) 1403–1410.
50 C.M. Sparks, Holcombe, J.A. and Pinkston, T.L., Sample Retention in the Transport Tubing Between an Electrothermal Vaporizer and an Inductively Coupled Plasma Mass Spectrometer, Appl. Spectrosc., 50, 86–90 (1996).
51 C.M. Sparks, J. Holcombe and T.L. Pinkston, Particle size distribution of sample transported from an electrothermal vaporizer to an inductively coupled plasmas mass spectrometer, *Spectrochim. Acta*, 48B (1993) 1607–1615.
52 D.C. Grégoire and R.E. Sturgeon, Background spectral features in

electrothermal vaporization inductively coupled plasma mass spectrometry: molecular ions resulting from the use of chemical modifiers, *Spectrochim. Acta*, 48B (1993) 1347–1364.
53 J.P. Byrne, D.M. Goltz, D.C. Grégoire and C.L. Chakrabarti, Vaporization and atomization of boron in the graphite furnace investigated by electrothermal vaporization inductively coupled plasma mass spectrometry, *Spectrochim. Acta*, 49B (1994) 433–443.
54 J.P. Byrne, D.M. Hughes, C.J. Chakrabarti and D.C. Grégoire, Mechanism of tungsten volatilization in electrothermal vaporization inductively coupled plasma mass spectrometry, *J. Anal. At. Spectrom.*, 9 (1994) 913–917.
55 R.St.C. McIntyre, D.C. Grégoire and C.L. Chakrabarti, Vaporization of radium and other alkaline earth elements in electrothermal vaporization inductively coupled plasma mass spectrometry, *J. Anal. At. Spectrom.*, 12 (1997) 547–551.
56 J.P. Byrne, R.St.C. McIntyre, M.E. Benyounes, D.C. Grégoire and C.L. Chakrabarti, Determination of chromium by electrothermal vaporization inductively coupled plasma mass spectrometry, *Can. J. Anal. Sci. Spectrosc.*, 42 (1997) 95–101.
57 D.M. Goltz, D.C. Grégoire and C.L. Chakrabarti, Mechanism of vaporization of yttrium and rare earth elements in electrothermal vaporization inductively coupled plasma mass spectrometry, *Spectrochim. Acta*, 50B (1995) 1365–1382.
58 J.P. Byrne and A.L. Carambassis, Vaporization and atomization of neodymium in graphite furnace atomic absorption spectrometry, *Spectrochim. Acta*, 51B (1996) 87–96.
59 D.M. Goltz, D.C. Grégoire, J.P. Byrne and C.L. Chakrabarti, Vaporization and atomization of uranium in a graphite tube electrothermal vaporizer: a mechanistic study using electrothermal vaporization inductively coupled plasma mass spectrometry and graphite furnace atomic absorption spectrometry, *Spectrochim. Acta*, 50B (1995) 803–814.
60 M.M. Lamoureux, C.L. Chakrabarti, J.C. Hutton, A.Kh. Gilmutdinov, Y.A. Zakharov and D.C. Grégoire, Mechanism of aluminum spike formation and dissipation in electrothermal atomic absorption spectrometry, *Spectrochim. Acta*, 50B (1995) 1847–1867.
61 J.P. Byrne, D.C. Grégoire, M.E. Benyounes and C.L. Chakrabarti, Vaporization and atomization of the platinum group elements in the graphite furnace investigated by electrothermal vaporization inductively coupled plasma mass spectrometry, *Spectrochim. Acta*, 52B (1997) 1575–1586.
62 D.M. Hughes, D.C. Grégoire, H. Naka and C.L. Chakrabarti, The vaporization of phosphorous and the use of chemical modifiers for the determination of phosphorus by electrothermal vaporization inductively

coupled plasma mass spectrometry, *Spectrochim. Acta*, 52B (1997) 517–529.
63 D.C. Grégoire and H. Naka, Mechanism of vaporization of sulfur in electrothermal vaporization inductively coupled plasma mass spectrometry, *J. Anal. At. Spectrom.*, 10 (1995) 823–828.
64 D.C. Grégoire, D.M. Goltz, M.M. Lamoureux and C.L. Chakrabarti, Vaporization of acids and their effect on analyte signal pulses in electrothermal vaporization inductively coupled plasma mass spectrometry, *J. Anal. At. Spectrom.*, 9 (1994) 919–926.
65 D.C. Grégoire, Application of isotope ratios determined by ICP-MS to earth science studies, *Prog. Anal. Spectrosc.*, 12 (1989) 433–452.
66 D.C. Grégoire and J. Lee, Determination of cadmium and zinc isotope ratios in sheep's blood and organ tissue by electrothermal vaporization inductively coupled plasma mass spectrometry, *J. Anal. At. Spectrom.*, 9 (1994) 393–397.
67 R.E. Sturgeon, D.C. Grégoire, S.N. Willie, J. Zheng and A. Kudo, Determination of ultratrace levels of heavy metals in arctic snow by ETV-ICP-MS, *J. Anal. At. Spectrom.*, 8 (1993) 1053–1058.
68 D.C. Grégoire, K.M. Ansdell, D.M. Goltz and C.L. Chakrabarti, Trace analysis of single zircons for rare earth elements, U and Th by electrothermal vaporization ICP mass spectrometry, *Chem. Geol.*, 124 (1995) 91–99.
69 D.C. Grégoire, N.J. Miller-Ihli and R.E. Sturgeon, Direct analysis of solids by ultrasonic slurry electrothermal vaporization ICP mass spectrometry, *J. Anal. At. Spectrom.*, 9 (1994) 615–610.
70 J.G. Williams, In: K.E. Jarvis, A.L. Gray and R.S. Houk (Eds), *Handbook of Inductively Coupled Plasma Mass Spectrometry*, Blackie, London, pp. 81–98, 1992.
71 J.M. Carey and J.A. Caruso, Electrothermal vaporization for sample introduction in plasma source spectrometry, *Crit. Rev. Anal. Chem.*, 23 (1992) 397–439.
72 I. Marawi, L.K. Olson, J. Wang and J.A. Caruso, Utilization of metallic platforms in electrothermal vaporization inductively coupled plasma mass spectrometry, *J. Anal. At. Spectrom.*, 10, (1995) 7–14.
73 D.M. Goltz, D.C. Grégoire and C.L. Chakrabarti, Characterization and evaluation of Mo, Ta, Re and W metal vaporization surfaces for electrothermal vaporization ICP mass spectrometry, *Can. J. Appl. Spectrocs.*, 41 (1996) 70–76.
74 V. Karanassios, P. Drouin and G.G. Reynolds, Electrically heated wire-loop in-torch vaporization (ITV) sample introduction system for ICP-AES with photomultiplier tube detection and ICP-MS, *Spectrochim. Acta*, 50B (1995) 415–423.
75 D.C. Grégoire and C.L. Chakrabarti, Atomization from a tantalum surface

in graphite furnace atomic absorption spectrometry, *Spectrochim. Acta*, 37B (1982) 611–623.
76 J.C. Bailar Jr., H.J. Emeleus, Sir R. Nyholm and A.F. Trotman-Dickenson (Eds), *Comprehensive Inorganic Chemistry*, Pergamon Press, Oxford, 3, 1973.
77 M. Hansen, *Constitution of Binary Alloys*, 2nd Edn, McGraw-Hill Book Co., Toronto, 1958.
78 A.M. Gunn, D.L. Millard and G.F. Kirkbright, Optical emission spectrometry with an inductively coupled radio-frequency argon plasma source and sample introduction with a graphite electrothermal vaporization device. Part I. Instrumentation assembly and performance characteristics, *Analyst*, 103 (1978) 1066–1073.
79 D.C. Grégoire and C.J. Park, Versatile spray chamber for alternate sample introduction techniques in ICP mass spectrometry, *Can. J. Appl. Spectrosc.*, 37 (1992) 115–118.
80 K. Yamanaka, Y. Kawaguchi, K. Kawabata and D. Potter, *Optimization studies in ETV-ICP-MS*, Twentieth Annual Meeting of the Federation of Analytical Chemistry and Spectroscopy Societies, October 17–22, Detroit, MI., Paper #336 (1993).
81 M.M. Lamoureux, A multi-technique strategy for study of some mechanisms of atom formation and dissipation in electrothermal atomic absorption spectrometry, Ph.D. thesis, Carleton University, pp. 233 1993.
82 M.M. Lamoureux, C.L. Chakrabarti, D.M. Goltz and D.C. Grégoire, Modification of a commercial electrothermal vaporizer for sample introduction into an inductively coupled plasma mass spectrometer. Part 1. Characterization, *Anal. Chem.*, 66 (1994) 3208–3216.
83 M.M. Lamoureux, C.L. Chakrabarti, D.M. Goltz and D.C. Grégoire, Modification of a commercial electrothermal vaporizer for sample introduction into an inductively coupled plasma mass spectrometer. Part 2. Performance evaluation, *Anal. Chem.*, 66 (1994) 3217–3222.
84 R.C. Hutton and P.M. Hulmston, Microvolume sampling by ETV-ICP-MS, 11th International Symposium on Microchemical Techniques, Wiesbaden, FRG, Aug. 28 to Sept. 1, 1989.
85 W. Slavin, A comparison of atomic spectroscopic techniques, *Spectroscopy*, 6 (1991) 16–21.
86 W. Slavin. *Graphite Furnace AAS, A Source Book*, Perkin-Elmer Corporation, Norwalk, CT, pp. 229, 1984.
87 D.L. Tsalev, V.I. Slaveykova and P.B. Mandjukov, Chemical modification in graphite furnace atomic absorption spectrometry, *Spectrochim. Acta Rev.*, 13 (1990) 225–274.
88 N. Jakubowski, I. Feldmann and D. Stuewar, Diagnostic investigations of aerosols with varying water content in inductively coupled plasma mass spectrometry, *J. Anal. At. Spectrom.*, 8 (1993) 969–977.

89 S.H. Tan and G. Horlick, Background spectral features in inductively coupled plasma mass spectrometry, *Appl. Spectrosc.*, 40 (1986) 445–460.
90 E. Björn, W. Frech, E. Hoffman and C. Lüdke, Investigation and quantification of spectroscopic interferences from polyatomic species in inductively coupled plasma mass spectrometry using electrothermal vaporization or pneumatic nebulization for sample introduction, *Spectrochim. Acta*, 53B (1998) 1765–1776.
91 J.W. McLaren, D. Beauchemin and S.S. Berman, Analysis of the marine sediment material pacs-1 by inductively coupled plasma mass spectrometry, *Spectrochim. Acta*, 43B (1988) 413–420.
92 A.Kh. Gilmutdinov, D.C. Grégoire, R.E. Sturgeon, C.L. Chakrabarti and A.E. Staroverov, Kinetics of carbon vaporization from a graphite furnace investigated by electrothermal vaporization inductively coupled plasma mass spectrometry, *Spectrochim. Acta*, 49B (1994) 1007–1026.
93 V. Karanassios and G. Horlick, Background spectral characteristics in direct sample insertion inductively coupled plasma mass spectrometry, *Spectrochim. Acta*, 44B (1989) 1361–1385.
94 D.L. Styris and D.A. Redfield, Perspectives on mechanism of electrothermal atomization, *Spectrochim. Acta Rev.*, 15 (1993) 71–123.
95 R.E. Sturgeon and C.L. Chakrabarti, Recent advances in electrothermal atomization in graphite furnace AAS, *Prog. Anal. At. Spectrosc.*, 1 (1978) 5–176.
96 D.R. Lide (Ed), *Handbook of Chemistry and Physics*, 74th Edn, CRC Press, Boca Raton, FL, 1993.
97 W. Frech, A. Cedergren and E. Lundberg, Vapour phase interference in electrothermal atomic absorption spectroscopy (AAS), Analytiktreffen., Nov. 8–12, Neubrandenburg Abstracts, p. 102 (1982).
98 G.F. Kirkbright and R.D. Snook, Volatilization of refractory compound forming elements from a graphite electrothermal vaporization device for sample introduction into an inductively coupled plasma, *Anal. Chem.*, 51 (1979) 1938–1941.
99 R.D. Ediger, *The Challenge of the Successful ETV-ICP-MS Analysis,* Nineteenth Annual Meeting of the Federation of Analytical Chemistry and Spectroscopy Societies, Philadelphia, PA, U.S.A., Sept. 20–25, paper #675 (1992).
100 V. Majidi and N.J. Miller-Ihli, Influence of graphite substrate on analyte signals in electrothermal vaporization inductively coupled plasma mass spectrometry, *Spectrochim. Acta*, 53B (1998) 965–980.
101 W.-C. Wei, C.-J. Chen and H.-M. Yang, Determination of boron using mannitol assisted electrothermal vaporization for sample introduction in inductively coupled plasma mass spectrometry, *J. Anal. At. Spectrom.*, 10 (1995) 955–961.
102 B. Wanner, P. Richner and B. Magyar, The role of modifiers in electro-

thermal vaporization inductively coupled plasma mass spectrometry (ETV-ICP-MS) for the determination of B, La and U, *Spectrochim. Acta*, 51B (1996) 817–827.
103 R.W. Fonseca and N.J. Miller-Ihli, Influence of sample matrix components on the selection of calibration strategies in electrothermal vaporization inductively coupled plasma mass spectrometry, *Spectrochim. Acta*, 51B (1996) 1591–1599.
104 D.C. Grégoire. Perspectives on research in electrothermal vaporization icp mass spectrometry, *Trends Appl. Spectrosc.*, 1 (1993) 313–324.
105 T. Kántor, Interpreting some analytical characteristics of thermal dispersion methods used for sample introduction in atomic spectrometry, *Spectrochim. Acta*, 42B (1988) 1299.
106 K. Grünke, H.-J. Stärk and H.M. Ortner, Removal of water and hydrogen from graphite tubes investigated by electrothermal vaporization inductively coupled plasma mass spectrometry, *Spectrochim. Acta*, 52B (1997)1545–1550.
107 D. Langer, G. Pineda and J.A. Holcombe, *Analyte Transport in ETV-ICP-MS: A Look at the Carriers*, 1998 Winter Conference on Plasma Spectrochemistry, Scottsdale, Arizona, USA, January 5–10, 1998, Paper M11.
108 R.W. Fonseca, N.J. Miller-Ihli, C. Sparks, J.A. Holcombe and B. Shaver, Effect of oxygen ashing on analyte transport efficiency using ETV-ICP-MS, *Appl. Spectrosc.*, 51 (1997) 1800–1806.
109 T. Kántor and S. Güçer, Efficiency of sample introduction into inductively coupled plasma by graphite furnace electrothermal vaporization, *Spectrochim. Acta*, 54B (1999) 763–772.
110 T. Buchkamp and G. Hermann, S*olid Sampling for Electrothermal Atomization Spectroscopy with Electrostatic Deposition of Particulates,* 3rd European Furnace Symposium, Prague, Czech Republic, June 14–18, 1998, Paper O6.
111 P. Barth, S. Hauptkorn and V. Krivan, Improved slurry sampling electrothermal vaporization system using a tungsten coil for inductively coupled plasma atomic emission spectrometry, *J. Anal. At. Spectrom.*, 12 (1997) 1351–1358.
112 U. Schäffer and V. Krivan, A graphite furnace electrothermal vaporization system for inductively coupled plasma atomic emission spectrometry, *Anal. Chem.*, 70 (1998) 482–490.
113 D.C. Grégoire, The effect of easily ionizable concomitant elements on non-spectroscopic interferences in inductively coupled plasma mass spectrometry, *Spectrochim. Acta*, 42B (1987) 895–907.
114 D.C. Grégoire, Influence of instrument parameters on non-spectroscopic interferences in inductively coupled plasma mass spectrometry, *Appl. Spectrosc.*, 41 (1987) 897–903.

115 J.-F. Alary and E.D. Salin, Quantitation of water and plasma diagnosis for electrothermal vaporization inductively coupled plasma mass spectrometry: The use of argon and argide polyatomics as probing species, *Spectrochim. Acta*, 53B (1998) 1705–1721.
116 G.E.M. Hall and J.C. Pelchat, Determination of palladium and platinum in fresh waters by inductively coupled plasma mass spectrometry and activated charcoal preconcentration, *J. Anal. At. Spectrom.*, 8 (1993) 1059–1065.
117 T.L. Pinkston, Multi-Element Determination of Trace Contaminants in Photoresist Materials by Inductively Coupled Plasma Mass Spectrometry, *Future Prac. Contam. Control, Proceedings Int. Symp. Cont. Control*, 11 (1992) 535–540.
118 T. Etoh, M. Yamada and M. Matsubara, Electrothermal vaporization ICP-MS (ETV-ICP-MS) determination of impurities in photoresist for very large scale integrated circuits (VLSI), *Anal. Sci.*, 7 (1991) 1263–1264.
119 P.K.G. Manninen, Determination of extractable organic chlorine by electrothermal vaporization inductively coupled plasma mass spectrometry, *J. Anal. At. Spectrom.*, 9 (1994) 209–211.
120 R.J. Bowins and R. McNutt, Electrothermal isotope dilution inductively coupled plasma mass spectrometry method for the determination of sub-ng ml^{-1} levels of lead in human plasma, *J. Anal. At. Spectrom.*, 9 (1994) 1233–1236.
121 M.D. Argentine and R.M. Barnes, Electrothermal vaporization inductively coupled plasma mass spectrometry for the analysis of semiconductor grade organometallic materials and process chemicals, *J. Anal. At. Spectrom.*, 9 (1994) 1371–1378.
122 W.Hub and H. Amphlett, Application of ETV-ICP-MS in semiconductor process control, *Fresenius J. Anal. Chem.*, 350 (1994) 587–592.
123 P. Richner, D. Evans, C. Wahrenberger and V. Dietrich, Applications of laser ablation and electrothermal vaporization as sample introduction techniques for ICP-MS, *Fresenius J. Anal. Chem.*, 350 (1994) 235–241.
124 K. Ruth, P. Schmidt, J. Coria and E. Mori, Silicon wafer surface analysis by electrothermal vaporization inductively coupled plasma mass spectrometry (ETV-ICP-MS), *Proc. Electrochem. Soc.*, 94 (1994) 565–572.
125 J. Wang, J.M. Carey and J.A. Caruso, Direct analysis of solid samples by electrothermal vaporization inductively coupled plasma mass spectrometry, *Spectrochim. Acta*, 49B (1994) 193–203.
126 R.E. Sturgeon and D.C. Grégoire, Electrothermal vaporisation inductively coupled plasma mass spectrometric detection of As, Sb, Se, Bi and Sn following preconcentration by in-situ collection of their hydrides, *Spectrochim. Acta*, 49B (1994) 1335–1345.
127 I. Marawi, J. Wang and J.A. Caruso, Graphite furnace hydride preconcentration and subsequent detection by inductively coupled plasma

mass spectrometry, *Anal. Chim. Acta*, 291 (1994) 127–136.
128 S.J. Santosa and S. Tanaka, Inductively coupled plasma mass spectrometry for sequential determination of trace metals in rain and river waters using electrothermal vaporization, *Anal. Lett.*, 28 (1995) 509–534.
129 W.-C. Wei and M.-H. Yang, Determination of phosphorus and arsenic in trichlorosilane by electrothermal vaporization inductively coupled plasma mass spectrometry with prior concentration by cuprous chloride, *Fresenius J. Anal. Chem.*, 353 (1995) 167–170.
130 F. Vanhaecke, L. Gelaude, L. Moens and R. Dams, Solid sampling electrothermal vaporization inductively coupled plasma mass spectrometry for the direct determination of Hg in sludge samples, *Anal. Chim. Acta*, 383 (1999) 253–261.
131 J.M. Ren, R. Rattray, E.D. Salin and D.C. Grégoire, Assessment of direct solid sample analysis by graphite pellet electrothermal vaporization inductively coupled plasma mass spectrometry, *J. Anal. At. Spectrom.*, 10 (1995) 1027–1029.
132 F. Vanhaecke, S. Boonen, L. Moens and R. Dams, Solid sampling electrothermal vaporization inductively coupled plasma mass spectrometry for the determination of arsenic in standard reference materials of plant origin, *J. Anal. At. Spectrom.*, 10 (1995) 81–87.
133 L. Moens, P. Verrept, S. Boonen, F. Vanhaecke and R. Dams, Solid sampling electrothermal vaporization for sample introduction in inductively coupled plasma atomic emission spectrometry and inductively coupled plasma mass spectrometry, *Spectrochim. Acta*, 50B (1995) 463–475.
134 F. Vanhaecke, G. Galbács, S. Boonen, L. Moens and R. Dams, Use of the Ar_2^+ signal as a diagnostic tool in solid sampling electrothermal vaporization inductively coupled plasma mass spectrometry, *J. Anal. At. Spectrom.*, 10 (1995) 1047–1052.
135 M. Komoda, K. Chiba and H. Uchida, Determination of trace impurities in silicon wafer surface by isotope dilution analysis using electrothermal vaporization inductively coupled plasma mass spectrometry, *Anal. Lett.*, 12 (1996) 21–25.
136 N. Berryman and U. Probst, Parameter optimization of electrothermal vaporization inductively coupled plasma mass spectrometry for oligoelement determination in standard reference materials, *Fresenius J. Anal. Chem.*, 355 (1996) 783–788.
137 S. Laly, K. Nakagawa, T. Arimura and T. Kimijima, Optimization of electrothermal vaporization inductively coupled plasma mass spectrometry conditions for the determination of iron, copper, nickel and zinc in semiconductor grade acids, *Spectrochim. Acta*, 51B (1996) 1393–1401.
138 H. Naka and D.C. Grégoire, Determination of trace amounts of sulfur in steel by electrothermal vaporization inductively coupled plasma mass spectrometry, *J. Anal. At. Spectrom.*, 11 (1996) 359–363.

139 D.W. Hastings, S.R. Emerson and B.K. Nelson, Determination of picogram quantities of vanadium in calcite and seawater by isotope dilution inductively coupled plasma mass spectrometry with electrothermal vaporization, *Anal. Chem.*, 68 (1996) 371–377.
140 J.S. Alvarado and M.D. Erickson, Determination of long-lived radioisotopes using electrothermal vaporization inductively coupled plasma mass spectrometry, *J. Anal. At. Spectrom.*, 11 (1996) 923–928.
141 M.P. Escobar, B.W. Smith and J.D. Winefordner, Determination of metallo-organic species in lubricating oil by electrothermal vaporization inductively coupled plasma mass spectrometry, *Anal. Chim. Acta*, 320 (1996) 11–17.
142 G. Chapple and J.P. Byrne, Direct determination of trace metals in seawater using electrothermal vaporization inductively coupled plasma mass spectrometry, *J. Anal. At. Spectrom.*, 11 (1996) 549–553.
143 G. Chapple, The Determination of Trace Metals in Seawater by Electrothermal Vaporisation Inductively Coupled Plasma Mass Spectrometry, M.Sc. Thesis, University of Technology, Sydney, Australia, 1998.
144 M.-J. Liaw and S.-J. Jiang, Determination of copper, cadmium and lead in sediment samples by slurry sampling electrothermal vaporization inductively coupled plasma mass spectrometry, *J. Anal. At. Spectrom.*, 11 (1996) 555–560.
145 S. Boonen, F. Vanhaecke, L. Moens and R. Dams, Direct determination of Se and As in solid certified reference materials using electrothermal vaporization inductively coupled plasma mass spectrometry, *Spectrochim. Acta,* 51B (1996) 271–278.
146 G. Galbács, F. Vanhaecke, L. Moens and R. Dams, Determination of cadmium in certified reference materials using solid sampling electrothermal vaporization inductively coupled plasma mass spectrometry, *Microchem. J.*, 54 (1996) 272–286.
147 B. Bitterli, H. Cousin and B. Magyar, Determination of metals in airborne particles by electrothermal vaporization inductively coupled plasma mass spectrometry, *J. Anal. At. Spectrom.*, 12 (1997) 957–961.
148 N. Berryman and T. Probst, Rapid determination of ^{90}Sr by electrothermal vaporization inductively coupled plasma mass spectrometry (ETV-ICP-MS), *Radiochim. Acta*, 76 (1997) 191–195.
149 L. Yu, R. Koirtyohann, M.L. Rueppel, A.K. Skipor and J.J. Jacobs, Simultaneous determination of aluminum, titanium and vanadium in serum by electrothermal vaporization inductively coupled plasma mass spectrometry, *J. Anal. At. Spectrom.*, 12 (1997) 69–74.
150 C.-C. Chang and S.-J. Jiang, Determination of copper, cadmium and lead in biological samples by electrothermal vaporization isotope dilution inductively coupled plasma mass spectrometry, *J. Anal. At. Spectrom.*, 12 (1997) 75–80.

151 C.-C. Chang and S.-J. Jiang, Determination of Hg and Bi by electrothermal vaporization inductively coupled plasma mass spectrometry using vapor generation with in-situ preconcentration in a platinum coated furnace, *Anal. Chim. Acta*, 353 (1997) 173–180.
152 C. Moor, P. Boll and S. Wiget, Determination of impurities in microamounts of silver alloys by electrothermal vaporization inductively coupled plasma mass spectrometry (ETV-ICP-MS) after in-situ digestion in the graphite tube, *Fresenius J. Anal. Chem.*, 359 (1997) 404–406.
153 B. Fairman and T. Catterick, Simultaneous determination of arsenic, antimony and selenium in aqueous matrices by electrothermal vaporization inductively coupled plasma mass spectrometry, *J. Anal. At. Spectrom.*, 12 (1997) 863–866.
154 D.C. Grégoire and M. Ballinas, Direct determination of arsenic in fresh and saline waters by electrothermal vaporization inductively coupled plasma mass spectrometry, *Spectrochim. Acta*, 52B (1997) 75–82.
155 S.J. Santosa, S. Tanaka and K. Yamanaka, Inductively coupled plasma mass spectrometry for the sequential determination of trace metals in sea water after electrothermal vaporization of their dithiocarbamate complexes in methyl isobutyl ketone, *Fresenius J. Anal. Chem.*, 357 (1997) 1122–1127.
156 S.J. Santosa, S. Tanaka and K. Yamanaka, Sequential determination of trace metals in sea water by inductively coupled plasma mass spectrometry after electrothermal vaporization of their dithiocarbamate complexes in methyl isobutyl ketone, *Env. Mon. Asses.*, 44 (1997) 515–528.
157 M.-J. Liaw, S.-J. Jiang and Y.-C. Li, Determination of mercury in fish samples by slurry sampling electrothermal vaporization inductively coupled plasma mass spectrometry, *Spectrochim. Acta*, 52B (1997) 779–785.
158 S.N. Willie, D.C. Grégoire and R.E. Sturgeon, Determination of inorganic and total mercury in biological tissues by electrothermal vaporization inductively coupled plasma mass spectrometry, *Analyst*, 122 (1997) 751–754.
159 S. Hauptkorn, V. Krivan, B. Gercken and J. Pavel, Determination of trace impurities in high-purity quartz by electrothermal vaporization inductively coupled plasma mass spectrometry, *J. Anal. At. Spectrom.*, 12 (1997) 421–428.
160 M.W. Hinds, D.C. Grégoire and E.A. Ozaki, Direct determination of volatile elements in nickel alloys by electrothermal vaporization inductively coupled plasma mass spectrometry, *J. Anal. At. Spectrom.*, 12 (1997) 131–135.
161 F. Vanhaecke, S. Boonen, L. Moens and R. Dams, Isotope dilution as a calibration method for solid sampling electrothermal vaporization inductively coupled plasma mass spectrometry, *J. Anal. At. Spectrom.*, 12 (1997) 125–130.
162 H.Th. Uggerud and W. Lund, Use of palladium and iridium as modifiers in

the determination of arsenic and antimony by electrothermal vaporization inductively coupled plasma mass spectrometry, following in situ trapping of the hydrides, *J. Anal. At. Spectrom.*, 12 (1997) 1169–1174.

163 D. Pozebon, V.L. Dressler and A.J. Curtius, Determination of arsenic, selenium and lead by electrothermal vaporization inductively coupled plasma mass spectrometry using iridium-coated tubes, *J. Anal. At. Spectrom.*, 13 (1998) 7–11.

164 D.M. Hughes, D.C. Grégoire, H. Naka and C.L. Chakrabarti, Determination of trace amounts of phosphorus in high purity iron by electrothermal vaporization inductively coupled plasma mass spectrometry, *Spectrochim. Acta*, 53B (1998) 1079–1085.

165 R.St.C. McIntyre, D.C. Grégoire and C.L. Chakrabarti, The direct determination of radium in geological materials by slurry sampling electrothermal vaporization inductively coupled plasma mass spectrometry, *Spectroscopy*, 13 (1998) 19–24.

166 N. Nonose and M. Kubota, Determination of metal impurities in sulfamic acid by isotope dilution electrothermal vaporization inductively coupled plasma mass spectrometry, *J. Anal. At. Spectrom.*, 13 (1998) 151–156.

167 K.-H. Lee, S.-J. Jiang and H.-W. Liu, Determination of mercury in urine by electrothermal vaporization isotope dilution inductively coupled plasma mass spectrometry, *J. Anal. At. Spectrom.*, 13 (1998) 1227–1231.

168 E.B. Buseth, G. Wibetoe and I. Martinsen, Determination of endogenous concentrations of the lanthanides in body fluids and tissues using electrothermal vaporization inductively coupled plasma mass spectrometry, *J. Anal. At. Spectrom.*, 13 (1998) 1039–1049.

169 D. Pozebon, V.L. Dressler and A.J. Curtius, Determination of copper, lead, bismuth and selenium(IV) in sea-water by electrothermal vaporization inductively coupled plasma mass spectrometry after on-line separation, *J. Anal. At. Spectrom.*, 13 (1998) 363–369.

170 E. Rosland and W. Lund, Direct determination of trace metals in sea-water by inductively coupled plasma mass spectrometry, *J. Anal. At. Spectrom.*, 13 (1998) 1239–1244.

171 K.-H. Lee, S.-H. Liu and S.-J. Jiang, Determination of cadmium and lead in urine by electrothermal vaporization isotope dilution inductively coupled plasma mass spectrometry, *Analyst*, 123 (1998) 1557–1560.

172 D. Pozebon, V.L. Dressler and A.J. Curtius, Determination of trace elements in biological materials by ETV-ICP-MS after dissolution or slurry formation with tetramethylammonium hydroxide, *J. Anal. At. Spectrom.*, 13 (1998) 1101–1105.

173 S.-F. Chen and S.-J. Jiang, Determination of cadmium, mercury and lead in soil samples by slurry sampling electrothermal vaporization inductively coupled plasma mass spectrometry, *J. Anal. At. Spectrom.*, 13 (1998) 1113–1117.

174 Y.-C. Li and S.-J. Jiang, Determination of Cu, Zn, Cd and Pb in fish samples by slurry sampling electrothermal vaporization inductively coupled plasma mass spectrometry, *Anal. Chim. Acta*, 359 (1998) 205–212.
175 J. Mierzwa and M.-H. Yang, Electrothermal vaporization inductively coupled plasma mass spectrometry for determination of metal impurities in slurries of aluminum oxide, *J. Anal. At. Spectrom.*, 13 (1998) 667–671.
176 S.-F. Chen and S.-J. Jiang, Determination of arsenic, selenium and mercury in fish samples by slurry sampling electrothermal vaporization inductively coupled plasma mass spectrometry, *J. Anal. At. Spectrom.*, 13 (1998) 673–677.
177 G. Galbács, T. Kántor, L. Moens and R. Dams, Mass spectrometric studies of thermal decomposition products of reference materials for use in solid sampling atomic spectrometry, *Spectrochim. Acta*, 53B (1998) 1335–1346.
178 J. Turner, S.J. Hill, H.E. Evans and B. Fairman, The use of ETV-IVP-MS for the determination of selenium in serum, *J. Anal. At. Spectrom.*, 14 (1999) 121–126.
179 J.W.H. Lam, R.E. Sturgeon and J.W. McLaren, Ultra-trace determination of Se in sediments by electrothermal vaporizer–inductively coupled plasma–mass spectrometry: User of the ETV as a thermochemical reactor, *Spectrochim. Acta*, 54B (1999) 443–453.
180 Y. Hu, F. Vanhaecke, L. Moens, R. Dams and I. Geuens, Determination of ruthenium in photographic materials using solid sampling electrothermal vaporization inductively coupled plasma mass spectrometry, *J. Anal. At. Spectrom.*, 14 (1999) 589–592.
181 J.B. Truscott, L. Bromley, P. Jones, E.H. Evans, J. Turner and B. Fairman, Determination of natural uranium and thorium in environmental samples by ETV-ICP-MS after matrix removal by on-line solid phase extraction, *J. Anal. At. Spectrom.*, 14 (1999) 627–631.
182 F. Vanhaecke, M. Verstraete, L. Moens, R. Dams and M. Nekkers, Determination of the palladium content in a solid plastic material by electrothermal vaporization ICP–mass spectrometry (ETV-ICP-MS), *Anal. Commun.*, 36 (1999) 89–92.
183 Y.-C. Li, S.-J. Jiang and S.-F. Chen, Determination of Ge, As, Se, Cd and Pb in plant materials by slurry sampling–electrothermal vaporization–inductively coupled plasma–mass spectrometry, *Anal. Chim. Acta*, 372 (1998) 365–372.
184 D. Pozebon, V.L. Dressler and A.J. Curtius, Determination of Mo, U and B in waters by electrothermal vaporization inductively coupled plasma mass spectrometry, *Talanta*, 47 (1998) 849–859.
185 H. Baumann, Solid sampling with inductively coupled plasma mass spectrometry—a survey, *Fresenius J. Anal. Chem.*, 342 (1992) 907–916.
186 U. Voellkopf, M. Paul and E. Denoyer, Analysis of solid samples by ICP mass spectrometry, *Fresenius J. Anal. Chem.*, 342 (1992) 917–923.

187 S.A. Darke and J.F. Tyson, Review of solid sampling introduction for plasma spectrometry and a comparison of results for laser ablation, electrothermal vaporization and slurry nebulization, *Microchem. J.*, 50 (1994) 310–336.
188 U. Kurfürst (Ed), *Solid Sample Analysis; Direct and Slurry Sampling using GF-AAS and ETV-ICP*, Springer Verlag, Berlin., 423 pp., 1998.
189 L.K. Olson, N.P. Vela and J.A. Caruso, Hydride generation electrothermal vaporization and liquid chromatography as sample introduction techniques for inductively coupled plasma mass spectrometry, *Spectrochim. Acta*, 50B (1995) 355–368.
190 A. Montaser (Ed), *Inductively Coupled Plasma Mass Spectrometry*, Wiley-VCH Publishers, NY, 964 pp., 1998.
191 P.P. Mahoney, S.J. Ray and G.M. Hieftje, Time-of-flight mass spectrometry for elemental analysis, *Appl. Spectrosc.*, 51 (1997) 16A–28A.
192 P.P. Mahoney, S.J. Ray, L. Gangqiang and G.M. Hieftje, Preliminary investigation of electrothermal vaporization sample introduction for inductively coupled plasma time-of-flight mass spectrometry, *Anal. Chem.*, 71 (1999) 1378–1383.
193 D.C. Grégoire, Electrothermal vaporization sample introduction for inductively coupled plasma mass spectrometry: Current status and future prospects, *Can. J. Anal. Sci. Spectrosc.*, 42 (1997) 1–9.

Chapter 4

Laser ablation for inductively coupled plasma-mass spectrometry

Detlef Günther and Jean-Michel Mermet

4.1 LASER ABLATION

The word 'ablation' originates from the Latin 'ablatus' (auferre, i.e. to carry away), which led to the French 'ablater' during the Middle Ages, then to 'to ablate' and 'ablation' in English. Ablation means a progressive and superficial destruction of a material by melting, fusion, sublimation, erosion, explosion, etc. Energy deposition is required for ablation and can be performed by using electrical discharges like arc and spark with electrically conducting materials or by using laser-material interaction with any type of materials, i.e. conducting or insulating. Laser ablation results in the formation of a microplasma at the surface of the target and in the production of fine particles and gaseous species. Several techniques are based on laser ablation, such as laser ablation coupled with ICP atomic emission spectrometry (LA-ICP-AES) or ICP mass spectrometry (LA-ICP-MS) and laser-induced plasma atomic emission spectrometry (LIP-AES). In LA-ICP-AES and LA-ICP-MS, ablation is performed in an ablation cell, and the ablation products are transported with a carrier gas stream to an ICP for further atomization, excitation and ionization. In LIP-AES, the detection is performed by observing directly the analyte line emission from the microplasma.

This chapter is mainly related to LA-ICP-MS [1–272].

4.2 DIFFERENT TYPES OF LASER ABLATION-BASED ANALYSIS

The different types of laser ablation-based analysis can be classified according to dimensional considerations.

Bulk analysis is aimed at replacing sample dissolution, because sample preparation can be complex, time-consuming and costly, or even either impossible to perform (e.g. SiC, zircon) or requiring more than a single dissolution procedure (e.g. Au and Ag). To perform bulk analysis, the target is generally scanned, and the signal is averaged. An alternative is the rotation of the target. There is no need for a sophisticated target observation system. Bulk analysis requires a fast, routine method, with the same analytical performance (LODs, accuracy and precision) as that obtained after sample dissolution. The key point is the availability of reference materials or certified reference materials (CRMs). They may be available in the case of metals and glasses, but the homogeneity is a limitation, because these materials were certified using a large amount of material, e.g. 1 g for sample dissolution. This may not give sufficient homogeneity in the mm-scale. The availability is more problematic with material such as ceramics and polymers. The use of different materials for calibration, for instance the NIST 612 glass CRM, or even the use of solutions can overcome this limitation when semi-quantitative results have to be obtained, but not when accurate results are required. The use of contained or added internal standards may be useful to improve both precision and accuracy.

Microanalysis is applied to inhomogeneous samples and requires a high lateral resolution, i.e. a small pit diameter (i.e. < 100 µm). This diameter corresponds to a small spot size of the focused laser beam, which is usually obtained by using optimized focusing systems, such as beam homogenizer, spatial filtering, etc. Requirements may be estimated to be in the range 5 µm to bulk for metals and alloys, 1–200 µm for geological samples, 5 µm for polymers, 1–10 µm for ceramics. (Microanalysis can be considered as a quasi non-destructive method, which is crucial in applications such as forensic sciences). The use of the word laser ablation microprobe (LAM) should be reserved for microanalysis only. A comparison of the characteristics of laser ablation for bulk analysis and microanalysis is given in Table 4.1.

In contrast to microanalysis, depth profiling does not require a high lateral resolution but a depth resolution in the order of 0.1 µm. This resolution may be obtained by using a flat distribution of energy across the laser beam at the target surface level. Moreover, it is critical to

TABLE 4.1

Comparison of the characteristics of laser ablation-ICP-MS for bulk analysis and microanalysis

	Bulk analysis	Microanalysis
Laser spot size	0.1–1 mm	5–100 μm
Laser energy	10–100 mJ	few mJ
Displacement	translation	usually single spot
Mass ablated	2 μg–2 mg	0.25 ng–2 μg
LODs	ppt–ppb	sub ppb– ppm

remove the material effectively and linearly and to avoid redeposition at the edge of the crater. Depth profiling may be applied when a transient change in concentration occurs in the solid, which is typically the case with multilayers. Requirements in depth resolution are 0.1 μm for metals and alloys, and in the range nm–μm for polymers and ceramics. Laser-based depth profiling cannot compete with conventional surface analysis methods in the nm range, but is certainly appropriate near 0.1 μm. Moreover, depth profiling can be conducted over several tens or hundreds of micrometers. Competitive methods are RF glow discharge for ceramics and polymers and DC glow discharge for conductive materials.

Mapping is certainly a growing field for laser-based technologies. Requirements are a lateral resolution in the range 3–50 μm, a depth resolution in the range 3–10 μm for level of impurities from ppm to %. Surfaces range from mm^2 to cm^2. It is expected that the maximum mapping procedure duration should be of the order of minutes or hours, depending on the scanned surface.

Most laser-based techniques are used in a laboratory, either for research or for routine analysis. However, there is a need for on-line analysis, which is usually applied to process control and which must be fast with respect to intrinsic process time, in order to save time and to increase the quality of manufactured products.

4.3 SPECIFIC USE OF LA-ICP-MS

When limits of detection (LODs) in the order of 1–10 ppb are required, the use of LA-ICP-MS is appropriate for bulk analysis. However, because of the risk of material deposition on the sampling cone orifice with subsequent signal and mass response drift, LA-ICP-AES, which allows the injection of a large amount of material without contamination, can be used as an alternative (particularly when multichannel detection is used). In this case LODs are only in the order of ppm. For bulk analysis, dedicated systems can be developed, without sophisticated observation and target displacement systems. For reasons similar to those for LA-ICP-AES, LIP-AES is also very suitable for bulk analysis. Several LIP-AES systems were introduced on the market several years ago, but it seems that currently, only one system is commercially available. This is used for the analysis of pig iron and blast furnace slags.

In contrast, because of the very low amount of ablated sample, LA-ICP-MS is most suitable for microanalysis. Powerful observation systems with large zooming facilities are necessary, with accurate positioning of the laser spot, ability to go from low to high magnification, and ability to view material removed during ablation. Because there is no transport of material, LIP-AES is very well suited for depth profiling and mapping, provided that the laser beam is modified according to the requirements, i.e. flat energy distribution for depth profiling, and high lateral resolution for mapping. Particularly for mapping, dedicated echelle grating-based dispersive systems need to be developed, based on the use of CCD or intensified CCD detection. Acquisition of entire spectra in the uv-visible range could then be obtained, leading to full flexibility in line and background selection, and to production of wavelength data base for elements of interest. Data processing such as time-correlated correction could be performed, and chemometrics could then be applied to provide element correlation and semi-quantitative analysis.

LA-ICP-MS may be necessary for depth profiling, when only low amounts of material can be ablated.

Selection of the adequate technology relative to the type of analysis is summarized in Table 4.2.

Clearly, LA-ICP-MS is unsurpassable when low LODs (ppb in a solid) and low amounts of probed materials (of the order of a few μm^3) are of concern. In particular, when a limited number of shots or even a single shot is performed in microanalysis, the technique can be considered as virtually non-destructive, which is important in applications such as forensic sciences and analysis of antique objects. Because laser ablation produces dry aerosols, the level of oxides and hydroxides can be significantly decreased, which can minimize polyatomic interferences, which are a severe limitation in ICP-quadrupole MS. Besides,

TABLE 4.2

Application of the various laser-based techniques and the types of analysis

Techniques	Type of analysis				Site
	Bulk	Micro-analysis	Depth profiling	Mapping	On-line
LA-ICP-AES	yes				yes
LA-ICP-MS	for low LODs	yes	for low LODs		
LIP-AES	yes	yes	yes	yes	yes

LA-ICP-MS keeps the capability for isotope abundance measurement, which is important in applications such as geochronology [e.g. 23,58,59, 72,97,99,126,132,139,169,171].

4.4 LASER PRINCIPLES

Lasers are characterized by monochromaticity, spatial and temporal coherence, laser beam with low divergence and high directionality, and high radiance. Only the last two characteristics are of concern for the ablation of solid materials, because a large amount of energy must be deposited onto the surface of the target.

Basically, a laser consists of three parts: energy input, amplification in an active medium, and light output. Lasers can be classified according to their active medium, gas, liquid and solid, and to the operating mode, i.e. continuous wave (CW) or pulse mode. Usually, the pulse mode is preferred for ablation, to obtain high-energy short pulses, i.e. to concentrate the energy over a short period of time. The first ever described laser, the ruby laser, was of the solid type. The most commonly used lasers for ablation are the Nd:YAG laser and the excimer laser.

The Nd:YAG laser belongs to the solid active medium laser type, while the excimer laser belongs to the gas type. A solid state laser uses ions in a crystalline matrix such as Cr^{3+} for the ruby laser and Nd^{3+} for the Nd:YAG laser. The crystal was Al_2O_3 for the ruby laser and is a garnet of yttrium and aluminium (YAG) for the Nd:YAG laser. This garnet is preferred over glass as in Nd:glass lasers, because of a higher thermal conductivity to improve thermal dissipation. The energy input is provided by a flash lamp. Both ruby and Nd:YAG lasers are conventional lasers because the rod is part of a resonator (or cavity) with a total reflection mirror on one side and a partially transparent mirror on the other side. The mirrors can be flat or spherical. The wavelength is 694 nm and 1064 nm for the ruby and Nd:YAG laser, respectively. The Nd:YAG laser can run in several pulse modes, the free-running mode (μs pulse duration) and the Q-switched mode (ns pulse duration). The latter type of pulses is produced by using Q-switching. The quality factor of a cavity, Q, is related to the cavity losses. A high Q resonator does not exhibit power loss. If the resonator is switched from no reflections (low Q) to near total reflection via a shutter (high Q), high population inversion and high energy storage are obtained. The pulse duration is shorter than 10 ns. Because of a cylindrical rod, the laser beam exhibits a circular shape.

The Nd:YAG is reliable and the only component that needs to be changed periodically is the flash lamp. The flash lamp lifetime can be longer than 10^6 pulses. The energy can be adjusted by modifying the Q-switch delay, the flash lamp voltage or by inserting neutral filters [34].

The excimer laser was first described in 1975 and is based on molecules that are stable only in an excited state. Their dissociation results in the emission of a photon related to the release of the molecule binding energy. Actually, the word excimer stands for excited dimer, which means that the molecule consists of two identical species (e.g. Xe_2), while the word exciplex stands for excited complex, where a noble gas atom combines with a halogen atom (e.g. ArF, KrF, or XeCl). Both excimer and exciplex lasers are used under the generic name of excimer laser. In contrast to solid state laser, the energy input is performed by means of an electrical discharge, which propagates along a rectangular cavity. The energy is stored in a high-voltage capacitor that is triggered by a thyratron and then discharged into the excimer gas load. The halogen atoms are ionized while the rare gas atoms are excited. They combine to obtain an ionized dimer. Because of the low amplification gain, a high power is required for the electrical discharge. As this power is delivered over a short period of time, the excimer laser is inherently working in the pulse mode, with pulse duration of the order of 10–20 ns. Actually, an excimer laser is not truly a laser since it does not exhibit spatial and temporal coherence. This is the reason why this effect is sometimes called superradiant. However, for ablation purposes, the excimer laser is very efficient. Moreover, wavelengths are obtained in the UV, i.e. 308 nm, 248 nm, 193 nm and 157 nm, for the XeCl, KrF, ArF and F_2 lasers, respectively. It should be noted that the F_2 laser is not truly an excimer laser because the F_2 molecule is stable. However, the gas discharge excitation is performed with a circuit similar to that used for excimer lasers. Because of a rectangular cavity, the laser beam exhibits a rectangular shape.

The discharge cavity is made of halogen resistant material and the gas affected by the discharge is recirculated by means of a fan in order to be replaced. However contaminants either in gaseous form or particulate form are produced, which reduces the gas lifetime and necessitates filtering and cleaning with a clean gas. The gas lifetime is usually expressed as the number of pulses to observe a decay to half of the maximum energy. Periodically, the gas mixture, i.e. halogen and rare gases diluted in a buffer gas such as Ne, must be totally replaced to obtain the initial conditions. When an excimer laser is operated at half of its maximum energy and with a repetition rate below the maximum value, a constant energy delivery can be obtained over a large number of pulses (several 10^6 pulses). In contrast, when the excimer laser is operated at its maximum energy and repetition rate, the delivered energy will decrease as a function of time. However, it should be mentioned that the lifetime of gas filling for an excimer laser has been significantly improved over the last few years, leading to reduced running costs of excimer lasers (308 nm–193 nm). Practically, the energy is adjustable via voltage of the discharge (e.g. 22–30 kV). Beam profile is usually rather flat in the horizontal axis and nearly Gaussian in the vertical axis.

4.4.1 Laser parameters

A laser is usually described by the wavelength, the energy, the operating mode, the pulse duration and repetition rate when the pulse mode is used, the laser beam dimensions and shape (circular or rectangular), and the energy distribution across the beam (e.g. Gaussian or flat). Typical characteristics are given in Table 4.3.

Practically, the energy range was 0.2–550 mJ, 0.9–70 mJ, 0.01–40 mJ, and 7.5 mJ for the Nd:YAG laser operating at 1064 nm, 532 nm, 266 nm and 213 nm, respectively. The range was 0.05–100 mJ, 20–250 mJ and 100 mJ for the excimer laser operating at 193 nm, 248 nm and 308 nm, respectively [e.g. 10,21,30,39,76,116]. Most of the ablation results were obtained for lasers with a pulse duration in the range 5–30 ns. However, some investigations were conducted with ps pulse duration in the 35–100 ps range [42, 122, 152].

Because the Nd:YAG laser provides an IR wavelength, it is necessary to add an optical device when a UV wavelength is required.

TABLE 4.3

Characteristics of the lasers currently used for laser ablation ICP-MS

Laser type	Wavelength (nm)	Maximum energy (mJ)	Beam size	Beam divergence	Pulse duration	Max. repetition rate
Ruby	694	100			1 ms	< 1 Hz
Nd:YAG	1064	25/500	dia. 3–9 mm	< 1 mrad	6–10 ns	10/20 Hz
	532	10/200				
	355	4/90				
	266	2/30				
	213	10				
XeCl	308	35/250	20–25 × 8–13 mm	< 3 mrad	8–30 ns	20/100
KrF	248	200/400				
ArF	193	200				
F_2	157	60				1

Frequency-doubling can be obtained by using non-linear optics such as KDP (potassium dideuterium phosphate) crystals for instance. Frequency-doubling, -tripling, -quadrupling and -quintupling can then be obtained, leading to 532 nm, 355 nm, 266 nm and 213 nm wavelengths. However, the frequency multiplication results in a decrease in the available energy. For instance, to obtain 7.5 mJ at 213 nm, an energy of 350 mJ is necessary at 1064 nm [161].

When comparing the Nd:YAG laser and the excimer laser, the advantages and limitations are summarized in Table 4.4.

4.4.2 Laser spot size

The laser energy can usually be adjusted. However, in terms of target–laser interaction, the irradiance and the fluence are the parameters to be considered. The irradiance is the power received per unit area (W m^{-2}), while the fluence is the energy per unit area (J m^{-2}), assuming a constant energy distribution across the laser beam. Practically, the irradiance is expressed in GW cm^{-2}, and an irradiance in the range 0.01–10 GW cm^{-2} is used for ablation. To obtain this type of irradiance, it is then necessary to focus the laser beam to obtain a small laser spot

TABLE 4.4

Advantages and limitations of Nd:YAG and excimer lasers

	Advantages	Limitations
Nd:YAG	reliability	need for complex optics for UV
	easy change of the flash lamp	lowest wavelength is 213 nm
	Gaussian or flat energy distribution	low energy in the UV
	compact size	
	competition	
	low diffraction in the UV	
	possibility to run ps laser	
Excimer	UV is obtained directly	use of halogen gases
	lowest UV wavelength (157 nm)	poor beam quality*
	high energy and high repetition rate	progressive decrease in max. energy**
	full automation of the gas filling	need for periodic cleaning
		size and cost
		limited competition

*Depending on the type of excimer and cavity.
**Without continuous refilling.

at the surface of the target. An energy of 0.1 mJ, with a pulse duration of 20 ns and a laser spot size of 5 μm corresponds to an irradiance of 25 GW cm^{-2}. For a spot size of 500 μm, it is necessary to have an energy of 100 mJ to obtain the same irradiance. Spot size is difficult to estimate, because the crater size is usually larger than that of the spot resulting from thermal diffusivity. Spot size must be measured by using a limited number of laser shots. Scanning electron microscopy, profilometer with a diamond sensor, and white light interferometry have been used for measuring the crater dimensions. The size of the spot will depend on the divergence of the beam, the diffraction, the laser beam diameter, the energy distribution across the beam and the quality of the focusing device, usually a lens, e.g. of the plano-convex type.

Moreover, energy distribution across the beam can be modified by using spatial filtering or by homogenizer arrays.

4.4.3 Laser beam processing

Because of the large variation of beam diameters (see Table 4.3) it is necessary to use different beam processing devices (focusing, spatial filtering, homogenizing).

4.4.3.1 Focusing devices

Two main categories of focusing objective are used—refracting and reflecting types, and within each category, a large range of models is available. The choice of objective must, therefore, be based upon analytical requirements (laser wavelength and power, magnification, focal spot size required, working distance, achromaticity, and cost). For very high resolution sampling, the user must be aware that the minimum limit of focal spot size of a simple objective, refracting or reflecting, is governed by diffraction as follows:

$$D = \frac{8\lambda F}{\pi W}$$

where λ is the laser wavelength, F is the focal length of the focusing system, and W is the laser beam diameter [1]. In practice, lens aberrations and thermal effects do not allow ablation pits with theoretical spot sizes to be produced. The resultant pit may be larger than theoretical due to heat conduction from the plasma formed at the ablation site. Working in the UV will decrease the diffraction limitation, which can be of the order of the μm in the UV. It should be emphasized that a Gaussian profile remains Gaussian after focalization. Smallest spot sizes are obtained with microscope lens, which can be a problem with UV wavelength, or with Cassegrain reflecting mirror. With UV wavelengths, it is possible to obtain spot diameters below 10 μm. A beam expander can be used to increase the laser beam diameter so as to obtain a smaller spot.

Another parameter is the depth of focus, L, which is related to the optical components used in the system.

$$L = \frac{8\lambda}{\pi}\left(\frac{F}{W}\right)^2$$

Whereas long-distance objectives allow the drilling of a deep hole without refocusing, objectives with high numerical apertures are rather limited in depth of ablation (e.g. diameter: depth of 1:0.8).

Refracting focusing lenses are potentially less robust than reflecting objectives because they are subject to damage through absorption by impurities or imperfections. For focusing UV lasers, UV grade synthetic fused silica or other UV transmitting materials must be used. However, simple single element lenses, while not providing high focusing or visual imaging performance because of the various lens aberrations, are available at low cost and can almost be considered consumable items (manufacturers are generally reluctant to guarantee lenses due to the high powers which are often used in laser applications). Standard cemented multiple element refracting lenses provide significantly improved focusing and imaging properties and work well for low power, high resolution sampling using IR and visible wavelengths only. However, the organic cements used in these lenses are extremely susceptible to laser damage. Air spaced compound lenses with very good focusing and imaging properties are now available with high laser power handling capability even in the UV.

Reflecting objectives provide a very interesting alternative to standard refracting lenses. With appropriate dielectric reflective coatings on a visible light-reflective substrate, they can be used for high power laser applications, while maintaining visual imaging capabilities, and are potentially long lived. They are capable of very high magnifications and have the added advantage of long working distances (distance between the lens or its mount and the point of focus) even at high magnifications (short focal length lenses). The absence of chromatic aberrations with these objectives (focusing of different wavelengths at different distances) provides very good visual imaging properties. Their main disadvantages are that, with a standard dielectric coating, they can be used only at one laser wavelength, they suffer an inherent transmission loss, they produce a strongly convergent beam with a small depth of focus and they are expensive.

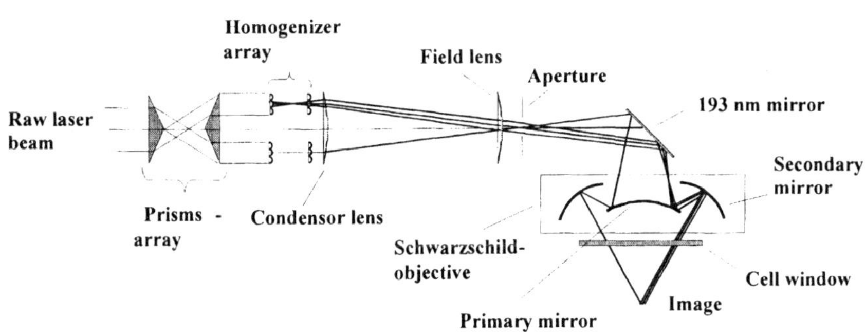

Fig. 4.1. Optical beam path for a excimer 193 nm laser for high resolution laser ablation analysis [116].

4.4.3.2 Focus positioning devices

The location of the laser–sample interaction point is one of the critical parameters. Using high magnification optics and a computerized *x,y,z* table, the focused position, and therefore any defocused position, can be adjusted. For highly transparent samples or bulk analysis, autofocus systems have been described [80].

4.4.3.3 Spatial filtering and homogenizing devices

In addition to focalization, laser beam can be processed using filtering via a mask to remove the edge of the beam with low energy and used to reduce the size of the spot, and homogenizing to obtain a flat-top energy distribution (Fig. 4.1), which has been used mainly for excimer lasers [39,116,119,153,182].

4.4.4 Interaction laser–material: UV laser vs IR laser

Although the interaction depends on the whole set of laser parameters, it appears that the wavelength plays a key role. With IR lasers, ablation is mainly obtained through the action of the microplasma which is formed by the surrounding Ar atmosphere, and the main process is plasma erosion. A consequence is that the focus distance is not highly critical. In contrast, with UV lasers, ablation is obtained by the direct action of the laser beam. The microplasma is formed by the particles ejected from the target surface. Therefore, the sharpness of focusing is crucial, and the absorption of the laser beam by the microplasma will modify the ablation efficiency. IR laser-based ablation is mainly related to thermal processes such as melting and vaporization, and therefore to the latent heat, the heat capacity and the heat of vaporization of the material, while UV laser-based ablation is related to the absorptivity of the material and the surface roughness, and to the chemical bonds of the materials, particularly with polymers. It is said that the use of UV lasers leads to so-called soft ablation because the thermal effects, and therefore fractionation and selective volatilization effects are minimized [e.g., 10, 40, 76, 124].

Besides, several materials, e.g. glass, are transparent to IR lasers. In this case, it is then necessary either to de-polish the surface to increase the absorptivity or to significantly defocus the laser beam. The alternative is to use a laser wavelength below 308 nm.

The influence of laser parameters discussed above on the ablation processes is summarized in Table 4.5.

TABLE 4.5

Brief description of the influence of laser parameters on ablation processes

Parameter	Controlled by	Influence on
Optical setup	Lens type/quality/combination	Spot diameter, crater geometry
Laser wavelength	Laser type, frequency frequency	Aerosol formation, particle size distribution, laser absorption
Laser energy	Flash lamp-voltage, tunable filters at constant flashlamp voltage	Amount of ablated material, pit size
Pulse width	Laser-type, Q-switch	Plume formation, pit geometry, transferred heat into the sample
Repetition rate	Recharge of the flashlamp circuit, heat-up of the active laser medium	Fractionation effects and heat transfer into the sample
Laser quality:		
Mode	Laser-resonance-cavity, intra cavity filter	Crater shape and structure
Beam homogeneity	Laser-type, intra cavity filter	Fractionation, particle formation
– beam divergence	Laser-cavity, beam correction	Defocusing of the focused laser spot
– beam diameter	Laser-rod (resonance cavity, filter	Necessary optical setup
–energy density	Laser-type	Amount of ablated material
Matrix:		
Thermal conductivity	Thermal properties of the matrix	Crater geometry, particle formation
Reflectivity/absorption	Optical properties of the matrix	Crater geometry

4.5 LASER ABLATION SYSTEMS

4.5.1 Specification

A laser ablation system consists of the laser unit, a laser beam processing device (e.g. lens), an ablation cell, and a transfer line to carry the ablation products into the torch of an ICP. In addition, depending on the application, laser systems can be equipped with simple or high magnification optics (TV camera, microscope) (Fig. 4.2).

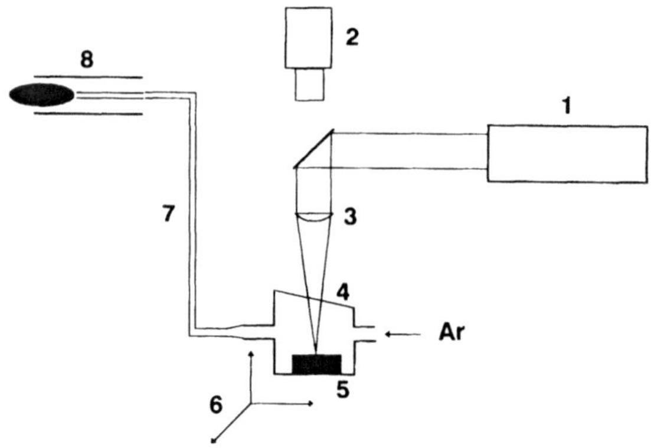

Fig. 4.2. Schematic drawing of a laser system: (1) laser, (2) observation system, (3) focusing device, (4) cell window, (5) sample, (6) *x-y-z* translation, (7) line transfer, (8) ICP torch.

Regardless of the type of analysis, a laser ablation system should have the following specifications:

- not sensitive to surface state (little or no physical sample preparation),
- any material shape,
- adjustment of laser parameters and beam processing according to the type of analysis,
- production of sub-micron particles in order to ensure a complete atomization in the ICP,
- no selective volatilization,
- no fractionation effects,
- no laser product transport effects,
- safety (laser beam),
- ease of use and maintenance.

4.5.2 Ablation cell and transportation system

The major task of the ablation cell is to position the sample with respect to the focused laser beam and to take up the ablation products. Therefore, the ablation cell specification should:

- be low dead volume,
- be able to deal with any sample shape and size,
- be easy change of the sample,

- have no memory effects (deposit on the cell walls) and short wash-out time,
- have a simply changeable window,
- possibly associate with an additional spray chamber,
- have programmable *x-y-z* translation with easy and reproducible focus and defocus distance (displacement of the target with respect to the laser beam or displacement of the laser beam).

A large variety of cells have been described in the literature and all of them are usable for laser ablation. This is not to say that there is no room for improvement in cell design, especially for specific applications. The cell must allow the laser energy to enter and be focused onto the sample, and contain ports for the entry and exit of the carrier gas that sweeps the ablated sample material to the ICP. A closed, gas-tight cell is generally used to avoid loss of sample or introduction of atmospheric gases. Good gas flow dynamics, which result in fast and efficient sample transportation (i.e., small chamber with high gas velocities), are an obvious design goal. However, since most mass spectrometers which are used for LA-ICP-MS have sequential quadrupole mass analyzers, it is desirable to have a cell (and transfer tubing) volume sufficient to allow dilution and mixing of the sample pulses so that the ICP-MS receives a fairly stable input of sample. Some study of this has been made [e.g., 45], but, no particular cell configuration has been widely adopted. Ablation cells described in the literature have been varied in volume from ca. 1 ml (for microanalysis) to more than 100 ml (bulk analysis). Thus, most cells typically reported a diameter sufficient to contain a small sample (a few cm) and meet the general gas flow dynamics criteria (Fig. 4.3). The literature suggests that smaller and larger cells also work well [2,46]. It is also important that the design allows rapid flushing of the cell to remove residual atmospheric gas following sample changing or otherwise a 'mixed gas' plasma will temporarily result. As typical argon supplies have concentrations of air in the part per million range, an appreciable time may be required to flush the sample cell to these concentrations. Therefore, an auto-sampler for laser ablation would be useful for large series of samples.

Since cell volume is apparently not a major design consideration for continuous ablation, the first practical requirement of the sample cell is that it must be able to accommodate the sample, and ideally several samples along with calibration and reference materials, to reduce the frequency of opening the cell, and thereby maintaining more constant ICP operating conditions. However, ablation of very small samples (e.g.

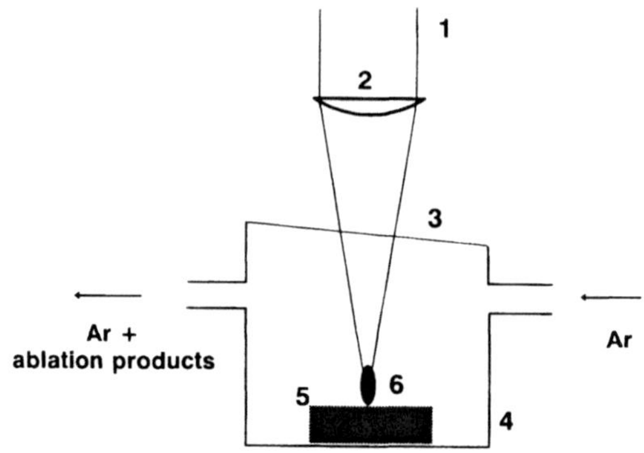

Fig. 4.3. Schematic drawing of an ablation cell: (1) laser beam, (2) focusing device, (3) cell window, (4) cell, (5) sample, (6) laser-induced microplasma.

fluid inclusions) may allow too much dilution of the sample when a large sample cell is used. In such a case, the cell volume should be kept as small as possible. For maximum flexibility, cells have also been described where the sample is not contained within the cell but forms the bottom of the cell [45] (Fig. 4.4). The cell must have a window which is transparent to the laser wavelength used (optical quality glass or, for UV lasers, synthetic fused silica). For high energy applications, the window is often mounted at a small angle to the incident laser beam to avoid harmful back-reflections into the laser [7]. However, this is not necessary when low energies are used in micro analytical applications. For transparent samples, which the operator wishes to observe using transmitted light, a second window, transparent to visible light, in the bottom of the cell is required. The cell should be designed to allow easy replacement of the windows in case of damage.

Ablated material is typically transferred to the ICP torch via plastic tubing. Tubing lengths are usually on the order of 1–2 m, with a length of almost 10 m having been reported [28,29]. Although difficult to measure with good accuracy, studies that have been made suggest that the efficiency of transportation ranges from 10–20% [43] to as high as 40% [45,47]. The remarkable simplicity and high efficiency of the transfer process is often very difficult for workers unfamiliar with laser ablation sample introduction to accept.

Some studies have been made concerning the length and internal diameter of the tubing used to transport the Ar and the sample to the ICP (e.g., Ref. [48]). The geometric considerations are obvious: a large

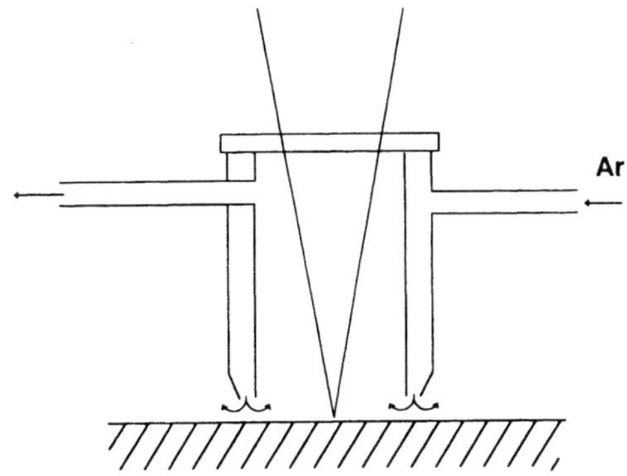

Fig. 4.4. Schematic drawing of the ablation cell from Ref. [45].

diameter has the advantage of an increased ratio of the gas volume to wall area, which should minimize the number of sample particles which collide with the wall and are potentially lost there. Conversely, a small diameter is desirable to increase the linear gas velocity (cm/s) thereby minimizing the time used to transport the sample from the cell to the ICP and reducing gravitational settling loss. In practice, as with sample cell design, any reasonable compromise results in reasonably efficient transfer. In the ETH system, readily available plastic tubing with a 6-mm (1/4-inch) outside diameter and a 3-mm (1/8-inch) inside diameter is used. Several plastic materials have been tested along with Cu tubing, and all performed comparably although the Cu tubing, occasionally touted to give higher transport efficiency, proved to be very inconvenient because of its lack of flexibility. Transfer tubing should be changed on a frequent basis (even daily) to minimize tube memory. For instance, for the determination of U/Pb isotopic ratios, the tubing should be pre-cleaned using nitric acid to remove surface Pb contamination. Some sample types may require more frequent tubing changes depending upon the volume of sample that is ablated. A need for a tubing change is often indicated by spurious spikes in the gas blank signal.

The transportation efficiency of the ablation product is related to the particle size distribution, the properties of the carrier gas, the volume of the ablation cell and the length of the transfer line. However, due to the large number of laser, cell and transportation parameters along with the huge variation from matrix to matrix, no definite figures can be given.

4.5.3 Ablation strategies

The three main types of analysis (micro-, bulk analysis, and depth profiling) need different sampling strategies. Microanalysis is high-resolution single or limited number of shots on the same position (e.g., single crater). Bulk analysis is low resolution multishot with translation or rotation of the sample (e.g., 10–100 µm s^{-1}). Translation patterns include line, raster or grid ablation. Depending on the translation speed in the case of continuous displacement and the laser spot size, crater overlaps can be obtained. An alternative is a discontinuous displacement, where a given number of shots is performed at a given location to obtain well-separated craters. For this, the practical crater depth usually remains within the depth of focus and therefore, the refocusing is not needed.

Depth profiling can be performed either on the same position or on a given area. In this case the crater depth can be larger than the depth of focus and the focus position needs to be continuously or periodically readjusted. In order to modify the fluence during the ablation the focus position can be significantly modified to change the spot size and therefore the ablation characteristics (see also Section 4.7).

4.6 SIGNAL ACQUISITION AND DATA PROCESSING

Since the signal, which is produced, is not stable, it is very important to be able to acquire and inspect the signal as a function of time, particularly where heterogeneous samples are being analysed. Without real-time data storage and display, the analyst is deprived of a wealth of important and useful information concerning the homogeneity of the ablated volume. Furthermore, the observation of signals during the ablation process is necessary in order to allow the operator to make changes, if necessary, to the laser operating parameters, including laser power and focus, during the ablation or to terminate an ablation if the signals indicate that the laser has penetrated a second phase. Acquisition of transient signals using single shot ablation is even more crucial.

Laser ablation produces a noisy, time-dependent signal (compared to conventional pneumatic nebulization) and needs a fast (possibly simultaneous) data acquisition system. Most of the so far used ICP-MS spectrometers in combination with laser ablation are quadrupole-filter based instruments and therefore sequential. Settling time, dwell time, points per peak, number of elements, time per sweep, sweeps per reading, and number of readings are all parameters (not independent) which are used to define the data acquisition procedure and must be optimally selected in order to produce the most efficient data acquisition. If no on-line signal observation and acquisition is available it may be useful to measure after the ablation process has stabilized. Other parameters which should be considered for transient data acquisition are documented in Ref. [107].

4.7 SAMPLE PREPARATION

Laser ablation is well-known for small or no sample preparation and almost every solid material as well as solutions can be ablated. However, this statement is not always valid for any sample and any type of analysis. Chemical and/or physical preparations have to be carried out. For example, powder analysis needs pelleting with the possible addition of internal standards. Even more complex are fusion products. Polishing to a few μm finish is required for a number of applications, such as depth profiling, single grain or inclusion analysis. Freeze-drying procedures have been used for biological samples to obtain powders.

Many techniques for the preparation of pellets of rock and other materials can be found in the literature, including both pressed powder and fusion procedures similar to those developed for XRF analysis of rock powder materials [12–27]. Binders are sometimes incorporated into the pellet. These include high molecular weight alcohols and plastic resins (such as phenols or polyethylene), and graphite and cellulose, which are mixed with the rock powder and pressed under high pressure [19,28–33]. Successful analyses of reference materials have been reported. Powdered samples of biological materials can be similarly prepared [12].

Fusion of samples to form a glass disk, although a more difficult sample preparation procedure, produces a more homogeneous material and has the potential to solve the representative sampling requirement of LA analysis. Fusions, both with fluxes (e.g., Li- or Na-tetraborate) or without fluxes, have been reported [14,23,24,26]. Fusion using a flux is a very much better established procedure, and both extensive XRF literature and sample preparation hardware are available. Fusion without fluxes to form glasses requires higher temperatures and thus there is potential for a possible increase in the loss of volatile elements. Fusion in an argon atmosphere at atmospheric or higher pressures has been used to reduce loss of volatiles. Since only a small weight of sample (typically a few mg) is fused, it is not always easy to obtain a representative sample of the powder. However, fusion without fluxes eliminates the disadvantages of sample dilution and potential contamination inherent in preparing pressed powder pellets using binders and glasses incorporating fluxes.

This section cannot be regarded as being comprehensive due to the fast expansion of laser ablation-based analysis and additional information can be found in the literature [12–243].

4.8 CALIBRATION

Laser ablation has been used for both qualitative and semi-quantitative analysis. However, there is a demand for more and more quantitative element concentration in smaller amounts of samples and for lower levels of concentration, which requires appropriate, precise and accurate calibration. Therefore, calibration procedure is still the major limitation for this solid sampling technique. Calibration requires that the background corrected signal from all or part of an ablation of a sample be referenced to that from a standard material. Background signals are usually obtained from measurement of a 'gas blank' (i.e., with no ablation), either performed as a separate acquisition or, preferably, as a precursor to initiating ablation. Three calibration strategies have been used: (1) external calibration alone, using a solid reference standard [19,88,110,174]; (2) external calibration to a solid reference standard, in conjunction with internal standardization [10,21,36,96]; (3) calibration using solutions nebulization with and without desolvation, and liquid ablation [37,44,48,49,101,111,115,165,178]. Additional external calibrations based on physical processes, such as acoustic waves, light scattering and crater volume have been described [148,162].

(1) *External calibration alone*: Calibration using external calibration alone has been used on occasion for preliminary investigations of sample compositions [e.g., 7]. However, because of the large differences in ablation yield that can result from fluctuations in laser output and even small differences in target matrix, it requires reference materials

that are closely matrix matched with respect to major element composition of the samples and, even then, it is barely more than a semi-quantitative technique.

(2) *Internal standardization*: The use of internal standards produces a much more robust calibration method by allowing a correction to be applied for differing ablation yield between sample and reference material. Additionally, it also partially corrects for matrix effects and signal drift in the ICP-MS. This is particularly important for bulk sampling techniques where vigorous ablation can result in heavy loading of the ICP and significant instrumental drift and matrix effects [28]. The most effective internal standards should have similar ablation characteristics to the elements being determined, and for most complete correction of ICP-MS drift and matrix effects, similar atomic mass. In practice, therefore, the internal standard is generally a major constituent, the concentration of which can be obtained from an analysis using an alternative technique (e.g., electron microprobe), or in some instances, be estimated with sufficient accuracy from the known stoichiometry of the material (e.g., trace element analysis of geological minerals). The lack of a need for matrix matching has the very important benefit that it allows use of easily obtainable glasses, either synthetic (e.g., the very widely used NIST SRM 612 [e.g., 10,54,129,130,157]) or prepared from well-characterised powders by fusing, to be used for calibration purposes.

(3) *Use of solution for calibration*: The introduction of reference materials using solution nebulization simultaneously with the laser ablated material has been reported [37, 44, 48, 55, 56]. In this procedure, the carrier gas flow from the laser cell is mixed, before entering the ICP torch, with an aerosol generated by standard solution nebulization. When ablation of the sample is performed, a blank solution is aspirated to maintain consistent conditions in the ICP. Internal standardisation is essential. Another method of calibration is the use of direct liquid ablation, which has only been used for quantification of fluid inclusions [68,74,91,101,115,141,164]. Solution-based calibration procedures offer the maximum flexibility in the composition of the reference material. However, due to the differences between dry and wet aerosol, mainly semi-quantitative data have been measured. Besides, the introduction of water into the ICP nullifies the significant advantages of reduced poly-atomic oxide ion backgrounds and interferences that are gained with a dry ICP. Therefore, desolvation of the wet nebulized material has been described to simulate dry plasma

conditions. Desolvation was applied to pneumatic nebulization and ultrasonic nebulization [44,111,178]. ETV is another method of desolvation [71,105].

It should be emphasized, that the use of a dry aerosol might result in changing the atomization and ionization processes, the formation of the Mach-disk and therefore, the ion-optics efficiency and the mass response.

4.9 ELEMENTAL FRACTIONATION

Elemental fractionation is one of the well-known, but not well understood problems in laser ablation-ICP-MS and mainly influences the accuracy of quantitative analysis. It has long been recognised that not all elements exhibit the same characteristics during ablation, sampling, transport, and ionization and that significant fractional ablation or fractionation of elements can occur. Fractionation means, that the stoichiometry of the laser products is not representative of the sample composition. Fractionation is a dynamic process, and the degree to which it occurs during an ablation is dependent on numerous processes, that are related to each other. Consequently, the investigation of separate parts is very difficult. Table 4.6 shows ablation parameters and their possible influence.

The number of interactive parameters and the difficulty of investigating single processes have led to a number of attempts to describe this phenomenon. The influence of the wavelength shows that the elemental fractionation can be improved using UV wavelength [10,76, 116,161]. In addition, output energy and the related plume and particle formation seem to make a major contribution to the elemental fractionation. However, for most of the experiments and suggestions to improve the ablation characteristics, laser parameters have been changed. The focus point of the laser on the sample surface, maintaining the sample in constant z position (active focus), moving the sample during the ablation (x,y movements of the sample) and slowly increased release of laser energy (so-called 'soft'-ablation) have been evaluated as strategies to try to reduce the elemental fractionation problem for different types of lasers and wavelengths. The fractionation is element-dependent [32,49,82,120,122,124,152,153] and seems to be related to thermal properties of the element in different matrices (melting point, boiling point, latent heat, etc.) and needs to be

TABLE 4.6

List of parameters, which possibly influence elemental fractionation

Parameter	Influence
Laser	
Wavelength	Dissipated heat into the sample; particle formation
Output energy/pulse energy	Amount of ablated material; crater depth and particle formation
Frequency	Particle density in and around the ablation spot
Pulse width	Dissipated heat into the sample; particle formation
Transportation system	
Cell volume	Plume expansion; particle agglomeration
Tube length and diameter	Particle separation; particle loss
Carrier gas	
Viscosity	Energy transfer onto the sample surface; plume formation; particle density
Flow	Transportation efficiency
Inductively Coupled Plasma	
Power	Excitation and ionization efficiency/spatial distribution
Carrier gas	Duration of particles in the analytical zone
Plasma gas	Air entrainment into the plasma and formation of polyatomic interferences
Vacuum	Space charge effects

determined for individual matrices. However, so far no efficient correction procedure generalized for a large range of matrices has been described.

4.10 FIGURES OF MERIT

As with all techniques, figures of merit are a function of numerous variables including instrumentation, operating conditions and elements being determined. Firstly, it must be noted that rapid technological developments in commercial ICP-MS instrumentation during the past decade have improved detection capabilities by more than three orders of magnitude. Secondly and even more significantly,

Table 4.7

Influence of the probed volume on relative LODs, assuming an absolute LOD of 1 fg and a sample density of 5 g cm^{-3}, and comparison with experimental values [116]

Absolute LOD	Relative LOD	Volume to be probed	Depth	Corresponding diameter	Practical LOD	Actual diameter
1 fg	1 ppm	200 μm^3	10 µm	5 µm	1 ppm	4 µm
	1 ppb	200 10^3 μm^3	10 µm	160 µm	1–10 ppb	80 µm

different applications use massively different sampling volumes. These two factors lead to LODs that, for any element, vary by many orders of magnitude between instruments and applications.

Spatial resolution: ablation pit size, both diameter and depth, can be varied, depending upon application, by a combination of adjustments of incident laser energy, number of laser shots and laser focus. Spot sizes greater than 0.5 mm down to 5 µm can readily be attained using a UV LA system [37] which, assuming a depth equal to the diameter, gives a million-fold difference in the volume ablated.

Precision: both precision and LODs are functions of signal and counting time, which in turn are dependent upon a large number of variables. However, various studies [e.g., 10,13,18,37] indicate that for standard multi-element analyses with pit sizes large enough to ensure that concentrations are well above LODs, precisions (1 r.s.d.) in the 2–15% range are typically achieved with use of internal standardisation.

Accuracy: for sets of elements with similar ablation behaviour and properly selected internal standards, a high degree of accuracy can be obtained without matrix matching. Accuracy for elements which fractionate significantly relative to the internal standard can be much poorer.

Limits of detection: like precision, for a given instrumental configuration, detection limits are largely a function of ablation volume and the counting time per element. LODs degrade when the amount of material sampled is decreased (higher spatial resolution) and when the number of elements determined is increased. Therefore, Table 4.7 contains an example of possible LODs related to the volume of ablated material.

4.11 APPLICATIONS

Laser ablation–ICP-MS has been used for a wide variety of applications in material science, biology, metallurgy, archaeology, medicine, and mainly geology. Due to the different parameters used for different studies, the original literature may contain the most useful and necessary structured information. Table 4.8 contains a (as good as possible) summary of laser-ablation works carried out in the last few years indicating the problem, the lasers and the calibration materials used. However, it should be mentioned that in addition to the table of application a number of publications from the 60s, 70s and 80s may also help to solve some of the problems.

4.12 SPARK AND ARC ABLATION

An alternative to laser ablation is based on the use of either spark or arc. Although the purpose is similar to laser ablation, the field of applications of spark or arc ablation is different from that of laser ablation, because these two techniques are inherent to conducting samples or non-conducting powders pressed with a conducting binder such as high-purity graphite. Moreover, both spark and arc are electrical discharges, which results in an interaction surface with the target that cannot be precisely defined in terms of space. Although a ceramic ring can be used to confine the discharge on the target surface, the spatial resolution is limited and is of the order of a few mm. Use of spark ablation is therefore restricted to bulk analysis. However, because a spark generator is simple to design, easy to use and cheaper than a laser system in terms of capital, running costs and maintenance, spark ablation has been used since the early works with ICP-AES [244,245]. Most of the applications of spark ablation were making use of ICP-AES [246–262]. Only a limited number of contributions have been published based on the use of ICP-MS [263–270].

Basically, a spark ablation cell is similar to the spark stand used in spark emission spectrometry, except that the ablated material is taken up by means of an argon carrier gas. The cell can be of the point-to-plane type, i.e. a counter electrode and a plane surface sample are used. In this case, change of the sample is very simple. Similar to spark emission spectrometry, the surface of the sample must be polished. An alternative is the point-to-point type, where both the counter electrode and the sample are in the form of an electrode. In this case, the dead volume of the cell is reduced [266]. The spark frequency is in the 25–2000 Hz range, while the analytical gap is the 3–4 mm range. Because a large amount of ablated material can be produced, a spray chamber can be added to remove the largest particles [204]. Depending on the material, the diameter of the particles is usually in the micrometer or submicrometer range [264–266]. Several spark ablation systems have been made commercially available, mostly to be used in conjunction with ICP-AES.

As for laser ablation, it is necessary to attain equilibrium to observe a stable signal that can be integrated over a given period of time. Repeatability is typically in the 1.5–3% RSD range [265–267], which is similar to laser ablation. LODs were in the 0.1–3 ppm range [265, 266], which can certainly be improved. A comparison between laser and spark ablation performed with ICP-AES for various alloys indicated that the limits of detection were rather similar [262]. Steel was the most commonly analyzed material [264,265,267]. However, brass material and Cu alloys were also studied [263]. More exotic samples such as fire assay beads of geological samples [266] and catalytic converter pressed with graphite [268] were also studied.

Arc ablation was only used by one research group [271,272]. Metallic samples [271] and non-conducting powders pelleted with graphite [272] were studied. In this case, LODs were in the 30–50 ppm range.

4.13 THE FUTURE

From this chapter it can be seen that most of the problems discovered earlier still exist and that a lot of work has been carried out to eliminate the early discovered problems of fractionation. It also can be seen that more and more work using lasers has been reported in the last few years and the number is still growing. New technologies, such as ps and fs lasers or lower UV wavelengths such as 213 nm, 193 nm, and 157 nm will possibly help to overcome some of the problems. Moreover, the recent introduction of time-of-flight mass spectrometers, which are highly suitable for the acquisition of transient signals, may be of benefit for microanalysis [104]. In any case, the availability of CRMs for a wide variety of solid samples would be highly beneficial to spread the use of LA-ICP-MS as a routine method.

TABLE 4.8

Applications based on the use of laser ablation ICP-MS

Elements	Matrix	Laser type	Comments	Calibration	Ref.
Multielement analysis	Tea leaves, milk powder, sediments	Ruby 694 nm	0.1–0.8 mJ	Bowen's Kale,NIST SRM 1549, GSD-1	[12] 1992
Li, B, Al, Si, Ti, Fe	Zeolites, glass fibres	Nd:Y AG 1064 nm		BCS 267, BCS 365 NBS 610-616	[14] 1992
Major, minor and trace elements	Carbonates, zircon, olivine, feldspar	Nd:Y AG 1064 nm	1 mJ, 2Hz	e.g. BCS-375/6 GSF JF-1/2	[8] 1992
Mg, Mn, Fe, Cu, Zn, Sr, Ba, U	Carbonates	Nd:Y AG 1064 nm	240 mJ, 2 Hz	GSJ Jls-1 BCS-CRM-393	[21] 1992
25 elements	Stalactite, fossil shells	Nd:Y AG 1064 nm	0.95 mJ, 1–10 Hz	Semiquant	[16] 1992
REE, Hf, Y, Th, U	Zircons	Nd:Y AG 1064 nm	240 mJ, 2 Hz	Fused glass of BCS CRM 388	[23] 1992
REE	USGS GXR-3	Nd:Y AG 1064 nm	320 mJ, 10 Hz	NBS 612	[60] 1992
Al, La, Ca, Sr, Mg, F, P	Fluorophosphate glass	Nd:Y AG 1064 nm	250 mJ, 10 Hz	Glass	[56] 1992
REE, Y, Zr, Nb, Hf, Th, U	Titanite, zircon, garnet, apatite, uranite	Nd:Y AG 1064 nm	1 mJ, 10 HZ	NBS 612	[10] 1992
29 elements	Glass, metals, alloys	Nd:Y AG 1064 nm	200-270 mJ, 10 Hz	Semiquant	[61] 1992
Al, Cu, In, Ni, Pb Sn, Co, Mn, V, W	Alloy	Nd:Y AG 532n m	4 mJ, 10 Hz	Semiquant	[43] 1993
REE, Hf, Ta Nb, Th, U	Rock powders	Nd:Y AG 1064 nm	240 mJ, 2 Hz	Fused glass	[22] 1993

continued

TABLE 4.8 (*continuation*)

Elements	Matrix	Laser type	Comments	Calibration	Ref.
$^{206}Pb/^{207}Pb$	Zircon grains	Nd:Y AG 1064 nm	5-6 mJ/shot, 4Hz	Zircon standard (TIMS)	[59] 1993
REE	Single mineral grains clinopyroxene, monazite	Nd:Y AG 266 nm	70 mJ, 10 Hz	Aqueous solutions:	[37] 1993
43 major, minor, trace elements	Whole rock powder	Nd:Y AG 1064 nm	144 mJ, <10 Hz	NBS 612	[13] 1993
$^{206}Pb/^{207}Pb$	Pitchblende, zircons	Nd:Y AG 1064 nm	320 mJ, 2,5, 10 Hz		[58] 1993
Major, traceand REE	Silicate rocks	Nd:Y AG 1064 ntn	600 mJ (max.) 20 Hz	Shale Sco-1 dolerite DNC-1	[18] 1993
Trace elements Group I, II, III, REE	Silicate rock powders	Nd:Y AG 1064 nm	160 mJ, 10 Hz	Diabase W-2	[62] 1993
Mo, Cu, Cr, Ni, Ca, V, As, Si, Fe	Steel, sediment	Nd:Y AG 1064 nm	90 mJ, 10 Hz	Sediment/ binder	[20] 1993
REE	Carbonates	Nd:Y AG 1064 nm	4 mJ/shot, 10 Hz	NIST 612	[54] 1994
Cd, Cr, Ni, Sr, Cu, Mn, Pb, Fe, Mg, Ca	Airborne particulates	Nd:Y AG 1064 nm	200 mJ, 10 Hz	Solution treated filter	[63] 1994
Pt, REE	Ferromanganese crusts	Nd:Y AG 1064 nm	500 mJ, 10 Hz	NOD-A-1, NOD-P-1	[15] 1994
REE	Glass bottle	Nd:Y AG 266 nm	40 mJ, 10 Hz	NIST 612	[11] 1994
REE	Soil	N&Y AG 1064 nm	0.2 mJ/shot, 10 Hz	Solutions / polyethylene mixed standard	[30] 1994
REE	Soil	Nd:Y AG 1064 nm	0.2 mJ/shot~, 10 Hz	Solutions	[57] 1994
12 minor and trace elements	Clinopyroxene, garnet, rutilC, glass	Nd:Y AG 266 nm	Approx. 1 mJ/ shot, 10 Hz	NIST 612	[35] 1994

Elements	Matrix	Laser type	Comments	Calibration	Ref.
Ta, Nb, Hf, Zr, Sc, V, Ga, Zn, Co	Spinel/melt	Nd:Y AG 266 nm	Approx. 1 mJ/ shot, 10 Hz	NIST 610	[38] 1994
REE + 22 traces	Biotite, cordierite, plagioclase, garnet	Nd:Y AG 1064 nm	120 mJ, 1-8 Hz	NIST 612	[39] 1994
26 elements	Steel and glass	Nd:Y AG 266 nm	0.5-60 mJ, 1–10 Hz	Solutions	[44] 1995
Sr, Y, REE	Hydrothermal calcites	Nd:Y AG 266 nm	Approx. 1 mJ, 10 Hz	NIST 610	[36] 1995
Cr, Mn, Co, Ni, Cu Cd, Zn, Pb	BCR141-144, BCR 146	Nd:Y AG 1064 nm		BCR 142	[70] 1994
Multielement	Volcanic eruption products	Nd:Y AG 1064 nm		NBS 1161-1168	[71] 1994
Cu	Copper	Nd:Y AG 532 nm		Ratios	[72] 1995
13 elements	Ti-rich amphibole Ti-rich basanitic glass	Nd:Y AG 1064 ntn		P-MT, HFA3	[73] 1995
Ba, Ca,Cs,Cu,K NaPb, Rb,Sr, Zn	Quartz, fluorite halite fluid inclusions	Nd:Y AG 266 nm	0.3-1 mJ, 10 Hz	NIST 612 epoxy resin inclusions	[74] 1995
Pb/U and 17 REE	Zircon, calcite dolomite	Nd:Y AG 1064 nm		NIST 612	[75] 1995
38 elements	Olivine, garnet, apatite, calcite, quartz etc.	Nd:Y AG 1064 and 266 nm		NIST 610	[76] 1995
Trace elements	Olivine, clinopyroxene	Nd:Y AG 1064 and 266 nm		NIST 612	[77] 1995
REE	Felsic glasses	Nd:Y AG 1064 nm		NIST 612/ 610, UDM-1 UTR-2, KN-9, KE 12	[78] 1995
20 elements	Garnet	Nd:Y AG 1064 nm		NIST 612	[79] 1995

continued

TABLE 4.8 (*continuation*)

Elements	Matrix	Laser type	Comments	Calibration	Ref.
Ca,Ce, Zr,Ba, Ti, Sr, Nb	Clinopyroxene, phlogopite	Nd:Y AG 266 nm	0.5-1 mJ, 10 Hz	NIST 612/ 610	[4] 1995
Chalcophile elements	Rock, soil, sediments	Nd:Y AG 1064 nm	550 mJ, 8Hz	spiked USGS PCC-1	[81] 1995
Chalcophile elements, litophile elements	NIST 612 Glass	Nd:Y AG 266	40 mJ, 10 Hz	NIST 610	[82] 1996
Cd, Zn, Pb, Cu	Sediments, shells	Nd:Y AG 1064 nm	200 mJ	Internal cal. method Synthetic $CaCo_3$ std.	[83] 1996
Trace elements	Silicate minerals	Excimer 308 nm		NIST 612	[84] 1996
REE, Y, Th, U	Granite, crustal protolith	Excimer 308 nm		NIST 612	[85] 1996
45 elements	Glass	Nd:Y AG 266 nm	0.5-1 mJ, 10 Hz	NIST 612/ 610	[86] 1996
Minor and trace elements	Shell	Nd:Y AG 266 nm	0.5-1 mJ	NIST 610	[87] 1996
Multielement	Leaves and their ashes	Nd:Y AG 1064 nm		Synthetic standards	[88] 1996
Trace elements	Plant ash, tree rings	Nd:Y AG 1064 and 266 nm		Synthetic cellulose Standards	[89] 1996
Trace elements	Fish bones	Nd:Y AG			[90] 1996
Sr, Rb trace elements	Fluid inclusions	Nd:Y AG 266 nm		Synthetic microcapill-ary fluid inclusion standards	[91] 1996
22 elements	Ceramics	Nd:Y AG 266 nm	0.5-2 mJ	Synthetic ceramics	[92] 1996
Sr, Ba, trace elements	Shell	Nd:Y AG 1064 nm		NIST 612	[93] 1996

Elements	Matrix	Laser type	Comments	Calibration	Ref.
Trace elements	Volcano glasses	Nd:Y AG 266 nm		NIST 612	[94] 1996
REE	Fish bone and teeth, apatite	Nd:Y AG 266 nm		Apatite reference material	[95] 1996
REE, Th, U	Garnet, pyroxene	Nd:Y AG 266 nm		NIST 612/ 610	[96] 1996
U/Pb	Zircon	Nd:Y AG 1064 nm		NIST 612	[97] 1996
REE, Th, U, Pb	Granites, pyroxene	Nd:Y AG			[98] 1996
Pb and U	Zircon, monazite	Nd:Y AG 1064 nm		Ratio (NIST 610, zircon UK-Z1)	[99] 1996
21 elements	Clinopyroxene, phlogopite	NdNA G 266 nm	0.05-1 mJ, 10 Hz	NIST 612/ 610	[100] 1996
Li, Na, Mg, K, Ca, Mn, Cu, Zn, Rb, Cs	Fluid inclusion	Nd:Y AG 266 nm		Ratios	[101] 1996
'Fingerprint' Hg, Au, Pb, Ca/P	Tooth	Nd:Y AG 1064 nm		Ratios	[102] 1996
Multi elements	Ceramic layers	Nd:Y AG 266 nm		Perovskite-type inhouse standards	[103] 1996
Si, Ti, Al Fe Mn Mg, Ca	geol. sample	ruby	100 mJ, 0.1 Hz	Cast Iron standard	[104] 1996
Co, Cu, Ag, Pb	NIST 1253c	Nd:Y AG 1064 nm	40 mJ, 10Hz	NIST 1253c copper standard	[105] 1996
Fingerprint	Gold, diamond, indicator minerals	Nd:Y AG 1064 nm		Ratio	[109] 1997
La, Sr, Mn and trace elements	Ceramics	Nd:Y AG 266 nm		Matrix-matched synthetic standards	[110] 1997

continued

TABLE 4.8 (*continuation*)

Elements	Matrix	Laser type	Comments	Calibration	Ref.
REE	Bentonite, soil	Nd:Y AG 1064 nm		Dried solutions	[111] 1997
Pb, Mg, Mn, Cu and Cd	Cellulose (tree rings)	Nd:Y AG 266 nm		Ratio	[112] 1997
Platinum group elements	Kamcite, taenite, plessite	Nd:Y AG 266 nm		Ratios, VIE iron meteorites	[113] 1997
As	Dust particles	Nd:Y AG 1064 nm	Long defocus	Standard filters	[114] 1997
Multielement	Glass	Excimer 193 nm	0.05-1 mJ	Solution	[115] 1997
Trace elements	Augite, hornblende	Excimer 193 nm	0.5-1 mJ, 10–100 Hz	NIST 612	[116] 1997
^{52}Cr, ^{53}Cr, V, Pb	SRM NIST	Nd:Y AG 532 nm	70 mJ, 15 Hz	Ratio NIST 1263/1767	[117] 1997
Trace elements	Geological materials	Nd:Y AG 1064 nm	10 Hz	NIST 612/ 610	[118] 1997
Impurities, Si, Cr, Fe, Co, Cu, Mo, Sn, Pb	Uranium matrix	Excimer 308 nm	100 mJ, 5 Hz	Uranium Standards CETAMA	[119] 1997
Pb, U / ratio	Zr-samples	Nd:Y AG 266 nm	0.2 mJ, 8 Hz	NIST 610	[120] 1997
Particules	Single crystal $NaNO_3$	Excimer 248 nm	250 mJ, 1Hz		[121] 1997
Ba, Ce, Nd, La,Al/ Zr Si, Ru, Ni, Na, Fe/ Zr	Glass, West Valley 205	Nd:Y AG 1064n m 532 nm 266n m	 2.4-3.0 μJ 1.6-400 μJ 1-300 μJ	–	[122] 1997
Radionuclids Tc, I, Th, U, Np	Synthetic cement standards	Nd:Y AG 266 nm	15 mJ, 10–20 Hz	Synthetic standards	[123] 1997
Trace Elements REE	Glass :NIST 610 Copper: NIST C1253	Nd:Y AG 1064 nm	40 / 70 mJ, 10 Hz	?	[124] 1997
Ti	NIST Al - standards	Nd:Y AG 1064 nm	20 mJ, 1 Hz	NIST	[125] 1997

Elements	Matrix	Laser type	Comments	Calibration	Ref.
Trace element REE	sintered Zr	Nd:Y AG 1064 nm	0.5 mJ 5 Hz	NIST 610	[126] 1997
Trace element Multielement	Scheelite	Excimer 193 nm	100mJ, 5Hz	NIST 610	[127] 1997
Heavy metals	tree	Nd:Y AG 1064 nm	120–140 mJ, 2Hz	?	[128] 1997
39 Elements REE, trace,major	NIST 614 /616 glass	Nd:Y AG 266 nm	20 mJ, 10Hz	NIST 612	[129] 1997
28 elements Cs– U	NIST 610/614 USGS BCR – 2G	Nd:Y AG 266 nm	2 mJ, 5 Hz	NIST 612 extern. Std. CaO intern. Std.	[130] 1997
Re, Au, Pd, Pt, Rh	NIST glass reference material 612-615	Excimer 193 nm	100 mJ, 5 Hz	External NIST612/ 613	[131] 1997
REE, U, Hf Ce, Eu	Zircons	Nd:Y AG 266 nm	2 mJ, 6.5 Hz	Synthetic standards	[132] 1997
Pb, Zn, Ag, Rb, Sr	Quartz	Nd:Y AG 1064 nm 266	5 mJ, 1–12 Hz	External standardi- sation Liquid standards	[133] 1997
Trace elements REE, Zr, Hf, Nb, Ta, Y	Silicate melt inclusion RGM - 1	Nd:Y AG 266 nm	0.25– 0.35 mJ 10 Hz	NIST 610	[134] 1997
Hf, Tb, Lu, Ta, Tm, U	Lithium tetraborate fusions	Nd:Y AG 266 nm	<30 mJ, 20 Hz	BCR-1, SY-2, BE-N, and others	[135] (1997)
Pb, Cu, As, Zn, U	Shells	Nd:Y AG 266 nm	1.5 mJ, 5 Hz	NIST 610	[136] 1997
Mg, Co, Al, Ni, Cu, Sr, Mo, Ba, La, Nb, Y	Silicon nitride Ceramics	Nd:Y AG 266 nm	1 mJ, 5Hz	NIST glasses	[137] 1997
REE	Granite	Nd:Y AG 266 nm	3 mJ, 10Hz	External, NIST 612	[138] 1997

continued

TABLE 4.8 (*continuation*)

Elements	Matrix	Laser type	Comments	Calibration	Ref.
REE, U, Pb, Sm,Nd	Geol. sample	Excimer 193 nm	10 Hz	NIST 612	[139] 1997
Trace elements	wastes	Nd:Y AG 266 nm	4 mJ, 1 – 20 Hz	NIST 612	[140] 1998
FE, Co, Ni, Zn, Pb, Cd Co	Solar grade silikon	Nd:Y AG 1064 nm	0.2 mJ, 4 Hz	NIST 610/ 614	[141] 1998
K, Ca, Mg, Pb	Sugar maple tree	Nd:Y AG 266 nm	-	Standard tree	[142] 1998
Major, minor, trace elements	Fluid inclusion	Excimer 193	100 mJ, 15 – 30 Hz	solutions NIST 612	[143] 1998
Multi elements	Airborne particulate samples	Nd:Y AG 1064 nm	100 mJ	NIST 1648 (urban particulates)	[144] 1998
Trace elements	Trees	Nd:Y AG 266 nm	15 mJ, 10 Hz	NIST	[145] 1998
Co	Serum	Nd:Y AG 1064 nm	300 mJ, 10 Hz	-	[146] 1998
Multi element	Alloy, brass, glass Fused ore NIST Std.	Nd:Y AG	60 mJ, 10 Hz	NIST	[147] 1998
Ca, Sr, Zn, Si	Glass, brass, ceramic soil (NIST)	Nd:Y AG 266 nm	0.2-1 mJ, 5 Hz	NIST, glass, brass, soil	[148] 1998
Trace element	Nuggets	Nd:Y AG 1064 nm	30–40 mJ (input), 10 Hz	Au 501 Standard NIST 610	[149] 1998
13 elements	Fish	Nd:Y AG 266 nm		Solution	[150] 1998
Multi element Isotope ratios	Glass samples	Nd:Y AG 266	10 mJ, 20 Hz	NIST	[151] 1998
Particle fractionation	NIST 613	Nd:Y AG 532 nm	0.9 –20 mJ, 5 Hz	–	[152] 1998
Element fractionation ?	NIST 612	Excimer 193 nm	-	-	[153] 1998

Elements	Matrix	Laser type	Comments	Calibration	Ref.
Pt	Silicon material	Nd:Y AG 266 nm	2.4 mJ, 10 Hz	Instrument calibration NIST 612	[154] 1998
Major, trace element REE	Mineral, rock	Excimer 193 nm	100 mJ, 5–10 Hz	NIST 612	[155] 1998
Multi element	NIST 612	Nd:Y AG 266 nm	3.5 mJ, 20 Hz	NIST 610	[156] 1998
Trace element REE	BCR – 2G	Nd:Y AG 266 nm	1–2mJ, 4 Hz	NIST 610	[157] 1998
Major-, minor-, elements	Filter samples Atmospheric particulate matter	Nd:Y AG 1064 nm	90 mJ	Standard filter	[158] 1998
Trace-, ultratrace elements	Doped GaAs crystal	Nd:Y AG 266 nm	20 Hz	NIST 610	[159] 1998
PGE, Au	Geol. materials UMT-1, WPR-1	Nd:Y AG 266 nm	0.01–4 mJ, 1–20 Hz	Synthetic standards	[160] 1998
Multi elements	NIST 612, Calcite fluorite	Nd:Y AG 266 nm 213 nm	35 mJ, 2, 5, 10 Hz 7.5 mJ	NIST 612	[161] 1998
Co–U	Sulfide mineral	Nd:Y AG 1064 nm	350–500 mJ	NIST 610/ 612/614	[162] 1998
Fingerprint Ti–U	Bulk samples of cannabis	Nd:Y AG 266 nm	350 mJ	NIST 610/ 612/614	[163] 1998
Minor elements	Fluid inclusions	Excimer 193 nm	3.5 mJ, 20 Hz	NIST 612	[164] 1998
Trace elements	KMO 1592	Nd:Y AG	2.8 mJ, 20 Hz	KMO 1592, solution	[165] 1998
REE	Zircon mineral standard 91500	Nd:Y AG 1064 nm 266 nm	2 mJ, 10 Hz 1.2 mJ, 10 Hz	NIST 612	[166] 1998
Major-, REE elements	Rocks		?	NIST 610	[167] 1998
PG elements	NiS Buttons	Nd:Y AG	2.5 mJ, 10 Hz		[168] 1998

continued

TABLE 4.8 (*continuation*)

Elements	Matrix	Laser type	Comments	Calibration	Ref.
Trace elements	Corals	Excimer 193 nm	50 mJ, 5 Hz	Wollastonite	[170] 1998
Os, Re, W, ratio	Iridosmine samples	Nd:Y AG 266 nm	2 mJ, 8 Hz	–	[171] 1998
Ca, Sr, Ba, Zn, Co, Cr, Mn, Al, Pb, Fe, Cd	Trees	Nd:Y AG 1064 nm	170 mJ, 10 Hz	Doped cellulose 13C internal standard	[172] 1998
Si, Al, Na, Mg, Li	Geochemical samples	Excimer 248 nm	20 mJ	SRM BCR 142	[173] 1999

REFERENCES

1 L. Moenke-Blankenburg, *Laser Microanalysis*. John Wiley, Toronto 1989.

2 L. Moenke-Blankenburg, Laser-ICP-Spectrometry, *Spectrochim. Acta Rev.* 15 (1993) 1.

3 S.A. Darke and J.F. Tyson, Interaction of laser radiation with solid materials and its significance to analytical spectrometry, *J. Anal. At. Spectrom.*, 8 (1993) 145.

4 B.J. Fryer, S.E. Jackson and H.P. Longerich, The design, operation and role of the laser ablation microprobe coupled with an inductively coupled plasma-mass spectrometer (LAM-ICP-MS) in the earth sciences, *Can. Min. J.*, 33 (1995) 303.

5 D.W Koppenaal, Atomic mass spectrometry, *Anal. Chem.*, 62 (1990) 303R.

6 D.W. Koppenaal, Atomic mass spectrometry, *Anal. Chem.*, 64 (1992) 320R.

7 A.L. Gray, Solid sample introduction by laser ablation for inductively coupled plasma source, *Analyst*, 110 (1985) 551.

8 N.G.J. Pearce, W.T. Perkins, I. Abell, G.A. Duller and R. Fuge, Mineral microanalysis by laser ablation inductively coupled plasma mass spectrometry, *J. Anal. At. Spectrom.*, 7 (1992) 53.

9 H.P. Longerich, S.E. Jackson, B.J. Fryer and D.F. Strong, The laser ablation microprobe-inductively coupled plasma-mass spectrometer, *Geosci. Can.*, 20 (1993) 21.

10 S.E. Jackson, H.P. Longerich, G.R. Dunning and B.J. Fryer, The application of laser-ablation microprobe-inductively coupled plasma-mass spectrometry (LAM-ICP-MS) to in situ trace-element determinations in minerals, *Can. Mineral.*, 30 (1992) 1049.
11 D. Günther, S. E. Jackson and H. P. Longerich, Laser ablation and arc spark solid sample introduction into inductively coupled plasma mass spectrometers, *Spectrochim. Acta.*, 54B (1999) 381.
12 S.F. Durrant, Multi-elemental analysis of environmental matrices by laser ablation inductively coupled plasma mass spectrometry, *Analyst*, 117 (1992) 1585.
13 L.S. Fedorowich, L.P. Richards, J.C. Jain, J. Fan and R. Kerrich, Rapid method for REE and trace element analysis using laser sampling ICP-MS on direct fusion whole-rock glasses, *Chem. Geol.*, 106 (1993) 229–249.
14 J. Franks, J. Marshall, I. Brown and L. Garden, Analysis of inorganic solids by laser ablation inductively coupled plasma spectrometry, *Anal. Proc.*, 29 (1992) 23.
15 D.-C. Garbe Schönberg and G.M. McMurty, In situ microanalysis of platinum and rare earths in ferro-manganese crusts by laser ablation-ICP-MS, *Fresenius J. Anal. Chem.*, 350 (1994) 264.
16 N. Imai, Microprobe analysis of geological materials by laser ablation inductively coupled plasma mass spectrometry, *Anal. Chim. Acta*, 269 (1992) 263.
17 N. Imai, Quantitative analysis of original and powdered rocks and mineral inclusions by laser ablation inductively coupled plasma mass spectrometry, *Anal. Chim. Acta*, 235 (1990) 381–391.
18 K.E. Jarvis and L.G. Williams, Laser ablation inductively coupled plasma mass spectrometry (LA-ICP-MS): a rapid technique for the direct quantitative determination of major, trace and rare-earth elements in geological samples, *Chem. Geol.*, 106 (1993) 251.
19 B. Magyar, H. Cousin and B. Aeschlimann, Combined use of X-ray fluorescence spectrometry and laser ablation inductively coupled plasma mass spectrometry in the analysis of soils, *Anal. Proc.*, 29 (1992) 282.
20 M. Paul, Analysis of solid samples by laser sampling ICP-MS, *At. Spectrosc.*, 15 (1994) 21.
21 N.G.J. Pearce, W.T. Perkins and R. Fuge, Developments in the analysis of carbonates by laser ablation inductively coupled plasma mass spectrometry, *Anal. Proc.*, 29 (1992) 280.
22 W.T. Perkins, N.G.J. Pearce and T.E. Jeffries, Laser ablation inductively coupled plasma mass spectrometry: a new technique for the determination of trace and ultra-trace elements in silicates, *Geochim. Cosmochim. Acta*, 57 (1993) 475.
23 W.T. Perkins, N.G.J. Pearce and R. Fuge, Analysis of zircon by laser ablation and solution ICP-MS, *J. Anal. At. Spectrom.*, 7 (1992) 611.

24 W.T. Perkins, R. Fuge and N.G.J. Pearce, Quantitative analysis of trace elements in carbonates using laser ablation inductively coupled plasma mass spectrometry, *J. Anal. At. Spectrom.*, 6 (1991) 445.
25 D. Potter and I. Abell, Direct bulk and feature analysis of electronic ceramic materials by laser ablation ICP-MS, *Anal. Sci.*, 7 (1991) 1239.
26 A.A. van Heuzen, Analysis of solids by laser ablation-inductively coupled plasma mass spectrometry. Matching with a glass matrix, *Spectrochim. Acta*, 46B (1991) 1803.
27 A.A. van Heuzen and J.B.W. Morinsk, Analysis of solids by laser ablation –inductively coupled plasma–mass spectrometry (LA-ICP-MS) II. Matching with a pressed pellet, *Spectrochim. Acta*, 46B (1991) 1819.
28 M. Broadhead, R. Broadhead and LW. Hager, Laser sampling ICP-MS: semiquantitative determination of sixty-six elements in geological sample, *At. Spectrosc.*, 11 (1990) 205.
29 M. Broadhead, Laser sampling ICP-MS: determination of rhenium in molybdenum concentrates, copper concentrates and other geological material, *At. Spectrosc.*, 12 (1991) 45.
30 H. Cousin and B. Magyar, Precision and accuracy of laser ablation ICP-MS of rare earth elements with external calibration, *Mikrochim. Acta*, 113 (1994) 313.
31 E.R. Denoyer, Analysis of powdered materials by laser sampling ICP-MS, In: G. Holland, and A.N. Eaton (Eds.) *Applied Plasma Source Mass Spectrometry*, The Royal Society of Chemistry, Cambridge, 1990, p. 199.
32 J.W. Hager, Laser Sampling ICP-MS: Quantitation Without Matrix Matched Standards. Laser Application Report-Perkin Elmer Sciex No. LAR-4.
33 T. Mochizuki, A. Sakashita, H. Iwata, Y. Ishibashi and N. Gunji, Direct analysis of silicon nitride powders by inductively coupled plasma mass spectrometry with a laser ablation technique, *Anal. Sci.*, 7 (1991) 151.
34 E.H. Piepmeier, *Analytical Applications of Lasers*, John Wiley, New York, 1986.
35 G.A. Jenner, S.F. Foley, S.E. Jackson, T.H. Green, B.J. Fryer and H.P. Longerich, Determination of partition coefficients for trace elements in high pressure-temperature experimental run products by laser ablation microprobe–inductively coupled plasma–mass spectrometry (LAM-ICP-MS), *Geochim. Cosmochim. Acta*, 57 (1993) 5099.
36 D.J. Kontak and S.E. Jackson, Laser ablation microanalysis of calcite cements from a Mississippi valley type Zn–Pb deposit, Nova Scotia, Canada: documentation of extreme rare earth element variation on macro- and micro-scales, *Can. Mineral.*, 33 (1995) 23.
37 S. Chenery and J.M. Cook, Determination of rare earth elements in single mineral grains by laser ablation microprobe-inductively coupled plasma mass spectrometry-preliminary study, *J. Anal. At. Spectrom.*, 8 (1993) 299.

38 I. Horn, S.F. Foley, S.E. Jackson and G.A. Jenner, Experimentally determined partitioning of high field strength and selected transition elements between spinel and basaltic melt, *Chem. Geol.*, 117 (1994) 193.
39 F. Bea, M.D. Pereira and A. Stroh, Mineral/leucosome trace element partitioning in a peraluminous migmatite (a laser ablation-ICP-MS study), *Chem. Geol.*, 117 (1994) 291.
40 C. Geertsen, A. Briand, F. Chartier, J-L. Lacour, P. Mauchien, S. Sjöström and J-M. Mermet, Comparison between infrared and ultraviolet laser ablation at atmospheric pressure—implications for solid sampling inductively coupled plasma spectrometry, *J. Anal. At. Spectrom.*, 9 (1994) 17.
41 J. Sneddon, Excimer laser-ablated plasmas for direct atomic emission spectroscopy of solids, *Spectroscopy*, 9 (1994) 34.
42 I. Ihleman, B. Wolff and P. Simon, Nanosecond and femtosecond-excimer laser ablation of fused silica, *Appl. Phys.*, A 54 (1992) 363.
43 Y. Huang, Y. Shibata and M. Morita, Micro laser ablation–inductively coupled plasma mass spectrometry. 1. Instrumentation and performance of micro laser ablation system, *Anal. Chem.*, 65 (1993) 2999.
44 E.F. Cromwell and P. Arrowsmith, Semiquantitative analysis with laser ablation inductively coupled plasma mass spectrometry, *Anal. Chem.*, 67 (1995) 131.
45 P. Arrowsmith and S.K. Hughes, Entrainment and transport of laser ablated plumes for subsequent elemental analysis, *Appl. Spectrosc.*, 42 (1988) 1231.
46 P. Mitchell, J. Sneddon and L.J. Radziemski, A sample chamber for solid analysis laser ablation/DCP spectroscopy, *Appl. Spectrosc.*, 40 (1986) 274.
47 S.A. Darke, S.E. Long, C.J. Pickford and LF. Tyson, Laser ablation system for solid sample analysis by inductively coupled plasma atomic emission spectrometry, *J. Anal. At. Spectrom.*, 4 (1989) 715.
48 L. Moenke-Blankenburg, M. Gäckle, D. Günther and J. Kammel, *Processes of Laser Ablation and Vapour Transport to the ICP*, Special Publication No. 85, TheRoyal Society of Chemistry, Cambridge, 1990, p. 117.
49 M. Thompson, S. Chenery and L. Brett, Nature of particulate matter produced by laser ablation—implications for tandem analytical systems, *J. Anal. At. Spectrom.*, 5 (1990) 49.
50 A.R. Date and A.L. Gray, *Inductively Coupled Plasma Mass Spectrometry*, Blackie, Glasgow and London, 1989.
51 A.J. Walder, I.D. Abell, I. Platzner and P.A. Freedman, Lead isotope ratio measurement of NIST 610 glass by laser ablation inductively coupled plasma mass spectrometry, *Spectrochim. Acta*, 48B (1993) 397.
52 G. Hieftje, Towards the next generation of plasma source mass spectrometers, *J. Anal. At. Spectrom.*, 7 (1992) 781.

53 P. Arrowsmith, Laser ablation of solids for elemental analysis by inductively coupled plasma mass spectrometry, *Anal. Chem.*, 59 (1987) 1437.
54 R. Feng, In situ trace element determination of carbonates by Laser probe inductively coupled plasma mass spectrometry using non-matrix matched standardization, *Geochim. Cosmochim.*, Acta 58 (1994) 1615.
55 M. Thompson, S.R. Chenery and L. Brett, Calibration studies in laser ablation microprobe–inductively coupled plasma atomic emission spectrometry, *J. Anal. At. Spectrom.*, 4 (1989) 11.
56 L. Moenke-Blankenburg, T. Schumann, D. Günther, H.-M. Kuss and M. Paul, Quantitative analysis of glass using inductively coupled plasma atomic emission and mass spectrometry, laser micro-analysis inductively coupled plasma atomic emission spectrometry and laser ablation inductively coupled plasma mass spectrometry, *J. Anal. At. Spectrom.*, 7 (1992) 251.
57 D. Günther, I. Leopold, H. Cousin and B. Magyar, Calibration Studies on Liquid and Laser-induced Aerosols for LA-ICP-MS, 1994 Winter Conference on Plasma Spectrochem. San Diego, Poster presentation 1994.
58 B. Fryer, S.E. Jackson and H.P. Longerich, The application of laser ablation microprobe–inductively coupled plasma–mass spectrometry (LAM-ICP-MS) to in situ (U)-Pb geochronology, *Chem. Geol.*, 109 (1993) 1.
59 R. Feng, N. Machado and J. Ludden, Lead geochronology of zircon by laser probe-inductively coupled plasma mass spectrometry (LP-ICP-MS), *Geochim. Cosmochim. Acta*, 57 (1993) 3479.
60 U. Voellkopf, M. Paul and E. Denoyer, Analysis of solid samples by ICP-mass-spectrometry, *Fresenius J. Anal. Chem.*, 342 (1992) 917.
61 P. Van de Weijer, W.L.M. Baeten, M.H.J. Bekkers and P.L.M.G. Vullings, Fast semiquantitative survey analysis of solids by laser ablation inductively coupled plasma mass spectrometry, *J. Anal. At. Spectrom.*, 7 (1992) 599.
62 L.G. Williams and K.E. Jarvis, Preliminary assessment of laser ablation inductively coupled plasma mass spectrometry for quantitative multi-element determinationin silicates, *J. Anal. At. Spectrom.*, 8 (1993) 25.
63 C. Lüdke, E. Hoffmann and J. Skole, Comparative studies on metal determination in airborne particulates by LA-ICP-MS and furnace atomization non-thermal excitation spectrometry, *Fresenius J. Anal. Chem.*, 350 (1994) 272.
64 R. Ediger, The Analysis of Nickel Alloys with Laser Sampling ICP-Mass Spectrometry. Laser Application Report—Perkin Elmer Sciex No. LAR-3.
65 C.T. Tye, J.S. Gordon, R.C. Hutton, P. Webb and J.E. Cantle, VG Instruments Canada, Analysis Of Uranium by Laser Ablation Inductively Coupled Plasma Mass Spectrometry. VG Sales Information Brochure. 1
66 L.E.J. Vandevelde and S. Van Winckel, Analysis by ICP/MS and laser ablation sampling; prospects and limits for metallurgical applications,

Revue Atb-Metallurgie, XXX (1990) 99.
67 H. Yasuhara, T. Okano and Y. Matsumura, Determination of trace elements in steel by laser ablation inductively coupled plasma mass spectrometry, *Analyst*, 117 (1992) 395.
68 T.J. Shepherd, S.R. Chenery and J.M. Cook, Chemical characterisation of metalliferous basinal brines by laser ablation microprobe-ICP-MS analysis of single fluid inclusions, *Terra Abst.*, 5 (1993) 468.
69 P.J. Potts, *A Handbook of Silicate Rock Analysis*, Blackie, New York, 1987.
70 H. Scholze, H. Stephanowitz, E. Hoffmann and J. Skole, Element analysis of inhomogeneous samples by laser ablation ICP-mass spectrometry, *Fresenius. J. Anal. Chem.*, 350 (1994) 247.
71 P. Richner, D. Evans, C. Wahrenberger and V. Dietrich, Applications of laser ablation and electrothermal vaporization as sample introduction techniques for ICP-MS, *Fresenius. J. Anal. Chem.*, 350 (1994) 235.
72 L.A. Allen, H.M. Pang, A.R. Warren and R.S. Houk, Simultaneous measurements of isotope ratios in solids by laser ablation with a twin quadrupole inductively coupled plasma mass spectrometer, *J. Anal. At. Spectrom.*, 10 (1995) 267.
73 C. Daple, D.R. Baker and S.R. Sutton, Synchrotron X-ray-fluorescence and laser ablation ICP-MS microprobes—useful instruments for analysis of experimental run-products, *Can. Min. J.*, 33 (1995) 481.
74 T.J. Shepherd and S.R. Chenery, Laser ablation ICP-MS elemental analysis of individual fluid inclusions: an evaluation study, Geochim. Cosmochim. Acta 59 (1995) 3997.
75 J.N. Ludden, R. Feng, G. Gauthier, J. Stix, L. Shi, D. Francis, N. Machado, G.P. Wu, Applications of LAM-ICP-MS analysis to minerals, *Can. Min J.*, 33 (1995) 419.
76 T.E. Jeffries, W.T. Perkins and N.J.G. Pearce, Comparison of IR and UV laser probe microanalysis inductively coupled plasma mass spectrometry in mineral analysis, *Analyst*, 120 (1995) 1365.
77 T.E. Jeffries, W.T. Perkins and N.J.G. Pearce, Measurements of trace elements in basalts and their phenocrusts by laser probe microanalysis inductively coupled plasma mass spectrometry (LPMA-ICP-MS), *Chem. Geol.*, 121 (1995) 131.
78 J. Stix, G. Gauthier and J.N. Ludden, A critical look at quantitative laser ablation ICP-MS analysis of natural and synthetic glasses, *Can. Min. J.*, 33 (1995) 435.
79 L.S. Fedorowich, L.C. Jain and R. Kerrich, Trace element analysis of garnet by laser ablation microprobe ICP-MS, *Can. Min. J.*, 333 (1995) 469.
80 H. Cousin, A. Weber, B. Magyar, I. Abell and D. Günther, An auto-focus system for reproducible focusing in laser ablation inductively coupled plasma mass spectrometry, *Spectrochimica Acta*, 50 B (1995) 63.
81 Xun Guo and F. E. Lichte, Analysis of rocks, Soils and Sediments for the

chalcophile elements by laser ablation inductively coupled plasma mass spectrometry, *Analyst*, 120 (1995) 2707.
82 H. P. Longerich, D. Günther and S. E. Jackson, Elemental fractionation in laser ablation inductively coupled plasma mass spectrometry, *Fresenius. J. Anal. Chem.*, 355 (1996) 538.
83 G. Schettler and N. J. G. Pearce, Metall pollution recorded in extinct Dreissena polymorpha communities, lake Breitling, Havel lake system, Germany: a laser ablation inducively coupled plasma mass spectrometry study, *Hydrobiologia*, 317 (1996) 1.
84 F. Bea, Residence of REE, Y, Th and U in granites and crustal protoliths: implications for the chemistry of crustal melts, *J. Petrol.*, 37 (1996) 521.
85 F. Bea, P. Montero, A. Stroh and J. Baasner, Microanalysis of minerals by an excimer UV-LA-ICP-MS system, *Chem. Geol.*, 133 (1996) 145.
86 A. Raith, J. Godfey and R.C. Hutton, Quantitation methods using laser ablation-ICP-MS evaluation of new glass standards, *Fresenius. J. Anal. Chem.*, 354 (1996) 163.
87 A. Raith, W.T. Perkins, N.J.G. Pearce and T.E. Jeffries, Environmental monitoring on shellfish using UV laser ablation ICP-MS, *Fresenius. J. Anal. Chem.*, 355 (1996) 789.
88 H. Scholze, E. Hoffmann, C. Lüdke and A. Platalla, Analysis of leaves by using the laser ablation-ICP-MS with isotope dissolution method, *Fresenius. Anal. Chem.*, 355 (1996) 892.
89 E. Hoffmann, C. Lüdke and H. Stephanowitz, Application of Laser ICP-MS in environmental analysis, *Fresenius. J. Anal. Chem.*, 355 (1996) 900.
90 S.L.M. Blaber, D.A. Milton, J. Pang, P. Wong, O. Boonteck, L. Nyigo and D. Lubim, The life history of the tropical shad tenualosa toli from Sarawak: first evidence of protandry in the clupeiformes, *Environ. Biol. Fishes*, 46 (1996) 225.
91 A.M. Ghazi, T.E. McCandless, D.A. Vanko and J. Ruiz, New quantitative approach in the trace elemental analysis of single fluid inclusions: applications of laser ablation inductively coupled plasma mass spectrometry, *J. Anal. At. Spectrom.*, 11 (1996) 667.
92 J. Westheide, J.S. Becker, R. Jager, H.J. Dietze and J.A.C. Broekeart, Analysis of ceramic layers for solid oxide fuel cells by laser ablation inductively coupled plasma mass spectrometry, *J. Anal. At. Spectrom.*, 11 (1996) 661.
93 H.A. Stecher, D.E. Krantz, C.J. Lord, C.W. Luther and K.W. Bock, Profiles of strontium and barium in Mercenaria mercenaria and spisula solidissima shells, *Geochim. Cosmochim. Acta*, 60 (1996) 3445.
94 N.J.G. Pearce, L.A. Westgate and W.T. Perkins, Developments in the analysis of volcanic glass shards by laser ablation ICP-MS: quantitative and single internal standard multielement methods, *Quart. Int.*, 34 (1996) 213.

95 R.S. Chenery, T. Williams, T.A. Elliott, L.P. Forey and L. Werdelin, Determination of rare earth elements in biological and mineral apatite by EPMA and LAMP-ICP-MS, *Mikrochim. Acta*, Suppl. (1996) 259.
96 M.D. Norman, N.J. Pearson, A. Sharma and W.L. Griffin, Quantitative analysis of trace elements in geological materials by laser ablation ICP-MS: instrumental operating conditions and calibration values of NIST glasses, *Geostand. Newslett.*, 20 (1996) 247.
97 D.J. Scott and G. Gauthier, Comparison of TIMS (U/Pb) and laser ablation microprobe ICP-MS (Pb) techniques for age determination of detrital zircons from paleoproterizoic metasedimentary rocks from northeastern Laurentia, Canada, with tectonic implications, *Chem. Geol.*, 131 (1996) 127.
98 F. Poitrasson, R.S. Chenery and D.J. Bland, Contrasted monazite hydrothermal alteration mechanisms and their geochemical implications, *Earth Planet. Sci. Lett.*, 145 (1996) 79.
99 N. Machado and G. Gauthier, Determination of Pb207/Pb206 ages on zircon and monazite by laser ablation ICP-MS and application to a study of sedimentary provenance and metamorphism in Southeastern Brazil, *Geochim. Cosmochim. Acta*, 60 (1996) 5063.
100 S.F. Foley, S.E. Jackson, B.J. Fryer, J.D. Greenough and G.A. Jenner, Trace element partition coefficients for clinopyroxene and phlogopite in a alkaline lamprophyre from Newfoundland by LAM-ICP-MS, *Geochim. Cosmochim. Acta*, 60 (1996) 629.
101 A. Moissette, T.J. Shepherd and S. R. Chenery, Calibration strategies for the elemental analysis of individual aqueous fluid inclusions by laser ablation inductively coupled plasma mass spectrometry, *J. Anal. At. Spectrom.*, 11 (1996) 177.
102 A. Cox, F. Keenan, M. Cooke and J. Appleton, Trace element profling of dental tissues using laser ablation inductively coupled plasma mass spectrometry, *Fresenius. J. Anal. Chem.*, 354 (1996) 254.
103 J.S. Becker, U. Breuer, J. Westheide, A.I Saprykin, H. Holzbecher, H. Nickel and H.J. Dietze, Trace and surface analysis of ceramic layers of solid oxide fuel cells by mass spectrometry, *Fresenius. J. Anal. Chem.*, 355 (1996) 626.
104 P.P. Mahoney, GangQiang Li and G.M. Hieftje, Laser ablation Inductively coupled plasma mass spectrometry with a time-of-flight mass analyser, *J. Anal. At. Spectrom.*, 11 (1996) 401.
105 W. Doherty, P.M. Outridge and D.C. Grégoire, Technique for the introduction of dry atomic vapours for improved optimization and diagnostic studies of laser ablation Inductively coupled plasma spectrometry, *J. Anal. At. Spectrom.*, 11 (1996) 1123.
106 N.J.G. Pearce, W.T. Perkins, J.A. Westgate, M.P. Gorton, S.E. Jackson, C.R. Neal and S.P. Chenery, A compilation of new and published major

and trace element data for NIST SRM 610 and NIST SRM 612 Glass reference materials, *Geostand. Newsletter*, 21 (1996) 115.
107 H.P. Longerich, S.E. Jackson and D. Günther, Laser ablation inducively coupled plasma mass spectrometry transient signal data acquisition and analyte concentration calculation, *J. Anal. At. Spectrom.*, 11 (1996) 899.
108 D. Günther, H.P. Longerich and S.E. Jackson, Effect of sampler orifice diameter on dry plasma inductively coupled plasma mass spectrometry (ICP-MS) backgrounds, sensitivities, and limits of detection using laser ablation sample introduction, *Fresenius J. Anal. Chem.*, 355 (1996) 771.
109 R.J. Watling, B.F. Lynch and D. Herring, Use of laser ablation inductively coupled plasma mass spectrometry for fingerprinting scene of crime evidence, *J. Anal. At. Spectrom.*, 12 (1997) 195.
110 J.S. Becker, J. Westheide, A.I. Saprykin, H. Holzbrecher, U. Breuer and H.J. Dietze, Mass spectrometric analysis of cermaic components for solid oxide fuel cells, *Mikrochim. Acta*, 125 (1997) 153.
111 D. Günther, H. Cousin, B. Magyar and I. Leopold, Calibration studies on dried aerosols for laser ablation-inductively coupled plasma mass spectrometry, *J. Anal. At. Spectrom.*, 12 (1997) 165.
112 S.A. Watmough, T.C. Hutchinson and R.D. Evans, Application of laser ablation inductively coupled plasma mass spectrometry in dendrochemical analysis, *Environ. Sci. Technol.*, 31 (1997) 114.
113 T. Hirata and R.W. Nesbitt, Distribution of platinum group elements and rhenium between metallic phases of iron meteorites, *Earth Planet. Sci. Lett.*, 147 (1997) 11.
114 C.F. Wang, S.L. Jeng and F.J. Shieh, Determination of arsenic in airborne particulate matter by inductively coupled plasma mass spectrometry, *J. Anal. At. Spectrom.*, 12 (1997) 61.
115 D. Günther, R. Frischknecht, H.J. Müschenborn and C.A. Heinrich, Direct liquid ablation: a new calibration strategy for laser ablation-ICP-MS microanalysis of solids and liquids, *Fresenius. J. Anal. Chem.*, 359 (1997) 390.
116 D. Günther, R. Frischknecht, C.A. Heinrich and H.J. Kahlert, Capabilities of an ArF 193 nm eximer Laser for LAM-ICP-MS micro analysis of geological materials, *J. Anal. At. Spectrom.*, 12 (1997) 939.
117 L.A. Allen, J.J. Leach, Ho-Ming Pang and R.S. Houk, Precise measurement of ion ratios in solide samples using laser ablation with a twin quadrupole inductively coupled plasma mass spectrometer, *J. Anal. At. Spectrom.*, 12 (1997) 171.
118 Zhongxing Chen, W. Doherty and D.C. Grégoire, Application of laser sampling microprobe inductively coupled plasma mass spectrometry to the in situ trace element analysis of selected geological materials, *J. Anal. At. Spectrom.*, 12 (1997) 653.
119 Ch. Leloup, P. Marty, D. Dall'Ava and M. Perderau, Quantitative analysis

for impurities in Uranium by laser ablation inductively coupled plasma mass spectrometry: Improvements in the experimental setup, *J. Anal. At. Spectrom.*, 12 (1997) 945.
120 T. Hirata, Soft ablations technique for laser ablation inductively coupled plasma mass spectrometry, *J. Anal. At. Spectrom.*, 12 (1997) 1337.
121 R.L. Webb, J.T. Dickinson and G.J. Exarhos, Characterization of particulates accompanying laser ablation of $NaNO_3$, *Appl. Spectrosc.*, 51 (1997) 707.
122 D. Figg and M.S. Kahr, Elementel fractionation of glass using laser ablation inductively coupled plasma mass spectrometry, *Appl. Spectrosc.*, 51 (1997) 1185.
123 M. Gastel, J.S. Becker, G. Küppers and H.J. Dietze, Determination of long-lived radionuclides in concret matrix by laser ablation inductively coupled plasma mass spectrometry, *Spectrochim. Acta*, 52 B (1997) 2051.
124 P.M. Outridge, W. Doherty and D.C. Gregoire, Ablative and transport fractionation of trace elements during laser sampling of glass and copper, *Spectrochim. Acta*, 52 B (1997) 2093.
125 S.A. Baker, B.W. Smith and J.D. Winefordner, Laser ablation inductively coupled plasma mass spectrometry with a compact laser source, *Appl. Spectrosc.*, 51 (1997) 1918.
126 L.P. Bédard, D.R. Baker and N. Machado, A new technique for the synthesis of geochemical reference samples for laser ablation ICP-MS analysis of zircon, *Chem. Geol.*, 138 (1997) 1.
127 P.J. Sylvester and M. Ghaderi, Trace element analysis of scheelite by excimer laser ablation inductively coupled plasma mass spectrometry (ELA-ICP-MS) using a synthetic silicate glass standard, *Chem. Geol.*, 141 (1997) 49.
128 C.D. Garbe-Schönberg, C. Reimann and V.A. Pavlov, Laser ablation ICP-MS analyses of tree-ring profiles in pine and birch from N Norway and NW Russia—a reliable record of the pollution history of the area?, *Environmental Geology*, 32 (1997) 9.
129 I. Horn, R.W. Hinton S.E. Jackson and H.P. Longerich, Ultra trace element analysis of NIST SRM 616 and 614 using laser ablation microprobe inductively coupled plasma mass spectrometry (LAM-ICP-MS): a comparison with secondary ion mass spectrometry (SIMS), *Geostandards Newsletter*, 21 (1997) 191.
130 Y. Lahaye, D. Lambert and St. Walters, Ultraviolet laser sampling and high resolution indctively coupled plasma mass spectrometry of NIST and BCR-2G glass reference materials, *Geostandards Newsletter*, 21 (1997) 205.
131 P.J. Sylvester and S.M. Eggins, Analysis of Re, Au, Pd, Pt, and Rh in NIST glass certified reference materials and natural basalt glasses by laser ablation ICP-MS, *Geostandards Newsletter*, 21 (1997) 215.

132 R.W. Nesbitt, T. Hirata, I.B. Butler and J.A. Milton, UV laser ablation ICP-MS: some applications in earth sciences, *Geostandards Newsletter*, 20 (1997) 231.
133 T.E. McCandless, D.J. Lajack, J. Ruiz and A.M. Ghazi, Trace element determination of single fluid inclusions in Quartz by laser ablation ICP-MS, *Geostandards Newsletter*, 21 (1997) 279.
134 R.P. Taylor, S.E. Jackson, H.P. Longerich and J.D. Webster, In-situ trace element analysis of individual silicate melt inclusion by laser ablation microprobe inductively coupled plasma mass spectrometry (LAM-ICP-MS), *Geostandards Newsletter*, 61 (1997) 2559.
135 M. Odgard and M. Hamester, Preliminary investigation into the use of a high resolution inductively coupled plasma-mass spectrometer with laserablation for bulk analysis of geological materials fused with $Li_2B_4O_7$, *Geostandards Newsletter*, 21 (1997) 245.
136 G.D. Price and N.J.G. Pearce, Biomonitoring of Pollution by cerastoderma edule from the British Isles: a laser ablation study, *Marine Pollution Bulletin*, 34 (1997) 1025.
137 S.C. Baker, M.J. Dellevecchia, B.W. Smith and J.D. Winefordner, Analysis of silicon nitride bearings with laser ablation inductively coupled plasma mass spectrometry, *Anal. Chim. Acta*, 355 (1997) 113.
138 K. Simon, U. Wiechert, J. Hoefs and B. Grote, Microanalysis of minerals by laser ablation ICP-MS and SIRMS, *Fresenius. J. Anal. Chem.*, 359 (1997) 458.
139 A. von Quadt, D. Günther, R. Frischknecht, R. Zimmermann and G. Franz, The evolution of pre-Variscan eclogites of the tauern window (eastern Alps): A Sm/Nd-, convential and Laser ICP-MS zircon U–Pb study, *Schweiz. Mineral. Petrog. Mitt.*, 77 (1997) 265.
140 M. Motelica-Heino, P. Le Costumer, J.H. Thomassin, A. Gauthier and O.F.X Donard, Macro and microchemistry of trace metals in vitrfied domestic wastes by laser ablation ICP-MS and scanning electron microprobe X-ray energy dispersive spectroscopy, *Talanta*, 46 (1998) 407.
141 P.L. Buldini, A. Mevoli and J.L. Sharma, LA-ICP-MS, IC and DPASV-DPCSV determination of metallic impurities in solar–grade silicon, *Talanta*, 47 (1998) 203.
142 S.A. Watmough, T.C. Hutchinson and D. Evans, The quantitative analysis of sugar maple tree rings by laser ablation in conjunction with ICP-MS, *J. Environ. Qual.*, 27 (1998) 1087.
143 D. Günther, A. Audétat, R. Frischknecht and Ch. Heinrich, Quantitative analysis of major, minor and trace elements in fluid inclusions using laser ablation inductively coupled plasma mass spectrometry, *J. Anal. At. Spectrom.*, 13 (1998) 263.
144 Chu-Fang-Wang, Su-Ling Jeng, Chien C. Lin and Pen-Chi Chiang, Preparation of airborne particulate standards on PTFE-membrane filter

for laser ablation inductively coupled plasma mass spectrometry, *Anal. Chim. Acta*, 368 (1998) 11.

145 S.A. Watmough, T.C. Hutchinson and R.D. Evans, Development of solid calibration standards for trace elemental aanalyses of tree rings by laser ablation inductively coupled plasma mass spectrometry, *Environ. Sci. Technol.*, 32 (1998) 2185.

146 J.L. Neilsen, A. Abildtrup, J. Christensen, P. Watson, A. Cox and C.W. McLeod, Laser ablation inductively coupled plasma mass spectrometry in combination with gel electrophoresis: a new strategy for speciation of metal binding serum proteins, *Spectrochim. Acta*, 53 B (1998).

147 Sang-Ho Nam, A. Montaser and E.F. Cromwell, Exploration of a helium inductively coupled plasma / Mattauch-Herzog mass spectrometry for simultaneous multielement analysis, *Appl. Spectrosc.*, 52 (1998) 161.

148 S.A. Baker, B.W. Smith and J.D. Winefordner, Investigation of light scattering for normalization of signals in laser ablation inductively coupled plasma mass spectrometry, *Appl. Spectrosc.*, 52 (1998) 154.

149 P.M. Outridge, W. Doherty and D.C. Gregoire, Determination of trace elemental signatures in placer gold by laser ablation inductively coupled plasma mass spectrometry as a potential aid for gold exploration, *J. Geochem. Exp.*, 60 (1998) 229.

150 D.A. Milton and S.R. Chenery, The effect of otolith storage methods on the concentrations of elements detected by laser ablation ICP-MS, *J. Fish Biol.*, 53 (1998) 785.

151 E. Schroeder, M. Hamester and M. Kaiser, Properties and characteristics of a laser ablation ICP-MS system for the quantitative elemental analysis of glasses, *Appl. Surf. Sci.*, 127–129 (1998) 292.

152 D.J. Figg, J.B. Cross and Ch. Brink, More investigations into elemental fractionation resulting from laser ablation inductively coupled plasma mass spectrometry on glass samples, *Appl. Surf. Sci.*, 127–129 (1998) 287.

153 S.M. Eggins, L.P.J. Kinsley and J.M.G. Shelley, Deposition and element fractionation processes during atmospheric pressure laser sampling for analysis by ICP-MS, *Appl. Surf. Sci.*, 127–129 (1998) 278.

154 R.E. Wolf, Ch. Thomas and A. Bohlke, Analytical determination of metals in industrial polymers by laser ablation ICP-MS, *Appl. Surf. Sci.*, 127–129 (1998) 299.

155 S.M. Eggins, R.L. Rudnick and W.F. McDonough, The composition of periotites and their minerals: a laser ablation ICP-MS study, *Earth Planet. Sci. Lett.*, 154 (1998) 53.

156 S.T. Shuttleworth and D.T. Kremser, Assessment of laser ablation and sector field inductively coupled plasma mass spectrometry for elemental analysis of solide samples, *J. Anal. At. Spectrom.*, 13 (1998) 697.

157 M.D. Norman, W.L. Griffin, N.J. Pearson, M.O. Garcia and S.Y. O'Reilly, Quantitative analysis of trace element abundances in glasses and

minerals: a comparison of laser ablation inducively coupled plasma mass spectrometry, solution inductively coupled plasma mass spectrometry, proton microprobe and electron microprobe data, *J. Anal. At. Spectrom.*, 13 (1998) 477.

158 S. Tanaka, N. Yasushi, N. Sato, T. Fukasawa, S.J. Santosa, K. Yamanaka and T. Ootoshi, Rapid and simultaneous multielement analysis of atmospheric particulate matter using inductively coupled plasma mass spectrometry with laser ablation sample introduction, *J. Anal. At. Spectrom.*, 13 (1998) 135.

159 J.S. Becker, S.S. Soman, T. Becker, V.K. Panday and H.J. Dietze, Trace and ultratrace analysis of gallium arsenide by different mass spectrometric techniques, *J. Anal. At. Spectrom.*, 13 (1998) 983.

160 E.K. Shibuya, J.E.S. Sarkis, J. Enzweiler, A.P.S. Jorge and A.M.G. Figueiredo, Determination of platinum group elements and gold geological materials using an ultraviolet laser ablation high-resolution inductively coupled plasma mass spectrometric technique, *J. Anal. At. Spectrom.*, 13 (1998) 941.

161 T.E. Jeffries, S.E. Jackson and H.P. Longerich, Application of a frequency quintupled Nd:YAG source (λ = 213 nm) for laser ablation inductively coupled plasma mass spectrometry analysis of minerals, *J. Anal. At. Spectrom.*, 13 (1998) 935.

162 R.J. Watling, In-line mass transport measurement cell for improving quantification in sulfide mineral analysis using laser ablation inductively coupled plasma mass spectrometry, *J. Anal. At. Spectrom.*, 13 (1998) 927.

163 R.J. Watling, Sourcing the provenance of cannabis crops using inter-element association patterns 'fingerprinting' and laser ablation inductively coupled plasma mass spectrometry, *J. Anal. At. Spectrom.*, 13 (1998) 917.

164 A. Audétat, D. Günther and Ch.A. Heinrich, Formation of magmatic-hydrothermal ore deposit: Insights with LA-ICP-MS analysis of fluid inclusions, *Science*, 279 (1998) 2091.

165 H.F. Falk, B. Hattendorf, K. Krengel-Rothensee, N. Wieberneit and S.L. Dannen, Calibration of laser ablation ICP-MS, Can we use synthetic standards with pneumatic nebulization?, *Fresenius J. Anal. Chem.*, 362 (1998) 468.

166 C.D. Garbe-Schönberg and T. Arpe, High resolution ICP-MS in fast scanning-mode: application for laser ablation analysis of zircon, *Fresenius J. Anal. Chem.*, 362 (1998) 462.

167 G. Suhr, H.A. Seck, N. Shimizu, D. Günther and G. Jenner, Infiltration of refractory melts into the lowermost oceanic crust: evidence from dunite- and gabbro-hosted clinopyroxenes in the bay of islands ophiolite, *Contrib. Mineral Petrol.*, 131 (1998) 136.

168 A.P. de S. Jorge, J. Enzweiler, E.K. Shibuya, J.E.S. Sarkis and A.M.G.

Figueiredo, Platinum-group elements and gold determination in NiS fire assay buttons by UV laser ablation ICP-MS, *Geostandards Newsletter*, 22 (1998) 47.
169 A.N. Halliday, Der-Chuen Lee, J.N. Christensen, M. Rehkämpfer, Wen Yi, X. Luo, Ch.M. Hall, Ch.J. Ballentine, Th. Pettke and C. Stirling, Applications of multiple collector ICP-MS to cosmochemistry, geochemistry and paleoceanography, *Geochim. Cosmochim. Acta*, 62 (1998) 919.
170 D.J. Sinclair, L.P.J. Kinsley and M.T. McCulloch, High resolution analysis of trace elements in corals by laser ablation ICP-MS, *Geochim. Cosmochim. Acta*, 62 (1998) 1889.
171 T. Hirata, M. Hattori and T. Tanaka, In-situ osmium isotope ratio analyses if iridosmines by laser ablation-multiple collector-inductively coupled plasma mass spectrometry, *Chem. Geol.*, 144 (1998) 269.
172 T. Prohaska, C. Stadelbauer, R. Wimmer, G. Stingeder, Ch. Latkoczy, E. Hoffmann and H. Stephanowitz, Investigation of element variability in tree rings of young Norway spruce by laser ablation ICP-MS, *The Science of the Total Environment*, 219 (1998) 29.
173 J. Nölte, F. Scheffler, S. Mann and M. Paul, The influence of true simultaneous internal standardization and background correction on repeatability for laser ablation and the slurry technique coupled to ICP emission spectrometry, *J. Anal. At. Spectrom.*, 14 (1999) 597.
174 W. Devos, Ch. Moor and P. Lienemann, Determination of impurities in antique silver object for authentication by laser ablation inductively coupled plasma mass spectrometry, *J. Anal. At. Spectrom.*, 14 (1999) 621.
175 Ching-Jer Chin, Chung-Fang Wang and Su-Ling Jeng, Multielement analysis of airborne particulate matter collected on PTFE-membrane filters by laser ablation inductively coupled plasma mass spectrometry, *J. Anal. At. Spectrom.*, 14 (1999) 663.
176 M. Motelica-Heino, O.F.X. Donard and J.M. Mermet, Laser ablation of synthetic geological powders using ICP–AES detection, effects of the matrix, chemical form of the analyte and laser wavelength, *J. Anal. At. Spectrom.*, 14 (1999) 675.
177 S. Rings, R. Sievers and M. Jansen, Analysis of the spatial distribution of the constituting elements in amorphous solids: laser ablation with ICP spectrometry, *Fresenius J. Anal. Chem.*, 363 (1999) 165.
178 J.J. Leach, L.A. Allen, D.B. Aeschlimann and R.S. Houk, Calibration of laser ablation inductively coupled plasma mass spectrometry using standard additions with dried solution aerosols, *Anal. Chem.*, 71 (1999) 440.
179 A.J. Campell and M. Humayun, Trace element microanalysis in iron meteorites by laser ablation ICP-MS, *Anal. Chem.*, 71 (1999) 939.
180 E. Hoffmann, C. Lüdke, J. Skole, H. Stepanowitz and G. Wagner, Studies on quantitative analysis of trace elements in single SiC crystals using laser ablation-ICP-MS *J. Anal. At. Spectrom.*, 14 (1999) 1679.

181 U. Schaltegger, C.M. Fanning, D. Günther, J.C. Maurin, K. Schulmann and D. Gebauer, Growth, annealing and recrystallization of zircon and preservation of monazite in high-grade metamorphism: conventional and in-situ U–Pb isotope, cathodoluminescence and microchemical evidence, *Contrib. Mineral. Petrol*, 134 (1999) 186.
182 H. Horn, Harvard University, personnal communication, 1998.
183 A. Audetat and D. Günther, Mobility and H2O loss from fluid inclusions in natural quartz crystals, *Contrib. Mineral Petrol*, 137 (1999) 1.
184 J.R. Bacon, J.S. Crain, L. Van Vaeck and J.G. Williams, Atomic mass spectrometry 1999, *J. Anal. At. Spectrom.*, 14 (1999) 1633.
185 S.A. Baker, M. Bi, R.Q. Aucelio, B.W. Smith and J.D. Winefordner, Analysis of soil and sediment samples by laser ablation inductively coupled plasma mass spectrometry, *J. Anal. At. Spectrom.*, 14 (1999) 19.
186 K.M. Beck, T. Sasaki and N. Koshizaki, Characterization of nanocomposite materials prepared via laser ablation of Pt/TiO_2 bi-combinant targets, *Chem. Phys. Lett.*, 301 (1999) 336.
187 J.S. Becker and H.J. Dietze, Determination of trace elements in geological samples by laser ablation inductively coupled plasma mass spectrometry, *Fresenius J. Anal. Chem.*, 365 (1999) 429.
188 M.C. Boiron, J. Dubessy, N. André, A. Briand, J.L. Lacour, P. Mauchien and J.M. Mermet, Analysis of mono-atomic ions in individual fluid inclusions by laser produced plasma emission spectroscopy, *Geochim. Cosmochim. Acta*, 55 (1991) 917.
189 C.A. Booth, D.A. Spears, P. Krause and A.G. Cox, The determination of low level trace elements in coals by laser ablation-inductively coupled plasma-mass spectrometry, *Fuel*, 78 (1999) 1665.
190 O.V. Borisov, X.L. Mao, A. Fernandez, M. Caetano and R.E. Russo, Inductively coupled plasma mass spectrometric study of non-linear calibration behavior during laser ablation of binary Cu–Zn Alloys, *Spectrochim. Acta*, 54 B (1999) 1351.
191 C.J. Bryant, R.J. Arculus and S.M. Eggins, Laser ablation-inductively coupled plasma-mass spectrometry and tephras: A new approach to understanding arc-magma genesis, *Geology*, 27 (1999) 1119.
192 Z.X. Chen, Inter-element fractionation and correction in laser ablation inductively coupled plasma mass spectrometry, *J. Anal. At. Spectrom.*, 14 (1999) 1823.
193 Y.S. Choi, S.C. Lee, H.S. Im and C.J. Park, Studies on the fundamental characteristics of laser ablation/low-pressure inductively coupled plasma, *Microchem. J.*, 63 (1999) 24.
194 G. Dobosi, G. Kurat, G.A. Jenner and F. Brandstatter, Cryptic metasomatism in the upper mantle beneath Southeastern Austria: a laser ablation microprobe-ICP-MS study, *Contrib. Mineral. Petrol.*, 67 (1999) 143.

195 S.F. Durrant, Laser ablation inductively coupled plasma mass spectrometry: achievements, problems, prospects, *J. Anal. At. Spectrom.*, 14 (1999) 1385.
196 W. Ertel, H.S. O'-Neill, P.J. Sylvester and D.B. Dingwell, Solubilities of Pt and Rh in a haplobasaltic silicate melt at 1300 degrees C, *Geochim. Cosmochim. Acta*, 63 (1999) 2439.
197 S.J. Fallon, M.T. McCulloch, R.van Woesik and D.J. Sinclair, Corals at their latitudinal limits: laser ablation trace element systematics in Porites from Shirigai Bay, Japan, *EPSL*, 172 (1999) 221.
198 Suarez, J. Fernandez, A.G. Gutierrez, G.A. Jenner and M.N. Tubrett, Crustal sources in Lower Palaeozoic rocks from NW Iberia: insights from laser ablation U-Pb ages of detrital zircons, *J. Geol. Soc.*, 156 (1999) 1065.
199 S.M. Glaser, S.F. Foley and D. Günther, Trace element compositions of minerals in garnet and spinel peridotite xenoliths from the Vitim volcanic field, Transbaikalia, Eastern Siberia, Lithos 48 (1999) 263.
200 D. Günther and C.A. Heinrich, Comparison of the ablation behaviour of 266nm Nd:YAG and 193 nm ArF excimer lasers for LA-ICP-MS analysis, *J. Anal. At. Spectrom.*, 14 (1999) 1369.
201 D. Günther and C.A. Heinrich, Enhanced sensitivity in laser ablation-ICP mass spectrometry using helium-argon mixtures as aerosol carrier, *J. Anal. At. Spectrom.*, 14 (1999) 1363.
202 M. Hassellov, B. Lyven and R. Beckett, Sedimentation field-flow fractionation coupled online to inductively coupled plasma mass spectrometry—New possibilities for studies of trace metal adsorption onto natural colloids, *Environmental Science & Technology*, 33 (1999) 4528.
203 C.A. Heinrich, D. Günther, A. Audetat, T. Ulrich and R. Frischknecht, Metal fractionation between magmatic brine and vapor, determined by microanalysis of fluid inclusions, *Geology*, 27 (1999) 755.
204 R.W. Hinton, NIST SRM 610, 611 and SRM 612, 613 multi-element glasses: Constraints from element abundance ratios measured by microprobe techniques, *Geostandards Newsletters*, 23 (1999) 197.
205 T. Hirata and T. Yamaguchi, Isotopic analysis of zirconium using enhanced sensitivity–laser ablation–multiple collector–inductively coupled plasma mass spectrometry, *J. Anal. At. Spectrom.*, 14 (1999) 1455.
206 E. Hoffmann, C. Ludke, J. Kurner, H. Scholze, E. Ullrich and H. Stephanowitz, Studies on the use of XeF2 as digestion reagent, *Fresenius J. Anal. Chem.*, 365 (1999) 592.
207 D.A. Ionov, M. Gregoire and V. Prikhod'ko, Feldspar-Ti-oxide metasomatism in off-cratonic continental and oceanic upper mantle, *EPSL*, 165 (1999) 37.
208 S.H. Jeong, O.V. Borisov, J.H. Yoo, X.L. Mao and R.E. Russo, Effects of particle size distribution on inductively coupled plasma mass spectrometry signal intensity during laser ablation of glass samples, *Anal.*

Chem., 71 (1999) 5123.
209 R.E. Jervis, P.V. Tan and G.J. Evans, Measurement of PM10/2.5 fractionated respirable particles in urban Toronto by INAA, PIXE, ICP-AES, and LAMS—A comparison, *Biological Trace Element Research*, 71 (1999) 223.
210 V.S. Kamentsky, R.C. Wolfe, S. M. Eggins, T.P. Mernagh and E. Bastrakov, Volatile exsolution at the Dinkidi Cu-Au porphyry deposit, Philippines: A melt-inclusion record of the initial ore-forming process, *Geology*, 27 (1999) 691.
211 K. Kawamoto, J. Takada, Y. Tanaka and M. Akaboshi, Determination of rare earth and other elements in standard biological samples by ICP-MS and activation analysis, *J. Radioanal. Nuclear Chem.*, 242 (1999) 527.
212 K.M. Lee, J. Appleton, M. Cooke, F. Keenan and K.K. Sawicka, Use of laser ablation inductively coupled plasma mass spectrometry to provide element versus time profiles in teeth, *Anal. Chim. Acta*, 395 (1999) 179.
213 X.R. Liang, X.H. Li, M. Sun, Y. Liu and X.L. Tu, Simultaneously in-situ analysis of trace elements and U-Pb and Pb-Pb ages for single zircons by laser ablation microprobe–inductively coupled plasma mass spectrometry, *Chem. Lett.*, (1999) 639.
214 X.R. Liang, X.H. Li, G.J. Wei, X.L. Tu and G.L. Wang, Uranium-lead age determination of single zircon grain by laser ablation microprobe–inductively couple plasma mass spectrometry, *Chinese J. Anal. Chem.*, 27 (1999) 1121.
215 F. Lochner, J. Appleton, F. Keenan and M. Cooke. Multi-element profiling of human deciduous teeth by laser ablation–inductively coupled plasma–mass spectrometry, *Anal. Chim. Acta*, 401 (1999) 299.
216 A.J.G. Mank and P.R.D. Mason, A critical assessment of laser ablation ICP-MS as an analytical tool for depth analysis in silica-based glass samples, *J. Anal. At. Spectrom.*, 14 (1999) 1143.
217 P.R.D. Mason, K.E. Jarvis, H. Downes and R. Vannucci, Determination of incompatible trace elements in mantle clinopyroxenes by LA-ICP-MS: A comparison of analytical performance with established techniques, *Geostandards Newsletters*, 23 (1999) 157.
218 J.M. Mermet, Fundamental principles of inductively coupled plasmas, In: S. Hill (Ed.), *Handbook of ICP Spectrometry*, Sheffield Academic Press, Sheffield, 1999. pp. 35–70
219 Y. Narita, S. Tanaka and S.J. Santosa, A study on the concentration, distribution, and behavior of metals in atmospheric particulate matter over the North Pacific Ocean by using inductively coupled plasma mass spectrometry equipped with laser ablation, *J. Geophys. Res. Atmospheres*, 104 (1999) 26859.
220 M. Odegard, Preparation of synthetic calibration materials for use in the microanalysis of oxide minerals by direct fusion in high-purity graphite

electrodes: preliminary results for quartz and rutile, *Geostandards Newsletters*, 23 (1999) 173.

221 N.J.G. Pearce, J.A. Westgate, W.T. Perkins, W.J. Eastwood and P. Shane, The application of laser ablation ICP-MS to the analysis of volcanic glass shards from tephra deposits: bulk glass and single shard analysis, *Global and Planetary Change*, 21 (1999) 151.

222 J.E. Reid, I. Horn, H.P. Longerich, L. Forsythe and G.A. Jenner. Determination of Zr and Hf in a flux-free fusion of whole rock samples using laser ablation inductively coupled plasma-mass spectrometry (LA-ICP-MS) with isotope dilution calibration, *Geostandards Newsletters*, 23 (1999) 149.

223 M.S. Roberts, P.L. Smart, C.J. Hawkesworth, W.T. Perkins and N.J.G. Pearce, Trace element variations in coeval holocene speleothems from GB cave, Southwest England, *Holocene*, 9 (1999) 707.

224 B. Schafer, R. Frischknecht, D. Günther and D.B. Dingwell, Determination of trace-element partitioning between fluid and melt using LA-ICP-MS analysis of synthetic fluid inclusions in glass, *Eur. J. Min.*, 11 (1999) 415.

225 D.J. Scott, U-Pb geochronology of the eastern Hall Peninsula, southern Baffin Island, Canada: a northern link between the Archean of West Greenland and the Paleoproterozoic Torngat Orogen of Northern Labrador, *Precambrian Res.*, 93 (1999) 5.

226 L.K. Sha and B.W. Chappell, Apatite chemical composition, determined by electron microprobe and laser-ablation inductively coupled plasma mass spectrometry, as a probe into granite petrogenesis, *Geochim. Cosmochim. Acta*, 63 (1999) 3861.

[227 T. Ulrich, D. Günther and C.A. Heinrich, Gold concentrations of magmatic brines and the metal budget of porphyry copper deposits, *Nature*, 399 (1999) 676.

[228 E. Vander Putten, F. Dehairs, L. Andre and W. Baeyens, Quantitative in situ microanalysis of minor and trace elements in biogenic calcite using infrared laser ablation–inductively coupled plasma mass spectrometry: a critical evaluation, *Anal. Chim. Acta*, 378 (1999) 261.

[229 G. Veinott, T. Northcote, M. Rosenau and R.D. Evans, Concentrations of strontium in the pectoral fin rays of the white sturgeon (Acipenser transmontanus) by laser ablation sampling–inductively coupled plasma–mass spectrometry as an indicator of marine migrations, *Can. J. Fish. Aquatic Sci.*, 56 (1999) 1981.

230 G.I. Veinott and R.D. Evans, An examination of elemental stability in the fin ray of the white sturgeon with laser ablation sampling-inductively coupled plasma-mass spectrometry (LAS-ICP-MS), *Transact. Am. Fish. Soc.*, 128 (1999) 352.

231 B. Wagner, S. Garbos, E. Bulska and A. Hulanicki, Determination of iron

and copper in old manuscripts by slurry sampling graphite furnace atomic absorption spectrometry and laser ablation inductively coupled plasma mass spectrometry, *Spectrochim. Acta*, 54B (1999) 797.
232 C.F. Wang, F.H. Tu, S.L. Jeng and C.J. Chin, The determination of silicon in airborne particulate matter by XRF and LA-ICP-MS, *J. Radioanal. Nuc. Chem.*, 242 (1999) 97.
233 B. Wanner, C. Moor, P. Richner, R. Bronnimann and B. Magyar, Laser ablation inductively coupled plasma mass spectrometry (LA-ICP-MS) for spatially resolved trace element determination of solids using an autofocus system, *Spectrochim. Acta*, 54B (1999) 289.
234 R. Wennrich, P. Morgenstern, G. Liebergeld and G. Werner, Influence of pellet composition in IR-laser ablation inductively coupled plasma atomic emission and mass spectrometry, *Chemia Analityczna*, 44 (1999) 523.
235 K.X. Yang, J.X. Zhou, X.D. Hou and R.G. Michel, Excimer laser ablation for microwave-induced plasma emission spectrometry, *J. Korean Phys. Soc.*, 35 (1999) 133.
236 P. Yang, T. Rivers, and S. Jackson, Crystal-chemical and thermal controls on trace-element partitioning between coexisting garnet and biotite in metamorphic rocks from western Labrador, *Can. Min.*, 37 (1999) 443.
237 G.M. Yaxley and V. Kemenetsky, In situ origin for glass in mantle xenoliths from southeastern Australia: insights from trace element compositions of glasses and metasomatic phases, *EPSL*, 172 (1999) 97.
238 R. Zhang, Y. Achiba, K.J. Fisher, G.E. Gadd, F.G. Hopwood, T. Ishigaki, D. R. Smith, S. Suzuki and G. D. Willett, Laser ablation mass spectrometry of pyrolyzed Koppers coal-tar pitch: A precursor for fullerenes and metallofullerenes, *J. Phys. Chem. B*, 103 (1999) 9450.
239 B. Gratuze, Obsidian characterization by laser ablation ICP-MS and its application to prehistoric trade in the mediterranean and the near east: sources and distribution of obsidian with the aegean and anatolia, *J. Archaelogical Science*, 26 (2000) 869.
240 P. Garofalo, A. Audetat, D. Günther, C.A. Heinrich and J. Ridley. Estimation and testing of standard molar thermodynamic properties of tourmaline end-members using data of natural samples, *Am. Mineral.*, 85 (2000) 78.
241 X.D. Cao, X.R. Wang and G.W. Zhao, Assessment of the bioavailability of rare earth elements in soils by chemical fractionation and multiple regression analysis, *Chemosphere*, 40 (2000) 23.
242 P. Stalder, P. Ulmer, A.B. Thompson and D. Günther, Experimental approach to constrain second critical end points in fluid/silicate systems: Near-solidus fluids and melts in the system albite-H_2O, *Am. Mineral.*, 85 (2000) 68.
243 X.S. Xu, S.Y. O'Reilly, W.L. Griffin and X.M. Zhou, Genesis of young lithospheric mantle in southeastern China: an LAM-ICPMS trace element

study, *J. Petrol.*, 41 (2000) 111.
244 H.G.C. Human, R.H. Scott, A.R. Oakes and C.D. West, The use of a spark as a sampling nebulising device for solid samples in atomic absorption, atomic fluorescence and inductively coupled plasma emisson spectrometry, *Analyst*, 101 (1976) 265.
245 R.H. Scott, Spark elutriation of powders into an inductively coupled plasma, *Spectrochim. Acta*, 33B (1978) 123.
246 J.Y. Marks, D.E. Fornwalt and R.E. Yungk, Application of a solid sampling device to the analysis of high temperature alloys by ICP-AES, *Spectrochim. Acta*, 38B (1983) 107.
247 A. Aziz, J.A.C. Broekaert, K. Laqua and F. Leis, A study of direct analysis of solid samples using spark ablation combined with excitation in an inductively coupled plasma, *Spectrochim. Acta*, 39B (1984) 1091.
248 P.M. Beckwith, R.L. Mullins and D.M. Coleman, Analysis of Eastern coal fly ash using separated sampling and excitation atomic emission spectrometry, *Anal. Chem.*, 59 (1987) 163.
249 A. Lemarchand, G. Labarraque, P. Masson and J.A.C. Broekaert, Analysis of ferrous alloys by spark ablation coupled to inductively coupled plasma atomic emission spectrometry, *J. Anal. At. Spectrom.*, 2 (1987) 481.
250 L.J. Prell and S.R. Koirtyohann, Analytical performance and transport studies using spark sampling followed by ICP excitation with a sequential spectrometer, *Appl. Spectrosc.*, 42 (1988) 1221-1228.
251 D.W. Golightly, A. Montaser, B.L. Smith and A.F. Dorrzapf, Spark ablation-inductively coupled plasma spectrometry for analysis of geological materials, *Talanta*, 36 (1989) 299.
252 G. Vujicic and I. Steffan, Spark ablation as sample introduction device for ICP-AES: Analysis of free cutting steels, *Mikrochim. Acta*, II (1990) 315.
253 A. Gomez Coedo, M.T. Dorado Lopez, J.L. Jimenez Seco and I. Gutteriez Cobo, Determination of aluminium, niobium and vanadium in low-alloyed steels by spark ablation coupled with inductively coupled plasma atomic emission spectrometry, *J. Anal. At. Spectrom.*, 7 (1992) 11.
254 A. Gomez Coedo, M.T. Dorado Lopez, I.G. Cobo and E.E. Baquero, Analytical system for the analysis of ferrovanadium using spark ablation coupled with inductively coupled plasma atomic emission spectrometry, *J. Anal. At. Spectrom.*, 7 (1992) 247.
255 I. Steffan and G. Vujicic, ICP-AES analysis of non conductive materials after spark ablation, *Spectrochim. Acta*, 47B (1992) 61.
256 A.G. Coedo, T. Dorado, E. Escudero and I.G. Coba, Boron determination in steels by inductively coupled plasma atomic emission spectrometry, Comparative study of spark ablation and pneumatic nebulization sampling systems, *J. Anal. At. Spectrom.*, 8 (1993) 827.
257 A.G. Coedo, T. Dorado, C.J. Rivero and I.G. Cobo, Study of X-ray fluorescence spectrometry and spark ablation inductively coupled plasma

atomic emission spectrometry for chromium determination in ferrochromium from bulk metal samples, *J. Anal. At. Spectrom.*, 8 (1993) 1023.
258 A.G. Coedo, M.T. Dorado and I.G. Cobo, Analysis of electro arc furnace flue dust by spark ablation inductively coupled plasma atomic emission spectrometry, *J. Anal. At. Spectrom.*, 9 (1994) 223.
259 C. Webb, C.B. Cooper III, A.T. Zander, J.T. Arnold, E.S. Lile and S.E. Anderson, Direct solid sampling for analysis with inductively coupled plasma using a novel electronic spark source, *J. Anal. At. Spectrom.*, 9 (1994) 263.
260 L. Moenke-Blankenburg, J. Kammel and T. Schumann, Microanalysis by laser ablation inductively coupled plasma–atomic emission spectrometry in comparison to spark ablation of certified and found minerals galena, sphalerite, and pyrite, *Microchem. J.*, 50 (1994) 374.
261 I.B. Brenner, A. Zander, S. Kim and C. Holloway, Multielement analysis of geological and related non-conducting materials using spark ablation and a sequential spectrometer, *Spectrochim. Acta*, 50B (1995) 565.
262 M. Gagean and J.M. Mermet, Comparison of ultraviolet laser ablation and spark ablation of metals and alloys for analysis by axially viewed inductively coupled plasma atomic emission spectrometry, *J. Anal. At. Spectrom.*, 12 (1997) 189.
263 K.A. Ivanovic, D.M. Coleman, F.W. Kunz and D. Schuetzle, Spark ablation of solid samples for analysis by inductively coupled plasma–mass spectrometry, *Appl. Spectrosc.*, 46 (1992) 894.
264 N. Jakubowski, I. Feldmann, B. Sack and D. Stuewer, Analysis of conductive solids by inductively coupled plasma mass spectrometry with spark ablation, *J. Anal. At. Spectrom.*, 7 (1992) 121.
265 A.G. Coedo, M.T. Dorado and B. Fernandez, Spark ablation as sampling device for inductively coupled plasma mass spectrometric analysis of low-alloyed steels, *J. Anal. At. Spectrom.*, 10 (1995) 859.
266 R.L. Van Hoven, S.H. Nam, A. Montaser, M.W. Doughten and A.F. Dorrzapf Jr, Direct solid sampling of fire assay beads by spark ablation inductively coupled plasma mass spectrometry, *Spectrochim. Acta*, 50B (1995) 549.
267 N. Jakubowski, I. Feldmann and D. Stuewer, Comparison of ICP-MS with spark ablation and GDMS for direct element analysis of conductive solids, *Spectrochim. Acta*, 50B (1995) 639.
268 O.V. Borisov, D.M. Coleman and R.O. Carter III, Determination of vanadium, rhodium and platinum in automotive catalytic converters using inductively coupled plasma mass spectrometry with spark ablation, *J. Anal. At. Spectrom.*, 12 (1997) 231.
269 R.D. Deutsch, G.M. Hieftje and J.P. Keilsohn, The use of a microarc atomizer for sample introduction into an inductively coupled plasma, *Appl. Spectrosc.*, 37 (1983) 101.

270 D.A. Rusak, R.L. Litteral, B.W. Smith and J.D. Winefordner, DC arc vaporization as a sample introduction technique for analysis of solids by ICP-OES, *Talanta*, 44 (1997) 1987.
271 S.J. Jiang and R.S. Houk, Arc nebulization for elemental analysis of conducting solids by inductively coupled plasma mass spectrometry, *Anal. Chem.*, 58 (1986) 1739.
272 S.J. Jiang and R.S. Houk, Elemental and isotopic analysis of powders by inductively coupled plasma-mass spectrometry with arc nebulization, *Spectrochim. Acta*, 42B (1986) 93.

Chapter 5

Direct sample insertion

Vassili Karanassios and Todd J. Wood

5.1 INTRODUCTION

The inductively coupled plasma (ICP) is a workhorse in elemental analysis and, ICP–atomic emission spectrometry (ICP-AES) and ICP–mass spectrometry (ICP-MS) are well established analytical techniques. Despite wide acceptance and applicability of ICP spectrometry, introduction of an initially ambient temperature sample into a hot, partially ionized gas phase plasma is a persistent problem and, to this day, it still remains an active area of research.

Although analytical samples can be introduced into an ICP as gases, liquids or solids, in routine elemental analysis, most samples are introduced as liquids using a pneumatic nebulizer. Arguably, there are three key limitations to pneumatic nebulization sample introduction. One, that a sample must be in liquid form. Since many samples occur naturally as solids, such samples must be converted to a liquid prior to analysis. This conversion (or sample digestion) results in sample dilution, it increases the potential for contamination from the digestion reagents, it is often labour-intensive and typically, costly. Since digestion usually involves use of corrosive acids, it also complicates waste-disposal of unused samples containing potentially hazardous elemental contaminants. Poor sample introduction efficiency (typically <3%) is another key limitation of pneumatic nebulization. In essence, over 97% of a hard-earned sample literally goes down the drain. No wonder then that sample introduction has been termed the "Achilles' heel" of atomic spectroscopy. Since pneumatic nebulization requires solution samples and volumes of 1–3 ml/min, the third limitation arises when there is only a small amount of sample available for analysis. Examples abound in biological, forensic and clinical analysis.

Direct sample insertion (DSI) devices were developed by Salin and Horlick [1] and almost concurrently by Sommer and Ohls [2,3], in part to address these limitations. In DSI-ICP (Fig. 5.1), a probe (e.g., graphite cup, wire-loop, a basket holding a pelletized solid) carrying a few μl of a liquid or mg/sub-mg amounts of a solid is inserted through the central tube of a modified torch into the plasma and the ICP is used for vaporization, atomization, excitation and ionization. Conceptually, the sample-carrying probe can be thought of as a sample holder, the DSI as a device that brings an initially ambient temperature sample in

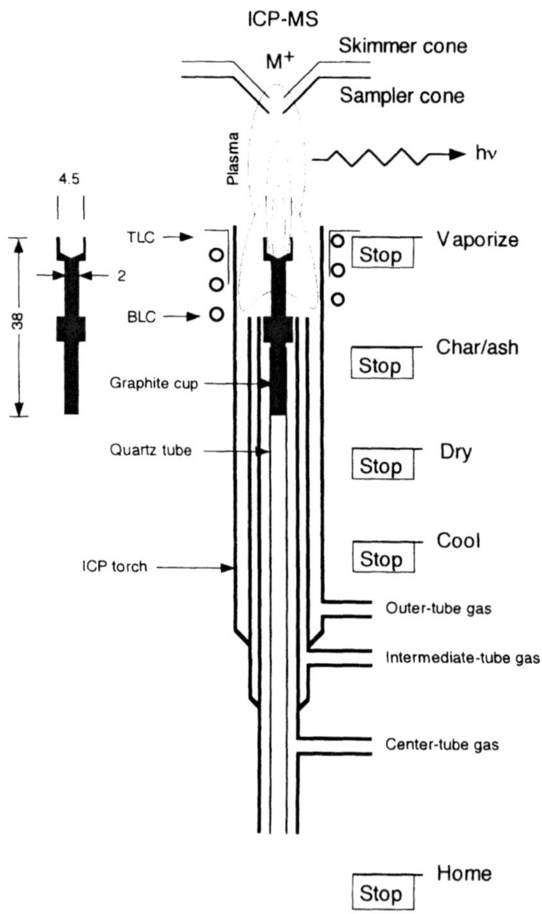

Fig. 5.1. DSI torch and graphite cup (illustration not to scale, dimensions in mm).

contact with the plasma (e.g., sample elevator technique or SET was an early acronym used in lieu of DSI [2]) and the ICP as an inert-gas "furnace" for sample-carrying probes. By taking advantage of the thermal gradient that exists below the plasma, some sample processing can take place *in situ*, for example, drying and charring/ashing are typically done inside the torch (Fig. 5.1).

Advantages of DSIs include high sample introduction efficiency, for instance, 100% of the sample enters the plasma, and a capability to provide analytical results from μl volumes of samples or from mg/sub-mg quantities of powders/solids with minimum sample pre-treatment. Over the last two decades, DSI-ICP has been refined and has been described in book chapters [4,5], reviews [6–8] and the primary literature [9–66].

This chapter is not a review in a strict sense of the word; it can be thought of as an overview and a practical guide to DSI-ICP spectrometry. The material may be particularly useful to researchers in this field or to those contemplating exploiting, exploring or further developing the technique. The material draws heavily from DSI-ICP-AES because most of the work reported in the literature has been carried out using such systems. However, some of the conclusions drawn using DSI-ICP-AES are equally valid for DSI-ICP-MS: for example, the effects of cup geometry and composition, drying and charring conditions, carbide formation, the temperature of a graphite cup in a plasma, volatilization enhancement using mixed-gas plasmas and *in-situ* chemical modification, pre-concentration or sample clean-up and slurry sampling, to name but a few. Furthermore, emphasis is placed on more recent work on DSI-ICP including some previously unpublished results.

Following an introduction, this chapter begins by discussing DSI instrumentation for ICP-AES and ICP-MS. To facilitate discussion and to enable comparisons, most of the experimental results included here have been acquired using an automated DSI-ICP-AES system which has been developed in the author's laboratory [9–11]. What follows is a description of the effect of operating conditions on analyte signal behaviour. Subsequently, the effect of mixed gases on emission signals is systematically evaluated. In most recent work with DSI-ICP, graphite cups which have been machined into different shapes appear to be the preferred sample-carrying probes, but their use makes carbide formation the key chemical limitation of graphite cup DSIs.

Chemical modification as a potential solution to the problem of carbide formation is the next main topic to be discussed. Additional advantages of DSIs, such as: an ability to resolve spectral interferences in the time-rather than the spectral-domain; the benefits arising from use of dried solution residues, such as reductions in spectroscopic and non-spectroscopic interference effects, and the capability to separate potentially troublesome matrix components of a sample or pre-concentrate analytes directly on graphite cups, or even use of graphite cups as *miniature graphite-beakers* that collect fractions eluting from a flow-injection system before inserting them into an ICP, are described next. The chapter concludes by discussing analytical figures-of-merit, analysis of powders or solids that can be converted to a powder as slurries and by listing applications of DSI-ICP.

5.2 INSTRUMENTATION FOR DSIs

The basic building blocks for a typical DSI-ICP-AES or DSI-ICP-MS system, are illustrated schematically in Fig. 5.2. These are: (1) the DSI-device, (2) the ICP (including the radio frequency generator and the

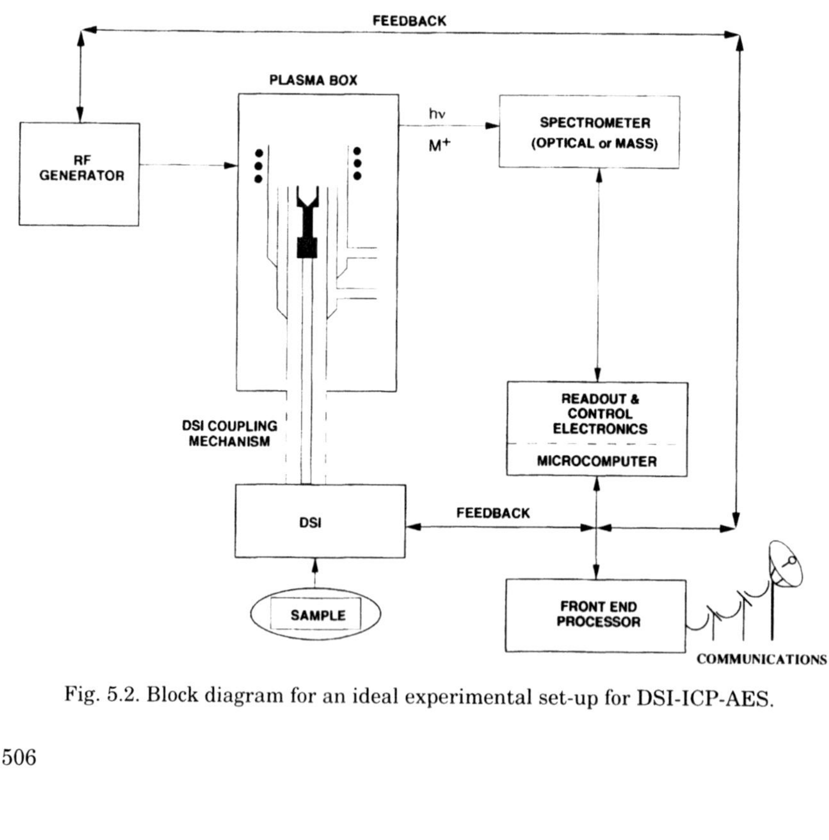

Fig. 5.2. Block diagram for an ideal experimental set-up for DSI-ICP-AES.

plasma box), (3) a spectrometer, (4) readout and control electronics and (5) a front end processor. The measurement end of the system shown in Fig. 5.2 consists of read-out and control electronics and a front end processor supports a graphical user interface. This end has been broken into two parts to indicate both functionality and actual implementation.

Although many of the building blocks shown in Fig. 5.2 are common to many ICP systems, parts of some of these blocks must be modified or tailored to DSIs to obtain optimum analytical performance. Typically, DSIs designed for AES work are different than those for MS work. These differences arise because most ICP-AES systems are operated vertically whereas commercially available ICP-MS systems are operated horizontally. As a consequence, for radially viewed ICP-AES systems, DSIs are mounted vertically and those for ICP-MS systems horizontally. DSI-mounting dictates the way samples are deposited into or onto a sample-carrying probe, how they are dried (e.g., externally or *in-situ*) and how they are retained in the sample-carrying probe (e.g., often necessitating use of boiler cups). At the heart of the system shown in Fig. 5.2 is a DSI-device. The device consists of a DSI-drive mechanism, of a modified ICP torch and of a sample-carrying probe. Design attributes and experimental parameters affecting analytical performance characteristics of DSI-ICP-MS are listed in Table 5.1 and described in detail in the following section.

5.2.1 DSI-drive mechanism and DSI-designs

An ideal drive mechanism must allow rapid and reproducible probe insertion or retraction. It must also allow the sample-carrying probe to stop on its way to the plasma for a programmable period of time for drying and charring/ashing (if necessary) and, during retraction, for a cooling period prior to bringing the DSI back to its home position (Fig. 5.1). In most cases, the mechanical stops shown in Fig. 5.1 indicate functionality rather than actual implementation.

Since DSIs are not commercially available, many types of manual, semi-automated and automated DSIs have been designed over the years. For example, some were designed for radially viewed ICP-AES [11–13,17], others were modified for axially viewed ICP-AES [41], and yet others for ICP-MS systems [11,13]. Furthermore, some were pneumatically driven [12,17], others were stepper-motor controlled [11,13,33] and others implemented the car aerial concept [64,67,68]. A

TABLE 5.1

Design attributes and experimental parameters and affecting analytical performance characteristics of DSI-ICP-MS

Plasma

- ICP torch
- Forward power
- Gas flow rates
- Gas composition

Spectrometer, readout electronics and software

- Analytical line(s) or *m/z*
- Sampling depth
- Scanning speed (ICP-MS)
- Custom software

DSI-device design attributes

- Design
  - Type (e.g., vertical, horizontal)
  - Drive mechanism (stepper motor, car aerial, pneumatic)
  - Probe insertion sequence
- Probe attributes
  - Type (e.g., metal or graphite cup, wire-loop)
  - Composition (e.g., graphite, metal, activated carbon)
  - Geometry (e.g., size, shape)
  - Physical condition (e.g., surface porosity, pyrolytic coating)

Run-time parameters

- Drying
  - Insertion position (with respect to BLC)
  - Drying time (s)
- Charring
  - Insertion position (with respect to BLC)
  - Charring time (s)
- Vaporization
  - Insertion position (with respect to TLC)
  - Observation time (s)
  - Integration time (s) for peak-area integration

Sample

- Type (e.g., liquid, solid/powder)
- Size (e.g., volume, mass)
- Element type (e.g., volatile, carbide forming; some examples are listed in Table 5.2)
- Analyte concentration-range
- Matrix composition

TABLE 5.2

Boiling points for Ti, V and W and the corresponding carbides

Element	T_b (°C)	Carbide	T_b (°C)
Ti	3287	TiC	4820
V	3380	VC	3900
W	5660	WC (WC_2)	6000

detailed summary can be found in Table 5.2 of Ref. [6] and more recently, in Ref. [8].

A recent trend in DSI-ICP is toward either simplification of the design, for example, using an inexpensive car aerial [67–69] to insert a sample-carrying probe into a plasma or toward stand-alone, computer-controlled and stepper-motor driven designs [11]. A DSI-device which was designed to be able to be used either vertically (for ICP-AES) or horizontally (for ICP-MS and, potentially with axially viewed ICP-AES systems) has been developed in our laboratories [11] and will be described in some detail.

This hybrid design has been constructed around a light, hollow aluminum frame with a top and a bottom cover (Fig. 5.3). Two steel rods are attached to the top and bottom covers of the aluminum frame and a sliding platform is placed inside the frame. The platform is guided by the steel rods and is attached to a timing belt. The timing belt is suspended between two timing belt gears and the bottom gear is attached to the stepper-motor using vibration absorbing rubber. A spring-loaded probe assembly is placed outside the aluminum frame and is attached to the sliding platform using a rotating shaft that acts as a pivot axis. One end of the probe assembly has a cam attached to it, the other has a hole into which a quartz tube is secured (i.e., locked into place). The graphite cup is placed into the other end of the quartz tube and the quartz tube (with the graphite cup on it) is inserted through the central tube of the torch (Fig. 5.1) into the plasma.

The stepper-motor is driven by one axis of a dual stepper-motor controller board which has been installed in a computer dedicated to the DSI. The DSI-computer can either stand alone or be stripped of its keyboard and screen and communicate serially with the controlling microcomputer. The benefits of making a computer a key component of the DSI-device are two-fold. First, the front end processor (Fig. 5.2) is

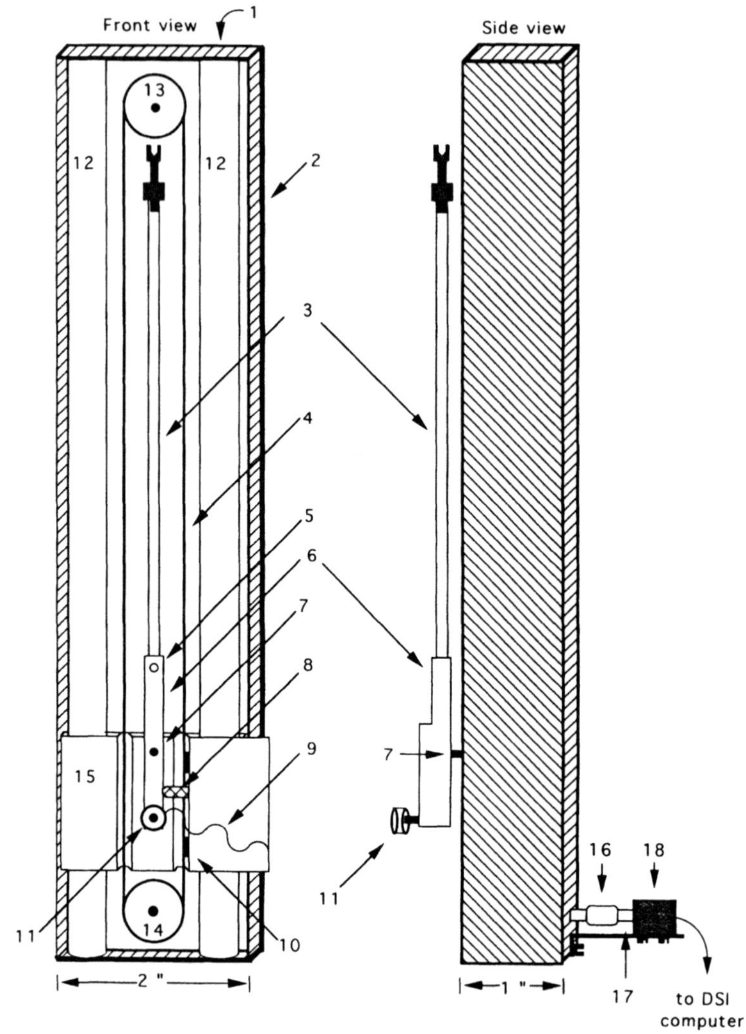

Fig. 5.3. Front and side views of the DSI-device. (1) Top aluminum cover; (2) hollow aluminum frame; (3) quartz rod; (4) timing belt; (5) quartz rod-securing screw; (6) pivoting probe support assembly; (7) rotating shaft and pivoting axis; (8) level adjusting screw and probe support assembly stop; (9) spring; (10) sliding platform-to-timing belt securing mechanism; (11) cam; (12) round steel rods; (13) top timing belt gear; (14) bottom timing belt gear; (15) sliding platform; (16) vibration absorbing rubber; (17) rotating shaft connected to stepper motor and to the bottom timing belt gear; (18) stepper motor. For clarity, the spring is not shown on side view. Illustration not to scale.

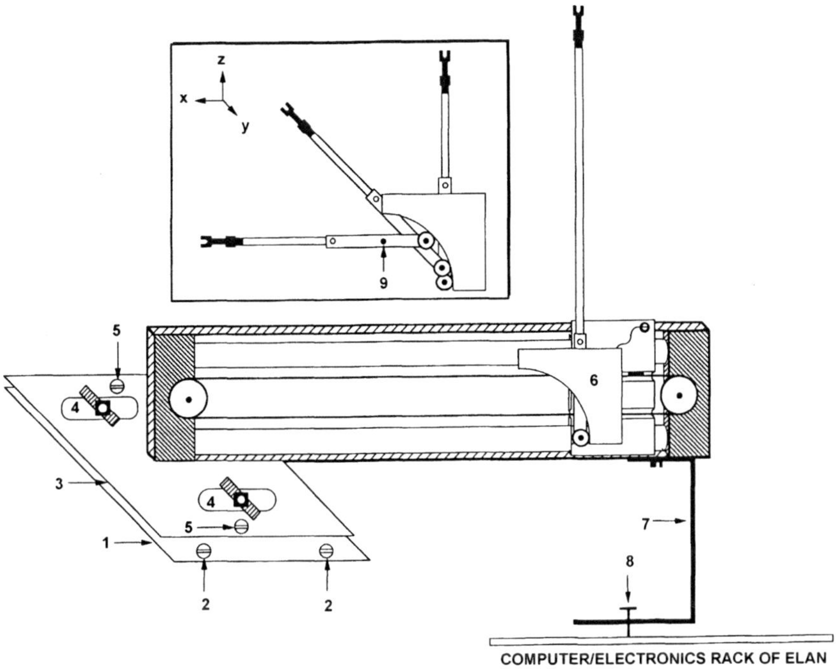

Fig. 5.4. Side view of DSI on an Elan 250 ICP-MS system. (1) Aluminum plate mounted on top of the matching network; (2) aluminum plate mounting screws; (3) DSI-support plate mounted on the side of the hollow aluminum frame; (4) slot-and-nut, x-y-z position adjusting and securing mechanism for alignment of the DSI with ICP torch (screw is attached on an aluminum plate which is mounted on the matching network); (5) level/height adjusting screws; (6) cam follower; (7) frame supporting the back-end of DSI; (8) height adjusting screw; (9) pivot axis. Illustration not to scale.

free to perform other tasks and second, the DSI-device becomes easier to attach to and to use with other ICP spectrometers.

For example, by simply extending the hollow aluminum frame, this DSI has been used [67] with a commercially available polychromator (JY-48, Instruments SA, Edison, N.J.). Also, by simply mounting an aluminum plate on top of the matching network, by mounting a level-adjustable support plate, a height-adjustable support frame and a cam follower as shown in Fig. 5.4, the DSI can be configured to operate horizontally for ICP-MS. When fully retracted, the sample-carrying probe of this device assumes a vertical position (parallel to the z-axis, Fig. 5.4) to allow easy addition of samples to the graphite cup. As the sample probe moves forward, it gradually assumes a horizontal position to allow probe insertion into the horizontal torch of the ICP-MS (parallel to the x-axis, Fig. 5.4).

Even with such a versatile DSI-design, some modifications to the ICP spectrometer may still be required. Modifications may be as simple as cutting access holes to an ICP-MS system [14] or as elaborate as re-writing software [17].

5.2.2 ICP torch

Several types of modified torches have been used with DSIs, as has been summarized elsewhere [8]. While the plasma can be operated with the bottom of the central tube open (Fig. 5.1), in some instances, a shutter mechanism has been placed to seal the bottom of the torch when the sample-carrying probe is retracted [14,33]. This was done to prevent the plasma from resting on the inner tubes of the torch, thus possibly causing torch meltdown.

In some cases, (arc) filament formation was noted during retraction between the plasma and the sample-carrying probe [12]. To prevent filament formation, either a "snuffer gas" was added to the central tube of the torch, or water-saturated Ar or He have been used as snuffer-gases [13,20,22]. For the torch shown in Fig. 5.1 and the DSI shown in Fig. 5.3, neither a snuffer gas nor a shutter were required [11].

5.2.3 Sample-carrying probes

There are three basic types: graphite probes (e.g., rods or cups with or without boiler cups), metal cups and metallic wire-based probes (e.g., wire-loops and wire baskets). Numerous probe geometries have been reported in the literature and examples are shown in Fig. 5.5. Inexpensive graphite probes, such as those used with dc arcs, are the most widely used in DSI-ICP. These are easy to machine using a lathe. To eliminate the possibility of contaminants boiling off the graphite, such supports are either 'burned off' in the plasma for 5–15 min (typical powers 1.5–2 kW) or are repeatedly inserted into the plasma following an acid wash [18]. Perhaps the key chemical limitation of graphite cups arises from the material itself because a number of elements form non-volatile carbides (some examples are listed in Table 5.2).

Metal cups fabricated from high melting point metals, such as Ta, Mo and W, have been used [12,15], mostly to eliminate carbide

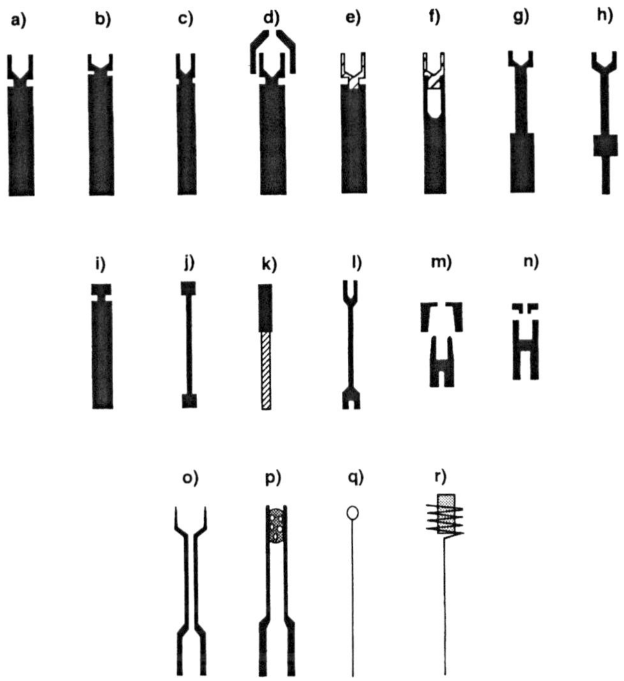

Fig. 5.5. A selection of sample carrying probes: (a–c) commercially available graphite cups [17]; (d) graphite cup with a boiler cup; (e) metal cup [17]; (f) metal cup on a hollow graphite rod [20]; (g, h) machined graphite cups (often referred to as a "long undercut"); (i, k) custom-made flat top rod; (j) glassy carbon stem [51]; custom made thin wall cups [34,57]; (m, n) graphite cup with a lid; (o, p) hollow stem graphite supports [41,42]; wire-loop [33] and (h) wire basket [36].

formation. However, metal cups are difficult to machine and, in all reported cases, only the actual cup has been made out of metal. Although the approach addresses the problem of carbide formation, inter-metallic compound formation, spectral interference from the probe material and potential vaporization of impurities in the metal must be considered when using them.

Loops made out of W or Ta provide an alternative to metal cups and have been used with microlitre volumes of samples [27,33]. Due to their low mass and high thermal conductivity, wire-loops heat very rapidly, thus solution residues on them vaporize very quickly. As a consequence, the transient emission signals last one or a few hundred milliseconds. This places new demands on the signal conditioning and data acquisition electronics of AES systems and further limits the multi-element capability of MS systems. Although spectral interference from the loop material has not been observed by ICP-MS [15], it has been reported by ICP-AES (most likely due to use of materials from different suppliers). Overall, the inability of wire-loops to handle solids or powders limits their analytical utility. Also, the use of wire baskets to support pelletized solids has been reported to be of questionable analytical utility [32,36].

5.2.4 Spectrometers

Due to the use of small amounts of sample, signals generated by DSI-ICP are transient by nature. For example, peak widths ranging from one to a few hundred milliseconds for wire-loop DSI to as many as 100 s for graphite cup DSIs, a medium volatility element (e.g., Fe, Co) and plasma powers around 1 kW. Such transient signals dictate spectrometer type, readout electronics hardware and software. These will be briefly discussed, first for ICP-AES and then for ICP-MS.

5.2.4.1 Spectrometers for DSI-ICP-AES

Due to the transient nature of analyte emission, it would be impossible to use a slew-scanning, sequential system for simultaneous multi-element analyses using DSI-ICP-AES. Although systems using photodiode arrays or area sensor detectors could be used with DSIs, such systems do not allow monitoring of temporal emission behaviour because the speed of readout electronics for such systems, at present, is relatively slow. Perhaps the only way to perform simultaneous multi-element determinations and, at the same time, monitor analyte emission temporal behaviour is by using a polychromator system with photomultiplier tube (PMT) detectors and data acquisition electronics and software capable of handling transient signals [6,8].

5.2.4.2 Spectrometers for DSI-ICP-MS

To our knowledge, DSIs have only been studied with quadrupole-based ICP-MS systems. In such systems, masses (m/z) are monitored using a peak hopping mode. In this mode, the quadrupole dwells on a peak for a user-defined period of time and then it hops to different m/z. The process is repeated until all desired m/z have been scanned. This sequential approach is similar to that used by slew-scanning optical emission spectrometers. Clearly, the width of the transient signal and the scanning speed of the quadrupole define the number of *m/z* (isotopes or elements) that can be monitored in a single analytical run. For a Perkin-Elmer/Sciex Elan 250, a graphite-cup DSI, high- to medium volatility elements (e.g., Zn, Cu) and plasma powers in the neighbourhood of 1.0 kW, the number of elements that may be scanned within each sample run is limited to 4 when a dwell time of 10 ms is used (Fig. 5.6). As dwell time increases or as signals get faster, the number of *m/z* that can be monitored decreases [14]. For example, for wire-loop DSI-ICP-MS, all one can hope to achieve is single ion monitoring. In such cases, multiple runs of the same sample are required in order to make multielement determinations, thus increasing analysis time. Time-of-flight DSI-ICP-MS or ion-trap DSI-ICP-MS have the potential to provide simultaneous multielement analysis capabilities from transient signals generated by DSIs but, at present, such a coupling has not been attempted.

Some quadrupole-based ICP-MS systems which have been introduced in the mid to late 90s are claimed to be able to better monitor faster transient signals (or more *m/z*), however, such systems have not been tested with DSIs. Readers interested in sample introduction systems that generate transient signals should closely evaluate the time resolved capabilities of ICP-MS systems before purchasing such an instrument. Custom software developed to fit specific needs may not necessarily be easy to write, as reported using an older Elan 250 [14].

5.3 EFFECT OF OPERATING CONDITIONS ON ANALYTE SIGNAL BEHAVIOUR

In DSI-ICP, analytical performance depends on a number of interdependent system parameters (Table 5.1) that vary in a complex manner. Key among them are insertion sequence (e.g., speed and reproducibility of insertion and final insertion position), applied forward power, cup geometry and composition and, to a lesser extent, gas flow rates and observation height [8].

For most studies on operating conditions, liquid standards and DSI-ICP-AES systems were used. But, direct and quantitative inter-comparisons may not necessarily be meaningful because different DSIs, ICPs and spectrometers were utilized for such studies, thus making the magnitude of these effects system-dependent [11–13,17]. To simplify

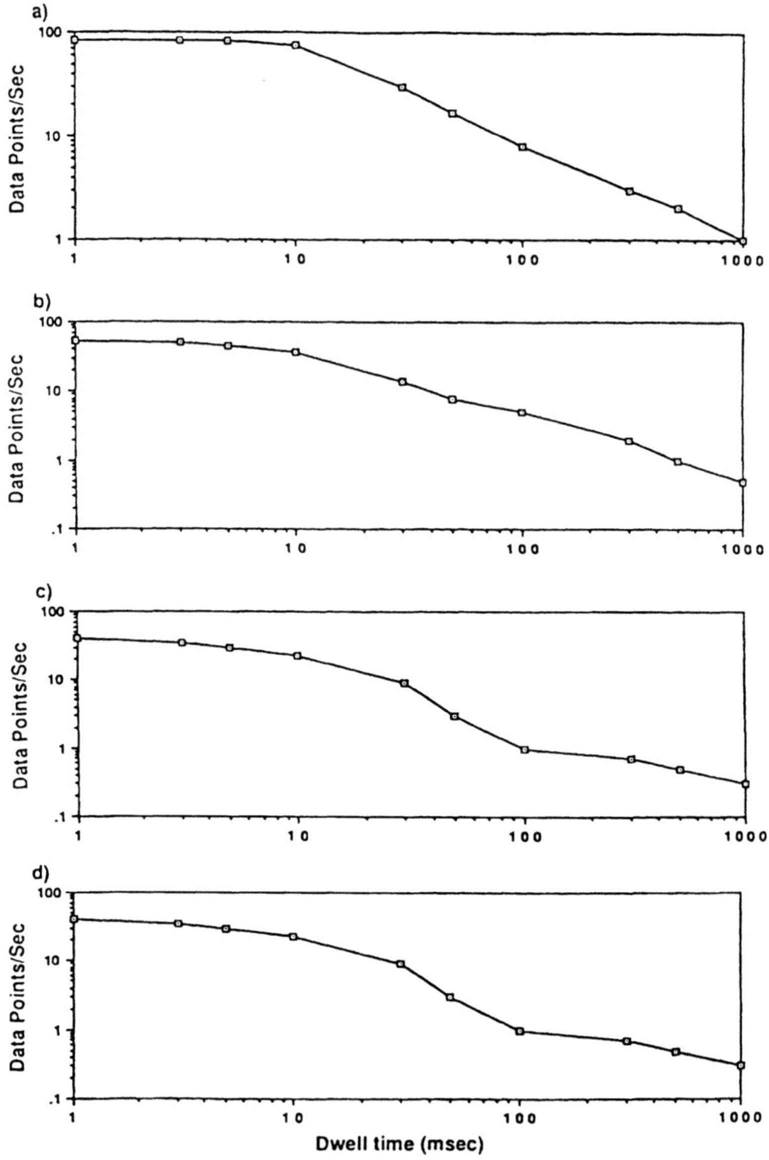

Fig. 5.6. Number of data points per second per number of masses scanned. (a) One mass scanned; (b) two masses scanned; (c) three masses scanned and (d) four masses scanned. Reproduced from Ref. [15] with permission.

discussion, to facilitate comparisons and to establish the extent of the effect of these parameters on analyte signals, the results presented in this chapter were obtained using the automated DSI-ICP-AES system described elsewhere [11] and µl volumes of diluted standard stock solutions.

5.3.1 Insertion sequence

The insertion sequence encompasses speed of insertion, reproducibility and final insertion position. Reproducible insertions are essential for reproducible analytical signals; for example, it has been reported that peak intensities may vary as much as 10% for a 1-mm change in insertion position [13]. Such insertions can be obtained using a stepper-motor drive. Rapid insertions are also essential because they produce sharp analyte signals which are well resolved from the change in background (due to cup insertion into the plasma; an example is shown in Fig. 5.7). As a consequence, rapid insertions provide improved analytical performance. Although the drop in background is not as

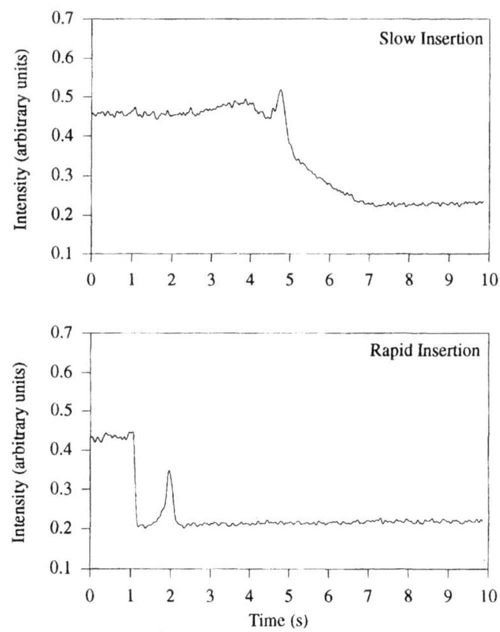

Fig. 5.7. Effect of insertion speed on analyte emission temporal behaviour (Cd, 100 pg).

noticeable in DSI-ICP-MS as it is in the example of Fig. 5.7, a slow insertion stretches out the transient signal. Commonly, in DSI-ICP sharper signals translate to improved detection limits.

Several strategies may be used to increase insertion speed. For instance, for the stepper-motor driven design of Fig. 5.3, because the probe is stopped close to the plasma for drying (and charring/ashing if necessary), the stepper-motor cannot attain its maximum speed over such a short distance. To further increase the speed of insertion, the following insertion sequence was applied. After drying and/or charring, the DSI was driven to its 'home' position (Fig. 5.1). Subsequently, it was driven at maximum speed to its final insertion position which, for this work, was when the top of the graphite cup was in alignment with the top load coil (TLC). At maximum speed, it takes approximately 2 s to drive the DSI from 'home' to TLC and insertion from the bottom load coil (BLC) to TLC takes a fraction of a second. This insertion sequence provided sharp analyte signals that were well resolved from the change in background (Fig. 5.7) and the signals are similar to those reported for a fast, one-shot insertion with a pneumatically driven DSI [17]. Although different final insertion positions have been used (e.g., –3 to +8 mm above TLC, Ref. [8]), for the last few years the top load coil has been the preferred final insertion position.

At the end of an analytical run, the graphite cup was stopped at about 14 cm from TLC for about 20 s to cool. This has been found to extend the useful life of the cup, otherwise, the hot cup gets oxidized by air and its surface degrades after several runs, thus necessitating frequent cup replacement. At the end of the cooling period, the DSI was retracted to its home position.

5.3.2 Sample probe geometry, composition and physical condition

Probe geometry and composition have been reported to exert a significant influence on analyte emission temporal behaviour [11,13,17,20, 33]. Although different types of sample-carrying probes have been used (Fig. 5.5), a key current trend is toward graphite cups of different geometries. Therefore, the following discussion will be restricted to such cups. Examples of the effect of cup geometry on analyte signals are shown in Fig. 5.8. From these results and a study with a CCD camera [42] it can be concluded that by lowering thermal mass, by better confining analytes in the centre of the plasma (e.g., using narrow

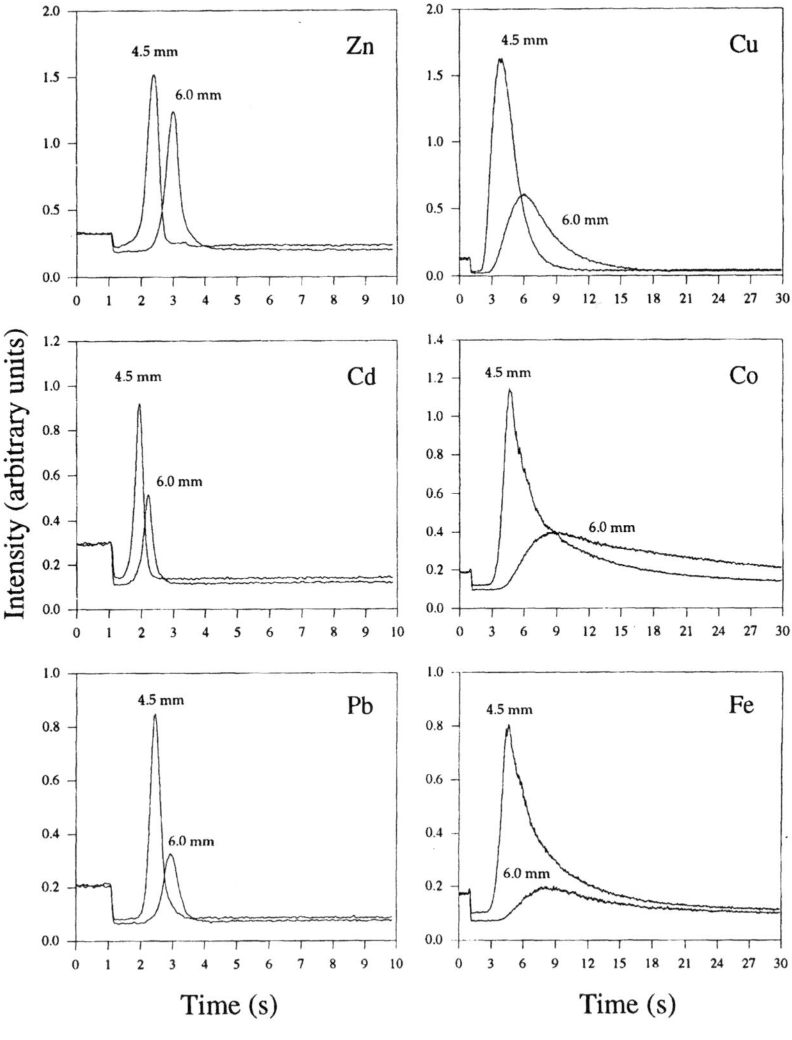

Fig. 5.8. Effect of graphite cup diameter on analyte emission temporal behaviour.

and thin wall cups) and by restricting heat conduction (e.g., using thin stems and long undercut graphite cups, Fig. 5.1), faster heating rates and higher final temperatures are reached. Thus sharper and more intense peaks are observed when such 'smaller' cups are used.

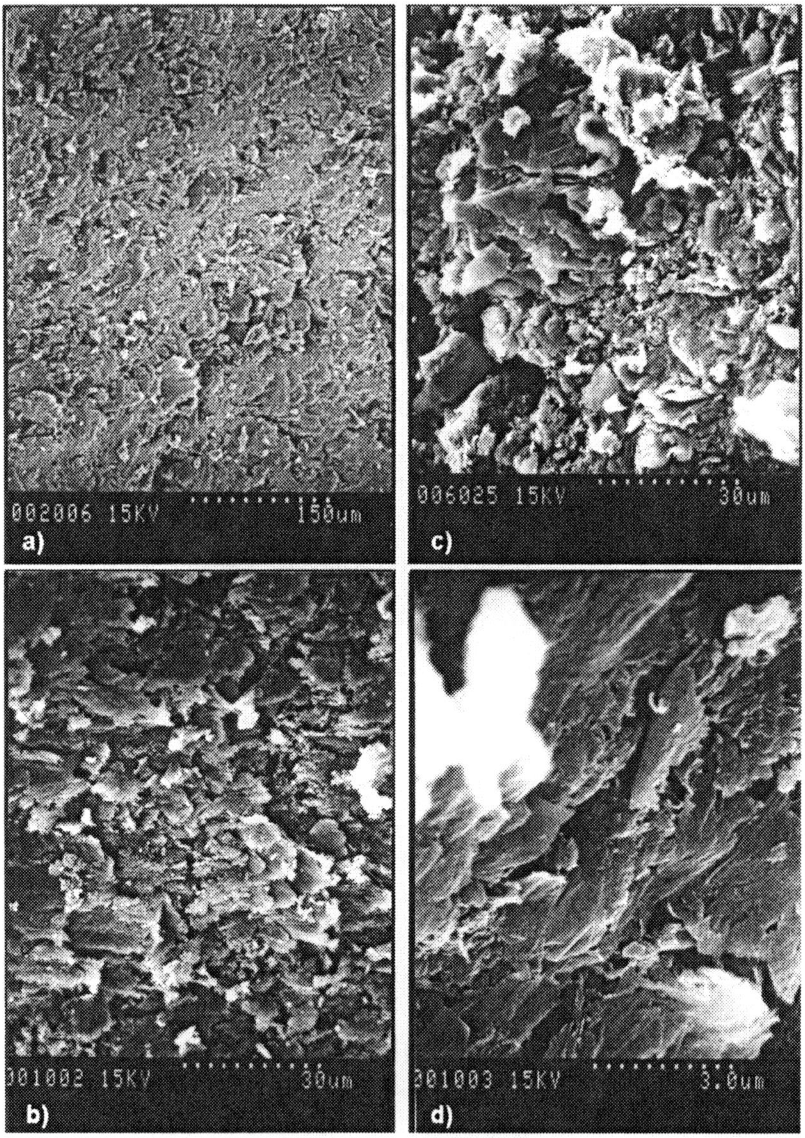

Fig. 5.9. Scanning electron micrographs (SEMs) of a graphite cup: (a) and (b) prior to insertion into a plasma; (c) and (d) after 10 insertions. Surface quality degrades and 'channels and crevices' are created (a) and (b) versus (c) and (d) and analytes entering these channels take longer to vaporize thus resulting broader peaks.

The physical condition of graphite cups also affects analyte signals, most notably peak shapes and reproducibility. For example, with clean water samples, as the number of insertions increases, the inner surface of the cup which is in contact with the sample becomes porous. An example is shown in Fig. 5.9. Thus, a sample may migrate into the graphite crevices and crevasses, hence causing peak broadening and signal degradation (e.g., poorer precision, in particular, when analytical determinations from peak height are attempted). This then raises the question, how many times may graphite cups be used before discarding them? The useful lifetime of graphite cups is a function of plasma power, plasma composition (e.g., $Ar–O_2$), of elements present in a sample (e.g., volatile or carbide-forming) and, of the sample matrix. In addition to signal degradation and, depending on concentration, carbide-forming elements may cause memory effects. Also, when mg amounts of solid samples are used, and depending on the volatility of the matrix, a residue may remain in the cup even after prolonged cup insertion into the plasma. We observed, for example, emission from the matrix even after a 3 min insertion into an ICP (typical insertions are 30 s). In both cases mentioned above, graphite cups must be discarded after a single use. Since graphite cups are relatively inexpensive, discarding them adds very little to the overall cost of an analysis. The useful lifetime of graphite cups also depends on plasma composition. For instance, with an $Ar–O_2$ (4%) mixed-gas plasma, graphite cups disintegrated after about five insertions (typically, 10–30 s each) and, with a 20% $Ar–O_2$ mixed-gas plasma graphite cups were totally consumed [22]. For completeness, with an all-Ar ICP or other mixed-gas plasmas ($Ar–H_2$, $Ar–N_2$, Ar–He), graphite cups used with non-carbide-forming elements and clean samples were re-usable for over 20 times before a drop in precision was noticed.

5.3.3 Sampling depth and gas flow rates

In DSI-ICP-MS, sampling depth (i.e., the distance between TLC and the sampler cone, Fig. 5.1) has been reported to be a more critical parameter defining peak height intensities than viewing height is in DSI-ICP-AES. For example, using a wire-loop DSI-ICP-MS system, Hall et al. [66] and Boomer et al. [30] reported that signal intensities go through a maximum.

Brief studies using DSI-ICP-AES indicate that outer, intermediate and central-tube gas flows have little influence on signal-to-background ratios (SBRs) [8]. Overall, gas flows for DSI-ICP-AES and DSI-ICP-MS are typically set to values commonly used for pneumatic nebulization sample introduction or they are dictated by reliable operation of the ICP. Central-tube gas flow, although it has been reported to exert very little influence on analyte emission temporal behaviour in DSI-ICP-AES, has been reported to affect drying and charring conditions [45,46,50]. For example, depending on the DSI design and coupling mechanism (Fig. 5.2), a small flow-rate prevents solvent condensation in the central tube of the torch. Following cup insertion into the plasma and upon retraction, the hot-cup vaporizes the condensate thus extinguishing the plasma. Relatively high flows tend to cool the cup, thus extending drying and charring time.

5.3.4 Drying and charring (or ashing)

Samples must be completely dried, otherwise the plasma may be extinguished. Samples may be dried externally or *in situ*. In recent literature, *in situ* drying is preferred. In this mode of operation, a sample-carrying cup is positioned below the BLC (Fig. 5.1) for 20–60 s. The actual time and distance from BLC (Fig. 5.1) depends on the DSI coupling mechanism (Fig. 5.2), on central-tube gas flow rate (if any) and on applied forward power. Often, drying is accompanied by an increase in reflected power and a change in plasma dimensions, both of which return to 'normal' when drying is complete. Reflected power has been used to mark the end of a drying step. Prolonged drying affects emission signal profiles but had no effect on integrated peak intensities when using a wire-loop DSI-ICP-AES system [33].

Typically, charring takes place *in situ* by stopping the sample-carrying probe closer to the plasma (Fig. 5.1) for a time period ranging between 30 s to 5 min. Similar to drying, the actual time and distance below the BLC depends on the DSI coupling mechanism (Fig. 5.2), on central-tube gas flow rate (if any), on applied forward power and on sample type. Charring is the preferred term in this chapter because ashing implies combustion which requires oxygen and, in most cases, samples are charred *in situ* in an inert atmosphere. Charring is necessary if the sample contains organic material, otherwise the plasma may be extinguished.

Samples in graphite cups have also been dried externally using ovens, IR lamps and hot plates, to name but a few. External drying allows batch operation and it reduces instrument time required per sample. Due to design constraints, external drying (e.g., using an IR lamp) has been used in DSI-ICP-MS [14]. External or *in situ* drying and charring also affects analyte signal behaviour. For example, Chan and Horlick [21] observed differences in peak shapes for samples dried and charred *in situ* versus those dried externally. These differences were attributed to differences in the chemical transformations of the sample that took place in the cup as the temperature was raised, and to differences in drying and charring rates.

5.3.5 Optimization

Although optimization strategies have not been employed in DSI-ICP-MS, a few comments will be made regarding alignment procedures and optimization of plasma and mass spectrometer parameters (e.g., ion lens voltages). Alignment of the centre of the graphite cup with the sampler cone was carried out using a pointing device inserted into a graphite cup [14]. To avoid tuning the MS part of ICP-MS by running solutions with the DSI (undoubtedly a lengthy process), it has been suggested that most of the optimization of the mass spectrometer may be done using solution nebulization sample introduction [39].

5.3.6 Plasma power and cup temperature

A set of experiments was carried out to test the effect of applied forward power on analyte signals; an example is shown in Fig. 5.10. Similar results have been obtained for other elements, as shown in Fig. 5.11. From these results it can be concluded that signal intensity increases and appearance times become shorter with an increase in applied power [18]. Although the increase in signal intensity varied from element to element [6,8,18] and from system to system, larger improvements were observed for medium volatility elements (e.g., Fe, Ni, Fig. 5.11). Since the 2.0 kW power provided the highest signal intensities and the best signal-to-background ratios, it was used to acquire all DSI-ICP-AES data included in the following sections. One would expect similar trends for DSI-ICP-MS as well.

Signal improvement at higher power levels can be attributed to the effect of plasma power on a spectral line and to an increase in the rate of volatilization. The heating rate and final (or equilibrium) temperature of a graphite cup in the plasma have been found to depend on cup geometry, cup composition and plasma power. The temperature of the

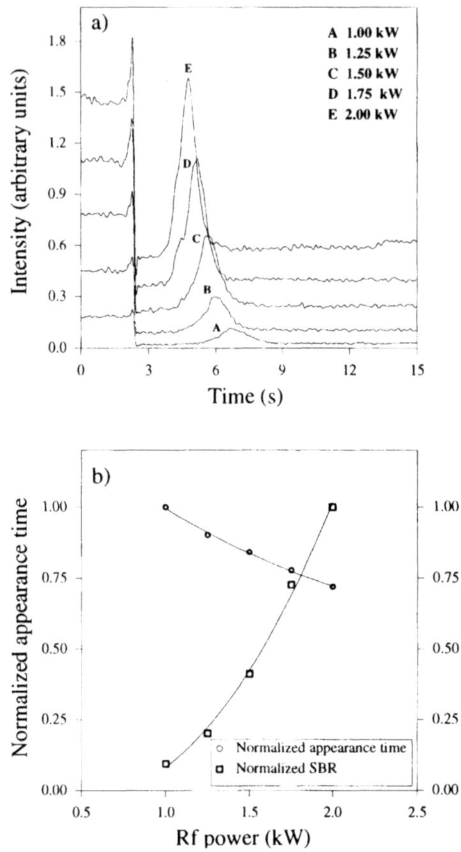

Fig. 5.10. Effect of applied forward power on analyte emission temporal behaviour (a) for Pb for different power setting and (b) appearance time (arbitrarily defined as the time at which the maximum of the peak appears) and signal-to-background ratio as a function of power.

external surface of a 4.5 mm long undercut graphite cup (Fig. 5.1) in the plasma has been measured as a function of applied forward power using an optical pyrometer, and the results are shown in Fig. 5.12. Measured values reported in the literature are between 1600–2200°C [63,51]. Cup temperatures in a plasma have been estimated by placing high melting point metals into a cup and by inserting the sample-carrying cup into the plasma. When significant emission was observed, the temperature of the cup was assumed to be the same as the melting

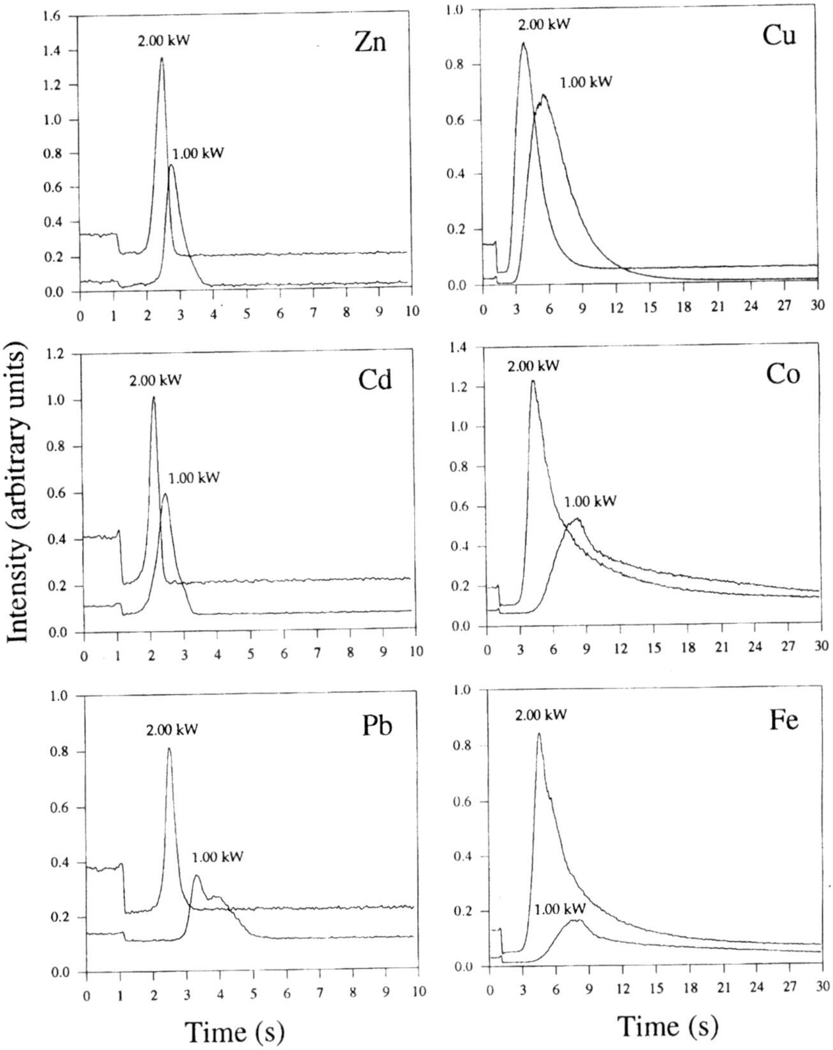

Fig. 5.11. Effect of applied forward power on analyte emission temporal behaviour for some high- and medium-volatility elements.

point of the material in it. With one exception in which a temperature of 1800°C was reported [59], such estimates ranged between 3000 and 3600°C [24,46,52,55,60,64]. Curiously, these results appear to be

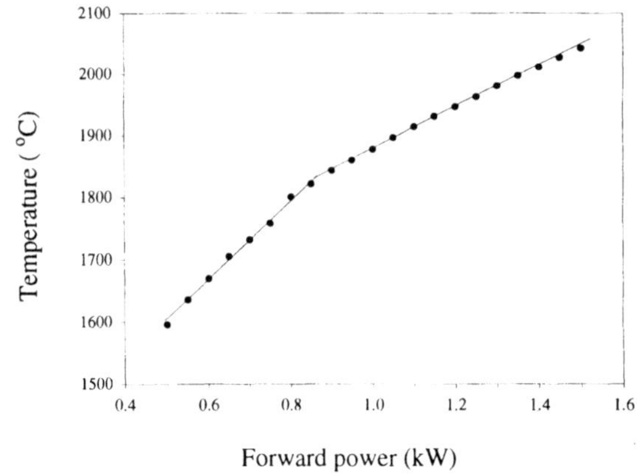

Fig. 5.12. Effect of applied forward power on graphite cup temperature (as measured using an optical pyrometer).

biased toward higher temperatures indicating that, in these types of experiments and in addition to volatilization, perhaps another mechanism may be operative.

Even at 2.0 kW, a signal was not observed for Ti, V and W; essentially analyte signals were indistinguishable from those obtained from blanks. These results indicate that these elements remain 'tied-up' in the cup, most likely in the form of a refractory carbide. The lack of analyte signals can be explained by considering the low temperatures of the cup in the plasma and the high vaporization temperature of Ti, V and W carbides (Table 5.2).

Based on the results shown in Fig. 5.11 and those reported elsewhere [18], it can be concluded that, similar to dc arc, analytes 'boil off' the cup and that those with the lowest boiling point will vaporize first, followed by analytes with a higher boiling point (sometimes this description is referred to as the frying pan theory). Thus, elements can be roughly ranked according to their volatility as highly volatile (e.g., Pb, Cd and Zn), medium- to low-volatility (e.g., Cu, Co, Fe) and as carbide-forming (e.g., Ti, V and W). By comparison, a different volatility sequence has been reported [13]. Moreover, a correlation between the boiling point of the element of interest and its appearance time could not be established. These observations suggest that, much like the dc arc, the volatility sequence does not depend on elemental volatilities alone and that a complex mechanism must be operative (and, the matrix may play a significant role as well).

Returning to the volatility sequence, can volatilization rates be improved, in particular, for low-volatility and carbide-forming elements? One way of increasing the rate of volatilization is by using higher power plasmas. Although DSIs have been used with 3–5 kW plasmas, since most commercial radio frequency (rf) generators typically deliver a maximum power of about 2.0 kW, wide applicability of this approach is limited. Alternatively, volatilization may be enhanced by modifying the properties of the plasma (e.g., heat capacity, thermal conductivity) using mixed gases. Although mixed-gas DSI-ICP data discussed in the next section were acquired using an AES system, the general conclusions drawn on volatilization enhancement are directly transferable to graphite-cup DSI-ICP-MS. It would be interesting, however, to see work documenting the actual magnitude of enhancements observed in mixed-gas DSI-ICP-MS.

5.4 MIXED-GAS DSI-ICP-AES

Enhanced volatilization may be obtained without further increasing rf power, but by modifying the bulk properties of the plasma (e.g., thermal conductivity, heat capacity, electrical conductivity), by altering fundamental parameters of the discharge (e.g., electron concentrations, gas kinetic temperatures), by improving energy transfer to the sample-carrying cup and to the vaporized analyte (e.g., due to higher thermal conductivity) and by making the chemical environment of the plasma oxidizing, reducing or halogenating using, for instance, Ar–He, Ar–N_2, Ar–H_2, Ar–O_2 or Ar–halocarbon mixed-gas ICPs. Depending on which gas is chosen, the physical and/or chemical properties of the plasma may be modified, thus providing the capability to control and alter, on demand, the environment a sample is experiencing. As a consequence, the plasma may be tailored to the application.

The use of mixed-gas ICPs with pneumatic nebulization sample introduction is not new and a comprehensive paper summarizing the literature and providing new insights into their effects has been published [70]. In addition, mixed gases have been found to augment the analytical performance of DSI-ICP-AES using Ar–O_2 [12,22], Ar–H_2 [12], Ar–air [22] mixed-gas plasmas. Also, Ar–N_2 [12,22] mixtures have been briefly tested for a small range of elements and Ar–N_2 [2,53,54] and Ar–O_2 [12,22] mixtures were used for elemental determinations in a variety of materials.

In this section, the effect of $Ar–H_2$, $Ar–O_2$, $Ar–N_2$ and Ar–He mixed-gas DSI-ICP-AES on analyte emission will be presented [71] and it will be systematically evaluated using representative examples of high-volatility, medium-to-low volatility and carbide-forming elements.

5.4.1 Experimental observations on mixed-gas DSI-ICP-AES

Mixed-gas plasmas are typically generated by first lighting an all-Ar plasma and by slowly mixing the foreign gas with the outer tube-gas flow (coolant). Upon introduction of the foreign gas, the plasma becomes brighter, thinner and develops an elongated tail, essentially becoming a 'pencil-thin' discharge. Reflected power also increases, thus necessitating a manual adjustment of the variable tuning capacitor in the matching network to bring it below 5 W [9,10]. Although significant spatial changes occur in the plasma upon introduction of a foreign gas and the magnitude of these changes is foreign-gas dependent, all data reported here were taken at a fixed observation height which was identical to that used for an all-Ar ICP and consequently, it may be sub-optimal for mixed-gas work. However, use of such compromise conditions preserved the simultaneous multi-element capability of the technique.

The amount of foreign gas added to the plasma was not optimized, the maximum percent mixed-gas composition that can be used with the automated DSI-ICP-AES system described in a previous section was determined by increasing the amount of foreign gas mixed with Ar until the plasma became unstable and was extinguished upon cup insertion. The maximum percentage of foreign gas that resulted in stable, continuous and reliable operation of the current configuration of the automated DSI-ICP-AES is listed in Table 5.3. The use of larger

TABLE 5.3

Mixed-gas composition

Mixed-gas	Composition (%)
$Ar–H_2$	6
$Ar–N_2$	5
$Ar–O_2$	4
Ar–He	15

percentages of foreign gases with DSI-ICP, for example, 20% O_2, has been reported [22] to require the addition of a 150 µF capacitor to the variable tuning capacitor in the impedance matching network. This was not done in work reported in this chapter in order to maintain the capability to operate the DSI-ICP-AES system [11] automatically, and to provide the ability to switch between different foreign gases relatively quickly.

5.4.2 Survey of analyte emission with mixed gases and µl volumes

To get an indication of the signals obtained by DSI-ICP-AES and mixed gases, the elements tested were grouped in three broad categories ranked according to volatility, and representative examples of high, medium and low volatility and carbide-forming elements were tested using Ar–He, Ar–N_2, Ar–H_2 and Ar–O_2 mixed-gases and 10 µl volumes of standard solutions [9,10,71,72]. Also, analyte emission signals were compared with those obtained using an all-Ar ICP, as briefly described in the next section.

The effect of an all-Ar and of Ar–He, Ar–N_2, Ar–H_2 and Ar–O_2 mixed-gas plasma on analyte emission signals for high-volatility elements, such as, Zn, Cd and Pb is shown in Fig. 5.13. As can be seen, analyte signals obtained using mixed-gas plasmas were narrower and more intense (by as much as five times in some instances, as compared to those obtained using an all-Ar ICP) and, with the exception of Ar–N_2, they showed shorter appearance times (arbitrarily defined as the time at which the peak maximum appears) than the all-Ar ICP. The shortest appearance time was for Cd and Ar–H_2 mixed-gas plasma, with the analyte signal appearing next to the rapid change in the background. This is analogous to what is observed when the cup is inserted slowly into the plasma (Fig. 5.7). Interestingly, the appearance time for Zn and the Ar–N_2 plasma was longer than that obtained using an all-Ar ICP.

Figure 5.14 shows analyte signals for medium- to low-volatility elements, such as Cu, Co and Fe using all-Ar and mixed-gas plasmas. Shorter appearance times and narrower and sharper analyte signals were observed with the mixed-gas plasmas compared to those obtained with the all-Ar plasma and the maximum improvement in peak heights was about seven times.

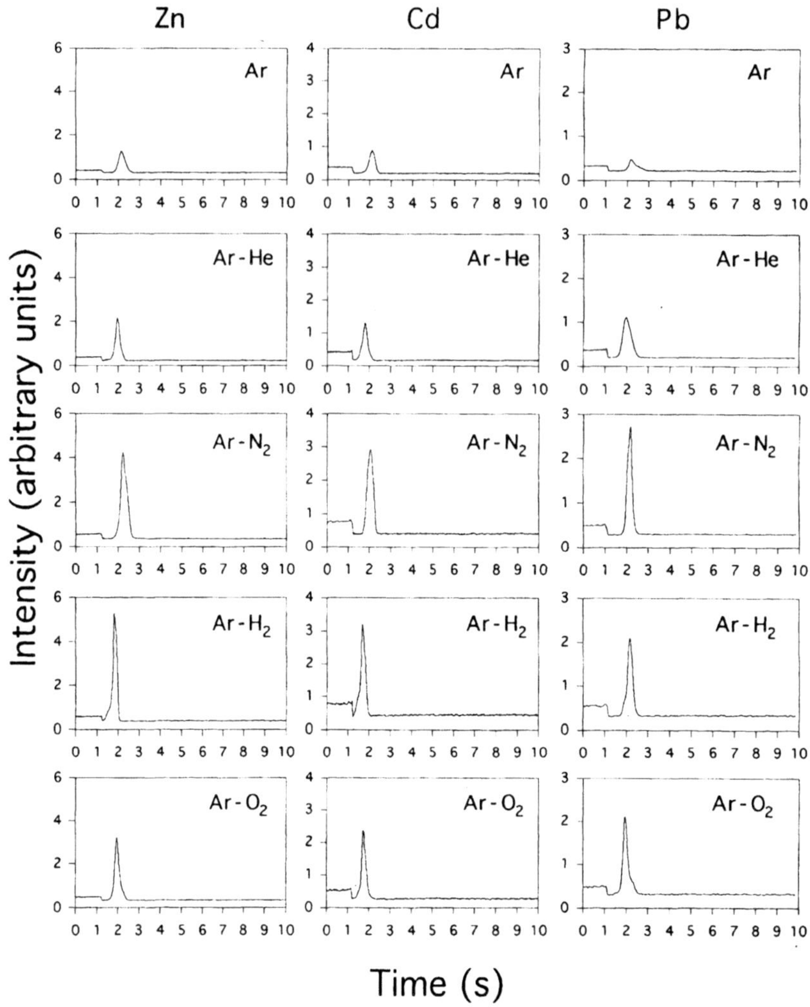

Fig. 5.13. Analyte emission temporal behaviour for Zn (1 ng), Cd (1 ng) and Pb (10 ng) using all-Ar and Ar–He, Ar–N_2, Ar–H_2 and Ar–O_2 mixed-gas DSI-ICP-AES.

As shown in Fig. 5.15, analyte signals were not observed for Ti and V when using an all-Ar ICP. Essentially, these elements have remained 'tied-up' in the cup. Similar results were obtained for W. Broad, low intensity signals were observed (Fig. 5.15) when using Ar–H_2, Ar–He and Ar–N_2 (not shown for clarity) plasmas, thus

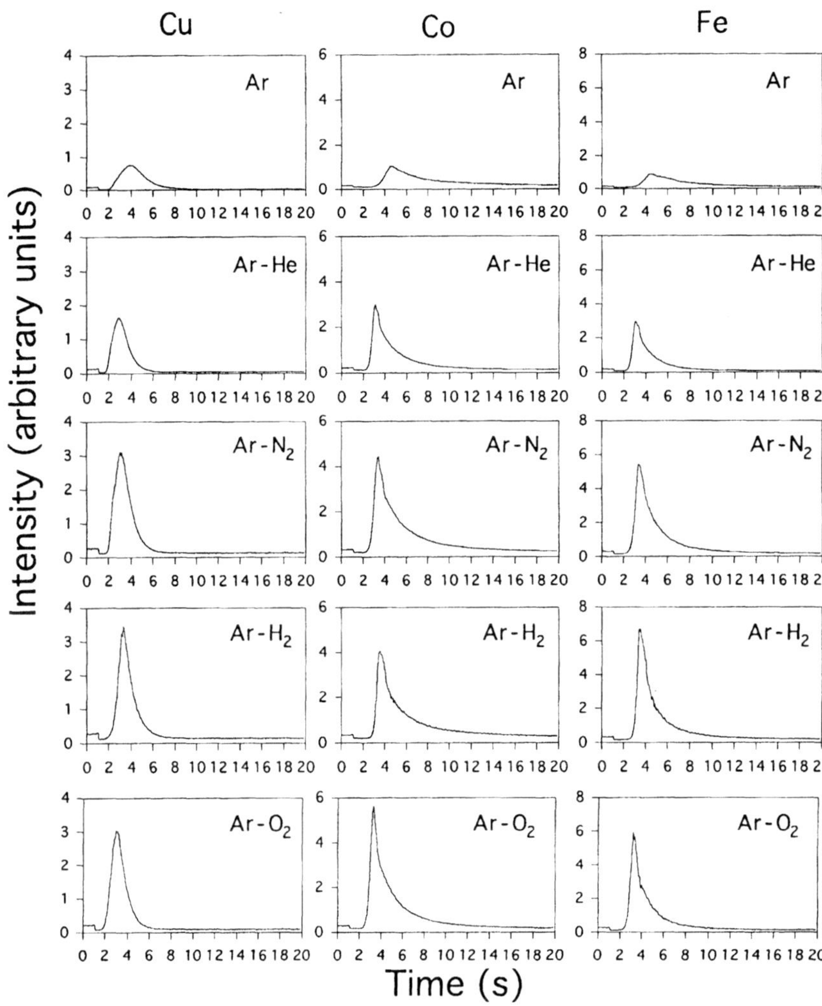

Fig. 5.14. Analyte emission temporal behaviour for Cu (30 ng), Co (1 μg) and Fe (1 μg) using all-Ar and Ar–He, Ar–N_2, Ar–H_2 and Ar–O_2 mixed-gas DSI-ICP-AES.

indicating incomplete volatilization with these gas mixtures. Although significant improvements in the sensitivities of these elements were observed for an Ar–O_2 ICP, the signals did not return to their post insertion background levels after a typical 30 s insertion. An indication of the amount of analyte remaining in the cup was obtained by

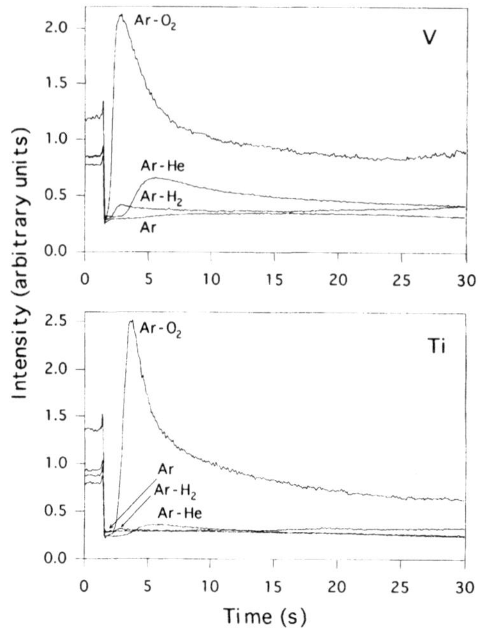

Fig. 5.15. Analyte emission temporal behaviour for V (100 ng) and Ti (100 ng) using all-Ar and Ar–He, Ar–H_2 and Ar–O_2 mixed-gas DSI-ICP-AES.

inserting a cup containing 10 ng of Ti into an Ar–O_2 plasma for 30 s and by re-inserting the same cup into an all-Ar ICP after addition of 10 ml of 0.25 M NaF, following a procedure described previously [19]. As shown in Fig. 5.16, a relatively intense signal was observed, thus indicating that, even after a 30 s insertion into a 4% Ar–O_2 plasma, some Ti remained in the cup. Alternatively, quantitative volatilization may be obtained by chemically modifying the environment of a sample.

5.5 CHEMICAL MODIFICATION

Chemical modification can be defined as the addition of a reagent (e.g., the modifier) to the sample or to the plasma in order to convert the matrix to a phase of higher thermostability (e.g., to delay or altogether suppress matrix volatilization) or, as is typical of DSIs, to increase the volatility of the analytes. Chemical modifiers are usually chlorinating

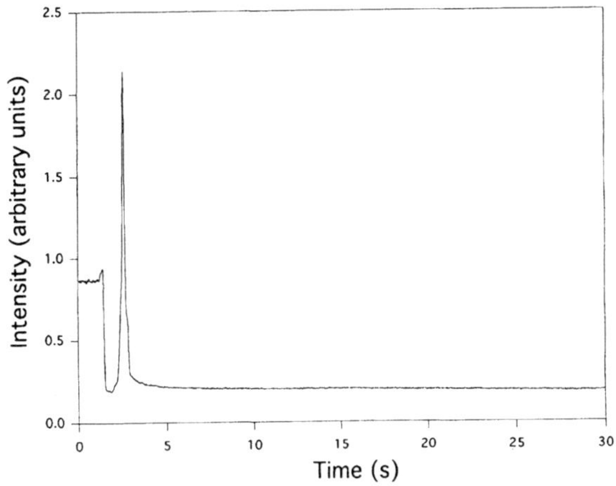

Fig. 5.16. Analyte emission temporal behaviour for Ti (10 ng) from an all-Ar ICP following chemical modification using of 10 μl of 0.25 M NaF [19].

or fluorinating reagents that improve volatilization of carbide-forming elements due to preferential formation of the corresponding high volatility halide. In general, chemical modification and graphite cup DSI-ICP appear to be tightly bound. A detailed summary of chemical modifiers used in DSI-ICP can be found in Table 5.5 of Ref. [6] and in Ref. [8].

Briefly, for solution sample residues, aliquots of an aqueous solution of a halide salt, such as AgCl, NaCl, KCl, $BaCl_2$, NH_4Cl, NaF, KF, BaF_2 are added to solution sample residues in a sample-carrying cup. For powders or solids that can be converted to a powder, the chemical modifier is mixed with the powder. For example [13,19,54], AgCl was used for the determination of trace impurities in U_3O_8, AgCl or $BaCl_2$ or NH_4Cl or PTFE for volatilization enhancement from powdered Al_2O_3 samples [53], AgCl for the determination of refractory elements in sediments [37], $BaCl_2$ or NH_4Cl or NaCl or NaF plus PTFE (poly-tetra-fluoro-ethylene) for determinations from powdered ceramic samples [52], and AgCl or LiF or SrF_2 for volatilization enhancement of lanthanide elements extracted from U or Pu [61]. In all cases, volatilization enhancement depended on sample type, thus indicating that the chemical modification reagent used must be tailored to the application.

Alternatively, chemical modification can be obtained by mixing Freon gases (e.g., CCl_2F_2, $C_2Cl_2F_4$) with Ar and by introducing this

Ar–halocarbon mixture to the plasma through the central tube of the torch. This approach has been briefly tested using flat-top graphite rods as sample-carrying probes [48]. More recently, Ar–Freon mixtures were delivered directly to the sample through a hollowed-out stem (Fig. 5.5) of a graphite cup [41]. This necessitated fabrication of new types of cups with a hole drilled right through them (the delicate and elaborate procedure is described in detail in Ref. [41]) and, the use of a 'plug' to prevent a sample from either plugging the hollowed-out stem or running through it. Overall, the approach complicated both gas delivery to the plasma and DSI probe design. It is worth noting that, at present, the discussion on Ar–Freon mixtures may be of historical interest to many readers because, for environmental reasons, production of Freons has been discontinued and their general use has been prohibited in many parts of the world.

To simplify modifier introduction to the plasma and using the reaction scheme shown in Fig. 5.17 as a guide [9–11,73,74], gaseous chemical modifiers were introduced through the intermediate gas-flow of the plasma torch rather than the central tube. Such *in situ* chemical modification via addition of a halogen-containing gaseous chemical modification reagent to the plasma was expected to force preferential formation of the corresponding high-volatility halide over the essentially non-volatile carbide and to offer several additional advantages. For example, since the modifier is added to the plasma rather than the sample, additional sample processing steps are not required and, as a result, analysis time is reduced. Also, the use of gaseous chemical modifiers eliminates stray light effects that may arise from high concentrations of the typical alkali–metal halide–salt chemical modifiers, reduces the possibility for contamination from the chemical modification reagents, and simplifies automation of chemical modification procedures, for instance by using a gas blender and a computer-controlled solenoid valve.

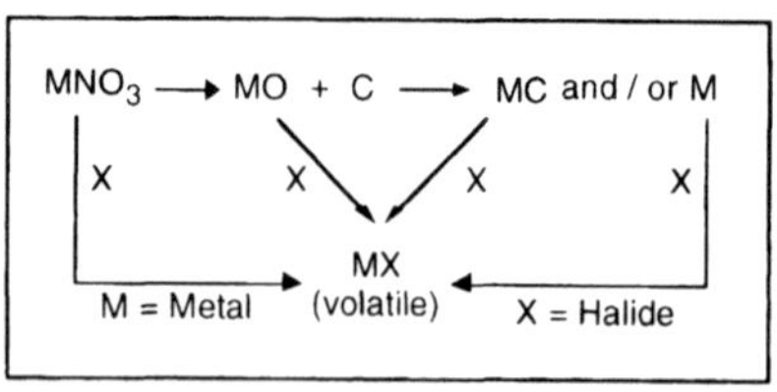

Fig. 5.17. Possible reaction pathways for volatile halide formation.

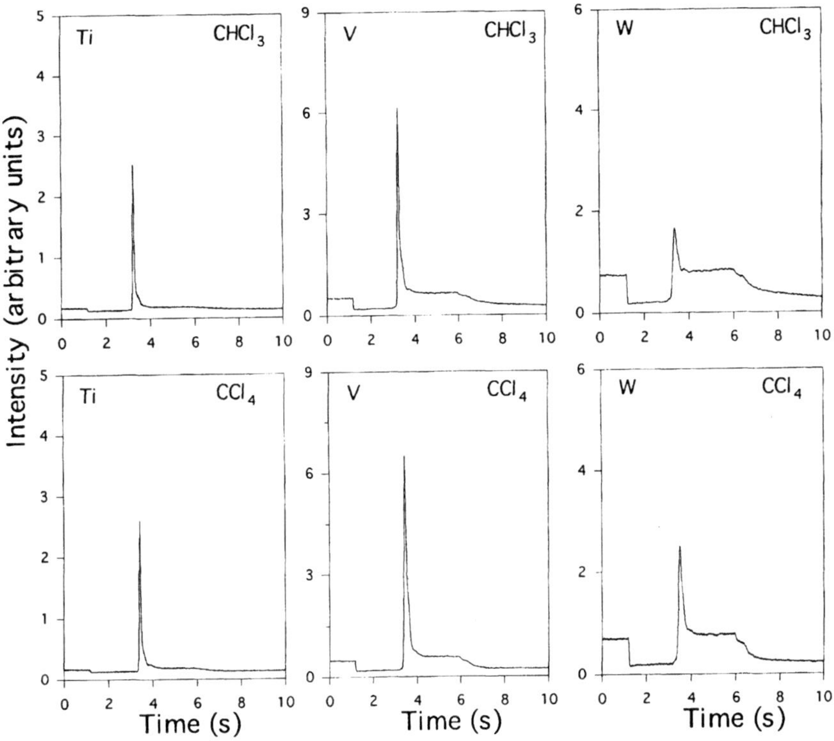

Fig. 5.18. Analyte emission temporal behaviour for Ti (10 ng), V (10 ng) and W (100 ng) using Ar–CCl_4 and Ar–$CHCl_3$ mixtures (see text for discussion).

To address the question, "would a chemical modifier introduced through the intermediate tube of the torch reach a dried solution residue at the bottom of a graphite cup in the plasma?", Ar was bubbled through a flask containing CCl_4 or $CHCl_3$ and the resultant Ar–halocarbon mixture was introduced into the ICP [9,10,73,74]. Examples are shown in Fig. 5.18. The increase in plasma background (for t between about 3.5 and 6 s) was attributed to several factors: among them, an increase in the auxiliary gas flow rate, an intense green and purple plasma emission and a slight axial contraction of the discharge. At about t = 6 s, Ar flow to the modifier-containing flask was terminated and plasma background began to decrease and, eventually, it reached the post-insertion background levels (e.g., between t = 1.5 and 3 s).

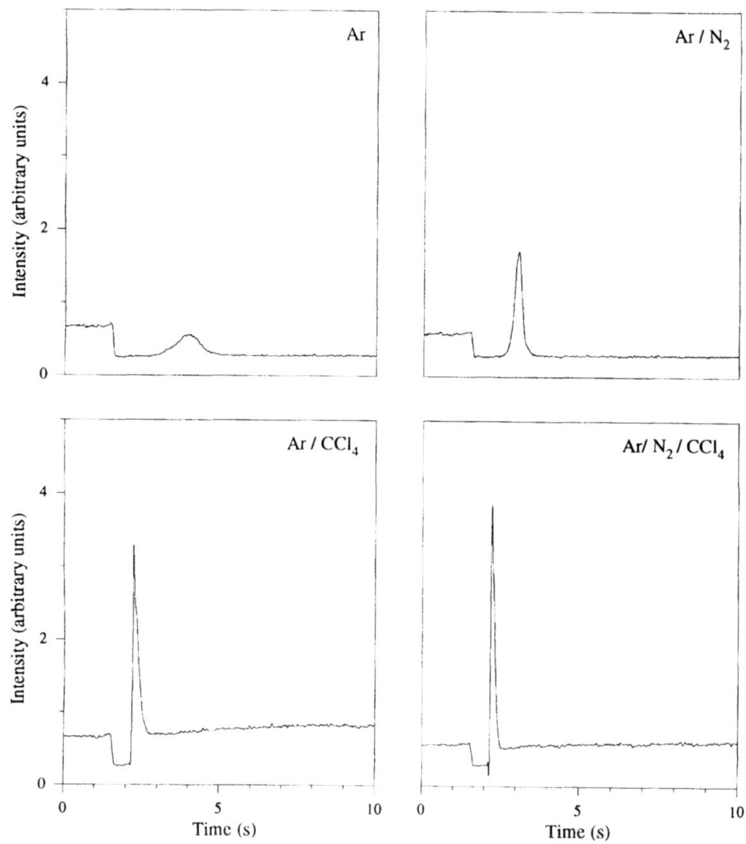

Fig. 5.19. Analyte emission temporal behaviour for 3 ng Pb and a ternary-gas mixed-gas plasma.

Despite its simplicity and the problems associated with carbon deposits on the torch, the approach proved the feasibility of *in situ* chemical modification (via the intermediate gas-tube). Further signal enhancements were obtained using ternary gas mixtures (Figs. 5.19 and 5.20) and initial results [10,11,73,74] indicate that in part, such mixtures facilitate energy transfer to the sample-carrying cup, thus further enhancing volatilization. Although successful, automation of ternary gas mixtures proved to be even more difficult than that required for the binary mixtures mentioned above.

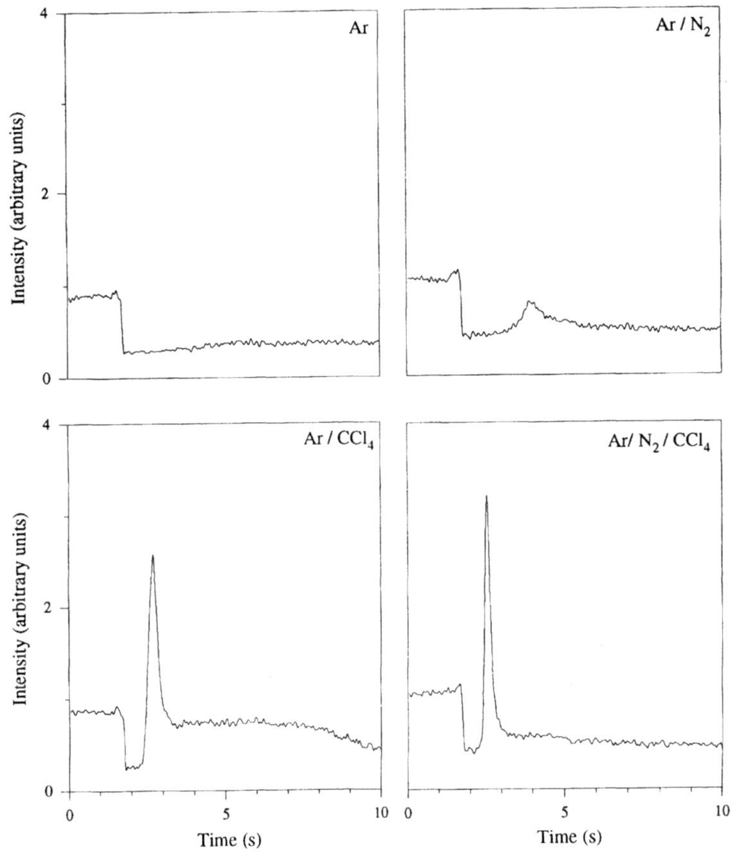

Fig. 5.20. Analyte emission temporal behaviour for 3 ng Co and a ternary-gas mixed-gas plasma.

To facilitate automation and to solve the problem of carbon deposits, SF_6 stored in gas cylinders and a computer-controlled solenoid valve were used. Sulphur hexafluoride was chosen because it was expected to eliminate the problems from carbon deposits and has a high degree of co-ordination of Fluorine. Also, sulphur has a relatively simple emission spectrum. However, when SF_6 was released into the ICP immediately prior to graphite cup insertion, the signal for W (WF_6 has a boiling point of 18°C) was on top of the changing background (due to graphite cup insertion) thus complicating background correction [11]. This problem was addressed by releasing the modifier after the

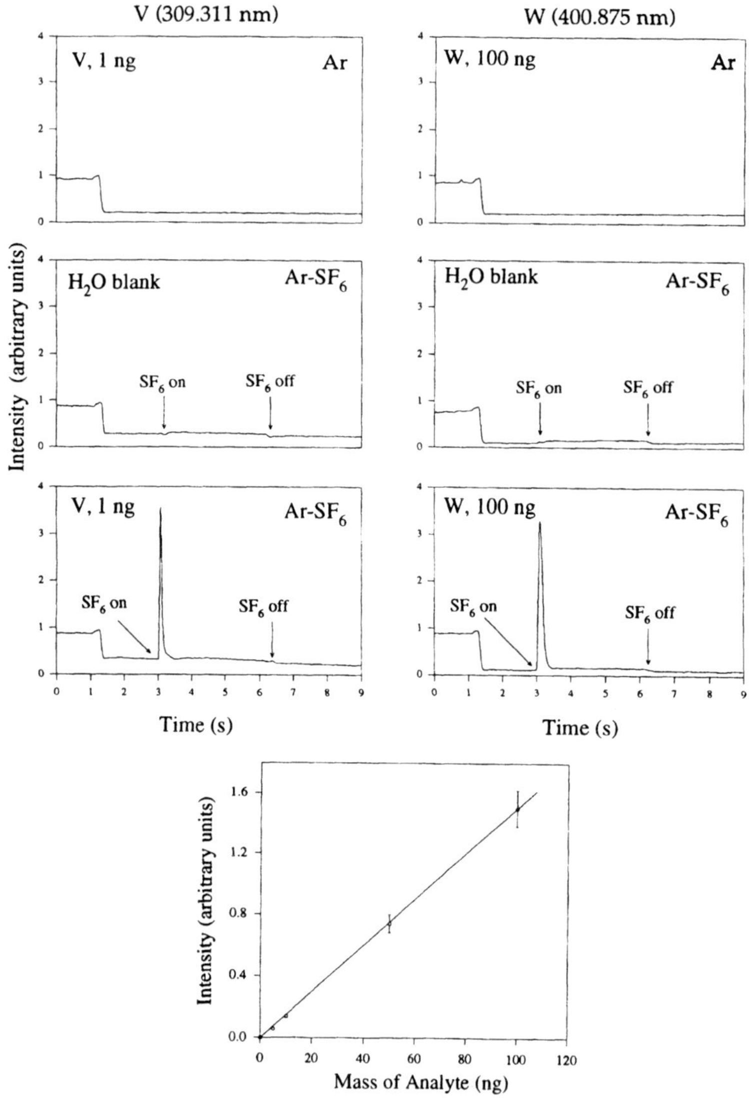

Fig. 5.21. Chemical modification using SF_6 and calibration curve for W (10 µl injections).

graphite cup was in the plasma for about 3 s (Fig. 5.21) using the computer-controlled solenoid valve [11].

5.6 FURTHER EXTENSION OF THE ANALYTICAL CAPABILITY OF THE ICP USING DSIs

In addition to the advantages mentioned in the introduction, the analytical capability of the ICP can be further extended by using DSIs. Some examples of innovative uses of DSIs to address some fundamental shortcomings of ICP spectrometry or to extend the analytical utility of the ICP are briefly described below.

5.6.1 Reductions of spectroscopic interference in ICP-MS

In ICP-MS, spectroscopic interference arises from overlaps of isobaric isotopes of elements; plasma background species; matrix induced background species; oxides; hydroxides; doubly charged species and overlaps with ArX^+ species [15,16]. Of the different types of spectroscopic interference, those due to background species are most prevalent. Such species arise from plasma gases, from air entrainment, from the sample and from the solvent (e.g., acidified water). Since DSI involves introduction of dried solution residues or powders/solids into a 'dry' plasma (i.e., one in which water is not being introduced), it was thought that solvent-induced background species should be significantly reduced or even eliminated. This is in stark contrast to pneumatic nebulization in which background spectral features of ICP-MS are dominated by solvent-induced species.

Background spectral features of a dry plasma were thoroughly investigated [15,16] and the results were compared to those obtained using a 'wet' plasma (i.e., by continuously aspirating distilled water using a pneumatic nebulizer). The results shown in Fig. 5.22 clearly demonstrate that the spectral background of ICP-MS was simplified considerably, that, in most instances, the magnitude of background spectral features was reduced and that, in some cases, spectral features were simply eliminated. Removal of background species enabled interference-free determinations for a number of isotopes in some cases, whereas in others, reduction of the magnitude of interfering species enabled determinations with substantially reduced interference. The remaining species containing oxygen, nitrogen and (some) carbon containing species which are still present despite the use of a dry plasma (Fig. 5.22) were thought to arise from air entrainment. This was confirmed by reducing sampling depth and by using an 'extended' torch (i.e., a torch fitted with a sleeve) [15].

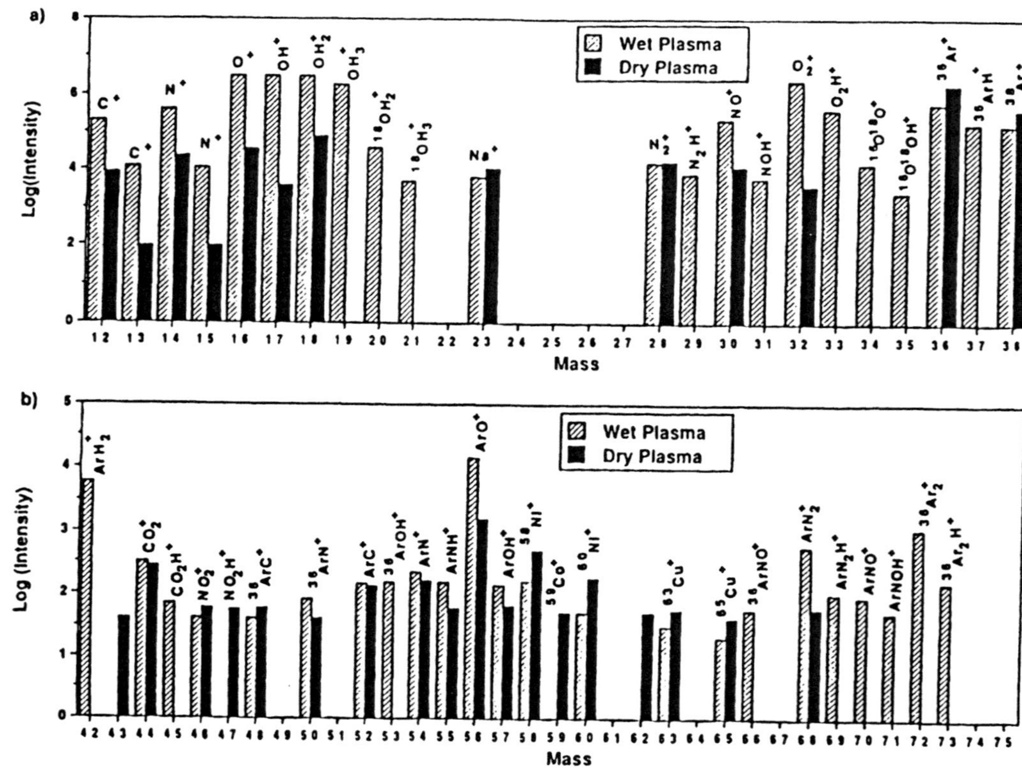

Fig. 5.22. Bar graph background spectra (logarithmic scale intensity axis) for the wet plasma (sampling depth 15 mm) and the dry plasma (extended torch) for (a) mass range 12–42 amu and (b) mass range 42–75 amu. Reproduced from Ref. [15] with permission.

However, when a DSI is used a foreign object (e.g., a probe) is inserted into the plasma. Therefore, a more realistic plasma background is obtained when the probe is inserted into the plasma. Such background spectra were also recorded [15]. For graphite cups, background spectra were nearly identical to those obtained using a dry plasma (as described above). Surprisingly, the intensities of $^{12}C^{+}$, $^{12}C^{16}O^{2+}$ and $^{40}Ar^{12}C^{+}$ were not much larger than those obtained with the dry plasma despite the presence of the graphite cup in the plasma. Thus, carbon deposits on the sample and skimmer cone and the ion optics were not expected to be a problem. In fact, it has been reported [30] that carbon deposits were not visible and that signal degradation was not observed when using graphite-diluted pelletized-solid standards and a wire-basket DSI, even after one week of operation. However, DSI-ICP-MS has not yet been tested with solution residues with high dissolved solids. Getting back to interference from probe material, Mo and Ta metal cups and W wire-loops added their respective ion or isotopic pattern to background spectra and, in addition, the isotope or isotopic pattern of impurities boiling off the metal cup (e.g., Sn in the Mo cup) [15]. Other than that, background spectra were 'clean' [15].

For the same plasma power, plasma background intensity (counts/s) was higher in the dry than the wet plasma and, $^{36}Ar^{+}$ and $^{38}Ar^{+}$ peaks were more intense indicating that, as expected, the plasma was hotter and brighter due to absence of water that cools the central channel. These results were acquired using an older Elan 250 ICP-MS system and it would be interesting to compare them with results obtained using modern ICP-MS systems with improved photon rejection capabilities.

Some examples of the successful reduction or elimination of spectral interference effects from oxides, hydroxides and poly-atomics are briefly discussed next [15]. Barium was chosen because its isotopes and combinations of associated species can potentially interfere with 29 *m/z*. Cerium was chosen because it is a good oxide former in the ICP and it is representative of rare earths. In both cases, oxide and hydroxide species were reduced or eliminated using DSIs. Another example involves an interference arising from solvent-induced polyatomic species and the effect of $^{35}Cl^{16}O^{+}$ on the nearly mono-isotopic $^{51}V^{+}$ is a classic one. Such interference can be eliminated by eliminating either chlorine (from the solvent) or oxygen (from the sample and the plasma). Due to use of a dried solution residue and a dry plasma, DSIs enabled interference-free determinations of V from chlorine-containing solutions. Overall, the simple idea that a reduction in the water content of the plasma would also reduce ArO^+, $ArOH^+$, MO^+ and MOH^+ has been verified. These conclusions transcend DSIs and have been useful with other sample introduction systems that use dry plasmas, such as, electrothermal vaporization or laser ablation.

5.6.2 Reductions of non-spectroscopic interference in ICP-MS

Non-spectroscopic inter-element interference effects can be thought of as analyte or matrix induced signal changes or, simply, as matrix effects [16]. Some of the most serious matrix effects occur when the determination of a low concentration of a light element (e.g., Zn) is attempted in the presence of a high concentration of a heavy element (e.g., U). Because Zn is a highly volatile element and U is a strong carbide-forming element, Zn was quantitatively volatilized from a graphite cup inserted into the plasma whereas U remained quantitatively in the cup (most likely tied up as a carbide). Thus, matrix effects were separated in the time domain. Although it was only demonstrated for U, because a large number of Lanthanides and Actinides are strong carbide formers, the approach is not expected to be an isolated one.

5.6.3 Pre-concentration and sample clean-up

Some liquid samples have complex matrices, for example, sea water, brines or mine tailings. For such samples, the matrix may have to be 'removed' to enable interference-free analytical determinations. In others, analytes are below the instrumental detection limit. In such cases, use of a different analytical technique or pre-concentration may be the only options. As briefly discussed below, DSIs offer unique possibilities for sample pre-concentration and clean-up.

Approaches to pre-concentration and clean-up using DSIs include solvent evaporation, electrodeposition, flow injection analysis (FIA) pre-concentration, chelation and pre-concentration on activated charcoal. The advantages of these methods have been recognized early in the development of DSIs. Pre-concentration [26,28] and sample clean-up (e.g., using artificial sea water) [47] using electrochemical methods were two of the first approaches used with DSI-ICP-AES. Evaporation methods, for example, by pipetting μl volumes onto a wire-loop, drying the sample and repeating the process several times was also tested [33]. Extraction and pre-concentration of analytes on cellulose and insertion of a graphite cup carrying analyte-loaded cellulose have also been described [53].

More recently, evaporation was re-visited, this time using a pneumatic nebulizer to deposit a sample aerosol on a graphite cup which, following a drying step, was inserted into an ICP of an AES system [38]. Using a 2 min deposition and sample volumes of 0.5 ml, concentration detection limits improved by a factor of 600 (Cu), 750 (Zn), 600 (Cd) and 570 (Pb) from clean standard solutions. It was expected that an increase in aerosol deposition time would improve concentration detection limits even further. However, since the matrix is not removed prior to deposition, the matrix is also pre-concentrated and this may cause problems if the matrix is complex. The aerosol deposition system described above was automated and tested using DSI-ICP-MS and the operational parameters affecting its analytical performance were established for eight elements [39]. The approach was also evaluated for samples in an ethanol matrix. A two minute deposition at 0.25 ml/min was used and detection limits improved by about two orders of magnitude. It was concluded that the system showed considerable promise for determinations from pristine waters or non-aqueous solvents.

To remove the matrix and pre-concentrate the sample, Salin and co-workers used Chelex-100 based FIA [44] and collected eluted fractions into a graphite cup which (following *in situ* or external drying) was inserted into the ICP of an AES system. For FIA, a 5 ml injection volume was used and FIA processing time was 6 min. Their detection limits turned out to be blank limited and improvements for what the authors called 'real blanks' were 700 (Pb), 80 (Zn) and 40 (Cu).

Matrix removal using batch pre-concentration on Chelex-100 resin and direct analysis of the analyte-containing resin using DSI-ICP-AES was also attempted [40]. In this case, matrix effects from the 'resin-ash' limited applicability of the approach. Trace metals have also been pre-concentrated on activated charcoal and by transferring sample-loaded charcoal into a DSI cup or by fabricating the cup itself from charcoal [43]. Several chelating agents were tested as well and 8-hydroxyquinoline appeared to be the most promising.

5.7.1 Liquid samples

Calibration curves are linear for 3–5 orders of magnitude and peak area integration is preferred. Non-linear calibration curves have been observed in some instances when using peak heights and this was attributed to changes in peak shape as a function of concentration [33].

For volatile elements, percent relative standard deviation (%RSD) range between 2 and 7% [6,8,11]. This precision reflects not only the capabilities of the technique but also that of manual pipetting. Automation improved %RSDs to <1% short term (30 runs) and to <2% long term (250 runs, 3 hours) [33]. These results were obtained using a sample-carrying probe which was permanently mounted on the DSI. The same authors examined the effect of changing sample-carrying probes for each run (thus potentially affecting positioning of the probe) and, as expected, precision worsened (up to about 5% RSD). This is important, for example, when carbide-forming elements are present in the matrix thus necessitating use of new graphite cups for each run.

Detection limits are listed in Table 5.4. Although they were obtained using DSI-ICP-AES, they have been included to provide an indication of the expected detection capability of DSI-ICP-MS. For example, detection limits of ICP-MS are in general, two to three orders of magnitude better than those of ICP-AES and similar improvements are expected when using DSIs. Comparisons of detection limits obtained by DSI-ICP with those obtained using a pneumatic nebulizer inevitably arise in the literature (and those obtained by DSIs are reported to be better than those obtained with a nebulizer), but such comparisons may be misleading because the pneumatic nebulizer cannot handle microlitre volumes of liquids or solids directly.

Accuracy has also been briefly tested with digested certified reference materials (CRMs) and standard reference materials (SRMs) and, in general, good agreement has been found between the certified and the determined values [6,8]. Signal reproducibility using mixed-gas plasmas ranged between 8.8% and 3.3% [9,10] and examples of three successive insertions for Pb, Cd and Zn in three different mixed-gas plasmas are shown in Fig. 5.23.

TABLE 5.4

Limits of detection[(a)] (see text for discussion)

Element	Range	References
Ag	40–0.4	13f, 33c, 49
Al	700–1030	19d, 45
As	220–700	33c, 56
B	1300–10,000	19d, 49b
Bi	42	56
Ca	5000–5960	19d, 45
Cd	1.2–21	11, 22f, 12e, 20, 33c, 45, 48
Co	20.5–13000	19d, 22f, 45, 48
Cr	20–110	27, 49d, 51, 49d
Cu	1.8–140	11, 20, 22f, 33c, 45, 48, 49b, 56
Fe	11–810	19d, 20, 22f, 45, 48, 56
K	1810	45
Mg	0.07–150	33c, 45
Mn	3–43	22f, 45, 48, 51
Mo	1000	49b
Na	5670	45
Ni	63.5–900	12e, 22f, 45, 48, 51, 56
P	640	45
Pb	21–140	11, 12e, 20, 33e, 45, 48, 51, 56
Sb	65	56
Se	200	33c
Sn	405	62, 65
Sr	30	20
Ti	6–890	11g, 19d, 49b
V	5–380	11g, 22f, 33c
W	750	11g
U	1900	49b
Zn	0.7–120	11, 12e, 20, 22f, 33c, 45
Zr	400–1100	19d, 49b

(a) All detection limits are expressed in pg. Concentration detection limits were converted to absolute detection limits by dividing by the pipetted volume (either 5 or 10 μl).
(b) Freon in Ar (0.1%).
(c) Wire-loop DSI.
(d) Chemical modification using 0.25 M NaF.
(e) W cup in an Ar–H_2 mixed-gas plasma.
(d) Ar–N_2 mixed-gas plasma.
(g) Chemical modification using SF_6.

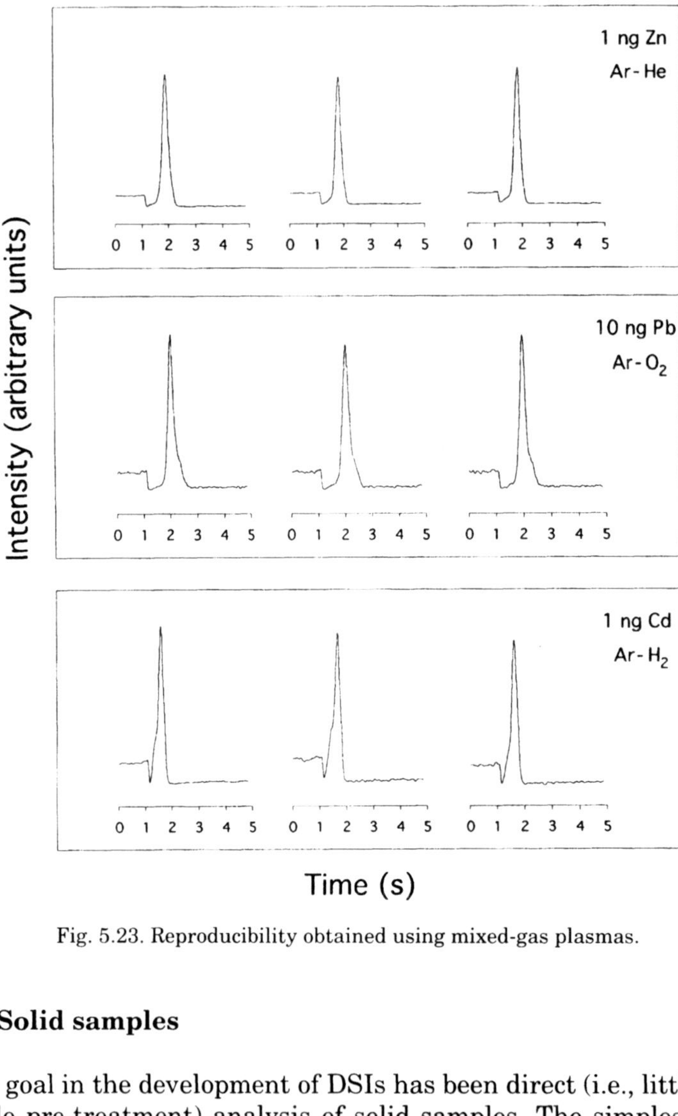

Fig. 5.23. Reproducibility obtained using mixed-gas plasmas.

5.7.2 Solid samples

A key goal in the development of DSIs has been direct (i.e., little or no sample pre-treatment) analysis of solid samples. The simplest of all seem to be chips [64] and alloys [22] which have been successfully analyzed by simply sample washing and drying for sample preparation, thus fulfilling this key goal. For other sample types, however, more elaborate method development proved to be necessary.

Powdered samples with an organic matrix, such as leaves of plants [46], hair [29], cellulose collectors [53] and motor-oils [12,22] all required a charring step. In some cases, a charring step was required even though the sample did not seem to contain a considerable amount of charrable material. This was particularly true for powders mixed with graphite and run as pellets [35,36]. Unless pellets were charred, gas released from the pellet would either extinguish the plasma or would shatter the pellet [35,36]. This is most likely due to use of large amounts of material: a total weight of 100–250 mg is typical of pellets.

For solids analysis of volatile elements in sediments [35,65], the emission signals tended to appear later, to split and tail and to be wider than signals obtained from dried solution residues. Dilution with graphite typically smooths out emission signals and prevents sample loss due to spluttering out of the cup [35,52,65]. Such an approach may not always be feasible because analytes may fall below the detection limit of the technique. In addition to careful packing of powders placed in a cup, boiler cups have been used to prevent loss of powdered samples during handling and insertion [12,60], thus alleviating the requirement to dilute the sample with graphite. Addition of graphite diluent was necessary, however, for some geological samples which tend to form globules [65,35].

With graphite cups, the weight of a solid sample varies typically between 1 and 10 mg. Invariably, the small amount of sample raises questions of homogeneity. To address such questions, larger sample masses may be used, for example 10–250 mg of a powder was mixed with graphite and it was pressed to form a pellet. Using this approach, and due to the large amount of matrix which is introduced into the plasma along with the analyte, matrix effects and spectral interference may become a problem. Plasma power may be used as an alternative and can be adjusted so that analytes vaporize before the matrix. This approach has been used successfully in the analysis of nickel alloys for volatile elements. Applied forward power reduced volatilization of matrix elements, such as, Ni, Al, Cr and Co, thus permitting interference-free determination [64].

In solids analysis by DSI-ICP, precision ranges between 1 and 25% and a typical value is around 10%. Although the causes of poor precision are neither well understood nor well documented at this point, it can be suggested that they are partially due to sample heterogeneity. As for calibration curves, in general they are linear. But, since analytes are in a solid matrix, instrument calibration is one of the remaining difficulties in direct elemental analysis.

There are three basic calibration methods. Using solid standard reference materials, using liquid standards and using standard additions. To facilitate discussion, examples from the literature will be presented along with some data obtained in our laboratory.

Standard reference materials have been used extensively for calibration and calibration curves were linear. Calibration curves using SRMs can be constructed using spectroscopically pure graphite as a diluent [35] or using multiple SRMs of similar composition. For example, we constructed a calibration curve for Pb using 1–2 mg of different lake sediment CRMs [11]. From the linearity of the calibration curve it was concluded that there is a relative independence from the matrix, as also observed by others for different sample types [1,11–13,19,21, 22,26,33,58]. Similar results are expected for DSI-ICP-MS, although some additional dilution with an inert diluent (e.g., spectroscopically pure graphite) may be required. The dynamic range was limited by the concentration range of the CRMs and it was extended somewhat by varying the weight of standards (from about 2 to about 6 mg). The approach provided a total of about one-order-of-magnitude dynamic range, which is sufficient in many cases since concentrations of analytes in solids, typically, do not vary considerably. It should be stressed, however, that applicability of this method depends on the availability of suitable calibration standards.

What if there are no such standards available? Calibration using liquid standards is one option. Umemoto et al. [58] demonstrated use of aqueous calibration of Cu and Zn in high purity Pb and Zn. We also tried this approach and a calibration curve for Pb was constructed using liquid standards and about 2 mg of a powdered lake sediment. Precise but not accurate analytical results were obtained as compared to the certified values. Liquid calibration was also used for direct determination of Pb in paint chips [10]. In this case, smooth signals and single peaks were obtained from both the liquid standard and about 2 mg of a paint chip (Fig. 5.24). As before, precise but not accurate analytical results were obtained as compared to Pb concentrations measured following acid dissolution. Although more work is required to test for a general applicability of the approach, this calibration method may be useful when rapid, semi-quantitative, order-of-magnitude compositional information is sufficient.

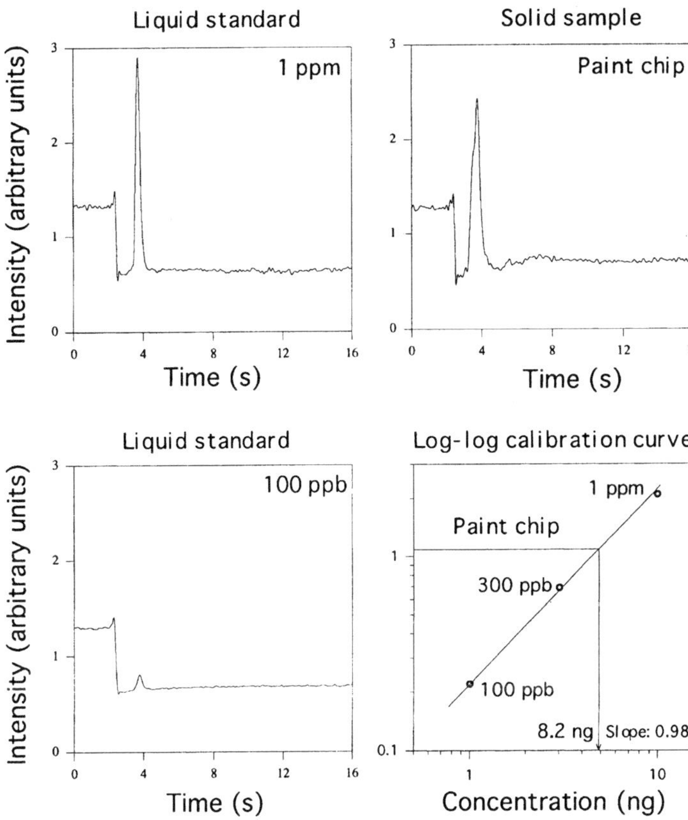

Fig. 5.24. Determination of Pb in paint chips.

The standard additions method offers another alternative. This calibration method has been used successfully with powders and DSI-ICP-AES and, in our case, it was tested for the determination of Pb in a lake sediment CRM (LKSD-4). However, we observed multiple peaks and erratic peak shapes and the concentration we determined (peak area integration) was about two times less than the certified values. The bias was attributed to Pb tied up in the residue that remained in the cup at the end of a 30 s insertion and to Pb entrapped within the silicate matrix. From the erratic peak shapes it was concluded that two vaporization mechanisms were operative: one from the aqueous standard addition (liquid standards were added onto the powdered solid); the other, from the solid (most likely because the analyte is tightly bound to the matrix). This was confirmed by running liquid standards, and Pb from the standard solutions was found to vaporize differently (e.g., peak shapes, appearance time) than from the solid. Similar observations have been reported by others. For example, Blain and Salin [35] found that peak profiles obtained from three different sediment SRMs (PACSS, MESS and BCSS) for Cd, Pb, Zn and Hg were similar, but their response factors were different. As a consequence, Cd had to be used as an internal standard for the other elements.

In direct elemental analysis of powders, concerns frequently arise due to use of small amounts of sample. There are difficulties, for example, in handling and weighing small amounts and difficulties in finding appropriate calibration standards. Questions of sampling and homogeneity also arise. As mentioned above, powdered solids can be diluted by mixing with an inert diluent, such as graphite. This is labour intensive and difficult to automate, it may further augment homogeneity problems (if mixing is not done properly) and it increases the potential for contamination from the dilution reagents and the mixing apparatus. These shortcomings were partially addressed by adapting slurry sampling to DSIs.

5.7.3 Slurry DSI-ICP-AES

Recently, slurry sampling for powders or solids that can be converted to a powder has been applied to DSI-ICP-AES [9-11, 73]. In slurry sampling, a powdered sample is mixed with distilled water (or a dilute acid) and, often, with a thixotropic agent (e.g., Triton X-100). The resultant suspension is stirred vigorously and µl volumes of the suspension are pipetted into a cup. Because a relatively large amount of sample can be used to make the slurry (e.g., 1 g), sample handling, weighing, sampling of the solid and homogeneity errors are reduced. Since distilled water is the usual diluent, dilution is simplified and contamination from the dilution reagents is no longer an issue. Also, the use of a pipette to deliver a slurry into a cup provides a straightforward and reproducible way of handling µg amounts of powders. Since smaller amounts of matrix are introduced into the plasma (vis à vis pellets or mg amounts of powders), the potential for interference from the matrix is also reduced. Furthermore, due to the use of aqueous solutions, slurry sampling facilitates automation of dispensing, dilution and of standard additions using commercially available technology for liquid sample-handling and delivery systems (e.g., an autosampler similar to that used in a graphite furnace). Despite these advantages, sampling and homogeneity may become again a concern due to use of microlitre volumes.

Slurry DSI-ICP-AES was evaluated [9–11,67,73,74] using a botanical SRM (Pb in pine needles), a geological CRM (Pb and Zn in lake sediments) and a biological SRM (Cu in bovine liver). Calibration curves were linear and the slopes of the log–log plot were reproducible. A charring time of 2 min was required for the biological SRM, otherwise multiple peaks were observed. With the exception of pine needles, precision was 4–6%. The relatively poor precision for the botanical SRM (13%) was attributed to the relatively large particle size of this sample which caused pipette tip clogging.

In general, slurry sampling proved to be a particularly attractive sample processing methodology for μg amounts of powdered samples, thus warranting further investigation. Overall, it can be concluded that it is advantageous to reduce sample size, for example, from mg used in pellets [35,36] to μg using slurries [11]. Although use of μg amounts raises questions of homogeneity, these questions can be addressed by simply running a sample multiple times. Perhaps more importantly, slurry sampling opens up the possibility for precise and accurate analytical results using liquid calibration standards. An example of such calibration is discussed below.

5.7.4 Benefits of tailoring the plasma environment to the application

Examples of the benefits arising from use of μg quantities of sample and from the ability to customize the ICP using mixed gases or chemical modification of the plasma environment are briefly presented in this section. The first example involves 0.2 mg of Al_2O_3 and a 20% $Ar–O_2$ plasma. According to Liu and Horlick [22], the mixed-gas plasma was the key operational feature that enabled determination of a number of elements in this sample.

Using a 15% Ar–He plasma, we were able to determine Co, Pb and Fe in pine needles SRM using the standard additions method by pipetting 10 μl of a liquid standard directly on the sample in a graphite cup. The measured concentrations compared favourably with the certified values. It is worth noting that, unless a mixed-gas plasma was used, erratic and multiple peaks were observed, thus making the method of standard additions unusable for these elements in this sample type.

Is it possible to use liquid calibration standards with slurries and DSI-ICP-AES? This question was briefly addressed using an all-Ar ICP and an Ar–O_2 plasma (4% O_2 was added to the outer tube of the torch) and, Pb and Zn in lake sediment CRMs and Cu in bovine liver SRM. The calibration curves obtained using liquid standards, slurries and an Ar–O_2 ICP were super-impossible, whereas those obtained using an all-Ar ICP were not [9–11,72]. Thus, it was concluded that liquid rather than solid standards may be used for calibration in slurry mixed-gas DSI-ICP-AES. These results are specific to these elements, matrices and operating conditions and more work remains to be done to test for a general applicability of the approach.

A final example of the benefits arising from micro-samples and from the ability to customize the ICP involves *in situ* chemical modification. For example, we were able to determine Ti (a strong carbide-forming element) in micro-samples of slurries of a lake sediment CRM using SF_6 gaseous chemical modifier [11,72,73]. Such determinations would have been impossible by graphite-cup DSI-ICP-AES and an Ar-only ICP operated at 2 kW.

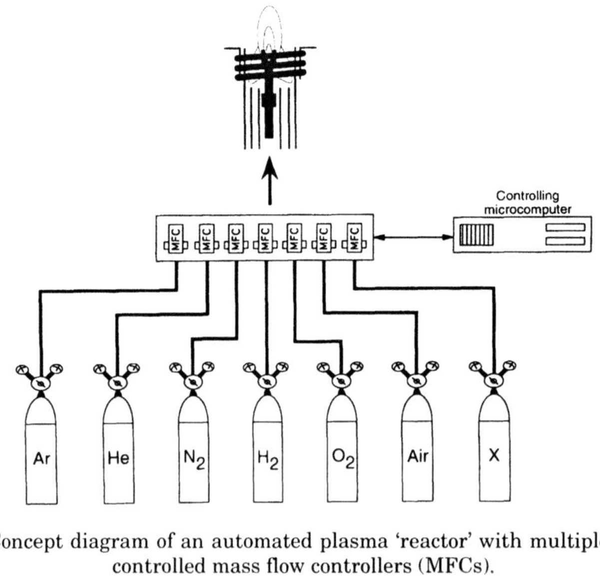

Fig. 5.25. Concept diagram of an automated plasma 'reactor' with multiple, computer-controlled mass flow controllers (MFCs).

5.7.5 The plasma as a programmable reactor

As mentioned in the introduction, an Ar-only ICP can be thought of as an inert furnace for sample-carrying probes, a DSI-device as a mechanism that brings an initially ambient temperature sample in contact with a high temperature, gas-phase plasma and a graphite cup as a mere sample holder. With computer-controlled addition of foreign gases to the plasma, the conceptual goal of a cup as a 'crucible' in which to carry chemical reactions in the plasma (rather than a plain 'sample holder') and of the plasma as a 'programmable reactor' whose environment can be easily tailored to the application, appear to be realizable. Automated foreign gas selection using, for example, an expert system is not hard to envision as well. A schematic illustration of a potential future implementation that embodies these concepts is shown in Fig. 5.25.

5.8 APPLICATIONS

For the most part, work with DSI-ICP-AES and DSI-ICP-MS carried out thus far has focused on the development and characterization of DSIs, on exploring and documenting their capabilities, on addressing their limitations (such as carbide formation), on further developing DSI-ICP, and on expanding the scope of the technique rather than the range of applications. However, there have been some application areas in which the capabilities of DSIs provided unique analytical solutions to practical analytical problems. For completeness and to provide interested DSI-ICP-MS readers with potential sources of information on analytical methodology in DSI-ICP, a list of sample types which have been analyzed or have been used to characterize DSI-ICP-AES is included in Table 5.5.

5.9 CONCLUSIONS

Overall, DSI-devices have extended the analytical capability and utility of the ICP. Although they are still being developed and characterized using ICP-AES, they may bring greater benefits to ICP-MS. In particular, a DSI-device coupled to a commercially available time-of-flight ICP-MS has the potential to offer true simultaneous multielement capabilities from small amounts of liquid or solid

TABLE 5.5

Sample types analyzed

Sample type	Elements	Reference
Analyzed as solids		
Botanical sample	Many	46
Botanical sample	Zn	12
Graphite	Many	1
Botanical samples	Many	12
Al_2O_3	Many	12
Synthetic silicate glass	Zn, Cd	65
Uranium oxide	Many	63, 60
SiO_2, Al_2O_3, graphite	Many	13
Sediments	Many	36
Sediment	Pb, Cd, Zn, Hg	35
Pb and Zn metal	Cu, Fe	58
$CaSO_4$	B	19
Cellulose collectors	Pb, Fe, Cr, Mn	55
Human hair	Pb, Cd, Zn, Hg, As, Cu	29
Nickel alloys	Cd, Zn, As, Se, Mg	14
Paint chips	Pb	
Analyzed as slurries		
Botanical samples	Pb, Zn	11
Biological samples	Cu	11
Sediments	Pb, Ti	11
Analyzed as liquids		
River water	As, Bi, Cd, Cu, Hg, In, Pb, Sb, Tl, Zn	39
River water	Mn, Pb, As, Au, Fe, Ni, Mo	66
Digested iron and steel	As, Sb	55
Digested botanicals	Cu, K, Mn, Pb	13
Digested botanical	Many	47
Synthetic sea water	Cd, Co, Cr, Cu, Mn, Ni, Pb, Zn	47
Synthetic sea water	Cd, Co, Cu, Ni, Pb, Zn	28
U/Pu nuclear fuel	Eu, Dy, Gd, Sm	61

samples. To quote, "*a great deal about DSI systems has been learned over the last decade and new generation DSI systems should now be developed from a synthesis of all this knowledge*" [6]. It was expected that such a DSI should be "*a very powerful system for sample introduction and processing for ICPs*" [6]. One such system that benefited greatly from the synthesis mentioned above and from a synergy between seemingly unrelated disciplines (i.e. plasma spectroscopy, software for acquisition of transient signals and for user interface design, DSI-device, optics, electronics hardware with some speculative gas-phase chemistry) has been described recently [11]. This system provided new capabilities: it enabled determinations previously thought impossible by DSIs, and it opened up the possibility of converting the ICP into a sample processor and into a programmable chemical reactor in addition to its traditional role as a signal generator.

In conclusion, DSIs offer unique ways for sample processing and clean-up and for introducing ambient temperature micro-samples into hot, gas-phase plasmas.

ACKNOWLEDGEMENTS

Financial assistance from the University of Waterloo and from the National Sciences and Engineering Council (NSERC) of Canada is gratefully acknowledged.

REFERENCES

1 E.D. Salin and G. Horlick, Direct sample insertion device for inductively coupled plasma-atomic emission spectrometry, *Anal. Chem.*, 51 (1979) 2284.

2 K. Ohls and D. Sommer, Analysis of liquid, solid and gaseous samples using high power nitrogen/argon ICP system, In: R.M. Barnes (Ed), *Developments in Atomic Spectrochemical Analysis*, pp. 321–336, Heyden, London, UK, 1981.

3 K. Ohls and D. Sommer, Analysis of liquid, solid and gaseous samples using a high power nitrogen/argon ICP system, Paper 78, International Winter Conference on developments in Atomic Plasma Spectrochemical Analysis, San Juan, Puerto Rico, Jan. 7–11, 1980.

4 C.W. McLeod, M.W. Routh and M.W. Tikkanen, Introduction of solids into plasmas, In: A. Montaser and D.W. Golightly (Eds), *Inductively Coupled Plasmas in Analytical Atomic Spectrometry*, VCH, NY, 1992.

5 A. Montaser, M.G. Minnich, J.A. McLean and H. Liu, Sample introduction in ICP-MS, In: A. Montaser (Ed), *Inductively Coupled Plasma Mass Spectrometry*, Wiley-VCH, NY (1998).
6 V. Karanassios and G. Horlick, Direct sample insertion for inductively coupled plasma spectrometry, *Spectrochim. Acta Rev.*, 13 (1990) 89.
7 H. Baoumann, Solid sampling with inductively coupled plasma–mass spectrometry: A survey, *Fresenius J. Anal. Chem.*, 342 (1992) 907.
8 R. Sing, Direct sample insertion for inductively coupled plasma spectrometry, *Spectrochim. Acta*, 54B (1999) 411.
9 V. Karanassios and T.J. Wood, Elemental analysis using an automated direct sample insertion–inductively coupled plasma–atomic emission spectrometry system, *Spectroscopy*, 11(8) (1996) 25.
10 V. Karanassios and T.J. Wood, An inductively coupled plasma as a programmable reactor applied to environmental micro-samples, *Proceedings, 1st Biennial International Conference on Chemical Measurement and Monitoring of the Environment*, R. Burk and R.E. Clement (Eds), 1 (1996) 321.
11 V. Karanassios and T.J. Wood, Development and characterization of an automated direct sample insertion–inductively coupled plasma–atomic emission spectrometry system, *Appl. Spectrosc.*, 53 (1999) 197.
12 W.E. Pettit and G. Horlick, An automated direct sample insertion system for the inductively coupled plasma, *Spectrochim. Acta*, 41B, 699–712 (1986).
13 Y. Shao and G. Horlick, Performance of a direct sample insertion system for the inductively coupled plasma, *Appl. Spectrosc.*, 40 (1986) 386.
14 V. Karanassios and G. Horlick, A computer-controlled direct sample insertion device for inductively coupled plasma–mass spectrometry, *Spectrochim. Acta*, 44B (1989) 1345.
15 V. Karanassios and G. Horlick, Background spectral characteristics in direct sample insertion–inductively coupled plasma–mass spectrometry, *Spectrochim. Acta*, 44B (1989) 1361.
16 V. Karanassios and G. Horlick, Elimination of some spectral interference and matrix effects in inductively coupled plasma–mass spectrometry using direct sample insertion techniques, *Spectrochim. Acta*, 44B (1989) 1387.
17 V. Karanassios and G. Horlick, A computer-controlled direct sample insertion device for inductively coupled plasma–atomic emission spectrometry device for ICP-AES, *Spectrochim. Acta*, 45B (1990) 85.
18 V. Karanassios, G. Horlick and M. Abdullah, Characterization of a direct sample insertion device for inductively coupled plasma–atomic emission spectrometry, *Spectrochim. Acta*, 45B (1990) 105.
19 V. Karanassios, M. Abdullah and G. Horlick, The application of chemical modification in direct sample insertion–inductively coupled plasma–atomic emission spectrometry, *Spectrochim. Acta*, 45B (1990) 119.
20 W. T. Chan and G. Horlick, An automated direct sample insertion system

for inductively coupled plasma–atomic emission spectrometry, *Appl. Spectrosc.*, 44 (1990) 380.
21 W.T. Chan and G. Horlick, Some signal characteristics in direct sample insertion–inductively coupled plasma–atomic emission spectrometry, *Appl. Spectrosc.*, 44 (1990) 525.
22 X.R. Liu and G. Horlick, Direct analysis of materials using direct sample insertion devices and mixed-gas inductively coupled plasma atomic emission spectrometry, *J. Anal. Atom. Spectrom.*, 9 (1994) 833.
23 X.R. Liu and G. Horlick, *In-situ* laser ablation sampling for inductively coupled plasma–atomic emission spectrometry, *Spectrochim. Acta*, 50B (1994) 537.
24 D. Sommer and K. Ohls, Direkte Probeneinführung in ein stabil brennendes, induktiv gekoppeltes Plasma, *Fresenius Z. Anal. Chem.*, 304 (1980) 97.
25 K. Ohls, Die Bestimmung kleiner Cadmium-Anteile in verschiedenen Materialien durch Festprobeneinsatz bei ICP-und flammendloser Atomabsorptionsspekrometrie, *Spectrochim. Acta*, 39B (1984) 1105.
26 E.D. Salin and M. Habib, Electrochemical pre-concentration and separation for elemental analysis by inductively coupled plasma–emission spectrometry, *Anal. Chem.*, 56 (1984) 1186.
27 E.D. Salin and R.L.A. Sing, Microsample liquid analysis with a wire-loop direct sample insertion technique in an inductively coupled plasma, *Anal. Chem.*, 56 (1984) 2596.
28 M.M. Habib and E.D. Salin, Controlled potential electrolysis coupled with a direct sample insertion device for multielement determination of heavy metals by inductively coupled plasma atomic emission spectrometry, *Anal. Chem.*, 57 (1985) 2055.
29 C.V. Monasterios, A.M. Jones and E.D. Salin, Determination of Cu and Zn in human head hair by inductively coupled plasma atomic emission spectrometry with a direct sample insertion device, *Anal. Chem.*, 58 (1986) 780.
30 D.W. Boomer, M. Powell, R.L.A. Sing and E.D. Salin, Application of a wire-loop direct sample insertion device for inductively coupled plasma–mass spectrometry, *Anal. Chem.*, 58 (1986) 975.
31 T.W. Williams and E.D. Salin, Hazards of naive approach to detection limits with transient signals, *Anal. Chem.*, 60 (1988) 725.
32 L. Blain, E.D. Salin and D.W. Boomer, Probe design for the direct insertion of solid samples in the inductively coupled plasma for analysis by atomic emission and mass spectrometry, *J. Anal. Atom. Spectrom.*, 4 (1989) 721.
33 R.L.A. Sing and E.D. Salin, Introduction of liquid samples into the inductively coupled plasma by direct insertion on a wire-loop by direct sample insertion on a wire-loop, *Anal. Chem.*, 61 (1989) 163.
34 P. Moss and E.D. Salin, Flow injection pre-concentration coupled with

direct sample insertion for inductively coupled plasma atomic emission spectrometry, *Appl. Spectrosc.*, 45 (1991) 1581.
35 L. Blain and E.D. Salin, Methodological solutions for the analysis of sediment samples by direct sample insertion–inductively coupled plasma–atomic emission spectrometry, *Spectrochim. Acta*, 47B (1992) 205.
36 L. Blain and E.D. Salin, Observations of pellet direct sample insertion for inductively coupled plasma–atomic emission spectrometry, *Spectrochim. Acta*, 47B (1992) 399.
37 L. Blain and E.D. Salin, Application of AgCl as a thermochemical additive for the determination of refractory elements in solid sediment samples by direct sample insertion–inductively coupled plasma–atomic emission spectrometry, *Spectrochim. Acta*, 47B (1992) 1471.
38 R. Rattray, J. Minoso and E.D. Salin, Rapid sample pre-concentration by aerosol deposition for the determination of trace elements by inductively coupled plasma spectrometry, *J. Anal. Atom. Spectrom.*, 8 (1993) 1033.
39 R. Rattray and E.D. Salin, Aerosol deposition direct sample insertion for ultra-trace elemental analysis by inductively coupled plasma mass spectrometry, *J. Anal. Atom. Spectrom.*, 10 (1995) 829.
40 R. Rattray and E.D. Salin, Chelation pre-concentration with resin analysis by direct sample insertion inductively coupled plasma spectrometry, *J. Anal. Atom. Spectrom.*, 10 (1995) 1053.
41 C.D. Skinner and E.D. Salin, Axial viewing and modified cup design for direct sample insertion inductively coupled plasma atomic emission spectrometry, *J. Anal. Atom. Spectrom.*, 12 (1997) 725.
42 C.D. Skinner and E.D. Salin, Images of direct sample insertion in an inductively coupled plasma, *J. Anal. Atom. Spectrom.*, 12 (1997) 1131.
43 C.D. Skinner, M. Cazagou, J. Blaise and E.D. Salin, Direct insertion of trace metals pre-concentrated on activated charcoal into an inductively coupled plasma, *Appl. Spectrosc.*, 53 (1999) 191.
44 H. Haraguchi, M. Abdullah and T. Hasegawa, Inductively coupled plasma-atomic emission spectrometry utilizing a direct graphite cup insertion technique, *Bull. Chem. Soc. Jpn.*, 57 (1984) 1839.
45 M. Abdullah, K. Fuwa and H. Haraguchi, Simultaneous multielement analysis of microliter volumes of solution samples by inductively coupled plasma–atomic emission spectrometry using a graphite cup direct insertion technique, *Spectrochim. Acta*, 39B (1984) 1129.
46 M. Abdullah and H. Haraguchi, Computer-controlled graphite-cup direct sample insertion device for direct analysis of plant samples by inductively coupled plasma atomic emission spectrometry, *Anal. Chem.*, 57 (1985), 2059.
47 M. Abdullah, K. Fuwa and H. Haraguchi, Electrochemical pre-concentration of trace metals for simultaneous multielement determination by inductively coupled plasma–atomic emission spectrometry utilizing

graphite-cup direct insertion technique, *Appl. Spectrosc.*, 41 (1987) 715.
48 G.F. Kirkbright and S.J. Walton, Optical emission spectrometry with an inductively coupled radio frequency argon plasma source and direct sample introduction from a graphite rod, *Analyst*, 107 (1992) 276.
49 G.F. Kirkbright and Z. Li-Xing, Volatilisation of some elements from a graphite rod direct sample insertion device into an inductively coupled argon plasma for optical emission spectrometry, *Analyst*, 107 (1982) 617.
50 Z. Li-Xing, G.F. Kirkbright, M.J. Cope and J.S. Watson, A microprocessor-controlled graphite rod direct sample insertion device for inductively coupled plasma optical emission spectrometry, *Appl. Spectrosc.*, 37 (1983) 250.
51 N.W. Barnett, M.J. Cope, G.F. Kirkbright and A.A.H. Taobi, Design consideration and temperature determination of an automated graphite furnace cup system used for direct sample introduction for ICP–optical emission spectrometry, *Spectrochim. Acta*, 39B (1984) 343.
52 G. Zaray, J.A.C. Broekaert and F. Leis, The use of direct sample insertion into a nitrogen–argon inductively coupled plasma for emission spectrometry—I. Technique optimization and application to the analysis of aluminium oxide, *Spectrochim. Acta*, 43B (1988) 241.
53 G. Zaray, P. Burba, J.A.C. Broekaert, The use of direct sample insertion into a nitrogen–argon inductively coupled plasma for emission spectrometry—II. determination of trace elements in high-purity aluminium after pre-concentration on cellulose collectors, *Spectrochim. Acta*, 43B (1988) 255.
54 M. Reisch, H. Nickel and M. Mazurkiewicz, Volatilization studies for Fe and Cd in their direct determination by inductively coupled plasma– atomic emission spectrometry using direct sample insertion or external electrothermal evaporation of powders, *Spectrochim. Acta*, 44B (1989) 307.
55 M. Umemoto and M. Kubota, Determination of As and Sb in iron and steels by inductively coupled plasma–atomic emission spectrometry using a graphite cup direct insertion device, *Spectrochim. Acta*, 42B (1987) 491.
56 M. Umemoto and M. Kubota, A study on detection limits of inductively coupled plasma–atomic emission spectrometry using a graphite cup direct insertion device, *Spectrochim. Acta*, 44B (1989) 713.
57 M. Umemoto and M. Kubota, Some characteristics of an inductively coupled plasma for atomic emission spectrometry with a graphite cup sample introduction, *Spectrochim. Acta*, 46B (1991) 1275.
58 M. Umemoto, K. Hayashi and H. Haraguchi, Direct determination of ultra trace copper and iron in lead and zinc metal by inductively coupled plasma–atomic emission spectrometry using the graphite cup direct insertion technique using the graphite cup direct insertion technique, *Anal. Chem.*, 64 (1992) 257.
59 A.G. Page, K.H. Madraswala, S.V. Goodbole, M.J. Kulkarni, V.S. Mallapurkar and B.D. Joshi, Carrier distillation ICAP approach for trace

metal assay of U_3O_8 powder, *Fresenius Z. Anal. Chem.*, 315 (1983) 38.
60 A.G. Page, S.V. Goodbole, K.H. Madraswala, M.J. Kulkarni, V.S. Mallapurkar and B.D. Joshi, Selective volatilization of trace metals from refractory solids into and inductively coupled plasma, *Spectrochim. Acta*, 39B (1984) 551.
61 P.J. Purohit, S.K. Thulasidas, N. Goyal and A.G. Page, Solid sampling approach in inductively coupled plasma–atomic emission spectrometry for the determination of four lanthanides with high sensitivity, *J. Anal. Atom. Spectrom.*, 12 (1997) 1317.
62 E.R. Prack and G.J. Bastiaans, Metal speciation by evolved gas/inductively coupled plasma emission spectrometry, *Anal. Chem.*, 55, (1983) 1654.
63 A. Lorber and Z. Goldbart, Convenient method for the determination of trace elements in solid samples using an inductively coupled plasma, *Analyst*, 110 (1985) 155.
64 C.W. McLeod, P.A. Clarke and D.J. Mowthorpe, Simultaneous determination of volatile trace elements in nickel–base alloys using a direct insertion probe and inductively coupled plasma–emission spectrometry, *Spectrochim. Acta*, 41B (1986) 63.
65 I.B. Brenner, A. Lorber and Z. Goldbart, Trace element analysis of geological materials by direct solids insertion of a graphite cup into an inductively coupled plasma, *Spectrochim. Acta*, 42B (1987) 219.
66 G.E.M. Hall, J.-C. Pelhat, D.W. Boomer and M. Powell, Relative merits of two methods of sample introduction in inductively coupled plasma mass spectrometry: electrothermal vaporization and direct sample insertion, *J. Anal. Atom. Spectrom.*, 3 (1988) 791.
67 V. Karanassios, Z. Zhang and T.J. Wood, Direct elemental analysis of solid samples, Paper 387, 21st FACCS Conference, St. Louis, MI, Oct. 2–7, 1994.
68 V. Karanassios and T.J. Wood, A direct sample insertion device for inductively coupled plasma–atomic emission spectrometry and inductively coupled plasma–mass spectrometry, Paper 51P, Pittsburgh conference, Chicago, IL, Feb. 27–Mar. 4, 1994.
69 L. Yin, Ph.D. thesis, University of Alberta, 1998.
70 N.N. Sesi, A. MacKenzie, K.E. Shanks, P. Yang and G.M. Hieftje, Fundamental studies of mixed-gas inductively coupled plasmas, *Spectrochim. Acta*, 49B (1994) 1259.
71 T.J. Wood and V. Karanassios, Mixed-gas plasmas in direct sample insertion inductively coupled plasma–atomic emission spectrometry, Paper TP55, Winter Conference on Plasma Spectrochemistry, Fort Lauderdale, FL, Jan. 8–13, 1996.
72 T.J. Wood and V. Karanassios, Development and characterization of an automated direct sample insertion–inductively coupled plasma–atomic emission spectrometry system, Paper 77, 41th International Conference on Analytical Sciences and Spectroscopy, Windsor, ON, Aug. 14–16,1995.

Index

ablation
— definition, 445
ablation cell, 458
ablation strategies, 462
accelerating and convergent electrostatic field, 30
acceleration potential, increase in, 92
acceleration region, 43
accelerator cone, 30
acetic acid
— enhancing effect of, 89
acetonitrile
— enhancing effect of, 89
acid L-cysteine, 304
acquisition time, 77
activated alumina, 308
active medium, 449
addition of O_2 to the plasma, 15
adsorption onto the walls of glass reaction vessels, 297
aerosol
— deposition, 543
— droplet size and plasma solvent loading, 94
— formation, 95
— ionic redistribution, 49,94
— loading, 49
— polydispersed, 46
— pre-vaporized, 58, 76
— production, 74
— transport efficiency, 95
— transport to the plasma, 45
aerosol carrier Ar flow
— addition of rare gases to, 162
aerosol carrier gas, 66
— addition of a molecular gas to, 148
— addition of a sheathing gas to, 151, 162
— addition of He to, 150,162
— addition of N_2 to, 150
aerosol carrier gas flow, 6, 242
— addition of a foreign gas to, 142
— addition of CH4 to, 160
— addition of CHF_3 to, 161
— optimum rate, 42, 157
— propan-2-ol with O_2 in, 156
AgCl, 533
air bubbles, 67,213
air entrainment, 539
air-segmentation, 213
air-segmented
— discrete introduction, 252
— water carrier, 250
all-He ICP, 147
all-He plasmas, 141
alternative high-efficiency nebulization system, 53
alumina instead of quartz for corrosive samples, use of, 28
ambipolar diffusion, 90
analysis
— of clean samples, 84
— of small-volume samples, 43
analyte emission, survey of, 529
analyte mass dependence, 91
analyte sensitivity, improving, 49
analyte signal enhancement, 242
analyte transport efficiency, 402

analytes that form volatile compounds, 55
anodic stripping voltammetry, 300, 306, 313
applications of limited dispersion FI systems with ICP-MS, 259
aqueous calibration, 548
aqueous carrier, 242
aqueous ethylation, 311
arc ablation, 469
archaeological applications, 97
Arctic snow, 417
argides, 394
^{36}Ar peaks, 541
ArF laser, 450
argon dimer, 369,409
argon-containing polyatomic species
— reduction of, 152
argon–nitrogen mixed-gas plasma, 301
Ar–H_2 mixed-gas, 527
Ar–halocarbon mixed-gas, 527, 535
Ar–N_2 mixed-gas, 527
Ar–O_2 mixed-gas, 521, 527
As signal enhancement in the presence of carbon, 285
ashing, 522
atom–ion equilibrium, 88
atomization, 5,56
autofocus, 456
automated sandwich FI manifold, 276
auxiliary complexing reagent, 302
auxiliary gas, 4
— addition of a molecular gas to, 151
average kinetic energy, 27
axial and radial dispersions, 219

Babington-type nebulizers, 44,48
back-extraction
— of analyte, 305
— into aqueous phase, 317
back-flow-prevention valve, 141
background
— decrease in, 157
— ions, 43
— originates from collisional processes in the ion beam, 29
— negligible, 38
— signals, 236, 464
— spectral features, 367, 539
back-pressure, 238,301
$BaCl_2$, 533
bacteria, 309
BaF_2, 533
bandpass , 32
barium, 541
batch extraction procedures, 317
battery-operated peristaltic pump, 294
Bayesian methods, 82
bead injection, 231
bed volume, 294
bench-top ICP-MS instrument, 24
beneficial ion-molecule reactions, 15
best sampling position, 12
best sensitivity, 146
bias factor, 97
binders, 463
biological applications, 97
biomass, 309
blanks in preconcentration, 300
blockage of the membrane, 57
body of the microcolumn, 293
boron, vaporization of, 381
boundary layer, 11
bubble gate, 252
bulk analysis, 446, 469
butanol, enhancing effect of, 89
by-pass valve, 287

calibration, 464, 548
calibration strategies, 134, 273
calibration using internal standardization, 43
carbide, 526, 534
carbide formation, 513
carbides, 364,394
carbon, 368

carbon deposits, 541
carbon-containing polyatomic species, 159,161
carbonyl vapour generation, 311
carrier, 389
carrier gas, 4
carrier stream, 214
— dispersion of the sample into the, 214
cavity-tuned, free-running rf generator, 309
CCl4, 535
centre-grounded coil, 25
cerium, 541
certified reference materials, 446
channel electron multiplier, 23
charge transfer, 7
— between background ions, 150
charge-transfer process, 162
charge-transfer reaction, 147
— with H_2, 16
charring, 522
CHCl3, 535
chelating/ion-exchange resin, 308
chelation, 131
Chelex-100, 543
chemical modification, 532, 552
chemical modifiers, 386
chemical reaction, 220
— on a short time scale, 278
chemical vaporization interferences
— removed on-line, 303
chlorine-containing polyatomic species
— reduction of, 152
circular shape of the laser beam, 450
cleanliness of the buffer, 302
clogging
— by salt deposition, 52
— of the sampler orifice, 78, 244
coagulation of droplets, 58
collector, 300
collision cell, 32
collisional ionization, 91
collisional reactions occurring in the boundary layer, 85
collision-cell technology, 32
collision-cell/DRC technology, 32
collision-induced dissociation, 16
common nebulization systems, 3
common systems with pneumatic nebulizers, 44
complete dissolution, 306
complete precipitation, 306
complexing agent immobilized on a solid support, 308
computer simulations, 154
computer-controlled autosampler, 238
computer-controlled commercial ICP-MS instrument, 24
computer-controlled pump, 229
computer-controlled sample injection valve, 238
concentration profile, 223
concentric nebulizers, 44
condensation reactions, 85
conducting binder, 469
conducting samples, 469
conespray nebulizer, 44, 94
connecting tubing between the exit of the spray chamber and the injector of the torch, 50
continuous aspiration, 240
continuous flow analysis, 213
continuous rinsing effect, 244
continuous vapour generation, 314
continuous wave, 449
controlled-dispersion flow analysis, 229
controlled-pore glass particles, 308
conventional ICP torches, 141
conventional tangential-flow torch, 28
conversion reactions, 270
cool conditions, 84
cool plasma, 71
— conditions, 93

cooled mini-spray chamber, 240
co-precipitation of the analyte, 312
correlated flicker noise, 75
counting statistics, 75
co-volatilization, 362
critical diameter, 404
cross-covariance function, 236
cross-flow nebulizers, 44, 63
cryogenic desolvation, 57,159
crystalline precipitates, 271
cup temperature, 523, 524
curve-fitting, 237
cyclonic spray chamber, 48, 231
cylindrical or planar stainless steel filters, 269

dead time, 129
— method for calculation of, 129
dead volume, extremely small, 62
degraded ion transmission, 26
demountable injector tube, 28
deposition and elution processes, 301
deposition and stripping potentials, 306
depth of focus, 454, 462
depth profiling, 446
depth resolution, 446
desolvating the aerosol, 26
desolvation, 5
— device, 56,76
— with high resolution ICP-MS, 83
desolvation system, 54,59
— efficient, 59
— involving Peltier coolers, 59
— with a reduced volume, 240
desolvation temperature
— optimal, 58
detection limits, 68, 71, 233
detection of transient signal, 43
detector's dead time, 24
differentially pumped interface, 11
diffusion coefficient, 221
digital filtering with discrete Fourier transform, 236
diluted seawater, 92
dilution effect, 219
dilution factor
— adjusting, 285
direct analysis, 133
direct injection high efficiency nebulizer, 63
direct injection nebulizer, 60
direct precipitation of the analyte, 312
direct sample insertion *see* DSI
direct sample introduction, 242
directionality, 449
discontinuous displacement, 462
discrete dynode electron multiplier, 24
discrete frequency noises, 74
dispersion
— definition and measurement, 217
dispersion coefficient, 217
dispersion processes, 218
displacement technique, 297
dissolution reagent, 271
dissolved salts
— high concentration of, 78
distribution of energy, 446
divergence, 449
donut-shaped spray chamber, 231
double-focusing instruments, 33
double-line FI manifold, 295
double-line manifold, 281
double-pass spray chamber, 231
doubly-charged ions, 83
— increased levels of, 148
doubly-charged ions, hydroxides and hydrides
— effects of, 82
drift correction procedures, 133
drift standard, 296
droplet size distribution, 49, 59, 80, 88
— of the aerosol, 7
dry aerosol, 50
— addition of H_2 to, 149

dry aerosols, 448
dry plasma, 539
drying, 522
drying step, 362
DSI torch and graphite cup, 504
DSIs
— advantages, 505
— designs, 507
— drive mechanism, 507
— instrumentation for, 506
— mounting, 507
dual-piston reciprocating pump, 45
duty cycle, 43, 235
dwell time, 77, 235, 515
dynamic reaction cell, 32

effect of a given concomitant element, 94
effective sampling diameter, 12
efficient elemental ion source for MS, 10
electrochemical cell, 317
electrodeposition, 542
electron number density, 85
electronic dilution, 284
electronically actuated valve, 228
electron-impact ionization, 7
electrostatic and a magnetic analyzer in series, 33
electrostatic interactions, 49
electrostatic ion lenses, 11
electrostatic lenses, 29
electrothermal vaporization, 347
electrothermal vaporization systems, 351
elemental fractionation, 466
elements prone to spectroscopic interferences, 71
energy, 451
energy distribution, 451
enhancement, 91
environmental applications, 97
enzymes, 309
ethanol
— enhancing effect of, 89
ethanol concentration, 49
evaporation, 133,543
evaporation column, 50
excimer lasers, 450, 456
exponential dilution system, 50
extended torch, 539
extent of mixing with the carrier gas, 144
external calibration, 464
— with matrix-matched standards, 133
— with standard solutions, 136
external drift monitor, 132
extraction, 131,543

F2 laser, 450
factorial optimization, 241
Faraday cups , 41
Fassel-type torch, 4
field sampling, 294
field-free flight tube, 43
figures of merit, 68
filterless collector, 300
filtration, 133
finer tertiary aerosol, 94
first vacuum stage, 95
fixed-volume rotary valve, 226
flame atomic absorption spectrometry, 1
flash lamp, 449
flat-top energy distribution, 456
flat-top peaks, 71,73
flexibility in terms of flow rate, 309
flexible tubing, 227
flow injection, 214
— principles of, 217
flow injection analysis, 542
flow rate of the carrier stream, 220
flow resistance, 222
flow restrictor, 281
fluctuations due to the peristaltic pump, 47
fluence, 452

fluxes, 463
focus positioning devices, 456
focused microwave heating systems, 305
focusing devices, 454
food-related applications, 97
formation of refractory oxides, 58
formation of the aerosol, 45
Fourier transform ion cyclotron resonance mass spectrometry, 16
fractionation, 456
free jet, 11
free-running 40-MHz generator, 138, 242
free-running mode, 450
freeze-drying procedures, 463
Freon, 387,533
frequency noise, 231
frequency-doubling, 452
frequency-quadrupling, 452
frequency-quintupling, 452
frequency-tripling, 452
fs lasers, 470
full mass spectra, 43
fused silica capillaries, 232
fusion of samples, 463
fusion products, 463

gas bubbles, 248
gas displacement pump, 45
gas flow rates, 521
gas lifetime, 451
gas proportioner or T-piece on a gas line, 141
gaseous or gas-segmented carrier, 248
gas-kinetic temperature, 150, 152
gas-pressurized reservoirs, 226
gate valve, 33
gelatinous precipitates, 271
generalized standard addition method, 259
geological applications, 97
glass collection chamber, 297
glass cyclonic spray chamber, 94
glass disk, 463
glycerol
— enhancing effect of, 89
good precision, 2
gradient calibration, 273
graphite cups, 518, 520, 521, 533
— composition, 523
— geometry, 523
— temperature, 523
graphite furnace atomic absorption spectrometry, 1
graphite probes, 512
grid ablation, 462

halide, 533
He ICP-MS, 165
heated extension, 76
heated microporous PTFE tubular membrane dryer, 55
heated spray chamber/membrane desolvation system, 66
heated spray chamber/Peltier cooling desolvation system, 241
heater/condenser system, 59
heat-transfer between the plasma and the aerosol particle surface, 138
heavy metal toxicology
— use of ICP-MS in, 97
heterogeneous nucleation, 405
heterogeneous samples, 249,462
hexapole collision cell , 32
high acid concentration, 244
high analyte concentrations, 285
high dissolved solids content, 244
high lateral resolution, 446
high mass resolution, 83
high resolution, 33
— highest achievable, 83
— highest available, 42
— highest setting, 83
high rf voltage in the plasma, 24
high selectivity, 309

high-efficiency nebulizer, 232
high-efficiency PN, 46
high-volatility elements, 529
holding coil, 230
hollow fibre membrane, 294
homogeneity, 446, 547, 550
homogeneous material, 463
human nutrition
— use of ICP-MS in, 97
hydraulic high-pressure nebulization, 63,65
hydride generation, 311
— with gas/liquid separation, 83
hydrological applications, 97

ICP torches, 512
ID analysis, 43,73,130,133
indirect detection of a variety of anions, 271
inductively coupled plasma atomic emission spectrometry, 1
inductively coupled plasma mass spectrometry, 1
industrial applications, 97
inert tubing, 227
inhomogeneous samples, 446
initial radiation zone, 6
injected sample volume, 220
injection volumes, 214, 229
insertion sequence, 517
insertion speed, 518
instrument warm-up time, 51
instruments using a low acceleration potential, 93
integration time, 75, 233
interface
— more efficient cooling of, 140
interfering oxide ions, 58
internal standard solution
— merging with the sample, 132
internal standardization, 55, 75, 132, 243, 253, 465, 468
internal standards, 446,463
— added on-line, 285
— careful selection of, 132
— extended range through the use of enriched isotopes, 132
interpolation, 274
ion beam
— widening of, 92
ion traps, 15
ionization, 5, 56
ionization equilibrium, change in, 88
ion-trap DSI-ICP-MS, 515
IR lasers, 456
irradiance, 452
IRZ position, 151
isobaric overlap, 78
isotope abundance measurement, 449
isotope dilution analysis, 253
isotope ratios, 130, 425
— measurement of, 43, 67
isotopic standard, 97

KCl, 533
KF, 533
kinetic discrimination, 312
kinetic energy distribution, 27
kinetic modelling, 131
knitted open tubular reactor, 224, 298
knotted reactor, 299, 309
— heating the, 306
KrF laser, 450

lab-on-a-valve concept, 321
laminar flow, 219
laminated magnet , 33
large dilution factors, 281
large dispersion systems, 220,2 81
laser ablation systems, 457
laser beam dimensions, 451
laser beam processing, 453
laser parameters, 451
laser spot, 452
laser spot size, 452

lens settings
— optimization of, 133
LiF, 533
limit of quantitation, 133
limited dispersion systems, 220,238
limits of detection, 447, 448, 468, 470, 545
line ablation, 462
linear calibration curves, 285
linear dynamic range, 68, 217, 235, 236
linear or polynomial interpolation versus time, 132
linear-flow torch, 28
liquid samples, 544
liquid–liquid extraction, 297, 305, 312
local thermal equilibrium, 8
long linear dynamic range, 2
long-term drift, 297
long-term stability, 78
loop approach, 309
low background, 2
— and high transmission, 71
low detection limits, 71
low frequency noises, 16
low resolution instruments, 81
low thermal conductivity of argon, 137

magnesium nitrate, 396
manifold design, 220
manual tuning, 140
manual univariate optimization, 32
mapping, 447
mass bias factor, 96
mass calibration, 42,77
mass discrimination, 95, 96, 129, 154, 277
— change in as a function of time, 130
— characteristics, 77
— decrease in, 163
— matrix induces additional, 97
— per mass unit, 96
mass resolution
— improved, 42
mass transfer, 213
— between sample film and subsequent liquid stream, 248
mass unit, 96
mass-flow controller, 6
materials not contaminated with the analyte, 294
mathematical mode for optimization of the FI parameters, 303
mathematical modelling, 241
mathematical models, 237
matrix effects, 52, 57, 64, 65, 86, 95, 257, 542
— observed in high-resolution instruments, 94
matrix interferences, 408
matrix matching, 55, 271, 465, 468
matrix removal, 543
matrix separation
— total or partial, 313
— with or without preconcentration, 272
matrix-matched standard solutions, 133
measurement conditions, 77, 233
measurement time, 68
mechanisms for the formation of polyatomic ions, 84
medium dispersion systems, 220, 267
— selected applications, 278
membrane dryer, 55
membrane drying tube, 59
membrane separator, 241
membrane separator interface, 59
memory effects, 50, 53, 57, 60, 160, 229, 250, 459
— of the enhancement effect, 90
merging manifold, 252
merging stream double-line manifold, 269

metal cups, 512
metal filaments, 374
metal oxide formation, 140
metal-free filter, 300
metallic wire-based probes, 512
metallurgical applications, 97
methanol, enhancing effect of, 89
method of standard additions, 55, 133, 226, 253
method of standard additions
— and internal standardization, 165
— with or without internal standardization, 134
microanalysis, 446, 448
microbore glass column, 294
microcolumn, 287
microconcentric nebulization, 63, 65
microflow USN, 58
micronebulizer, 244
microplasma, 445
microporous PTFE tube, 298
microwave-assisted procedures, 135
miniaturization, 215
— instrumentation, techniques, 321
minimal treatment, 133
Mistral, 59
mixed Ar–He ICP, 147
mixed Ar–He plasmas, 141
mixed-gas Ar–O_2 plasma, 241
mixed-gas DSI-ICP-MS, 527
mixed-gas plasmas, 141, 146, 241
— better ionization sources for MS, 147
— extended capabilities with, 137
— require higher rf power than an all-argon plasma, 139
— with only 2% N_2, 142
mixing chamber, 218,281
mixing coil, radius of, 221
moderate dilution, 271
molecular-gas plasmas, 138
monodisperse dried microparticulate injection, 63
Monte Carlo simulation, 134
multi-collector instruments, 42
multicomponent analysis of a full mass spectrum, 131
multi-detector instruments, 77
multielemental capability, 2, 68
multiposition valve, 230
multivariate calibration, 131
multivariate optimization techniques, 84
multivariate simplex optimization, 15

NaCl, 533
NaF, 533
Nafion membrane dryer, 232
narrower size distribution, 46
natural water
— analysis of, 314
— samples, 314
Nd:YAG laser, 449
nebulization
— continuous, 156,216
— efficiency, 15, 151, 217, 250
— increase in, 89
— reduced, 162
nebulizer, 5
— high-efficiency, 240
— position of, 94
Neoprene pump tubes, 227
neutral filters, 450
NH4Cl, 533
nickel alloys, 547
nickel nitrate, 398
nitrides, 394
noise, 236
— amplitude spectra, 12
— in the ICP, 74
— major source of, 75
— reducing, 49
— reduction of, 90
non-spectroscopic interferences, 86,542
normal analytical zone, 6

normal FI manifold, 218
nuclear applications, 97
numerical apertures, 454

O_2, effect on the electric field, 139
off-line column separation, 314
off-line method of standard additions, 274
offset ion optics, 30
Omega lens, 30
on-line addition of an internal standard, 253
on-line analysis, 447
on-line dilution, 275
on-line dissolution, 321
on-line electrolytic dissolution of Al alloys, 321
on-line extraction manifold, 240
on-line isotope dilution analysis, 259,275
on-line microwave digestion of slurries, 321
on-line sample pretreatment, 270
on-line separation of the analytes from the matrix, 307
on-line standard additions, 253,274
operating mode, 451
operating radio frequency of the generator, 27
optimization, 523
— of the plasma and MS operating conditions, 240
optimum aerosol carrier flow rate, 42
optimum sampling position, 29
organic precipitants, 320
organic solvents, 15, 244
organic-based samples, 165
— sample introduction of, 159
other or all gas flows
— addition of a foreign gas to, 163
overall transport efficiency, 52
oxide interferences, 82
oxide ratio, 66
oxides, 394
oxygen-containing polyatomic species, 159
— reduction of, 152

palladium nitrate, 400
partial least square regression, 131
particle coagulation, 49
particle size distribution, 52
particle size of the column material, 294
peak area, 217
peak height, 217
peak hopping, 77, 234, 514
— increased reliability, 38
peak width, 219,236
— at half maximum, 249
— increases with m/z, 38
PEEK, 227
pelleting, 463
Penning ionization, 7
peristaltic pump, 16, 45, 222, 226, 238
— fluctuations due to, 47
— noise induced by, 48
pervaporation device, 296
phase separator, 297
photon stop, 29
physical carriers, 386
physical properties of the analyte, 57
piezoelectric micropump, 226
plasma
— addition of a foreign gas to, 138
— addition of O_2 to, 140,310
— as a programmable reactor, 553
— improved stability of, 154
— overall size reduced, 146
— shrinkage of, 142, 144
plasma (outer) gas, 4
plasma ignition, 4
plasma interface, 33
plasma operating conditions
— change in, 133
plasma power, 523
plasma sampling position, 7,76

plasma solvent loading, 49,80
plasma torches, 28
plasma volume
— reduction in, 149
pneumatic nebulizers, 44, 221
polishing, 463
polyatomic ions, 369
polyatomic species
— reducing interferences from, 154
polyatomic species in ICP-MS
— formation of, 152
polymeric supports, 308
porous membranes, 298
position of the IRZ, 14,76
position of the tip of the IRZ, 14
positioning, 448
powder analysis, 463
powdered samples, 547
Powell's optimization algorithm, 241
precipitation set-up, 271
precipitation/dissolution, 299,306,312
precise mass calibration, 42
precision, 468
preconcentration, 231,287, 542
— on a microcolumn, 287, 301, 308
pre-derivatization, 304
pre-evaporation, 77
pressure wave, 412
probe geometry, 518
process control, 278
programmable x-y-z translation, 459
propanol
— enhancing effect of, 89
ps lasers, 470
PTFE, 533
pulsations, 213
— minimizing, 227
pulse damper, 45
pulse mode, 449,450
pulseless pump, 45
pumping rate, 222
pyrolysis step, 362

Q-switch delay, 450
Q-switched mode, 450
quadrupole-based instruments, 15, 16, 96
qualitative analysis, 464
quartz bonnet, 4
quasi steady-state signal, 240

radiance, 449
radiatively heated spray chamber combined to a two-stage condenser, 59
radioactive samples, 244
radioactive waste samples, 281
ramp scanning, 77,234
random background, 233
random noise, 75
rapid analysis with high sampling frequency, 278
rapid isotopic analysis, 2
raster ablation, 462
rate at which the potential is ramped, 306
reaction coil, 297
reaction gas, 32
reactions where the change in concentration is a function of time, 281
reactor volume, 220
readout time corresponding to the tailing section of the FI peak, 284
reagent consumption, 230
real-time data storage, 462
reciprocating positive displacement pumps, 226
recirculating loop manifold, 283, 285
recombination with electrons, 91
rectangular shape of the laser beam, 451
recycling nebulization system, 60
reduced dispersion systems, 220, 286
reduced-volume spray chamber, 240
reducing CO^+, 59
reductive precipitation, 304

REE, 419
refocusing, 462
refractory elements, 533
refractory materials, 136
refractory oxides
— formation of, 49,82
relative freedom from chemical interferences, 2
relative standard deviation, 544
renewable surface technique, 231
repetition rate, 451
replicate standard additions, 255
residence time
— longer, 84
— in the plasma, 6
resolution instruments, 71
resolution of about 1 amu, 30
response surfaces, 241
reverse-FI manifold, 218,274
rinsing solution, 53
robust ICP conditions, 87
robust plasma conditions, 84, 241
roughing pump system, 141

sample-carrying probes, 512
sample clean-up, 542
sample injection valve, 225, 238
sample injection volume, 224
— decreasing, 285
sample introduction, 214
— efficiency, 310
— system, 3, 86, 94
sample preparation steps, 135
sample solutions containing organic solvents, 78
sample transport efficiency, 246
sampler, 10
samples
— with a high ionic content, 313
— with high levels of dissolved solids, 272
sampling analytes in the field, 307
sampling depth, 521
sampling frequency, 213, 224
sampling position, 12
— shorter, 139
sampling rate, 213
saturation ratio, 404
scanning mode of the MS, 234
scanning rate, 235
Scott spray chamber, 47, 50, 63, 231
seawater, 87, 305, 314
— analysis of, 157
— anaysis of, 301
— dissolved CO_2 in, 305
— use as a mixed physical carrier, 389
secondary discharge, 25,85,146
— avoiding, 26
— formation of, 154
— intensity of, 148
secondary electron multiplier, 24
sector field instruments, 96
sector-based instruments, 15
sediments, 547, 550
segmented flow injection, 248, 249
selective precipitation, 131, 320
— of the analyte, 299
selective retention of the analyte on a microcolumn, 287
semi-quantitative analysis, 464
sensitivity, 72, 157, 221
separation in time of unextracted and extracted species, 312
separation of the analyte from the matrix, 286, 310
sequential injection, 230, 298
SF6, 537
sheathing device, 57, 144, 150
sheathing gas, 162, 163
shielded ICP, 27
ShieldTorch, 27, 71
short-term precision of ICP-MS, 74,76
shrinking of the plasma, 5
signal drift, 52
signal enhancement, 94
— extent of, 153

signal suppression, 245,409
simple spectra, 2
simplex optimization, 31, 84, 139, 157, 160, 161, 241
— using several criteria, 140
simultaneous full mass spectrum, 43
simultaneous multiple injection, 282
simultaneous nebulization of solutions, 297
single shot, 448
single shot ablation, 462
single-line FI manifold, 219, 223
single-point peak hopping, 233
size of the droplets, 6
skimmer, 10
slider valves, 226
slurries of reference soils, 136
slurry, 135, 244, 423
— aerosol, 136
— analysis, 135
— nebulization, 136, 231
— on-line electrolytic dissolution of, 321
— on-line microwave digestion of, 321
— particles, 136
— particles, evaporation of, 158
— preparation of, 135
— sampling, 550
small preconcentration, 308
smoothing, 236
sodium chloride, 395
soft ablation, 456,466
solenoid-driven diaphragm pump, 226
solid samples, 135,546
solids analysis, 413,423
solution nebulization, 367
solution-based calibration, 465
solvent evaporation, 49,542
solvent load in the plasma, 26, 58
solvent loading
— decrease in, 82
soot deposition, 78
sorbent extraction, 309
source of drift, 51
source of fluctuation, 52
space charge effects, 30, 91, 95
— deleterious, 92
— minimized by reducing the charge density of the beam, 93
— within the ion optics, 93
spark ablation, 469
spatial resolution, 468
speciation, 13
— of the analyte, 57
— of the analyte and the sample matrix, 53
specific interferences, 161
spectral fitting, 81
spectroscopic interferences, 78, 272, 277, 539
— corrections for, 133
— eliminating from polyatomic ions, 165
— from polyatomic ions, 80
— source of, 83
speed of acquisition, 43
spikes in the gas blank signal, 461
spray chamber, 5, 44, 47
— cyclonic, 48,60
— dead volume of the, 218
— disposable, 60
— generating vapour within, 296
— gradual heating of, 50
— heating of, 50
— temperature of, 49, 94
SrF2, 533
stabilization of unstable species, 311
stable isotope dilution (ID) analysis, 55, 134
stable isotope ratios, 129
stable isotope tracer studies, 131
stable mixed-gas plasmas
— formation of, 156
standard additions, 549
standard ion source, 3

standard reference materials
— use of ICP-MS in the production of, 97
starving conditions, 45
statistical uncertainty, 130
steady-state signal, 213
stoichiometry, 466
stopped-flow condition, 225
storage capability, 15
storage of species, 308
straight tubes, experiments with, 220
structured background, 233, 234, 236
Sturman-Masters spray chamber, 48
successive dilutions, 283
— of a single standard solution, 285
supersonic expansion, 11
suppression, 91
surfactant-based organised assemblies, 311
survival of polyatomic species, 86
SynchroScan, 38
synthetic sea water, 52
syringe, 228
syringe pumps, 226

tandem injection, 272
TEFLON, 227
TEFZEL, 227
ternary gas mixtures, 536
thermal conductivity
— and electrical resistivity, 145
— of the plasma gas, 6
— of the vapour, 6
thermal pinch effect, 145,149
thermospray, 63
thermospray nebulizer, 232
thick films on hydrophobic materials, 298
thioglycollic acid, 305
three-dimensional coiled tube, 224
time-based sample loading and elution, 287
time-of-flight analyzer orthogonal to the ICP-sampling interface, 43
time-of-flight based instruments, 15, 77
time-of-flight DSI-ICP-MS, 515
torch designs, 4
torch shield, 139
total flexibility, 309
toxic elements
— likely sources of, 131
T-piece, 141, 142
trace determination of Si, 28
transfer lens system, 30
transient signals, 214, 462, 514
translation speed, 462
transport efficiency, 13, 59, 402, 407, 460
— changes in, 95
tube length, 220
tungsten
— vaporization of, 383
turbulent flow, 220
Turner interlaced induction coils, 26

ultrasonic nebulizer with desolvation, 53
undesirable diffusion effects, 245
uniform mass response, 72
univariate optimization, 140
— of the plasma and MS operating parameters, 14
univariate searches, 31
uranium, 542
urine, 52, 165
— diluted, 320
UV lasers, 456
UV photoreactor to photochemically transform an inert species, 305
UV wavelengths, 470

vanadium, 542
vaporization, 5
— mechanisms of, 375
vaporization step, 362
vapour generation, 295, 302, 310
void volume, 231

volatility differences between the analyte compounds in the sample and standards, 45
volatility sequence, 526,527
volatilization, incomplete, 531

washout process, 243
washout time, 53, 57
water-cooled spray chamber, 158
water-removing device, 296
wavelength, 451, 456, 466
wet aerosol, 5
wet and dry aerosols, 58
whole blood, 249
wider skimmer orifice, 11
width of the entrance and exit slits, 42
wine samples, 286
wines
— analysis of, 89
wire-loops, 513

XeCl laser, 450

zinc, 542
zircons, 419
zone of silence, 11
zone-sampling manifold, 274
zone-sampling of Na standard solutions, 271
zone-splitting, 282
zooming facilities, 448

Printed in the United States
200461BV00001B/109/A

9 780444 899514